WITHDRAWN

Statistical Data Analysis Handbook

FRANCIS J. WALL, Ph.D.

McGRAW-HILL BOOK COMPANY

New York   St. Louis   San Francisco   Auckland   Bogotá
Hamburg   Johannesburg   London   Madrid   Mexico
Montreal   New Delhi   Panama   Paris   São Paulo
Singapore   Sydney   Tokyo   Toronto

Library of Congress Cataloging in Publication Data

Wall, Francis J.
　　Statistical data analysis handbook.

　　Bibliography: p. B.1
　　Includes index.
　　1. Mathematical statistics—Handbooks, manuals, etc.
I. Title.
QA276.25.W35　1986　　519.5　　85-8942
ISBN 0-07-067931-2

1234567890　DOC/DOC　89321098765

ISBN 0-07-067931-2

The editors for this book were Harold B. Crawford and Dennis
Gleason, the designer was Mark E. Safran, and the production
supervisor was Thomas G. Kowalczyk. It was set in Times Roman
by University Graphics, Inc.

Printed and bound by R. R. Donnelley & Sons Company.

**to
Jeanie**

---

and in memory of my friend,
Robert Proper, M.D.

Contents

Preface

As our society grows in complexity, so do the problems that must be solved. Pertinent data are often incomplete and initially seldom in a form that specifically supports decision making. The use of statistical methods to transform relevant data into usable summaries is data analysis.

Data analysis problems can be approached from two different directions. One of these I will call the "abstract" direction. In this approach, the data analyst is trained in statistical theory and methodology before being confronted with problems that must be analyzed for commerce and industry, government and research. These problems are, unfortunately for the data analyst, far more complex than the problems used in training.

The other direction I will call the "pragmatic." In this approach, the data analyst is presented with a problem and told to solve it. An employer is indeed fortunate if the data analyst recognizes the importance of statistics in solving data analysis problems. One or two or three courses in statistics would appear useful, and often elementary courses are the basis for employment as data analysts. Just as often, these analysts have little or no idea how to use the statistical methodology they have been taught. The problem is that the methodology they do know was presented from the abstract viewpoint, not from the pragmatic viewpoint. The learning experience required to apply abstract concepts to pragmatic problems is usually painful and not always successful.

I have become acutely aware of this problem through my work as a consultant in statistical analysis. Every consultant in statistics knows people in industry and government who have completed degrees in statistics without learning anything about using the methodology for data analysis problems. Thus my experience has been that not all statisticians are data analysts. It has also been my experience that the more I can teach a client's employees about data analysis, the more I am understood and needed. This book is written for these people.

I believe that data analysis stands in relation to statistics as medicine stands in relation to microbiology, biochemistry, anatomy, and physiology. These latter are the academic subjects that a medical doctor must apply just as statistics is the academic subject that a data analyst must apply. The problem is this: Most students of statistics will become data analysts, not statisticians. Unfortunately they are not trained to apply statistical methodology as a medical doctor is trained to apply microbiology and biochemistry.

Dr. Ross Wood summarized all this for me in one sentence. He said "Statistics is a science; data analysis is an art." I hope this book is about art.

I believe many people who have had statistical methodology courses can analyze data. Of course, the more statistical methodology they know, the better analysts they can become. This book is one attempt to show them how statistics can be used to analyze data.

In writing this book, I have assumed readers will have had one or more elementary courses in statistics. I have tried to show how elementary statistics can be used in data analysis. Part IV, Supporting Topics, is included not to teach methodology but as a source of review and as an opportunity to present to the reader certain broad concepts from the data analysis viewpoint.

This material has been used successfully for nonacademic training of data analysts whose backgrounds in statistics ranged from no courses to three or four. While not recommended for independent study without some previous formal training in statistics, much of the material in this book is suitable for independent study after only one or two elementary statistics courses. A serious reader completing this book should acquire a much better understanding of how statistics is used in the analysis of data.

It is not my intention that this book replace any elementary course in statistics. It is, rather, my intention to show data analysts how to use statistical methods. Some of the better analysts whom I have known did not begin their training in a department of statistics or mathematics. They began, like the founders of statistics, with practical problems to solve. This book is an effort to show others how, in this sense, to be better analysts.

I want to thank Milton Jenkins, Ross Wood, George Brown, Paul Littleton, and others for reading the manuscript and making valuable suggestions that improved the final version. Errors, omissions, and unclear explanations are my responsibility, however, not theirs.

FRANCIS J. WALL

CHAPTER 1

The Data Analysis Viewpoint

INTRODUCTION

A data analysis problem always begins with data. The data may already be collected or they may be only conceptual. If they have not been collected, the data analyst may be able to influence how they are collected and recorded. A plan for collecting and recording data should not be confused with experimental design. Experimental design is a branch of statistics and is not a subject considered in this book.

Data, and the innate curiosity of humans, generate a need to understand the data. An understanding of the data and, it is hoped, an understanding of the situation that generated the data are developed through an analysis. An understanding of the real world is the ultimate purpose of a data analysis.

An illustration is never as complex as a real problem and real problems can never be made as simple as most illustrations. Nevertheless, the problems of data analysis must be explored with illustrations and examples, so the examples used in this book are all taken from actual data analysis problems. Real problems always require background information to understand them. Such information contributes to a better understanding of the data analysis principles involved.

The principal tools of the data analyst are statistical procedures, but this is not a statistics book in the usual sense. A statistics text is organized to teach in a systematic manner a particular field of human knowledge. In contrast to this approach, this book is organized to show the data analyst how to use statistics in some common data analysis problems. The distinction between the data analysis and the statistics viewpoints will be elaborated in the next section.

DATA ANALYSIS AND STATISTICS

Data analysis and statistics are contrasted in this section by describing the viewpoint and purpose of each, some of the ways that concepts are presented, and a few miscellaneous ideas designed to set the stage for a study of data analysis.

Merriam-Webster's *Ninth New Collegiate Dictionary* defines *statistics* as "a branch of mathematics dealing with the collection, analysis, interpretation, and presentation of masses of numerical data." Kendall and Buckland (1957) define *statistics* as "the science of collecting, analyzing, and interpreting data." Neither dictionary defines *data analysis,* and perhaps for a good reason. Their definitions of statistics are good definitions of data analysis, but it may be difficult to convince someone confronted with a real data analysis problem that they are the same.

Most statistics books and courses begin with a study of distributions and their properties. Then procedures are developed for analyzing data that are from these distributions. Examples and problems are chosen to illustrate particular theory and concepts. Often the examples and problems used are simple and unreal but necessary for elementary courses.

In statistics, everything is under control. The characteristics of data can be defined; mathematical relationships can be proven. The statistician then hopes that these "laboratory" creations have some relationship to the data analysis problems encountered or that the problems encountered can be simplified to something statistical theorists have investigated.

The data analyst approaches problems from the opposite direction. Seldom, if ever, is the distribution of the data known. More often than not the sample distribution does not even resemble anything studied in statistics. Nothing is under control; no assumptions are valid. Finally at some point, the data analyst uses a statistical procedure while hoping that the deviations from the necessary conditions are not so great as to invalidate the analysis completely.

The purpose of data analysis is to understand nature. Jerzy Neyman helped shape statistics in the early days of the science. He summed up the data analyst's problem: "A strict accordance between practical work and a corresponding theory is never possible and yet all our life is based on constant practical application of inapplicable theories." The founders of statistics as a science were more data analysts than many statisticians are today. Professor Neyman recognized the problems.

So what is data analysis? Data analysis is an attempt to apply statistics to practical problems. It is a constant effort to transform data analysis situations into situations covered by statistical theory and to adapt statistical methodology for use with practical problems. No assumptions about data are permissible without an analysis. Finally, data analysis is not exact; it requires good judgment and much experience.

MEASUREMENT SCALES

Measurement is the assignment of numbers to the characteristics of objects or events in such a way that the numbers satisfy the conditions necessary for an analysis. Different assignment rules lead to different kinds of measurement. From the data analysis viewpoint, all assignment rules result in only four kinds of measurements: nominal, ordinal, interval, and ratio. Each has specific characteristics that distinguish it from the others and, consequently, different statistical methods are required. Data with the characteristics typical of one of these kinds of measurements are said to be measured in that scale and to be that kind of data.

Data analysts must be able to recognize the four kinds of data and it is not always possible to do so by simply looking at the numbers. The scale achieved for a particular measurement depends on the mathematical relationships appropriate for the characteristic being measured and not on the zeal and enthusiasm of the researcher or data collector. Consequently, it is always the responsibility of the data analyst to determine the *true* measurement scale achieved so that appropriate statistical procedures may be used.

The most important consequence of the different measurement rules is the meaning that can be given to the numbers assigned to the characteristics. For two scales, nominal and ordinal, the numbers are nothing more than names, even though ordinal scale names may convey some information about relative size of the characteristic measured. Interval and ratio data are quite different in the sense that they convey information about the magnitude or intensity of the characteristic measured.

Statistical procedures applicable to interval and ratio data generally are called *parametric* methods while statistical procedures applicable to nominal and ordinal data generally are called *nonparametric* methods. It is always possible to transform interval and ratio data into ordinal and even nominal data. Then nonparametric methods, when they exist, can be used for analyzing characteristics of objects and events originally measured in the ratio and interval scales.

The interval and ratio scales differ in one principal aspect. The ratio scale has a true zero while the interval scale does not. An illustration will make the distinction clear. Temperatures and intelligence quotients (IQs) are typical interval data, whereas distances and dollars are typical ratio data. The usual test to distinguish between ratio and interval data, i.e., to determine if the data have a true zero, is to consider the meaning of ratios of the data. Ratios of interval scale data make no sense, whereas they do for ratio scale data. For example, a temperature of 80°F is not twice as hot as 40°F, but according to the 1974 edition of the *Rand McNally Road Atlas,* it is three times as far from Albuquerque to Birmingham (1266 miles) as it is from Albuquerque to Denver (422 miles). Zero on the Fahrenheit temperature scale is an arbitrary point while zero on a distance scale is a true zero.

Although ratios of interval scale data are meaningless, ratios of intervals on an interval scale are meaningful. Consider two temperature changes, the first from 50° to 80° and the second from 80° to 95°. These are intervals on an interval scale. The first interval, 30°, is twice as large as the second interval, 15°. Such a statement concerning the ratio of two intervals has meaning since intervals on an interval scale can be truly zero.

The difference between the interval and ratio scales can be described more formally. If characteristics of objects and events can be associated with numbers on a ratio scale, then all of the operations of arithmetic may be performed meaningfully with the numerical values assigned to the objects or events themselves. If characteristics of objects and events can be associated with numbers on an interval scale but not on a ratio scale, then all arithmetic operations except division may be performed meaningfully with the numerical values assigned to the objects themselves. All arithmetic operations are meaningful for intervals, i.e., differences, on the interval scale.

The simplest mathematical relationship of concern to data analysts is equality. Characteristics of objects or events can be classified either equal or not equal. This relationship is symbolized by the familiar "equals" sign, $=$. If x and y are equal, i.e., $x = y$, then $y = x$. Furthermore, if $x = y$ and $y = z$, then $x = z$.

Measurement in its simplest form consists of determining which objects or events have equal characteristics and which do not. When objects or events are assigned to two or more classes or categories that have no natural order, the measurement is in the nominal or classificatory scale. Perhaps the simplest two classes that can be defined are (1) has a characteristic, and (2) doesn't have the characteristic. A classification of household pets as dog and not dog is an example. Acceptable product and not acceptable product is another example.

The classes in a nominal scale are often identified for calculation purposes with numbers or letters, but the natural order of the numbers or the accepted order of the letters must not imply any order to the classes. For example, the designation of classes dog and not dog as 1 and 2 is no more meaningful then designating the classes as 2 and 1. This interchangeability of nomenclature is a distinguishing feature of nominal scale measurement.

Nominal scales are not limited to two classes. There can be as many classes as necessary to distinguish the particular objects or events being studied. For example, household pets might be classified as barkers, meowers, chirpers, and others. Production might be classified acceptable, unacceptable but reworkable, and unacceptable and not reworkable. Sometimes distinguishable classes of objects or events are equivalent for specific problems. For example, other household pets certainly places the distinguishable classes slitherers, squeakers, etc. into a single category.

Characteristics of objects or events can be ordered when members of one class are larger, or smaller, than members of another class. This relationship is symbolized by the familiar "greater than" sign, $>$. The symbol for "less than" is $<$.

Ordinal measurement consists of assigning objects or events, on the basis of their characteristics, to ordered classes or categories. The nominal scale for classifying production into acceptable, unacceptable but reworkable, and unacceptable and not reworkable categories can become an ordinal scale if a monetary value is associated with each category. The categories, in order of increasing value, would be unacceptable and not reworkable, unacceptable but reworkable, and acceptable. Ordinal scale measurement makes no assertion about the difference between contiguous classifications, just that they are orderable.

Characteristics of objects or events are ranked by sorting the observations according to their magnitude and then assigning numbers called ranks to the observations that correspond to their positions in the sorted arrangement. In general, sorting is from least to greatest, and for N observations the integers from 1 through N are used as ranks. Ranking defines an ordinal scale that has as many classes or categories as there are objects or events to classify. Sometimes it is not possible, particularly when the original measurement is in the ordinal scale, to distinguish between two or more objects or events. Such objects or events are tied and special analysis procedures may be required.

Many nonparametric procedures are based on ranks. These procedures are equally applicable to observations measured originally with an ordinal scale and to ranks obtained by transforming ratio and interval data.

Table 1.1 summarizes the four kinds of measurement. The reader will find additional information on measurement scales in Stevens (1946) and in Siegel (1956).

TABLE 1.1 Measurement Scales

Scale	Defining relationships
Ratio	1. Equivalence 2. Greater than 3. Ratio of any two intervals has meaning 4. Ratio of any two scale values has meaning
Interval	1. Equivalence 2. Greater than 3. Ratio of any two intervals has meaning
Ordinal	1. Equivalence 2. Greater than
Nominal	1. Equivalence

SOME DEFINITIONS

Many of the practical problems that were the early inspiration for the development of statistical methods originated in agricultural and demographic research. Names given various aspects of data were naturally descriptive of these fields or else they were names that the data analyst could easily explain. Some of the names are less than appropriate for generalists to use nowadays but data analysts are stuck with them. So to avoid confusion in later chapters, some simple definitions are given here.

Units: The objects or events studied. In planned experiments, units are called *experimental units* or *subjects*.

Variables: The characteristics of the objects or events studied.

Data: Numbers or symbols assigned to characteristics of objects or events. These numbers may be descriptive or simply classificatory. Symbols serve to identify objects or events considered equivalent for an analysis.

Populations or treatment groups: These are synonymous terms for identifying large actual or hypothetical groups of units. In surveys, retrospective studies, and censuses, populations may actually exist. Treatment groups as populations exist only hypothetically. An investigator creates treatment groups by assigning homogeneous units to two or more groups that are treated differently in an experiment. Populations and treatment groups should always be carefully delineated in a data analysis problem as an aid to understanding and interpreting results.

Sample: A subset of a population.

Statistic: A summary value calculated from a sample of observations. See Kendall and Buckland (1957).

Strata or blocks: These are synonymous terms for identifying major subclassifications of units in a population. When units are not sufficiently homogeneous, it is sometimes possible to create subgroups that are distinctively more homogeneous than the full group. Subgrouping is based usually on a natural characteristic of the units such as location or time factors or on attributes of the units themselves. Generally, these subgroups are called strata in censuses, surveys, and retrospective studies while they are called blocks in all prospective studies, i.e., planned experiments. Their purpose is always to identify a source of differences in data and to provide an analysis mechanism for removing the effect of the differences identified.

Covariates: These are variables in the ratio and interval measurement scales that can be identified with an important source of differences in the data. Age is an often-used covariate in studies involving humans, machines, animals, etc. The differences in the data resulting from the covariates can be removed with appropriate analysis methodology.

Strata or blocks and covariates are used to make an analysis more precise so that the investigator can see smaller data differences. No report of an analysis is complete without a full discussion of any use made of strata or blocks and covariates and the effect of their use on the analysis.

SOME TYPES OF STUDIES

It is presumptuous of me to think I can classify data analysis problems. There are, nevertheless, specific study characteristics that data analysts should recognize because they determine the statistical procedures that should be used and the conclusions that can be made. For this discussion, studies will be classified in two ways: first, the method by which populations occur, and second, the method by which samples are selected from the populations.

Populations occur in two ways, natural assignment and random assignment. In random assignment, the investigator takes a group of homogeneous units, assigns them randomly to two or more subgroups, and then treats the subgroups differently, thus creating separate populations. For practical purposes, the subgroup is a sample from a hypothetical population. Only the sample has been created, but as much more of the population as the investigator might want could be created. The only limitation is the size of the original homogeneous group.

In natural assignment, the investigator does not create the populations. They occur naturally on the basis of one or more characteristics of the units themselves. For example, male and female is a natural assignment characteristic. Smokers and nonsmokers is another, even though an investigator might someday try to assign smoking habits.

Sampling occurs in two ways, natural selection and random selection, and the investigator may elect to use all available data. In random selection, the investigator delineates the population in some way and then, usually with the aid of random numbers, selects a portion of the population for study. See chapter 12, Random Sampling, for further discussion of this subject.

In natural selection, the sampling mechanism is not determined by the investigator. Units are in samples naturally on the basis of one or more characteristics of the units themselves. For example, a medical investigator who wants to compare brain tissue of stroke victims with brain tissue of people who have not had a stroke will use autopsied tissue from a group of stroke victims and a group of nonstroke victims. The natural selection mechanism is death and then the decision to autopsy.

Population assignment and sample selection are identified for study types in table 1.2. The distinction between the types of studies is ignored generally in statistical literature.

Retrospective is often associated with any study that uses already available data such as census, production, hospital, and business records. Prospective is associated generally with any study for which the data are not available at the time plans are made but which can be made or will become available in the future. These are broader definitions of the terms than that implied in table 1.2. The problem is not the availability of data when the study is planned but rather the assignment of subjects to populations and the selection of samples from the populations. This determines the generalizations that can be made with the results.

TABLE 1.2 Types of Studies

Population assignment	*Sample selection*		
	Random	*Natural*	*All*
Natural	Survey	Retrospective	Census
Random	Prospective with subsampling	Prospective with natural subsampling	Prospective, e.g., planned experiment

Since most, but not all, studies called prospective employ random assignment, while most, but not all, studies called retrospective use natural assignment to populations, the relative merits of prospective and retrospective studies are often confused. Data analysts should either not use the names at all or they should be used in the slightly restricted and more precise sense presented by table 1.2.

The data analysis issue here concerns which statistical procedures are appropriate for each type of study and then the interpretation of the results. Investigators and management always want results to be as widely applicable as possible, so it is the analyst's responsibility to report the results in ways to avoid misinterpretation.

In general, any randomization extends the applicability of results. For example, survey results can be immediately extended to the individual populations, and with proper weighting factors to the group of populations represented by the individual samples. Care must be taken not to extend results to populations not represented by the samples studied.

All results from comparing populations created by random assignment can be extended to the hypothetical populations. Results may not be extendable to similar populations created from other homogeneous groups without first demonstrating the comparability of the homogeneous groups. Prospective studies with natural subsampling can create interpretation problems unless the cause for the natural subsampling is completely independent of the population differences being studied. Data analysts can sometimes get around these problems by qualifying extensions of the results; other times the necessary qualifying statements so limit extensions that the results are useless.

Interpretation problems arise most often in retrospective studies. Results from these studies should always be qualified with statements describing the natural assignment and natural selection process but they seldom are. Even when these studies are done properly, investigators tend not to describe the studies sufficiently for proper interpretation.

Interpretation is a difficult problem. There are no specific rules and few generalities. This section of chapter 1 introduces problems and pitfalls; it doesn't solve them. One purpose for the rest of this book is to help the data analyst resolve such problems.

HYPOTHESES

Every data analysis problem is the result of an investigator or management trying to document the existence of desirable or undesirable differences or relationships between two or more things. The *research hypothesis* is the purpose for the research and should be stated precisely for every analysis.

A research hypothesis stated as the equality of two or more things is a deception. Equality, outside of formal mathematics, is impossible to prove. In the real world, it is always possible to show inequality simply by refining the measurement techniques.

This inability to demonstrate equality and the universal ability to demonstrate inequality provides the basis for all data analyses. Every statistical test is a test of the *null* hypothesis, i.e., a test of the hypothesis of zero differences or equality, in contrast to the research hypothesis, which is a statement of expected or anticipated differences. The statistical test is a use of the data to see if differences can be detected based on the measurements available for analysis. Since differences can always be detected by properly refining the measurement techniques, the analysis is an attempt to see differences in data measured by accepted or customary techniques. Differences that are evident under these conditions are called *statistically significant.* When extraordinarily precise measurements or very sophisticated analysis techniques are required to support a research hypothesis, I have found that the differences or relationships demonstrated are of no value in the real world.

The data analysis problem does not end with statistically significant differences. In some respects, it begins there because it is then that the investigator, management, or both must decide the real-world importance of the difference. They naturally want statistically significant differences to be as important as possible, and the main way differences are important is for them to have wide applicability. The data analyst, with an understanding of statistics, must help interpret the results properly for correct applicability.

ASSUMPTIONS VS. CONDITIONS

All statistical tests that are of interest in data analysis are dimensionless ratios. One member of the ratio, usually the numerator, is a measure of the differences that are the motivation of the analysis. The other member of the ratio is a measure of data repeatability without the difference being present.

Statistical tests are each applicable only in very special situations. These situations are described by the *assumptions* that are associated with each statistical test. The use of the term assumption in statistical texts is unfortunate. It implies that making an assumption is the "right and proper thing to do," that making the assumptions makes the methodology applicable, that somehow making the assumptions solves problems. In reality, an assumption creates problems. Statisticians may have the experience and training to evaluate the consequences of making assumptions, but most managers and researchers, untrained in statistics or data analysis, do not.

Assumptions in statistics become conditions in data analysis. The data analyst knows that a statistical procedure is not valid unless all the associated conditions, i.e., statistical assumptions, are true. Every condition necessary for the chosen statistical procedure must be tested statistically to determine if the conditions for its applications are likely to be true or not.

Sometimes, when the conditions for a particular statistical procedure are not true, the data can be transformed within the original measurement scale so that the necessary conditions are "probably" true. The transformed data are then analyzed using the selected procedure. Many problems are associated with the analysis of transformed data, but sometimes it is the best an analyst can do.

Often a better procedure is to transform the data to a lower measurement scale. The rank transform is an example. This creates a new distribution for the data at the same time the conditions for an analysis of the transformed data are satisfied. Transformations to a lower measurement scale result in a loss of information but they are the salvation of many difficult situations. Consequently, both kinds of transformations will be discussed whenever appropriate.

Sometimes data don't quite satisfy the conditions for a particular statistical test and a transformation is undesirable. The data analyst must do something; the problem can not be ignored. What is done depends on the analyst's experience. Statisticians try to help through investigations of the robustness of statistical procedures. Unfortunately texts report these investigations with such comments as: "when h is large enough, this statistic has approximately a standard normal distribution." Such texts are not written for data analysts who need to know what "large enough" means.

UNDERSTANDING

A data analyst uses statistics but data analysts are not all statisticians. Conversely, statisticians are not all data analysts, either. What then is a data analyst? What does a data analyst do?

The data analyst is anyone who is confronted with the problem of interpreting data. The principal tools used for the analysis are statistical methods, but data analysts recognize that any particular statistical procedure chosen may be inappropriate for the data one seeks to understand. Inappropriate use of statistical procedures is reduced by constantly checking the conditions that must be true for use of the procedure to be valid.

Assuming there is an appropriate statistical procedure for a given analysis situation, the data analysis eventually results in an estimate of a probability assuming that the null hypothesis is true, i.e., that the data analyzed differ from each other only because of random effects. At the same time, the analyst does not overvalue these probabilities since statistical significance is the result of an arbitrary definition of what will be considered significant. There is nothing magical or sacred about the 0.05 probability level commonly used in analyses; significance can be defined at any probability level one chooses. Hence, conclusions about the real world the analyst seeks to understand are dependent not only on the appropriate use of statistical procedures but on the definition chosen for significance.

Frequent errors in data analysis include

1. Concluding that differences exist when none do.
2. Concluding that differences don't exist when they do.
3. Using statistical procedures when the necessary conditions are not true.
4. Answering the right questions with the wrong data.
5. Answering the wrong questions with the right data.

The data analyst knows that one or more of these errors may be present in each and every analysis. Consequently, analyses do not produce blacks and whites but only shades of gray; not truths but, it is hoped, understandings. Ultimately the real decisions about the data are the analyst's to make.

GETTING STARTED

Two purposes for a data analysis are the comparisons of two or more samples and the exploration of the relationship between two or more variables. The first step in any data analysis is to organize the data in tables that can be easily explained to the investigator, management, or both. This usually means listing the data for each sample separate from the other samples and for each relationship explored from any others that might also be under study. Every example in this book begins with this step. It is not enough to do this internal to a computer because often the analyst will "see" differences and relationships in the data that no amount of testing can ever find. Sometimes sorted data are even more useful, especially data for the study of a relationship.

AN OVERVIEW OF THIS BOOK

To this point, chapter 1 has defined a viewpoint for the remainder of the book. The chapter would not be complete without a brief description of what the reader can expect.

This book is divided into five parts. Part I discusses the simple comparison of two or more independent samples as well as dependent samples. Specifically, the analysis of completely randomized designs is discussed in chapter 4 under independent samples, while the analysis of randomized block designs is covered in chapter 5 as dependent samples. Part II presents simple relationships of two or more variables with brief introductions to somewhat more complex relationships. As explained in part II, all relationships are limited to linear regressions. Part III discusses simple comparison of the relationships introduced in part II. Chapter 10 presents the comparison of two regression relationships, while chapter 11 covers the comparison of more than two.

Part IV stands apart from the rest of the book. Here a number of supporting topics are discussed from the data analysis viewpoint. This part is intended for reference as the need arises and as suggested in parts I, II, and III.

The basic structure of chapters 2 through 11 is the same. A common data analysis problem is introduced and the analysis of such problems in each measurement scale is discussed. Only a few carefully chosen examples are used in each chapter so that the reader can get a feel for data analysis rather than individual statistical procedures. Examples are presented, insofar as possible, from the researcher or management viewpoint, i.e., why should the data be analyzed? Results of analyses are presented, insofar as possible, from the data analyst's viewpoint, i.e., what summary or analytical results would be useful and how should they be presented? Emphasis is placed on simple data analysis reports that assist researchers and management in making decisions, but at the same time do not make decisions for them. Data analysis is not omnipotent; it is only one of many tools necessary for intelligent decisions.

Part V contains statistical tables. Most statistical tables are abominable. In order to make them brief, they are unintelligible to everyone except specialists. Since data analysts are generalists, an attempt has been made to present tables for generalists. Whenever possible, statistical tables have been extended, reformatted, or both for easier use. This has made some more bulky, for which no apologies are offered.

All of the analyses presented can be done with nothing more than a hand-held pocket calculator, preferably one that is programmable. Two old Hewlett-Packard 25's were used in preparation of this manuscript. A Digital Equipment Corporation 11/23 and FORTRAN were used for some problems, more to create orderly tables than as a necessary calculation tool, even though the calculations associated with the tables were also done with the 11/23. In all cases, numbers were rounded off as they were entered in the manuscript, not during calculations. This may make it difficult for a reader to duplicate results exactly step by step in the examples without carrying all possible digits from the beginning.

PART I

Simple Comparisons

Simple comparisons of two or more things are considered in part I. The comparison of drug responses to placebo responses or to other drug responses are examples of the comparison of two things. The lengths of stay for patients with pneumonia in each of a group of hospitals can be the subject for the comparison of more than two things. Tread-wear for automobile tires can be compared for two to four brands by using the brands simultaneously on each of several cars.

Once the things for comparison are identified, data are needed on each. The drug-placebo comparison requires that the effect of the drug be measured several times and that the effect of the placebo be measured several times. Patients are assigned to one of two groups. One group is given the drug while the other is given the placebo. Then their responses are measured. The drug and placebo responses are most likely all of the data that exist for the hypothetical populations created especially for this study.

The hospitalization comparison requires that the lengths of stay for patients with pneumonia be determined for several patients in each of the group of hospitals. Probably all pneumonia patients between two pre-specified dates would be used for this study, but if some of the hospitals are much larger than others, the records in the larger hospitals might be sampled in contrast to using all records for the smaller hospitals. The tread-wear comparison requires that tread-wear be measured for some specified time period or a specified number of miles, or both.

A sample is a group of data. The measured drug responses would be a sample, as would the measured placebo responses. The lengths of stay for each of the hospitals would be a sample. All of the tires of a particular brand would be a sample. Strictly speaking, from the statistical viewpoint a sample is part of a population. It would be difficult, however, to specify precisely the population represented by any one of these samples.

Most populations are only conceptual for the data analyst. For example, the group of measurements on the response to a drug is a sample. The conceptual population may be the group of all human responses to the drug or it may be more restricted, e.g., all humans with a particular disease entity, or humans under 6 years of age, or females.

A specific population for the lengths of stay in a hospital is hard to conceptualize. It might be some vague group such as "all the possible pneumonia patients that the hospital might ever have," but that doesn't convey much meaning. It might be "all possible pneumonia patients so long as conditions remain unchanged at the hospital," but conditions do change constantly so that definition has no real meaning either.

Such definitions of populations are designed to give some legitimacy to the data analyst's use of statistical methods. While these conceptual populations may sometimes be important, the data analyst seldom must define them precisely or even be aware of their existence until interpreting the analysis for the real world.

The drug-placebo and tread-wear studies are examples of prospective studies, while the hospitalization study is an example of a retrospective study. Prospective studies are planned and then the measurements are made, whereas in retrospective studies, the data are probably obtained from records already in existence.

Data for a retrospective study are usually obtained by some deliberate selection process. In the hospitalization example, the sample is simply all pneumonia patients between two specified dates. It is not clear just what population is sampled here. The population of all hospital patients does not seem appropriate, since only pneumonia patients were selected for the sample. All pneumonia patients for some longer time period does not seem an appropriate population, because seasonal and treatment modalities could cause time trends within the population rendering it not homogeneous unless such trends were first removed. All pneumonia patients in several hospitals would not be an appropriate population unless the patients in the hospitals all had the same chance of being chosen for the sample. Nevertheless, the data analyst, together with the investigator or manager, should describe to the best of their abilities the populations they think were sampled so that users of their research will not associate the results with inappropriate populations. Populations for prospective studies are usually hypothetical, i.e., created especially for the study, and the samples used in the analysis are all of the populations that really exists.

A good data presentation facilitates every analysis. The simplest and best way to present data for the comparison of two or more things is in a table with different columns devoted to the different things to be compared. Table I.1 illustrates the concept. Each column should be labeled and contain the sample of data associated with that label. For generality, the data in table I.1 are denoted by $x_{i,j}$, where the subscripts i and j identify datum i in sample j.

Data tables such as table I.1 can be classified in two ways. First, the table may present two samples or more than two samples. Second, the data in one sample may be associated in some way with data in the other samples or they may be independent. The first classification seems trivial, but since there are special statistical methods for two samples, the classification is necessary. The second classification is not trivial, as the following demonstrates.

Suppose that we want to study human skin sensitivity to a drug. First, we need a group of human volunteers to be the subjects for the study. The drug and a placebo are applied to small areas of the subject's skin, and after an appropriate time, the erythema is rated. The data can be presented in a table like table I.2.

TABLE I.1 Data Presentation Example

Thing A	Thing B	Thing C	Thing D
$x_{1,1}$	$x_{1,2}$	$x_{1,3}$	$x_{1,4}$
$x_{2,1}$	$x_{2,2}$	$x_{2,3}$	$x_{2,4}$
$x_{3,1}$	$x_{3,2}$	$x_{3,3}$	$x_{3,4}$
$x_{4,1}$	$x_{4,2}$	$x_{4,3}$	$x_{4,4}$
$x_{5,1}$		$x_{5,3}$	$x_{5,4}$
$x_{6,1}$			$x_{6,4}$
			$x_{7,4}$
			$x_{8,4}$

TABLE I.2 Associated vs. Independent Example

Drug	Placebo
$x_{1,1}$	$x_{1,2}$
$x_{2,1}$	$x_{2,2}$
$x_{3,1}$	$x_{3,2}$
$x_{4,1}$	$x_{4,2}$
$x_{5,1}$	$x_{5,2}$
$x_{6,1}$	$x_{6,2}$

Suppose the drug is applied to one arm of each subject while the placebo is applied to the other arm. Then for each datum in the drug sample of table I.2 there is associated with it a datum in the placebo sample. The two data should logically be on the same line. The data in table I.2 would then represent two measurements for each of six subjects—one measurement for drug and one measurement for placebo.

In contrast to the above, suppose that either the drug or the placebo, but not both, is applied to the skin of each subject. Then the data in table I.2 would represent one measurement for each of 12 subjects. No datum in the placebo sample has any special relationship with any datum in the drug sample. The samples are not associated; they are independent.

Association can occur for all kinds of reasons and is not limited to two samples. Since analysis procedures are different for associated and non-associated data, it is an important distinction.

In part I, *comparison* means the comparison of sample averages through an appropriate statistical test. The statistical test yields information on the consistency of the data with the null hypothesis that no differences exist between the populations. Of course, differences exist between the samples and it is the data analyst's responsibility to decide if the apparent differences are meaningful or simply due to poor measurement or other random or chance influences.

This introduction to simple comparisons has presented two ways to classify data tables. Table I.3 introduces some very useful statistical terminology for identifying the four kinds of problems. If these names are new to the reader, their meanings will become clear later in part I.

TABLE I.3   Classification of Data Tables

	Two samples	*More than two samples*
Independent or not associated	Independent samples Chapter 2	Completely randomized Chapter 4
Associated	Paired samples Chapter 3	Randomized blocks Chapter 5

CHAPTER 2

Two Independent Samples

The comparison of two things using independent samples is discussed in this chapter.

RATIO AND INTERVAL MEASUREMENTS

Example 2.1

Claudication is a medical term meaning "limping" or "lameness." The term is derived from Tiberius Claudius Drusus Nero Germanicus, the emperor of Rome from A.D. 41 to 54, whose distinguishing physical characteristic was limping or lameness. Intermittent claudication is a complex of symptoms characterized by pain or discomfort in one or both legs during walking. Typically, the pain is not present when the patient is at rest. Patients report pain, tension, and weakness in their calves, thighs, and/ or buttocks that begins and intensifies on walking. The symptoms disappear after a period of rest but of course recur with additional exercise.

Intermittent claudication is an atherosclerotic disease wherein insufficient blood is supplied to the affected muscles. It is quite debilitating, especially for persons whose jobs require them to stand or walk for even short periods of time. Any pharmaceutical that can relieve these symptoms would be a valuable aid for the patient.

There is no good way to measure the severity of intermittent claudication, mainly because individual tolerance to pain is so different and because a single individual's tolerance is believed to differ considerably from one day to another. As they walk on a treadmill, patients report that the first pain experienced can be tolerated for a few minutes. The pain usually intensifies rapidly, and soon the patient must get off the treadmill to sit down. The typical measure is the distance the patient walks before experiencing "rapidly increasing pain." Intermittent claudication is not usually treated if the patient can walk for 30 minutes on a treadmill inclined at 7° and running at 2 miles per hour.

Pharmaceuticals for these symptoms are tested by comparing the improvement in walking distance for patients taking the drug with the improvement for patients taking a placebo. Twenty subjects, all intermittent claudication patients, walked on a treadmill until they experienced rapidly increasing pain. The distance was recorded. Each patient was individually assigned a medication, i.e., the drug or the placebo, that was taken for 1 month before the walk was repeated. The data in table 2.1 are the differences in the distances for the two walks. Consequently, these data are measurements on a ratio scale. A negative datum means the patient walked less after taking the individually assigned medication than before taking it.

TABLE 2.1   Data for Example 2.1

Drug	Placebo
377	−79
871	334
1290	−182
475	−665
167	−380
−88	88
602	188
−88	−88
202	−132
17	
−290	

Two populations were created for this study. One consisted of patients who took the drug while the other consisted of patients who took the placebo. The samples are all that exist of the two populations. It is convenient to think of the data as random samples from populations of patients, but they aren't. First, the populations don't exist except for the sample, and second, if they did, the patients were not selected at random from the populations. The data analyst will have to decide if useful hypothetical populations can be defined and then if the patients can be considered random samples from those populations. These are difficult problems.

The principal data analysis issue is the following: Did the drug enable patients to walk further? In other words, is the average for subjects assigned to the drug greater than that for the subjects assigned to the placebo? If it is and if the difference is so large that chance occurrence can be ruled out, then perhaps the difference is due to the active drug. The decision must ultimately be made and defended by the data analyst on the basis of the data analyst's analysis.

The research hypothesis can now be stated more formally as: The drug group is expected to improve more than the placebo group is expected to improve. The corresponding null hypothesis, the hypothesis that will be tested statistically, is: The improvement for the drug group will be no more than the improvement for the placebo group. These hypotheses are stated here so that the objective of the analysis is clearly understood.

The statistical test for comparing these two samples is a form of Student's t-test. Details of this particular test are presented here, although statistical tests are discussed in more general terms in part IV.

To test the hypothesis that no difference exists between two samples, compute

$$t = \frac{\bar{x}_1 - \bar{x}_2}{\mathrm{sd}\sqrt{1/n + 1/m}}$$

where

$$\mathrm{sd}^2 = \frac{\sum_{i=1}^{n}(x_{i1} - \bar{x}_1)^2 + \sum_{i=1}^{m}(x_{i2} - \bar{x}_2)^2}{n + m - 2}$$

is a pooled estimate of the variance and n and m are the two sample sizes.

The quantity

$$\Sigma(x - \bar{x})^2$$

stands for the sum of the square of the difference between each datum in a set of data and the average of that set. As a symbol, it is great for explanations, but it is virtually useless in calculations. One could calculate the sum this way but an easier way can be developed with a little algebraic manipulation. First, square the quantity within the parentheses and sum the parts separately.

$$\Sigma(x - \bar{x})^2 = \Sigma(x^2 - 2x\bar{x} + \bar{x}^2)$$
$$= \Sigma x^2 - \Sigma 2x\bar{x} + \Sigma \bar{x}^2$$

Now for the data set, the average x is a constant, so that

$$\Sigma 2x\bar{x} = 2\bar{x}\,\Sigma x$$

and

$$\Sigma \bar{x}^2 = \bar{x}^2\,\Sigma 1$$

These new forms are possible because the sum of a constant times a set of data is that constant times the sum of the set. Furthermore, the sum of 1 for each datum in the data set is just the size n of the set. Next, substitute n for $\Sigma 1$ and $\Sigma x / n$ for $\bar{x}$ in these two quantities to get

$$\Sigma 2 x \bar{x} = 2 \left(\frac{\Sigma x}{n} \right) \Sigma x = \frac{2\,(\Sigma x)^2}{n}$$

and

$$\Sigma \bar{x}^2 = \left(\frac{\Sigma x}{n} \right)^2 n = \frac{(\Sigma x)^2}{n}$$

Then on collecting terms

$$\Sigma (x - \bar{x})^2 = \Sigma x^2 - \frac{(\Sigma x)^2}{n}$$

This is the calculation form used extensively in data analysis. Variations appear in many chapters throughout this book and each is developed very much as this one has been.

Now three conditions must be true for this test to be valid.

1. The samples must be random.
2. The samples must be from normal distributions.
3. The variance for the first population must be equivalent to the variance for the second population.

It is not sufficient in data analysis to assume these conditions are true; they can all be tested and at least the last two should be tested. Tests for normality and tests for the equality of variances are discussed in chapter 15. In some situations, the first condition can also be tested, but the data analyst can probably do as well by comparing the way the samples were obtained with characteristics of random samples as discussed in chapter 12.

The standard deviations for the drug and placebo samples are 467.5 and 298.4, respectively. The standard statistical test for the difference of two independent variances is an F-test. The test statistic is the ratio of the larger of the two variances to the smaller.

$$F = \frac{467.5^2}{298.4^2} = 2.45$$

Since there is no prestudy reason why one variance should be larger than the other, this test is said to be two-tailed. This means that the apparent probability of a larger F-statistic must be doubled. Using table D in part

V, the probability of an F-statistic with 10 and 8 degrees of freedom this large or larger in a two-tailed test is approximately 0.23, assuming the null hypothesis that the population variances are not different and observed differences are just chance variation. The variances for the drug and placebo samples are obviously different, but from the F-test it is reasonable to consider the population variances equivalent for the analysis. The population variances could indeed be equal and the observed sample difference or something larger will occur 23% of the time due to chance variation alone.

Lilliefors' (1967) procedure, described in chapter 15, is used in this analysis to test normality of the two populations. Table 2.2 presents the calculations for this particular sample. Since the variances for the two populations can be considered equivalent, the two samples have been combined, or pooled, to provide a larger number of degrees of freedom for the Lilliefors test statistic.

Column 1 identifies the sample, while the data are displayed in column 2. The standard deviations of the two populations have been accepted as equivalent but the averages have not. So the first step is to subtract the sample averages from each of the data. These differences are recorded in column 3. Now that the difference between sample averages has been removed, the pooled standard deviation of the data in column 3 is computed. The degrees of freedom for the estimate is the sum of the degrees

TABLE 2.2 The Lilliefors Test for Normality for Example 2.1

						Probabilities			
ID	y	$y - \bar{y}$	z	ID	z	Normal	Sample	da	db
D	377	56	0.139	D	−1.524	0.064	0.050	0.014	
D	871	550	1.370	P	−1.404	0.080	0.100	−0.020	0.030
D	1290	969	2.414	D	−1.020	0.154	0.150	0.004	0.054
D	475	154	0.383	D	−1.020	0.154	0.200	−0.046	0.004
D	167	−154	−0.385	D	−0.759	0.224	0.250	−0.026	0.024
D	−88	−409	−1.020	P	−0.693	0.244	0.300	−0.056	−0.006
D	602	281	0.699	D	−0.385	0.350	0.350	0.000	0.050
D	−88	−409	−1.020	D	−0.297	0.383	0.400	−0.017	0.033
D	202	−119	−0.297	P	−0.200	0.421	0.450	−0.029	0.021
D	17	−304	−0.759	P	−0.075	0.470	0.500	−0.030	0.020
D	−290	−611	−1.524	P	0.034	0.514	0.550	−0.036	0.014
P	−79	23	0.057	P	0.057	0.523	0.600	−0.077	−0.027
P	334	436	1.086	D	0.139	0.555	0.650	−0.095	−0.045
P	−182	−80	−0.200	D	0.383	0.649	0.700	−0.051	−0.001
P	−665	−563	−1.404	P	0.473	0.682	0.750	−0.068	−0.018
P	−380	−278	−0.693	D	0.699	0.758	0.800	−0.042	0.008
P	88	190	0.473	P	0.722	0.765	0.850	−0.085	−0.035
P	188	290	0.722	P	1.086	0.861	0.900	−0.039	0.011
P	−88	14	0.034	D	1.370	0.915	0.950	−0.035	0.015
P	−132	−30	−0.075	D	2.414	0.992	1.000	−0.008	0.042

NOTE: The Lilliefors test statistic is 0.095. Refer to chapter 15 for details.

of freedom for the two samples.

$$\text{Drug average} = \quad 321.4$$

$$\text{Placebo average} = -101.8$$

$$\text{Pooled standard deviation} = \quad 401.2$$

Columns 4 and 6 contain the standardized deviations of the data from the individual sample averages; these are often called z-scores. They are sorted in column 6 with the samples identified for the sorted data in column 5. Columns 7 and 8 display normal and sample probabilities of the standardized deviations or smaller numbers. Finally, columns 9 and 10 display the differences between the probabilities. da is the difference between the normal and sample probabilities on the same line, e.g., 0.014 = 0.064 − 0.050. db is the difference between the normal probability on the same line and the sample probability on the preceding line, e.g., 0.030 = 0.080 − 0.050. The Lilliefors test statistic is the largest absolute datum in column 9 or 10, 0.095. The probability of a Lilliefors statistic this large or larger when chance is the only cause for sample deviations from the normal distribution is considerably more than 0.20, the limit of table M in part V. Consequently, it is reasonable to assume the samples came from normal distributions.

The conditions for a valid t-test of the sample differences have been established. Patients were assigned randomly to the two medications, drug and placebo; the sample standard deviations indicate the populations can have the same standard deviation; and the distribution of the sample data is not significantly different from the normal distribution.

The basic purpose for this analysis is to determine if the drug patients improve more than placebo patients. If the drug is in fact no more effective than the placebo, this is equivalent to chance fluctuations being the only cause for the differences between the samples. No difference and placebo improvement are equivalent alternatives. Consequently, there is a prestudy reason why one sample average should be larger than the other, resulting in this being a one-tailed t-test. The t-statistic can now be calculated.

$$sd = 401.2$$

$$t = \frac{321.4 - (-101.8)}{401.2\sqrt{\frac{1}{11} + \frac{1}{9}}} = 2.346$$

The probability of a t-statistic this large or larger when chance fluctuations in the data are the only cause for sample differences, using table B in part V, is approximately 0.017. This is a small probability, and so we may reasonably conclude that the drug improves walking distance on the treadmill. Unless the data analyst is also a physician, the medical evaluation of the difference should be left to a medical doctor. Many statistically significant differences are not useful or economically important.

Sometimes investigators and/or managers will give enough thought to a research problem to recognize that statistically significant differences may not be useful. The difference must also make sense economically, i.e., it must be marketable; important from an engineering, medical, psychological, or economic viewpoint; worth changing a production system for, etc.; otherwise, who cares? From the statistical viewpoint, this is a decision theory problem, but nothing that precise or elegant will be considered here. Decision theory is not easily explained to investigators and managers, yet these data analysis clients usually have a good understanding of differences that are economically important.

The data analyst can often help to determine an economically important difference while discussing the purpose for a data analysis. The research hypothesis in these situations should reflect this difference. Suppose, in example 2.1, that an improvement in walking distance of less than 100 feet would not be marketable. This doesn't seem like much improvement, but for homemakers and the retired elderly it could represent substantially more freedom of movement. The research hypothesis would then be: The drug group is expected to improve at least 100 feet more in walking distance than the placebo group is expected to improve. The corresponding null hypothesis would be: The improvement in walking distance for the drug group will be no more than the improvement for the placebo group plus 100 feet. Again these hypotheses are stated formally so that the objective of the analysis is clearly understood.

The formula for the t-statistic can be modified to accommodate this new feature. To test this hypothesis, compute

$$t = \frac{\bar{x}_1 - \bar{x}_2 - d}{\text{sd}\sqrt{1/n + 1/m}}$$

The difference d may have either a positive or a negative sign, depending on how the null hypothesis is stated. All quantities in this equation except d are the same as those presented earlier for example 2.1. So, from this point on, the procedure is the same.

$$t = \frac{321.4 - (-101.8) - 100.0}{401.2\sqrt{\frac{1}{11} + \frac{1}{9}}} = 1.792$$

If the effect of the drug is in fact the same as the placebo plus 100 feet, i.e., the null hypothesis is true and the observable differences are due to chance fluctuations only, then the probability of a t-statistic this large or larger is approximately 0.046. This is less than the usually accepted 0.05 probability for significance, but the result of this test is hardly the kind of result on which to base momentous decisions. A distance of 115 feet instead of 100 feet would push the probability above 0.05, for example, or a small error or two in the data as presented in table 2.1 could do the same thing. Consequently, the investigator and/or management should be encouraged to repeat the study at another time and location.

Since the null hypothesis in this part of example 2.1 is that the drug patients walked the same as the placebo patients plus 100 feet, the reader might find it more rational to subtract 100 feet from each drug response or to add 100 feet to each placebo response and then use the t-statistic without the d constant. The results will be the same in any case.

For a particular population parameter, confidence limits define a range of values, or an interval, in the measurement scale such that there is a given probability called the *confidence* that the interval will include the unknown, fixed parameter. Confidence limits are customarily 95% limits, even though the analyst is free to choose any confidence level desired. Formulas for confidence limits are given in chapter 19. The formula for the lower confidence limit for the difference between the two hypothetical populations is

$$\text{cl} = \bar{x}_1 - \bar{x}_2 - t \cdot \text{sd} \sqrt{\frac{1}{n} + \frac{1}{m}}$$

All quantities on the right except t have already been defined and calculated for the sample. Choose t for $n + m - 2$ degrees of freedom and the desired confidence level, $1 - \alpha$, where α is the probability that the confidence interval will *not* include the parameter.

For example 2.1, let $\alpha = 0.05$; then t for a one-sided confidence interval from table B in part V is 1.734. The lower 95% confidence limit for the difference is 110.4. Since the principal data analysis issue was whether or not the drug enabled patients to walk further, this confidence limit is a lower bound on the difference, i.e., the analyst can be 95% confident that 110.4 feet is less than the difference between the means of the two hypothetical populations.

The notion of a population is useful even though a precise definition of a population is not always possible. This example is such a situation. The hypothetical drug and placebo populations might be all patients who would ever take the respective agents, who could participate in such studies, etc. The definitions for populations are quite nebulous in most studies. Nevertheless, the analyst can be very confident (95%) that the hypothetical drug population has an average walking distance 110.4 feet greater than the average for the hypothetical placebo population, assuming appropriate population definitions.

A data analysis is only one aid in understanding a situation. This analysis indicates that the observed drug and placebo difference is probably not due to chance and that there is a high probability 110 feet is less than the population difference. Most management would never make a decision on the basis of this analysis alone. The medical significance of the observed difference and the limit for that difference must be decided. This is a subjective decision outside the scope of either data analysis or statistics.

For patients who only walk around a house or yard, this increase may be very desirable, but for postal workers and police who do much walking, this increase may not be medically worthwhile. Analysts should never expect management to base decisions solely on statistical results. To do so implies an emotional attachment to the results which can only decrease the analyst's objectivity and usefulness.

Reports must always present the results of an analysis in simple terms. Probabilities and test statistics are not universally understood, so they should not be used as substitutes for clear, nontechnical explanation of analytical results. Test results should be verbalized and presented with the associated probabilities. Averages, standard deviations, and confidence limits for the population expected values and the difference between two expected values should be reported either in the body of the report or in an appendix. The statistical tests used to compare the two sample variances and to test for normality should be identified and the results verbalized, but the test statistics themselves shouldn't necessarily be in the body of the report. Descriptions and details of these ancillary tests can more appropriately be placed in an appendix.

The results of this analysis can be safely summarized as follows. The drug and placebo samples appear to be from normally distributed populations. The standard deviations are sufficiently alike to justify the use of a *t*-test to test the differences between the sample averages. The expected values for the populations appear to be different. It is probably reasonable to conclude that other patients with intermittent claudication assigned these treatments would show similar differences. Care must be taken that the patients, before assignment to either the drug or the placebo groups, are like the first ones used. The patients of another physician, from different racial or ethnic backgrounds, in a different age group, etc. might not respond in the same way.

Suppose that the samples are not from normal distributions or that the sample standard deviations cannot be considered equivalent. What can the analyst do? Here are two alternatives.

1. Transform the data within the ratio and/or interval scale.
2. Transform the data to a lower measurement scale, i.e., ordinal or nominal.

Both of these alternatives will be considered.

PARAMETRIC TRANSFORMATIONS

Parametric transformations are transformations within the ratio and interval measurement scales. Their purpose is to change the shape of non-normal sample distributions to shapes that appear and/or test more nearly normal. The analyst must first determine the shape of the sample distribution. This is best done by preparing a histogram of the data with at least five to seven cells. Consequently, a sample of less than 15 to 20 data is probably too small for preparing a histogram as a prelude to choosing a transformation.

Parametric transformations are seldom determined as the result of analysis; they are usually selected arbitrarily. In practice, their use is more art than science. Optimum transformations might be possible for very large samples but samples are seldom very large. Instead, the analyst begins with the histogram, uses good judgment, and then experiments. Sample distributions with long tails toward larger measurements can have their tails shortened by using some form of the simple logarithmic transformation, i.e.,

$$y' = \ln (a + by)$$

where the y are the original data and the y' are the transformed data. Similarly, sample distributions with long tails toward smaller measurements can have their tails shortened by using some form of the simple exponential transformation, i.e.,

$$y' = e^{a+by}$$

where e is the exponential function. There are many other transformations, some of which are noted in chapter 14, but these two are probably used more than all the others combined.

Parametric transformations present several problems. An appropriate transformation must be selected and then the coefficients (e.g., a and b in the above transformations) that will make the transformation work best must be determined. The analyst must be able to defend these decisions when questioned by management or the investigator. Last, but not least, the transformation should be appropriate and necessary for similar samples in other studies. Problems that data analysts encounter in using parametric transformations are discussed in some detail in chapter 14.

Once the data have been transformed, the transformed data are used for all calculations and tests of hypotheses. The results such as averages, standard deviations, and confidence limits are usually reverse-transformed before reporting to management, but reverse transformation can create interpretation problems. The use of parametric transformations is illustrated in example 2.2.

Example 2.2

The data in table 2.3 are similar to the data in table 2.1. They are differences in the distances for two walks. The principal data analysis issue is also the same: Did the drug enable patients to walk further?

TABLE 2.3   Data for Example 2.2

Drug	Placebo	Drug	Placebo
324	70	202	−100
408	89	173	649
669	356	−34	123
2	39	1041	237
266	−27	113	23
1286	165	242	932

The standard deviations for the drug and placebo samples are 408.5 and 300.9, respectively, and the F-statistic for comparing the variances is 1.84. The probability of an F-statistic with 11 and 11 degrees of freedom this large or larger in a two-tailed test, assuming the null hypothesis that the population variances are not different and the observed sample differences are just chance fluctuations, is approximately 0.34. Like example 2.1, the sample standard deviations are obviously different, but it is reasonable to proceed with the analysis assuming equivalency of the population standard deviations.

The data in the two samples are combined as in example 2.1 for a Lilliefors test of normality. The Lilliefors statistic is large, 0.223, indicating the data shouldn't be considered normally distributed. The data minus the respective sample averages are then used to prepare the histogram in figure 2.1. The long tail toward larger measurements indicates that a logarithmic transformation can be tried.

The log transform has two parameters and the question is: What values should be used? The simplest transformation is always the best, so the natural logarithm of $y + 110$, i.e., $\ln (y + 110)$, will be tried first. The constant 110 is chosen to eliminate negative arguments for which the natural logarithmic function is not defined. The Lilliefors test statistic for the transformed data is 0.149. This is smaller than the test statistic for the untransformed data, but it can be made even smaller by using the transformation $\ln (y + 200)$.

Using $\ln (y + 200)$, the Lilliefors test statistic 0.115 is now reasonably small. The variances of the transformed data should be compared just to ascertain that the transformation has not had an adverse effect. The new drug and placebo standard deviations are 0.6694 and 0.6576. The F-statistic, 1.036, is also very small. The remainder of the analysis is like example 2.1.

$$\text{sd} = 0.6635 \qquad t = 1.361$$

The probability of a t-statistic this large or larger is approximately 0.095, assuming the null hypothesis that the population expected values are not different is true and that the sample difference is just chance vari-

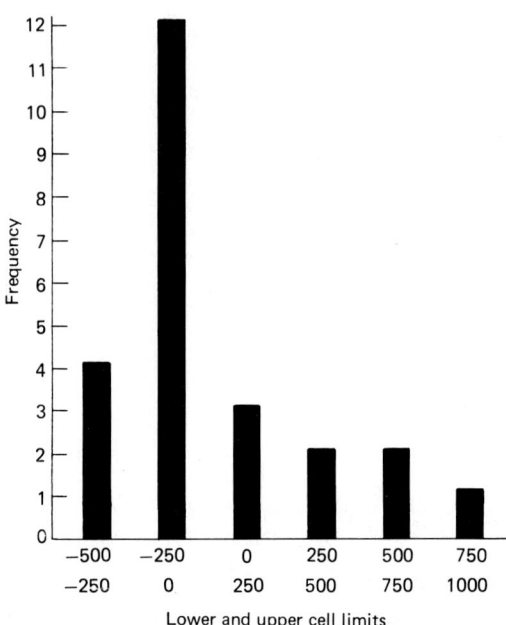

FIGURE 2.1 Histogram for example 2.2; data responses minus sample means.

ation. This probability is much too large to attach any statistical significance to the difference between the sample averages.

This analysis can be summarized as follows. The original data do not appear to be from normal distributions, so the function $\ln (y + 200)$ was used to transform the data to approximately normally distributed samples. The sample averages and standard deviations in the transformed measurement scale are

	Average	Standard deviation
Drug	6.1822	0.6576
Placebo	5.8134	0.6694

The standard deviations are not significantly different and a *t*-test of the difference, 0.3688, between the averages indicates the difference is probably due to chance. The lower 95% confidence limit for the difference between the expected values in the transformed scale is -0.0963.

Again the principal data analysis issue was whether or not the drug enabled patients to walk further. The confidence limit is a lower bound on the difference, i.e., the analyst can be 95% confident that -0.0963 is less than the difference between the expected values of the hypothetical populations in the transformed measurement scale.

The reverse transformation presents some interesting problems. The analyst may want to report the averages of the raw data or to report the reverse transformation of the averages of the transformed data. These will not be the same, of course.

	Average raw data	*Reverse transform of the average transformed data*
Drug	391	284
Placebo	213	135

The reverse transform of the difference between the averages of the transformed data 0.3688 and the 95% confidence limit for the population difference -0.0963 have no real meaning in the original measurement scale. A meaningful difference in the original measurement scale can be the difference between the reverse transforms of the averages, i.e.,

$$284 - 135 = 149$$

Confidence limits that perhaps have meaning can be obtained as follows from the drug and placebo averages.

The drug average exceeds the placebo average, but the confidence limit indicates that the placebo average can exceed the drug average by 0.0963 with a probability of 0.05. To base a confidence limit in the original measurement scale on the drug average, first add 0.0963 to the drug average. This results in a kind of upper limit on the placebo average, so perform the reverse transform and subtract it from the reverse transform of the drug average.

Drug average	6.1817
Confidence limit on difference	0.0963
Upper placebo limit	6.2780
Reverse transform of drug average	284
Reverse transform of placebo limit	333
Confidence limit in the original measurement scale	-49

The confidence limit indicates that the drug average can be less than the placebo average by 0.0963 with a probability of 0.05. To base a confidence limit in the original measurement scale on the placebo average, first subtract 0.0966 from the placebo average. The result is a kind of lower limit for the drug average, so subtract the reverse transform of the placebo average from the reverse transform of the drug limit.

Placebo average	5.8142
Confidence limit on difference	0.0963
Lower drug limit	5.7179
Reverse transform of drug limit	104
Reverse transform of placebo average	135
Confidence limit in the original measurement scale	−31

Now these are not the same and the data analyst must decide which to use. Leaving it to the investigator and/or management to decide will only create confusion for the client and disrespect for the analyst. Before leaving this, the reader should note that these "limits" for the difference in expected values of the drug and the placebo populations in the original measurement scales have been derived from the confidence limit for the difference in the transformed measurement scale. Confidence limits for the expected values of the separate populations in the transformed scale provide an alternative method for defining a confidence limit for the difference in the original measurement scale.

NONPARAMETRIC TRANSFORMATIONS

Nonparametric transformations are transformations that move data from one measurement scale to another. The original sample distribution and statistics have no meaning in the new measurement scale. The distribution of the transformed data is always known, conceptually simple, but sometimes difficult to use. Nonparametric transformations are discussed in chapter 14.

Two of the most important reasons for transforming ratio and interval measurements to the ordinal scale are (1) unequal sample standard deviations and (2) nonnormal sample distributions. Neither of these conditions is uncommon in data analysis. The two samples in table 2.4 are

TABLE 2.4   Data for Example 2.3

Drug	Placebo	Drug	Placebo
705	190	15	127
962	359	475	480
11	679	584	546
311	270	281	332
1413	−116	119	313

similar to the data used for examples 2.1 and 2.2 except that these samples are from populations that have different variances and so will be used to illustrate nonparametric transformation from the ratio measurement scale down to the ordinal measurement scale.

Ordinal Measurements

The transformation from the ratio and/or interval measurement scale to the ordinal scale is done as follows. Sort the two samples together while retaining sample identity. Now replace the data with ranks: the smallest datum is replaced with 1, the next smallest with 2, and so on. The largest datum is replaced with N where $N = n + m$, the sum of the sample sizes. Now put the data back into the original samples. The samples in table 2.4 have been transformed in this manner to produce the data in table 2.5.

TABLE 2.5 Transformed Data for Example 2.3

Drug	Placebo	Drug	Placebo
18	6	3	5
19	12	13	14
2	17	16	15
9	7	8	11
20	1	4	10

A transformation to ranks has some special implications. The difference between data adjacent on an ordinal measurement scale is always the same. An average rank has no particular meaning on the original interval scale, so a test of sample differences can no longer be made in terms of averages. Instead the test for sample differences involves the complete samples. The null hypothesis tested is that the two samples are from distributions that occupy the same space on the ratio and/or interval scale vs. the hypothesis that one is shifted to the right or the left.

Strictly speaking, this null hypothesis requires that the shape, and therefore the variances, of the two distributions be the same. Theoreticians have made this amply clear, but in practice, data analysts use the rank transform deliberately whenever sample variances are unequal. Pratt (1964) has shown that this practice can be a serious mistake, particularly when statistical significance at a given α level is rigidly interpreted. Data analysts, nevertheless, continue to use rank procedures in this situation, perhaps because the theoreticians have provided no useful alternatives.

Consider the situation when there is no difference between the samples. The data consist of the ranks 1 through N. Samples under this assumption can be simulated by drawing numbers from a hat. The first n numbers drawn from the hat can be the ranks for one sample, while the $m = N - n$ numbers remaining in the hat can be the ranks for the second sample. The numbers are drawn without replacement.

Now if the simulation conditions do not bias the draws from the hat, the ranks assigned each sample will be about the same but mixed, of course, in different ways for each simulation. The most extreme result would be for ranks 1 through n or ranks $m + 1$ through N to be drawn for the first sample. This is a very unlikely event. Absolutely every possible arrangement of N ranks assigned to two samples can be delineated and any summary statistic that measures the sample differences can be studied.

Wilcoxon (1945) proposed using the sum of the ranks as a measure of sample differences. The Wilcoxon rank sum test, as usually presented in statistical texts, is no exception to the general rule that nonparametric procedures can be a bit tedious. The tedium with this procedure arises from the fact that tables for expected values of the rank sum are large and difficult to arrange succinctly. To reduce the table size, only a portion of it is published, with instructions for obtaining the remainder as needed. The modified Wilcoxon procedure presented here permits greatly simplified tables and, it is hoped, increased understanding, thereby reducing the tedium.

Readers may know this general procedure as the Mann-Whitney test. Mann and Whitney (1947) proposed a test statistic that is equivalent to the Wilcoxon statistic, except for a constant, when there are no ties. Since Wilcoxon's paper predates the paper of Mann and Whitney by 2 years, the procedure is given Wilcoxon's name for this book.

Example 2.3

Consider as an example the data in table 2.5. Designate the number of data in the drug sample n and the number in the placebo sample m even though the samples contain equal numbers of data. Since Wilcoxon's rank sum procedure does not require sample sizes to be equal, using m and n permits a general presentation.

The two samples can be summarized.

	Drug	Placebo
Sum of ranks	112	98
Sample size	$n = 10$	$m = 10$

Either of the two sums is a Wilcoxon statistic. For the modified procedure presented here, the expected value is subtracted from the rank sums. The average rank for the ranks from 1 to $N = n + m$ is $(N + 1)/2$ so the expected value for the drug sample of n is this average multiplied by n, i.e.,

$$\frac{n(n + m + 1)}{2}$$

and for the placebo sample of m, it is

$$\frac{m(n + m + 1)}{2}$$

Now $n = m = 10$ in this example, so the expected value for both samples is 105. The test statistics for the modified procedure are

	Drug	*Placebo*
Test statistic, W	+7	−7
Sample size	$n = 10$	$m = 10$

$W = +7$ is the statistic for testing the research hypothesis that the drug responses have greater ranks than the placebo responses, implying that the patients given the drug walked further than the patients given the placebo. Comparison with table E in part V shows that the probability of a deviation as large as or larger than $+7$, assuming the null hypothesis is true, i.e., that the two hypothetical populations are equivalent, is 0.315.

When either sample size exceeds 20, the extent of table E, a normal theory approximation to the test can be used. The test statistic is

$$W^* = \frac{W_1/n - W_2/m}{\sqrt{\mathrm{var}(R)(1/n + 1/m)}}$$

where W_1 and W_2 are the test statistics for drug and placebo as defined above, n and m are the sample sizes, and $\mathrm{var}(R)$ is the common variance calculated for the ranks, i.e.,

$$\mathrm{var}(R) = \frac{\sum_{i=1}^{N} (R_i - \bar{R})^2}{N - 1}$$

where $N = n + m$.

The statistic W^* has a distribution that approaches the standard normal distribution asymptotically as m and n tend to infinity. Technically, both m and n should be greater than 20 before the normal approximation is used. In practice, the normal approximation is used when either is greater than 20 simply because the tables for W do not go beyond that point.

While the samples in example 2.3 have less than 20 data each, they can be used to demonstrate the normal theory approximation. The sum of the 20 ranks is 210 and the sum of the ranks squared is 2870. The variance is computed from these sums using $n + m - 1 = 19$ for the degrees of freedom. Then

$$W^* = \frac{\%_{10} - (-\%_{10})}{\sqrt{35(\%_{10} + \%_{10})}} = 0.529$$

This is compared to the standard normal distribution, table A in part V. If the null hypothesis is true, so that the only reason W^* differs from zero is chance, then the probability of a W^* as large as or larger than 0.529 is 0.298, which is in agreement with the probability of W using table E.

Ties in ranks can exist because measurements in ratio or interval scales cannot always resolve two very close quantities. When there are ties among the $m + n$ observations, average the ranks that would normally be assigned to a set of equal observations and assign this average rank to each in the set. Repeat this procedure for all sets of equal observations and then use the average ranks to compute W, W^*, and the variance of R in the standard normal approximation. The variance computed with average ranks will always be slightly less than the variance computed from an equivalent number of untied ranks.

The nonparametric transformation does not eliminate the need for an estimate of the sample difference or for confidence limits for the population difference in the ratio and/or interval scales. Hodges and Lehmann (1963) suggested that an estimate of the sample difference could be obtained as follows.

Pair each datum in one of the samples with each datum in the other sample and compute the difference. If the two sample sizes are m and n, there will be m times n of these differences. The Hodges-Lehmann sample difference is the median of the pair differences.

Consider the data in table 2.4. There are 10 data in each sample, so there are 100 differences. These differences are presented sorted in table 2.6. The median is 118.5, i.e., the average of the two middle differences.

A confidence interval for the difference between the populations can be obtained in the following way. Select the confidence level $1 - \alpha$ for the interval. If equal tails are desired, divide α by 2. Find W in table E for α or $\alpha/2$ and the sample sizes. This procedure was first defined as a graphical procedure by Moses in chapter 18 of Walker and Lev (1953) and may also be found in Moses (1965).

TABLE 2.6 100 Pair Differences for Example 2.3

−668	−265	−71	127	348	700
−664	−259	−71	131	373	734
−560	−255	−51	143	392	772
−535	−240	−48	154	394	821
−531	−235	−32	159	397	835
−469	−213	−21	162	416	867
−465	−204	−8	184	427	933
−427	−199	−5	205	435	1054
−398	−194	−2	225	457	1078
−368	−179	11	225	482	1081
−361	−175	26	235	515	1100
−348	−169	38	252	578	1143
−344	−151	41	271	591	1223
−321	−116	91	283	603	1286
−317	−112	104	285	630	1529
−302	−95	116	314	649	
−298	−78	121	346	692	

Let the median rank be the rank of the median difference if there is an odd number of differences, and let it be the average rank of the two differences averaged for the median if there is an even number of differences. The confidence limits are the differences whose rank is the median rank plus and/or minus W from table E. This will often result in fractional ranks and the rule is to round these toward the median rank.

Consider the sample differences in table 2.6 and find the lower 0.95 confidence limit. W from table E is 23 ($\alpha = 0.05$) and the median rank is 50.5. The rank of this confidence limit is

$$50.5 - 23 = 28$$

rounded toward the median rank. The confidence limit is then −175.

To appreciate the simplicity of the modified procedure presented here, the reader should compare it to the procedure presented in other books.

Nominal Measurements

Data can originate in the nominal scale and, of course, data measured in the ratio and interval scales can be transformed to the nominal scale. Measurement in the nominal scale is simple classification, so classes must be defined for the transformation. The analyst is free to determine classes in any appropriate way, but the definition often determines the outcome. Ratio or interval data transformed to the nominal may not completely lose the "greater than" defining relationship, so in some respects the transformed data are "grossly" measured ordinal data rather than true nominal data.

Consider the data in table 2.1. Positive numbers indicate improvement while negative numbers indicate a deterioration. This is then a natural classification to use for nominal scale measurements. Table 2.7 presents the data in example 2.1 transformed to this particular nominal scale (1). These positive and negative signs can be counted for a contingency table as follows:

	Drug	Placebo
Plus	8	3
Minus	3	6

The research hypothesis is that more people improved on the drug than on the placebo. The null hypothesis that can be tested is: The odds for improvement on the drug are the same as the odds for improvement on the placebo.

The chi-square test for goodness of fit is used to test the null hypothesis for contingency tables. Using the simplified formula with Yates' correction presented in chapter 15, the chi-square test statistic is 1.72. The probability of chi-square 1.72 or larger when chance is the only cause for differences in improvement is about 0.193 (table C in part V). This does not agree well with the previous analysis of these data in this chapter. Much so-called information is lost by transforming the data to the nominal scale. This probability is too large to reject the null hypothesis, although in example 2.1 the null hypothesis was rejected.

Suppose for medical or promotional reasons, the classification scale is not simple improvement vs. deterioration but that the scale is improvement by 200 feet vs. improvement by less than 200 feet and deterioration.

TABLE 2.7 Example 2.1

Nominal scale 1		Nominal scale 2	
Drug	Placebo	Drug	Placebo
+	−	+	−
+	+	+	I
+	−	+	−
+	−	+	−
+	−	−	−
−	+	−	−
+	+	+	−
−	−	−	−
+	−	+	−
+		−	
−		−	

Table 2.7 also presents the data in example 2.1 transformed to this nominal scale (2). The contingency table now becomes

	Drug	*Placebo*
Plus	6	1
Minus	5	8

and chi-square is 2.42. The probability of a chi-square 2.42 or larger when chance is the only cause for differences in improvement is about 0.127. While this probability is still large, it illustrates the importance of choos-

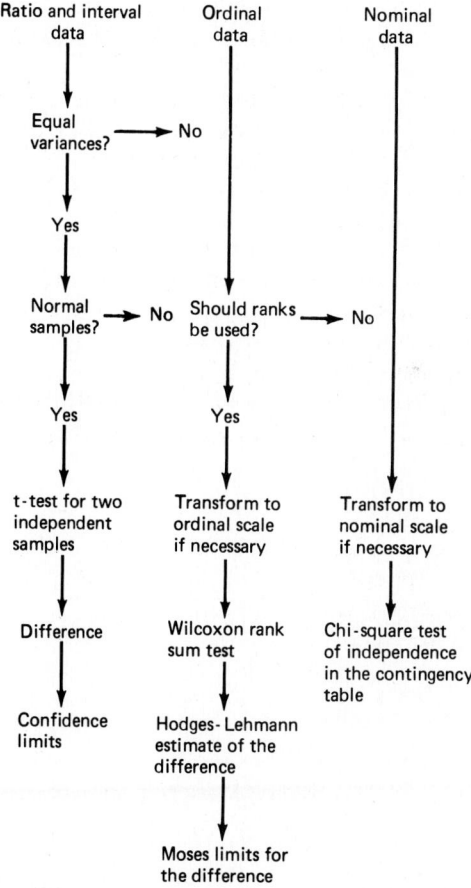

FIGURE 2.2 Analysis sequence flowchart for chapter 2.

ing meaningful nominal class definitions. Arbitrary definitions can result in completely wrong decisions.

SUMMARY

There is no procedure for comparing two independent samples that is correct to use all of the time. The analysis sequence suggested in this chapter is not universally accepted by data analysts. Statistical purists can find many legitimate faults with it, but until theoreticians develop something better, data analysts, who must help investigators and managers understand data, are stuck with it or with something very similar.

Figure 2.2 is a flowchart that summarizes the analysis sequence suggested here.

CHAPTER 3

Two Dependent Samples

The comparison of two things using dependent samples is discussed in this chapter. The data in two dependent samples can be classified two ways, in contrast to the data in two independent samples which can be classified in only one way. One way is obvious, i.e., the two samples. The other way is not so obvious. It associates each datum in one of the samples with a single datum in the other sample. The associated data are called *paired observations*.

A new concept is introduced in this chapter, i.e., the notion of variability, or difference, in data arising from more than one identifiable source. Variability can arise from many sources in a single problem, but in this chapter, variability from only two sources is considered. The two sources will be associated with the two ways that the data can be classified. Secondary classifications for data usually identify undesirable or nuisance sources of differences that make it more difficult, or even impossible, to find meaningful differences between the two samples.

There are many ways that data in two samples can be associated. Measurements made before and after some event are common ways that association arises. For example, the measurements might be miles per gallon before and after automobile engine tune-ups, or they might be white cell counts before and after administration of antibiotics. Here the association arises from two observations on the same experimental units.

Associations in space and time are also common. The difference in yields for two varieties of apples may be the research problem. The primary classification for the data will be the two varieties, while the secondary classification might be the location of the trees (space) or it might be the harvest year (time).

Suppose several orchardists grow two varieties of apples but the orchards are in different locations. Yields from a single variety grown in widely differing locations may exhibit greater differences than the differences between varieties grown in the same location. Thus location can introduce undesirable differences into the data, differences that may make it impossible to see varietal differences. The primary classification for yields per tree or per acre would be the two varieties. The secondary classification would be locations of the orchards. Each yield for one variety is associated with one yield for the other variety by being grown in the same location. An analysis procedure that can remove undesirable location differences before examining the difference between varieties would certainly be useful for the data analyst to have.

Now consider a single orchardist who has been growing two varieties of apples for a number of years and who now wants to plant a new orchard or to replant an old one. A new orchard requires several years to become productive, so to maximize future profits, the orchardist compares the annual yields for the two varieties over several years. Like the location example, the primary classification for the yields would be the two varieties, but the secondary classification would be years. Yields from different years for one variety may exhibit greater differences than the differences between the varieties. Such year-to-year differences would certainly be undesirable and a nuisance.

The single orchardist has a real problem: Which variety should be planted? It is not sufficient to simply compare the average yields over the years assuming independent samples. While this will certainly tell which was most productive, a test of the null hypothesis of no differences may indicate there are only chance differences between the varieties. The choice might then be made on some other factor, e.g., selling price. The procedures described in this chapter can make the test more sensitive to varietal differences by removing the undesirable geographic location or time differences.

No matter how such studies as these are planned, undesirable variability can be associated often with the secondary classification. Normally an engine tune-up study will employ several different motors. The undesirable variability will be associated with the motors, their differences, or differences in the effectiveness of the tune-ups. The problem cannot be avoided by using the same motor repeatedly. Each time a motor is tuned it is older, and hence different from previous times observations were made.

The situation is similar for white cell counts before and after therapy. If more than one experimental unit, i.e., animal or human, is used for the study, then the undesirable variability is associated with the experimental unit differences. Again the problem cannot be avoided by using a single experimental unit repeatedly. Each time an experimental unit is given antibiotic therapy, it becomes a different unit in many ways, e.g., older, perhaps less responsive, from times it was previously treated.

The research hypothesis always concerns the two samples and is a statement of some anticipated difference. The null hypothesis tested is that no difference exists between the two populations after removing differences associated with the secondary classification. The idea is this. The data can be classified in two ways. One classification, the two samples themselves, identifies the primary research interest. The other classification identifies something that introduces undesirable differences into the data, i.e., variability that can make a statistical test less sensitive. These undesirable differences are removed in the analysis before making a statistical test. This is a simple idea but a powerful tool in the hands of a resourceful data analyst.

RATIO AND INTERVAL MEASUREMENTS

Example 3.1

In 1972, the U.S. Congress mandated the establishment of Professional Standards Review Organizations (PSROs). According to the authorizing legislation, PSROs review services provided in hospitals and other inpatient facilities under the Medicare, Medicaid, and the Maternal and Child Health and the Crippled Children's programs. PSROs have the responsibility of assuring that federally funded health services are medically necessary, meet professionally recognized standards of care, and are appropriately provided in the most economical settings. Over the years, the name and direction of this program have been changed, but the basic purpose of the original and subsequent legislation has remained the containment of the cost of medical services.

The PSRO program is based on the concept that physicians are the most appropriate individuals to assess the quality of medical care and that local peer review is the most effective means for ensuring appropriate use of health care resources and facilities. Every PSRO governing body, composed primarily of physicians, is required to create an administrative structure capable of efficiently and effectively carrying out the purposes of the organization. An executive director and a medical director are usually the key individuals charged with these responsibilities. They in turn hire and supervise other PSRO personnel for program management, data management, and review functions. Patterns of medical practice and care that deviate from norms are reviewed and programs for bringing the deviations in line with norms are carried out. One easily defined and measured variable that PSROs have studied extensively is lengths of stay in hospitals. There are all sorts of reasons why lengths of stay can be too long and a few good reasons why they can be too short. Most are traceable to physician and/or hospital inefficiencies. Consequently, they have been a natural for PSRO study.

Table 3.1 presents data for two dependent samples. The samples contain the average lengths of stay in hospitals for the patients of 11 physicians. One sample contains averages for the period July through December while the other contains averages for the same months a year later. Between the two periods, the PSRO conducted a campaign to reduce lengths of stay. The principal data analysis issue is the following: Was the campaign successful? To state the issue another way: Were the lengths of stay shorter in the second period than in the first?

TABLE 3.1   Data for Example 3.1: Average Lengths of Stay in Hospitals (in days)

Physician	July–December, year 1	July–December, year 2	Difference
1	14.3	10.9	3.4
2	12.9	10.5	2.4
3	12.2	13.9	−1.7
4	10.4	7.9	2.5
5	14.4	12.2	2.2
6	9.7	9.2	0.5
7	9.9	11.8	−1.9
8	13.2	9.2	4.0
9	10.8	9.9	0.9
10	10.2	11.4	−1.2
11	17.5	13.3	4.2

These data can be classified in two ways, by time period and by physician. The principal data analysis issue defines the primary classification, i.e., time periods. Physicians are the secondary classification and are associated with such factors as different practices, different case loads, different towns, and all of the other factors that make one physician's work different from another. Each datum in the first sample is paired with a single datum in the second sample. The data pairs are, of course, identified with physicians. Differences between physicians are secondary to the differences between time periods. They are unavoidable, and for the data analyst they are an absolute nuisance.

The research hypothesis is that the lengths of stay for the second year were shorter than they were for the first year, i.e., the PSRO campaign to reduce hospital stays for these 11 physicians was successful. The reason for the study was to examine this research hypothesis. The null hypothesis that will be tested statistically in the analysis is that no difference exists between the two populations. The purpose of the secondary classification, i.e., physicians, is to identify and to permit removal of differences that would otherwise interfere with a comparison of the time periods. To illustrate this, the samples in table 3.1 are first compared as independent samples and then as dependent samples.

The reader should use the Lilliefors procedure to verify that the samples are from normal distributions and a two-tailed F-test to verify that the population standard deviations may be considered equivalent. The sample averages and standard deviations are

	First year	Second year
Average	12.32	10.93
Standard deviation	2.44	1.83

If the samples are considered to be independent, the t-statistic for a test of the null hypothesis is 1.51. The probability of a t-statistic this large or larger when the null hypothesis is true is approximately 0.076.

Physician differences are removed by computing the change in the length of stay for each physician. The differences, one for each physician, are then the subject of the analysis. Now this may seem like a trivially simple procedure, but it is equivalent to the more sophisticated procedure used in chapter 5 for removing secondary differences from more than two dependent samples. Thus the null hypothesis that no differences exist between the two samples after removing differences associated with physicians can be rephrased as follows: The expected value of the pairwise differences, one for each physician, is equivalent to zero.

The statistical test for comparing two dependent samples is also a form of Student's t-test, specifically, the form for one sample. The one sample is, of course, the pair differences. Additional details of this test are presented in chapter 15.

To test the restated null hypothesis, first compute the physician differences as in table 3.1. Then compute

$$t = \frac{\overline{D}}{\text{sd}\sqrt{1/n}}$$

where $\overline{D}$ is the average physician difference and

$$\text{sd}^2 = \frac{\sum_{i=1}^{n}(D_i - \overline{D})^2}{n-1}$$

is the estimate of the variance computed from the n physician differences.

This is a simpler form of the t-statistic than that used in chapter 2, and the conditions that must be true for validity of the t-test are also simpler. They are

1. The sample must be random.

2. The sample must be from a normal distribution.

Note that the variances for the two dependent samples do not have to be equal since the *t*-test is a one sample test of the physician differences. As will be emphasized repeatedly, it is not sufficient in data analysis to assume that necessary conditions are true. The last condition above can be tested.

Calculations for the Lilliefors test are presented in table 3.2. The test statistic is 0.187. Assuming that the sample is from a normal population and that the only differences are due to chance variation, the probability of a Lilliefors statistic this large or larger for a sample of 11 is greater than 20%. This is a large probability, therefore the sample can be assumed to be from a normal population and a valid *t*-test of the paired differences can be made.

TABLE 3.2 The Lilliefors Test for Normality for Example 3.1

						Probabilities			
ID	y	y − ȳ	z	ID	z	Normal	Sample	da	db
1	3.4	2.0	0.901	7	−1.475	0.070	0.091	−0.021	
2	2.4	1.0	0.452	3	−1.385	0.083	0.182	−0.099	−0.008
3	−1.7	−3.1	−1.385	10	−1.161	0.123	0.273	−0.150	−0.059
4	2.5	1.1	0.497	6	−0.399	0.345	0.364	−0.019	0.072
5	2.2	0.8	0.363	9	−0.220	0.413	0.455	−0.042	0.049
6	0.5	−0.9	−0.399	5	0.363	0.642	0.545	0.096	0.187
7	−1.9	−3.3	−1.475	2	0.452	0.674	0.636	0.038	0.129
8	4.0	2.6	1.170	4	0.497	0.690	0.727	−0.037	0.054
9	0.9	−0.5	−0.220	1	0.901	0.816	0.818	−0.002	0.089
10	−1.2	−2.6	−1.161	8	1.170	0.879	0.909	−0.030	0.061
11	4.2	2.8	1.259	11	1.259	0.896	1.000	−0.104	−0.013

NOTE: The Lilliefors test statistic is 0.187. Refer to chapter 15 for details.

The research hypothesis is that the campaign to reduce lengths of stay was successful, i.e., that the lengths of stay in the second period are shorter than the lengths for the first period. The null hypothesis is that there are no differences between the two periods. The null hypothesis, restated in terms of the physician differences, is that the expected value for the pair differences is not different from zero.

As the research hypothesis is stated, this is a one-tailed *t*-test. The *t*-statistic can now be calculated.

$$\text{average difference} = 1.39$$

$$sd = 2.23$$

$$t = \frac{1.39}{2.23\sqrt{\frac{1}{11}}} = 2.07$$

The probability of a *t*-statistic this large or larger is approximately 0.032, assuming the null hypothesis is true and that the observed difference from no change in length of stay is due to chance variation only. While small, this probability may not be small enough for management to conclude that the campaign to reduce lengths of stay was successful.

Removing the physician differences increased the t-statistic from 1.51 to 2.07 and decreased the p-value from 0.076 to 0.032. The comparison of the two time periods was made more sensitive by removing the differences in the data that can be associated with the physicians. Since the primary purpose of every analysis is the examination of sample differences under reasonable conditions, this demonstrates the usefulness of utilizing secondary classifications whenever possible. A report comparing the two time periods should state clearly the use made of secondary classifications.

It can be argued that this should be a two-tailed test. The research hypothesis is correct because the purpose of the campaign was to reduce lengths of stay; but if the campaign had the opposite effect, i.e., increased lengths of stay, then management would be alarmed. Hence a two-tailed test might be in order.

Just as with independent samples, it may occasionally be necessary to compare sample differences with a given quantity, which for convenience can be called d. Usually such given quantities are the result of an investigator and/or management decision concerning what is important to their goals. While a data analyst may be able to help a client select these arbitrary quantities, their selection is not really a data analysis or statistical problem.

The research and null hypotheses must reflect the arbitrary difference. Suppose, for example, the PSRO management considered the lengths of stay so large during the first time period that an average reduction less that a half day could not be considered important. The research hypothesis would then be that the lengths of stay for the second year should be a half day or more shorter than they were for the first year. This may not seem like much for a single patient, but one-half day for each Medicare-Medicaid–financed hospitalization in a PSRO area is an enormous dollar value. The null hypothesis becomes: There is no difference between lengths of stay during the first year and lengths of stay during the second year plus a half day. Contrast the null and the research hypotheses. The null hypothesis is a statement that no differences exist, while the research hypothesis is a statement of the difference anticipated.

To test the null hypothesis, compute

$$ t = \frac{\overline{D} - d}{\text{sd}\sqrt{1/n}} $$

where $\overline{D}$ is the average of the pair differences and $d = 0.5$ days. The t-statistic is then 1.32. Assuming that the null hypothesis is true and that the observed differences are due to chance variation only, the probability of a t-statistic this large or larger is about 0.110. The quantity d may have either a positive or a negative sign, and care must be exercised in determining which is appropriate in a particular situation. Because of the way t-tables are prepared, the research hypothesis should be stated so that $\overline{D} - d$ is positive.

Confidence limits are discussed in chapter 19. The formula for the lower confidence limit for the difference between the two hypothetical populations is

$$\text{cl} = \overline{D} - t \cdot \text{sd} \sqrt{\frac{1}{n}}$$

All quantities on the right of this equation except t have been defined. Choose t for $n - 1$ degrees of freedom and the desired confidence level, $1 - \alpha$, which is usually 95%.

For example 3.1, let $\alpha = 0.10$; then t for a one-sided 90% confidence interval is 1.372 and the confidence limit is 0.47. This is a lower limit for the reduction in lengths of stay, i.e., whatever the effect of the campaign on the lengths of stay, the analyst can be 90% confident that 0.47 of a day is less than the decrease achieved. The 89.1% confidence limit is exactly one-half day. If the only influence on the medical care system as practiced by these 11 physicians was the campaign to reduce lengths of stay, then the campaign can be credited with this reduction.

Any particular confidence interval calculated should include the true, or population, average with a probability of $1 - \alpha$. This does not mean that any given interval will include a proportion, $1 - \alpha$, of the population of sample differences or of future sample averages. Intervals with these properties are tolerance intervals, not confidence intervals. Tolerance intervals are discussed in Eisenhart, Hastay, and Wallis (1947) and in many journals. The results of the analysis can be summarized as follows.

Data in the two samples can be paired on physicians. Consequently, a procedure appropriate for two dependent samples was used to compare the average lengths of stay for the two time periods. The difference between the sample averages was 1.39 days. The probability of a difference this large or larger, assuming the null hypothesis is true and the observed difference is due to chance alone, is approximately 0.032. While this probability is not very small, the minimum set by management as economically important, i.e., one-half day, is the 0.891 confidence limit for the reduction achieved by the campaign. Thus the significance of the decrease might be accepted with caution.

Central Limit Theorem Implications

Statisticians often ignore completely the possibility that pair differences might not be normally distributed, almost as if taking differences preordained normality. A moment's reflection should reveal the fallacy in the assumption, even though there seems to be an element of truth to it from the data analysis viewpoint.

In its simplest form, the central limit theorem is that any sum tends toward a normal distribution with unknown average and variance. The pair differences qualify as a sum of two samples, even though one sample is given a minus sign. No matter what distribution the original samples had, the difference will be more nearly normally distributed than either original. If the originals were reasonably bell-shaped, then the difference may be very close to the normal. All of this is very problematical from the data analysis viewpoint. So the prudent analyst will always check the differences for normality.

The central limit theorem in action can be easily seen in the data for example 3.1. These are "average" hospital lengths of stay for 6-month periods. Lengths of stay have a lower bound of zero and a very long tail toward long stays. Figure 3.1 is a histogram of the lengths of stay for phy-

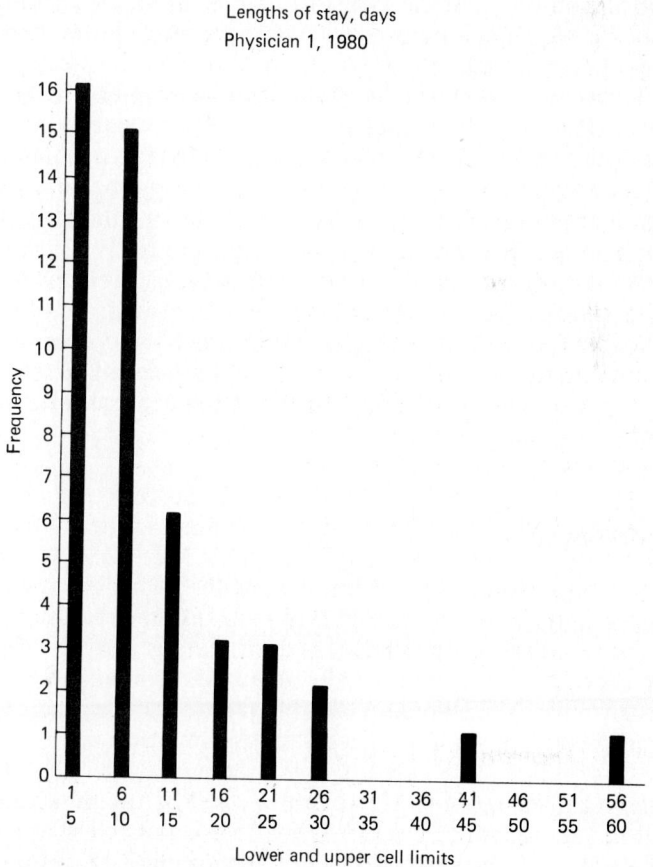

FIGURE 3.1   Histogram for example 3.1.

sician 1 during the second year and illustrates the anormality of the raw data. Nevertheless, averages over 6 months test normal. This the reader has verified.

TRANSFORMATIONS

Parametric transformations are not used very much with two dependent samples, although except for tedium, there is probably no reason for not using them. Use of parametric transformations with two dependent samples could cause a great debate. Should the sample data be transformed or should the differences be transformed? Probably the differences should be transformed because it is the differences that must be normally distributed for the t-test, not the data from which the differences are computed. Of course, if the samples were normal, the differences would most likely be normal too.

The most useful parametric transformations are the logarithmic and square root for reducing long tails toward larger measurements, whereas the exponential will reduce long tails toward smaller measurements. Many of the interpretation problems discussed in chapter 2 will certainly exist for two transformed dependent samples. So the analyst must weigh the advantages and disadvantages of staying in the ratio or interval measurement scale against moving into the ordinal or even the nominal scale.

Nonparametric transformations are much more common for two dependent samples than are parametric. There are many different ways to transform ratio or interval data to ordinal. A particular transformation is usually associated with a particular nonparametric procedure or test and vice versa. The rank transformation introduced in chapter 2 for two independent samples is just one of many rank transformations. A different transformation, one specifically for two dependent samples, is introduced in this chapter.

Ordinal Measurements

A common nonparametric procedure or test for comparing two dependent samples is the Wilcoxon signed rank test. Like the rank sum for two independent samples, this procedure is a bit tedious because of the way tables for expected values are usually arranged. A modified procedure that is easier to use is presented in this chapter.

The transformation for the signed rank is done as follows. Compute the data pair differences. Discard any differences that are zero and recount the number of pairs. This count, n, will be the number of data pairs used in subsequent calculations. While it is possible that all of the pair differences have the same sign, more likely some are positive while others are negative. Ignore the sign temporarily and rank the differences by absolute value. Substitute ranks for the differences and then assign the signs of the differences to the ranks.

The ranking procedure for two dependent samples is not as simple as the procedure for independent samples. Table 3.3 demonstrates the procedure using the dependent sample data in table 3.1. The first four columns are identical to the columns in table 3.1. Since none of the differences are zero, none of the data pairs are discarded. The data in the fourth column are sorted without regard to sign and placed in the fifth column. Sorting without regard to sign places +0.9 before −1.2 in column 5. Column 6 contains the ranks with the signs of the differences from column 5.

TABLE 3.3 Data for Example 3.1: Average Lengths of Stay in Hospitals (in days)

			Differences		
Physician	*July–December, year 1*	*July–December, year 2*	*Not sorted*	*Sorted*	*Signed*
1	14.3	10.9	3.4	0.5	1
2	12.9	10.5	2.4	0.9	2
3	12.2	13.9	−1.7	−1.2	−3
4	10.4	7.9	2.5	−1.7	−4
5	14.4	12.2	2.2	−1.9	−5
6	9.7	9.2	0.5	2.2	6
7	9.9	11.8	−1.9	2.4	7
8	13.2	9.2	4.0	2.5	8
9	10.8	9.9	0.9	3.4	9
10	10.2	14.4	−1.2	4.0	10
11	17.5	13.3	4.2	4.2	11

Column 6 in table 3.3 immediately suggests possible test statistics: sum of the positive ranks, sum of the negative ranks, and the sum of all the ranks. All of these are used in statistical literature to the confusion of the data analyst. The sum of the positive ranks is the sum of the ranks where the first measurement in the pair is greater than the second. The sum of the negative ranks is the sum of the ranks where the second measurement in the pair is greater than the first. Either is a rank sum statistic.

Consider the situation when there is no difference between the two samples. The positive rank sum and the negative rank sum should be about equal, i.e., near one-half of the total sum of absolute values of the signed ranks. If there are n nonzero differences, this is $n(n + 1)/4$ for the positive and $-n(n + 1)/4$ for negative rank sums.

Now consider the two extreme situations:

1. The data in the first sample are all pairwise larger than the data in the second sample.

2. The data in the first sample are all pairwise smaller than the data in the second sample.

In the first situation, the sum of the positive ranks would be the sum of the positive integers from 1 through n, i.e., $n(n + 1)/2$, and the sum of the negative ranks would be zero. In the second situation, the sum of the positive ranks would be zero while the sum of the negative ranks would be the sum of the negative integers from -1 through $-n$ or $-n(n + 1)/2$.

From the above, it seems reasonable to expect the average sum of the positive or negative ranks to be midway between their extremes and that is precisely what happens. The modified Wilcoxon signed rank procedure presented here uses the sum of the positive ranks minus its expected value or the sum of the negative ranks minus its expected value as the test statistic. The expected value for the sum of positive ranks is

$$\frac{n(n + 1)}{4}$$

and the expected value for the sum of negative ranks is

$$-\frac{n(n + 1)}{4}$$

It is now irrelevant which signed rank sum is used because the modified signed rank statistic, i.e., the signed rank sum minus its expected value, will be the same for both sums. If the difference was originally defined as sample 1 minus sample 2, then a positive modified rank sum indicates sample 1 has larger data while a negative modified rank sum indicates sample 1 has smaller data.

The signed ranks in table 3.3 can be summarized now.

	Positive ranks	Negative ranks
Sum	+54	−12
Expected	+33	−33
Test statistic W	+21	+21
$n = 11$		

Table F of part V presents upper-tail probabilities for the test statistic for n from 2 through 30. Comparison with this table shows that the probability of a deviation from expectation this large or larger is 0.034, assuming the null hypothesis of no differences is true and the observed differences are due to chance causes only. This is in good agreement with the probability determined for the ratio and interval data using Student's t-statistic.

Sometimes the research hypothesis is that one sample is different from the other without the direction of that difference being specified. There are two alternatives to the research hypothesis in this situation. The first sample may be larger than the second or it may be smaller than the second. Such tests are, of course, two-tailed, and the probability obtained from table F in part V should be doubled.

Estimates are usually needed in the ratio and interval scales for the difference between the samples and for confidence limits for the difference between populations. Hodges and Lehmann (1963) proposed the following procedure for an estimate of the differences.

Using the nonzero pairwise sample differences D, compute Walsh (1949) averages

$$\frac{D_i + D_j}{2}$$

for every pair of differences such that $i \leq j$. If n is the number of nonzero pairs, then there will be $n(n + 1)/2$ Walsh averages. Now sort these and use the median as the estimate of the difference between the two samples.

Consider example 3.1. There are 11 nonzero differences in table 3.3. Table 3.4 presents sorted the 66 Walsh averages that can be computed from these 11. The estimate for the difference is the median, i.e., the average of the two middle Walsh averages, which is 1.425.

TABLE 3.4 Walsh Averages for Example 3.1

−1.90	−0.15	0.70	1.45	2.35	3.20
−1.80	0.15	0.75	1.50	2.35	3.20
−1.70	0.25	0.85	1.50	2.40	3.25
−1.55	0.25	0.90	1.55	2.45	3.30
−1.45	0.30	1.05	1.65	2.45	3.35
−1.20	0.35	1.10	1.70	2.50	3.40
−0.70	0.40	1.15	1.95	2.55	3.70
−0.60	0.50	1.15	2.15	2.80	3.80
−0.50	0.50	1.25	2.20	2.90	4.00
−0.40	0.60	1.35	2.25	2.95	4.10
−0.35	0.65	1.40	2.30	3.10	4.20

A confidence interval for this estimate of the sample difference can be obtained in the following way. Select the confidence level $1 - \alpha$ for the interval. If both upper and lower limits are needed, divide α by 2. Find W in table F for α or $\alpha/2$ and the sample size. The confidence limits are the Walsh averages whose rank is the median rank plus and/or minus W.

This will often result in fractional ranks and the rule is to always round these toward the median rank.

Consider the 66 Walsh averages for example 3.1 and find the lower 0.897 confidence limit, i.e., $\alpha = 0.10$. The median rank is 33.5 and W from table F is 15. The rank of this confidence limit is

$$33.5 - 15 = 19$$

rounded toward the median rank. The confidence limit is 0.50. To find the upper 0.90 confidence limit, the factor from table F is given a positive sign.

The reader should compare this procedure with the procedure presented in chapter 2 for a nonparametric estimate and confidence limit in the two-independent-sample problem. The similarity should make both easier to remember and use.

When the sample sizes are greater than 30, a normal theory approximation to the test can be used. The test statistic is:

$$W^* = \frac{W}{\sqrt{\text{var}(W)}}$$

where $W =$ the test statistic already defined
$\quad$ $\text{var}(W) = \Sigma R^2/4$
$\quad\quad\quad$ $R =$ the ranks from 1 to n

The statistic W^* has a distribution that approaches the standard normal distribution asymptotically as n tends to infinity. The approximation is acceptable for data analyses with n above 20 and quite good for n above 30, the limit of table F.

Since the samples in example 3.1 have only 11 data, another problem has been chosen to illustrate use of the above formula and to introduce another normality test.

Example 3.2

The Hamilton psychiatric rating scale for depression is a widely used measure of depression arranged so that a higher score indicates deeper depression. Twenty-four patient characteristics are rated, most on a five-point scale, and then these characteristic ratings are summed. *Suicide* is one of the characteristics rated. The score and a description of each category are:

(0) absent
(1) feels life is not worth living

(2) wishes he or she were dead or has any thoughts of possible death to self

(3) suicide ideas or gestures

(4) attempts of suicide (any serious attempt rates 4)

Retardation is another characteristic rated. The scores and category descriptions for this characteristic are:

(0) normal speech and thought

(1) slight retardation at interview

(2) obvious retardation at interview

(3) interview difficult

(4) complete stupor

Hopelessness is a characteristic usually included in the depression scale even though it was not part of the original group. The scores and descriptions for the five categories in this characteristic are

(0) not present

(1) intermittently doubts that "things will improve" but can be reassured

(2) consistently feels hopeless but accepts reassurances

(3) expresses feelings of discouragement, despair, and pessimism about the future which cannot be dispelled

(4) spontaneously, continuously, and inappropriately asserts, "I'll never get well" or its equivalent

The data presented in table 3.5 are Hamilton depression scores for 32 patients before and after treatment with an antidepressant. The research hypothesis is that treatment with the antidepressant should lower the Hamilton scores.

The before-therapy minus after-therapy differences for example 3.2 are given in column 4 of table 3.5. The first issue is the normality of these data. Table 3.6 presents the calculations for the Lilliefors test of normality. The test statistic is 0.160. Assuming that these data are from a normal population and the observable deviations from normality are just chance variation, then the probability of a Lilliefors statistic larger than 0.160 is approximately 0.045. This probability is much too small to justify using parametric methods.

The Shapiro-Wilk (1965, 1968) test for normality will be demonstrated using the data in example 3.2. The only condition for using this test is that the data be a random sample. The hypothesis tested is that the sample is from a normal distribution with unspecified average and variance.

The Shapiro-Wilk procedure requires the calculation of three sum-

TABLE 3.5   Data for Example 3.2: Hamilton
Depression Scores

Patient number	Before therapy	After therapy	Difference, D
245	32	9	23
274	37	5	32
442	30	37	−7
548	38	13	25
850	34	3	31
867	34	9	25
868	34	9	25
869	39	9	30
870	32	32	0
871	36	47	−11
872	34	35	−1
880	25	11	14
881	26	17	9
882	33	16	17
886	27	11	16
887	24	11	13
890	33	6	27
891	28	18	10
892	29	10	19
893	30	15	15
895	26	12	14
896	31	16	15
897	28	15	13
898	28	13	15
900	34	7	27
901	32	8	24
902	28	29	−1
903	27	10	17
905	27	20	7
906	32	41	−9
908	28	34	−6
910	31	16	15

mations after first sorting the data in ascending order. These summations
are

$$D = \sum_{i=1}^{n} (y_i - \bar{y})^2$$

$$c_1 = \sum_{i=1}^{k} a_i y_{n-i+1}$$

and

$$c_2 = \sum_{i=1}^{k} a_i y_i$$

TABLE 3.6 The Lilliefors Test for Normality for Example 3.2

ID	y	$y - \bar{y}$	z	ID	z	Normal	Sample	da	db
						\multicolumn{2}{c}{*Probabilities*}		
245	23	9.2	0.757	871	−2.054	0.020	0.031	−0.011	
274	32	18.2	1.501	906	−1.889	0.029	0.063	−0.033	−0.002
442	−7	−20.8	−1.723	442	−1.723	0.042	0.094	−0.051	−0.020
548	25	11.2	0.922	908	−1.641	0.050	0.125	−0.075	−0.043
850	31	17.2	1.419	902	−1.227	0.110	0.156	−0.046	−0.015
867	25	11.2	0.922	872	−1.227	0.110	0.188	−0.078	−0.046
868	25	11.2	0.922	870	−1.145	0.126	0.219	−0.093	−0.061
869	30	16.2	1.336	905	−0.566	0.286	0.250	0.036	0.067
870	0	−13.8	−1.145	881	−0.401	0.344	0.281	0.063	0.094
871	−11	−24.8	−2.054	891	−0.318	0.375	0.313	0.063	0.094
872	−1	−14.8	−1.227	887	−0.070	0.472	0.344	0.128	0.160
880	14	0.2	0.013	897	−0.070	0.472	0.375	0.097	0.128
881	9	−4.8	−0.401	895	0.013	0.505	0.406	0.099	0.130
882	17	3.2	0.261	880	0.013	0.505	0.438	0.068	0.099
886	16	2.2	0.178	893	0.096	0.538	0.469	0.069	0.101
887	13	−0.8	−0.070	910	0.096	0.538	0.500	0.038	0.069
890	27	13.2	1.088	896	0.096	0.538	0.531	0.007	0.038
891	10	−3.8	−0.318	898	0.096	0.538	0.563	−0.024	0.007
892	19	5.2	0.426	886	0.178	0.571	0.594	−0.023	0.008
893	15	1.2	0.096	903	0.261	0.603	0.625	−0.022	0.009
895	14	0.2	0.013	882	0.261	0.603	0.656	−0.053	−0.022
896	15	1.2	0.096	892	0.426	0.665	0.688	−0.022	0.009
897	13	−0.8	−0.070	245	0.757	0.776	0.719	0.057	0.088
898	15	1.2	0.096	901	0.840	0.799	0.750	0.049	0.081
900	27	13.2	1.088	867	0.922	0.822	0.781	0.041	0.072
901	24	10.2	0.840	548	0.922	0.822	0.813	0.009	0.041
902	−1	−14.8	−1.227	868	0.922	0.822	0.844	−0.022	0.009
903	17	3.2	0.261	900	1.088	0.862	0.875	−0.013	0.018
905	7	−6.8	−0.566	890	1.088	0.862	0.906	−0.045	−0.013
906	−9	−22.8	−1.889	869	1.336	0.909	0.938	−0.028	0.003
908	−6	−19.8	−1.641	850	1.419	0.922	0.969	−0.047	−0.016
910	15	1.2	0.096	274	1.501	0.933	1.000	−0.067	−0.035

NOTE: The Lilliefors test statistic is 0.160. Refer to chapter 15 for details.

where y = the sample data sorted
k = $n/2$ rounded down
a = Shapiro-Wilk factors from table N1 of part V.

The test statistic is then

$$\pi = \frac{(c_1 - c_2)^2}{D}$$

Table N2 in part V is used to determine the probability of a π smaller than an observed π.

Most test statistics are defined so that significance increases as the test statistic increases, i.e., the probability of the test statistic decreases as the test statistic increases. The Shapiro-Wilk statistic is one that behaves in the opposite way. This test statistic was defined so that significance increases as the test statistic decreases, i.e., the probability of the test statistic decreases as the test statistic decreases. Consequently, the analyst must be particularly careful using table N2 of part V.

Table 3.7 illustrates the Shapiro-Wilk procedure for example 3.2. Column 1 identifies the patient and column 2 presents the sorted pairwise differences. The Shapiro-Wilk factors have been copied into column 3 from table N1 of part V. The products are given in column 4. The three summations required for the test statistic are

$$D = 4534.22 \qquad c_1 = -4.38 \qquad c_2 = 60.63$$

TABLE 3.7   Shapiro-Wilk Test for Normality for Example 3.2

Patient number	Differences	Factors	Products
871	−11	0.4188	−4.6068
906	−9	0.2898	−2.6082
442	−7	0.2462	−1.7234
908	−6	0.2141	−1.2846
872	−1	0.1878	−0.1878
902	−1	0.1651	−0.1651
870	0	0.1449	0.0000
905	7	0.1265	0.8855
881	9	0.1093	0.9837
891	10	0.0931	0.9310
887	13	0.0777	1.0101
897	13	0.0629	0.8177
895	14	0.0485	0.6790
880	14	0.0344	0.4816
910	15	0.0206	0.3090
898	15	0.0068	0.1020
896	15	0.0068	0.1020
893	15	0.0206	0.3090
886	16	0.0344	0.5504
903	17	0.0485	0.8245
882	17	0.0629	1.0693
892	19	0.0777	1.4763
245	23	0.0931	2.1413
901	24	0.1093	2.6232
868	25	0.1265	3.1625
867	25	0.1449	3.6225
548	25	0.1651	4.1275
900	27	0.1878	5.0706
890	27	0.2141	5.7807
869	30	0.2462	7.3860
850	31	0.2898	8.9838
274	32	0.4188	13.4016

The test statistic is

$$\pi = \frac{(-4.38 - 60.63)^2}{4534.22} = 0.932$$

Assuming that the data are from a normal population and that the sample differs from normality only because of chance, then the probability of a test statistic this small or smaller is approximately 0.059. This probability, like the probability for the Lilliefors statistic, is too small to justify a parametric test.

A histogram of the data for example 3.2 is presented in figure 3.2. Note that there are no differences in the +1 to +5 cell so that the sample is bimodal, a fact that rules out use of a parametric transformation.

The research hypothesis is that the Hamilton depression scores before therapy are greater than they are after therapy. Accordingly, the pair differences have been defined as the before-therapy minus after-therapy scores. These differences, sorted without regard to the sign, are presented in column 5 of table 3.8. Column 1 identifies the patient and column 6 presents the signed ranks. Note that tied ranks have been given the average rank for the tied group and that zero differences have been discarded.

The signed ranks can be summarized as follows.

	Positive ranks	Negative ranks
Sum	470	−26
Expected	248	−248
Test statistic W	222	222
$n = 31$		

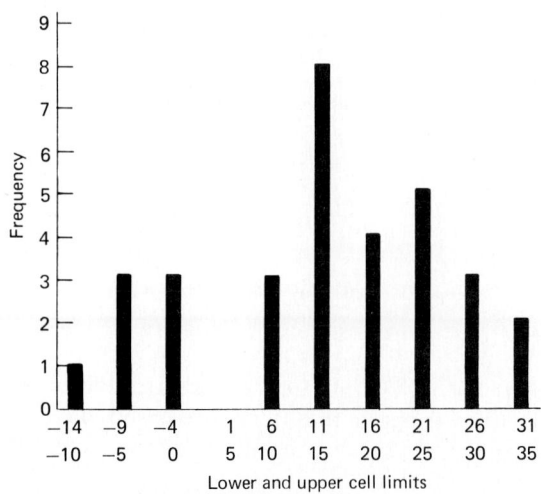

FIGURE 3.2 Histogram for example 3.2.

TABLE 3.8   Data for Example 3.2: Hamilton Depression Scores

Patient number	Before therapy	After therapy	Differences Not sorted	Differences Sorted	Signed ranks
245	32	9	23	0	
274	37	5	32	−1	−1.5
442	30	37	−7	−1	−1.5
548	38	13	25	−6	−3
850	34	3	31	−7	−4.5
867	34	9	25	7	4.5
868	34	9	25	−9	−6.5
869	39	9	30	9	6.5
870	32	32	0	10	8
871	36	47	−11	−11	−9
872	34	35	−1	13	10.5
880	25	11	14	13	10.5
881	26	17	9	14	12.5
882	33	16	17	14	12.5
886	27	11	16	15	15.5
887	24	11	13	15	15.5
890	33	6	27	15	15.5
891	28	18	10	15	15.5
892	29	10	19	16	18
893	30	15	15	17	19.5
895	26	12	14	17	19.5
896	31	16	15	19	21
897	28	15	13	23	22
898	28	13	15	24	23
900	34	7	27	25	25
901	32	8	24	25	25
902	28	29	−1	25	25
903	27	10	17	27	27.5
905	27	20	7	27	27.5
906	32	41	−9	30	29
908	28	34	−6	31	30
910	31	16	15	32	31

The test statistic W^* is

$$W^* = \frac{222}{\sqrt{2601.4}} = 4.35$$

Assuming that the null hypothesis is true and the observable difference between the samples is just due to chance alone, the probability of a W^* as large or larger than this asymptotically normal statistic (table A in part V) is less than 0.00002.

An estimate in the ratio or interval scale of the difference between the samples is obtained from the $n(n + 1)/2$ sorted Walsh averages. Table 3.9 presents these averages for example 3.2. The median is 15.0.

Confidence limits for the difference between samples are based on the normal approximation used for the test. There is an even number of Walsh averages so the median rank is the average of the two middle

TABLE 3.9 Walsh Averages for Example 3.2

−11.0	2.0	5.5	8.5	11.5	13.5	15.0	17.0	19.5	21.0	23.0	26.0
−10.0	2.5	6.0	9.0	11.5	13.5	15.0	17.0	19.5	21.0	23.0	26.0
−9.0	2.5	6.0	9.0	11.5	14.0	15.0	17.0	19.5	21.0	23.0	26.0
−9.0	2.5	6.0	9.0	12.0	14.0	15.0	17.0	19.5	21.0	23.0	26.0
−8.5	3.0	6.0	9.0	12.0	14.0	15.5	17.0	19.5	21.0	23.0	26.5
−8.0	3.0	6.0	9.0	12.0	14.0	15.5	17.0	19.5	21.0	23.0	27.0
−7.5	3.0	6.0	9.0	12.0	14.0	15.5	17.0	19.5	21.0	23.0	27.0
−7.0	3.0	6.5	9.0	12.0	14.0	15.5	17.5	19.5	21.0	23.5	27.0
−6.5	3.0	6.5	9.0	12.0	14.0	15.5	17.5	19.5	21.0	23.5	27.0
−6.0	3.0	6.5	9.0	12.0	14.0	15.5	17.5	20.0	21.0	23.5	27.0
−6.0	3.0	6.5	9.5	12.0	14.0	15.5	17.5	20.0	21.0	23.5	27.5
−6.0	3.0	6.5	9.5	12.0	14.0	15.5	18.0	20.0	21.0	23.5	27.5
−5.0	3.0	6.5	9.5	12.0	14.0	15.5	18.0	20.0	21.0	23.5	27.5
−5.0	3.0	7.0	9.5	12.0	14.0	15.5	18.0	20.0	21.0	23.5	27.5
−4.0	3.5	7.0	9.5	12.0	14.5	15.5	18.0	20.0	21.0	23.5	27.5
−4.0	3.5	7.0	10.0	12.0	14.5	16.0	18.0	20.0	21.5	24.0	28.0
−3.5	3.5	7.0	10.0	12.0	14.5	16.0	18.0	20.0	21.5	24.0	28.0
−3.5	3.5	7.0	10.0	12.0	14.5	16.0	18.5	20.0	21.5	24.0	28.0
−2.0	3.5	7.0	10.0	12.0	14.5	16.0	18.5	20.0	21.5	24.0	28.0
−1.0	4.0	7.0	10.0	12.5	14.5	16.0	18.5	20.0	21.5	24.0	28.5
−1.0	4.0	7.0	10.0	12.5	14.5	16.0	18.5	20.0	22.0	24.0	28.5
−1.0	4.0	7.0	10.5	12.5	14.5	16.0	18.5	20.0	22.0	24.0	28.5
−1.0	4.0	7.0	10.5	12.5	14.5	16.0	18.5	20.0	22.0	24.5	28.5
−1.0	4.0	7.0	10.5	12.5	14.5	16.0	18.5	20.0	22.0	24.5	28.5
−0.5	4.0	7.0	10.5	12.5	14.5	16.0	19.0	20.0	22.0	24.5	29.0
0.0	4.0	7.0	10.5	12.5	14.5	16.0	19.0	20.0	22.0	24.5	29.0
0.0	4.0	7.5	10.5	13.0	14.5	16.0	19.0	20.0	22.0	24.5	29.5
0.5	4.0	7.5	11.0	13.0	15.0	16.0	19.0	20.0	22.0	24.5	29.5
0.5	4.0	7.5	11.0	13.0	15.0	16.0	19.0	20.0	22.0	25.0	30.0
1.0	4.0	8.0	11.0	13.0	15.0	16.0	19.0	20.0	22.0	25.0	30.5
1.0	4.5	8.0	11.0	13.0	15.0	16.5	19.0	20.5	22.0	25.0	31.0
1.0	4.5	8.0	11.0	13.0	15.0	16.5	19.0	20.5	22.5	25.0	31.0
1.5	4.5	8.0	11.0	13.0	15.0	16.5	19.0	20.5	22.5	25.0	31.5
1.5	4.5	8.0	11.0	13.0	15.0	16.5	19.0	20.5	22.5	25.0	32.0
1.5	4.5	8.0	11.0	13.0	15.0	16.5	19.0	20.5	22.5	25.0	
1.5	4.5	8.0	11.0	13.0	15.0	16.5	19.0	20.5	22.5	25.0	
2.0	4.5	8.0	11.5	13.0	15.0	17.0	19.0	20.5	22.5	25.0	
2.0	5.0	8.0	11.5	13.0	15.0	17.0	19.0	20.5	22.5	25.5	
2.0	5.0	8.0	11.5	13.5	15.0	17.0	19.5	20.5	22.5	25.5	
2.0	5.0	8.0	11.5	13.5	15.0	17.0	19.5	20.5	23.0	25.5	
2.0	5.0	8.5	11.5	13.5	15.0	17.0	19.5	20.5	23.0	26.0	
2.0	5.5	8.5	11.5	13.5	15.0	17.0	19.5	21.0	23.0	26.0	

ranks, i.e., 248.5. The confidence limits are the Walsh averages whose ranks are this median rank plus and/or minus

$$z_\alpha \sqrt{\text{var}(W)}$$

where z_α is the standard normal deviate for the confidence level $1 - \alpha$ desired.

For example 3.2, let $\alpha = 0.10$; then z for a two-sided confidence interval is 1.645. The ranks for the Walsh averages that define this confidence interval would be

$$248.5 - 1.645 \sqrt{2601.4} = 164.6$$

and

$$248.5 + 1.645 \sqrt{2601.4} = 332.4$$

Some writers advise simply rounding these to the nearest integer. The conservative data analyst will round these away from the median rank, i.e., use ranks 164 and 333 for this problem. The analyst can then report that the interval 11.5 through 19.5 includes the true sample difference with a 90% confidence.

Nominal Measurements

Nominal scale measurement is simply the classification of observations. Data measured in the ratio and interval scales can always be transformed to the nominal scale, and of course data can originate in the nominal scale. Classes appropriate for the problem are chosen for transforming ratio or interval data to the nominal measurement scale, but the choice of classes can influence the outcome of an analysis. An example of this problem will be given. When the nominal classes can be ordered then ordinal measurement has been achieved, which is usually the case with ratio or interval data transformed to the nominal, and procedures appropriate for ordinal measurements should be used even though the number of classes might be small.

A popular test for the analysis of two dependent samples in the nominal scale is Fisher's sign test. As usually presented, this procedure is not very useful to the data analyst. A small but important modification will greatly increase its usefulness and illustrate the effect of choosing different definitions for the classes.

Like the ratio, interval, and ordinal scale procedures, the pairwise differences are computed first. Then the differences are classified into one of two categories: plus or minus, depending on the sign of the difference. Of course if the measurements originate in the nominal scale, this difference is determined subjectively, but plus or minus signs can still be assigned to the pairs even though they may not refer to a greater than or a less than status. They may simply mean that for a data pair, the datum from the first sample is preferred over the datum from the second sample or vice versa.

If the two samples are nearly alike, the number of positive signs will be very near the number of negative signs. If one sample is generally larger than or preferred over the other, then the number of positive signs will not be near the number of negative signs. The most numerous sign will depend on how the pairwise differences are defined.

The research hypothesis is again that one of the samples is in general larger than the other, or that one is preferred over the other, or that one is simply different from the other. The null hypothesis that can be tested statistically is that there is no difference between the two samples, i.e., the data in the two samples differ only randomly. Assuming this is correct, then the probability of a positive sign for a given data pair is one-half. The probability of B or more positive signs in n pairwise comparisons is given by the binomial distribution. An example will make this clear.

Consider example 3.1. The number of positive signs is distributed as the binomial with parameters n and p. For this example, n is 11, and assuming the null hypothesis is true, $p = 0.5$. There are eight positive signs (table 3.3) in this sample, so the test statistic B is 8. The probability of seeing 8 or more positive signs in 11 differences can be determined from table H in part V. Assuming the null hypothesis is true and the observed difference in the number of positive and negative signs is due to chance alone, then this probability is 0.1133. Based on this nominal measurement analysis, the data analyst can report that the first year lengths of stay are in general greater than the second year lengths of stay. The p-value for this nominal scale difference, however, is a large 0.1133. The analyst might want to advise management or the researcher to do other studies before making a decision that will significantly influence future actions.

Table H focuses on the most numerous sign. If the research hypothesis focuses on the least numerous sign, then the pairwise difference should be redefined.

Research hypotheses are always a statement of the difference anticipated or desired by management or a researcher and so simple differences as presented in table 3.3 are seldom the principal interest in a data analysis. As already discussed in example 3.1, a change less than a half day may be considered unimportant. The Fisher sign test can be modified slightly and made more useful by taking pairwise differences plus or minus a constant. Using one-half day as the constant, table 3.10 was prepared from table 3.1. The zero difference is discarded and the number of nonzero differences is used to enter the binomial table. There are now seven positive signs. Under the same assumptions, the probability of 7 or more positive signs in 10 differences is 0.1719 from table H. The choice of class definitions has had an effect on the results of the analysis.

TABLE 3.10 Data for Example 3.1: Average Lengths of Stay in Hospitals (in days)

Physician	July–December, year 1	July–December, year 2	Differences − 0.5
1	14.3	10.9	2.9
2	12.9	10.5	1.9
3	12.2	13.9	−2.2
4	10.4	7.9	2.0
5	14.4	12.2	1.7
6	9.7	9.2	0.0
7	9.9	11.8	−2.4
8	13.2	9.2	3.5
9	10.8	9.9	0.4
10	10.2	11.4	−1.7
11	17.5	13.3	3.7

When n is about 20 (for $p = 0.5$), a normal approximation can be used. The test statistic is

$$B^* = \frac{B - E(B) - 0.5}{\sqrt{\text{var}(B)}}$$

where $E(B) = np$
$\text{var}(B) = np(1 - p)$

(The quantity 0.5 is a correction for continuity and should not be confused with the half day discussed in the preceding paragraph.)

The approximation of the distribution of B^* by the normal deteriorates as p, the unknown population parameter, differs from 0.5, but improves as the sample size increases. The problem can usually be avoided by formulating the hypotheses so that an equal number of plus and minus differences are expected, i.e., so that p is expected to be 0.5. For example 3.1 where $p = 0.5$, $E(B) = n/2$, $\text{var}(B) = n/4$, and

$$B^* = \frac{8 - 11/2 - 0.5}{\sqrt{11/4}} = 1.206$$

Assuming the null hypothesis that the two populations are not different and the observed difference between positive and negative signs is due to chance alone, the probability of B^* can be determined by comparing 1.206 to percentiles of the normal distribution. Using table A in part V, the probability of 8 or more plus signs in 11 differences is 0.1139. This agrees very well, even when $n = 11$, with the binomial probability.

Table H gives exact probabilities for specific counts of the binomial variable. The reader may find table G more familiar and/or more useful where counts for specific probabilities are given for the special case of p = 0.50. To demonstrate use of this table, consider example 3.1 where there were 8 positive signs in 11 differences. The line for $n = 11$ indicates that the probability of observing 8 or more signs alike in 11 observations is less than or equal to 0.25. Likewise, this line indicates that the probability of there being 9 or more signs alike in 11 observations is less than or equal to 0.10. Taken together, this means that the probability of observing 8 or more signs alike in 11 observations is between 0.25 and 0.10, of course assuming the null hypothesis of no difference between the populations and that the observed difference is due to chance alone. This is not a precise p-value, but for some situations the reader may find it quicker and easier to explain to the investigator or to management.

SUMMARY

Just as with two independent samples, there is no procedure for comparing two dependent samples that is correct to use all of the time. The analysis sequence suggested in this chapter is probably more universally accepted by data analysts for dependent samples than the sequence sug-

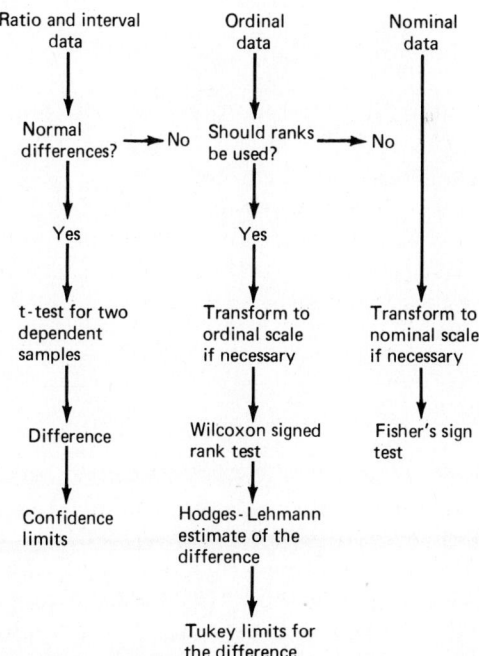

FIGURE 3.3   Analysis sequence flowchart for chapter 3.

gested in chapter 2 is accepted for independent samples. Perhaps that is because there are fewer alternatives in the dependent case. Statistical purists can find legitimate faults with the sequence suggested here but data analysts, who must help investigators and managers understand data, will have to use this sequence or something similar until theoreticians develop something better.

Figure 3.3 is a flowchart that summarizes the sequence suggested here.

CHAPTER 4

More Than Two Independent Samples

Simple comparisons of more than two independent samples are considered in this chapter. Lengths of stay for patients with pneumonia in several hospitals are an example of such data. The comparison of a new drug with a standard drug and a placebo is another example. The principal data analysis issue in both situations concerns differences between the things represented by the samples.

A casual observation may have alerted someone to the possibility of abnormal lengths of stay for patients with pneumonia in a particular hospital. To determine if this is true, lengths of stay for this hospital must be compared with lengths of stay for several other hospitals that are similar in size and patient characteristics. Such a study would be retrospective. Prospective research would require that pneumonia patients be assigned at random to the hospitals in the study, a procedure that is completely unreasonable, of course. Merely planning the study before patients are admitted would not constitute prospective research. The sample of patients for each hospital would probably be all people hospitalized for pneumonia between two arbitrary dates. Thus both the population and the sample would be naturally occurring groups.

Sample distributions for lengths of stay in hospitals often have long tails toward longer stays, i.e., are positively skewed, so that a test of sample differences depending on normality would not be valid. Lengths of stay in a hospital for any particular disease may tend to form natural subsamples that can be identified with physicians since the decision to hospitalize and then when to discharge patients is the result of each physician's judgment. Consequently, it may be better to concentrate attention initially on average lengths of stay for physicians within a hospital rather than on patients. If the number of patients for each physician is 10 or more, averaging for each physician could solve the normality problem also. Table 4.1 is a dummy table without actual measurements to illustrate how these data might be presented for analysis.

Assuming the data analyst finds the suspected hospital different from the others, the next logical step would be to compare the lengths of stay for the different physicians using that hospital. These individual length-of-stay data, as already noted, would probably not be normally distributed. Methods for analyzing such data are discussed later in this chapter.

The comparison of a new drug with an old drug and a placebo is less complex. This would most likely be a prospective study using samples that included all of three populations created especially for the study. Table 4.2 is a dummy table without actual measurements to illustrate how these data might be presented for analysis. The principal data analysis issue would be the comparison of a new product with an old product and a placebo.

TABLE 4.1

Hospital			
1	*2*	*3*	*4*
MD 60	MD 31	MD 79	MD 63
MD 86	MD 58	MD 59	MD 34
MD 08	MD 93	MD 17	
	MD 39		

NOTE: MD = medical doctor.

TABLE 4.2

Active drugs		
New	*Old*	*Placebo*
sn 31	sn 12	sn 72
sn 95	sn 65	sn 44
sn 09	sn 92	sn 15
sn 56	sn 11	sn 05
	sn 81	sn 66

NOTE: sn = subject number.

RATIO AND INTERVAL MEASUREMENTS

Example 4.1

The third most frequent cause of death in humans (in 1981) is heart and brain infarcts. In both of these events, a portion of the heart or the brain is denied the necessary supply of oxygen because of an obstruction in the arterial blood to the affected area. This causes necrosis, i.e., a small amount of local tissue dies, resulting in debilitation or even death for the victim. These incidents occur most often in individuals with hypertension, a medical term meaning abnormally high blood pressure and all the systemic conditions that accompany high blood pressure. Consequently, a diagnosis of hypertension always results in measures to reduce the blood pressure, especially the diastolic pressure.

Two different blood pressures are of interest to the physician. These are the systolic and the diastolic pressure and when measured (in milli-meters of mercury, mm Hg), are usually entered in a patient's records as systolic/diastolic, e.g., 140/80. Now these are measured as follows. The physician or other medical examiner attaches a band around the arm. The band, called a cuff, contains a balloon that when inflated will com-pletely cut off the blood flow to the lower arm. The pressure is then slowly released while the examiner listens with a stethoscope for the first sound of blood getting through the restriction of the cuff. As might be expected, this sound is a faint "squirt, squirt" synchronous with the heartbeats. The heart at that moment is producing a pressure just sufficient to overcome the restriction of the cuff. The air pressure in the cuff is the systolic pres-sure, the highest pressure in the arterial system at the measurement point. As the examiner continues to release the air in the cuff, the squirting sound becomes louder and louder and finally ceases to be a squirt. It becomes rather a "thump, thump" that fades as the pressure is released. When it disappears, the cuff is having no further effect. This is the dia-stolic pressure, the pressure below which the arterial system does not fall.

Hypertension is treated in a number of ways, such as diet, diuretics, antihypertensive drugs, etc. The object for any treatment of hypertension is to reduce the blood pressure, particularly the diastolic pressure. In example 4.1, an antihypertensive was given to five groups of people that differed in age and socioeconomic characteristics. The data analysis issue was this: Did it tend to reduce diastolic blood pressure about the same amount in all five groups or did some groups respond differently from others?

Table 4.3 presents data from these five studies. The first number in

TABLE 4.3   Diastolic Pressures for Example 4.1

		Study		
1	*2*	*3*	*4*	*5*
108, 82	90, 76	124, 102	104, 82	114, 94
108, 100	98, 82	110, 90	105, 84	105, 94
92, 78	102, 86	112, 82	105, 80	110, 96
102, 86	90, 74	118, 96	102, 82	100, 80
102, 88	110, 84	114, 96	106, 86	110, 92
120, 100	105, 82	126, 90	102, 82	98, 86
120, 100	104, 86	126, 92	108, 84	112, 90
105, 92	98, 86	120, 92	108, 82	102, 86
98, 88	98, 86	122, 86	104, 82	110, 96
108, 88	104, 92	116, 96	108, 85	100, 80
108, 90	105, 85		106, 84	100, 88
100, 88	106, 88		110, 90	115, 92
106, 94	102, 84		104, 82	98, 82
100, 82			106, 80	108, 84
95, 76			100, 76	
88, 88				

each data pair is the diastolic blood pressure before beginning medication. The second is the diastolic pressure after taking the drug for 9 weeks. Since the principal issue is equality in the reduction in blood pressure, the reduction computed for each patient is presented in table 4.4.

TABLE 4.4   Reductions in Diastolic
Blood Pressure in Example 4.1

		Study		
1	2	3	4	5
26	14	22	22	20
8	16	20	21	11
14	16	30	25	14
16	16	22	20	20
14	26	18	20	18
20	23	36	20	12
20	18	34	24	22
13	12	28	26	16
10	12	36	22	14
20	12	20	23	20
18	20		22	12
12	18		20	23
12	18		22	16
18			26	24
19			24	
0				

An analysis of variance, abbreviated ANOVA for tables, is a procedure for dividing a total sum of squares into identifiable components. The total sum of squares for this example is

$$\sum_{j=1}^{m} \sum_{i=1}^{n_j} (y_{ij} - \bar{\bar{y}})^2$$

where $\bar{\bar{y}}$ is the average of the data in all five samples and the summation is over all 68 data. The analysis of variance procedure divides this total sum of squares into two components. The first component can be identified with and is a measure of differences *between* the five samples. The other component can be identified with and is a measure of differences between the data *within* the samples. The size of this second component is not influenced by any average difference between the samples since it measures only the differences between data within samples.

These two component sums of squares are used to calculate mean squares or variances identified with the differences between and the differences within the samples. If the mean square for data between samples is about the same size as the mean square for data within samples, then any apparent differences between the samples are probably unimportant. Conversely, a mean square for between samples that is much larger than the mean square for within samples will imply important between-sample differences. This comparison of the components is formalized in an *F*-statistic.

Chapter 20 discusses the analysis of variance in more general terms and actually presents the algebra for dividing the total sum of squares for problems like this most elementary example into components. The formula for the sum of squares between samples is

$$\sum_{j=1}^{m} n_j(\bar{y}_j - \bar{y})^2 = \sum_{j=1}^{m} \frac{\left(\sum_{i=1}^{n_j} y_{ij}\right)^2}{n_j} - \frac{\left(\sum_{j=1}^{m}\sum_{i=1}^{n_j} y_{ij}\right)^2}{N}$$

and for within samples

$$\sum_{j=1}^{m}\sum_{i=1}^{n_j} (y_{ij} - \bar{y}_j)^2 = \sum_{j=1}^{m}\sum_{i=1}^{n_j} y_{ij}^2 - \sum_{j=1}^{m} \frac{\left(\sum_{i=1}^{n_j} y_{ij}\right)^2}{n_j}$$

where N is the sum of the n_j. Some useful intermediate summations are presented in table 4.5 for the analysis of variance presented in table 4.6.

The *F*-statistic will be large, but before making an *F*-test, certain conditions must be true. These conditions are

1. The samples must be independent.
2. The samples must be random.

TABLE 4.5 Sample Statistics for Example 4.1: Reductions in Diastolic Blood Pressure

	\multicolumn{6}{c}{*Sample*}					
	1	*2*	*3*	*4*	*5*	*Total*
Σy	240	221	266	337	242	1306
Σy^2	4154	3973	7524	7635	4426	27,712
n	16	13	10	15	14	68

$$\frac{240^2}{16} + \frac{221^2}{13} + \frac{266^2}{10} + \frac{337^2}{15} + \frac{242^2}{14} = 26,187$$

$$\frac{1306^2}{68} = 25,083$$

TABLE 4.6   The ANOVA Table for Example 4.1

Source of variation	df	ss	ms
Between samples	4	1104.13	276.03
Within samples	63	1524.99	24.21
Total	67	2629.12	

3. The samples must be from normal distributions.

4. The variances of the samples must be homogeneous.

Homogeneity is a technical term which, as used here, implies that the five samples are from populations that have equivalent variances even though they may have different averages.

The reader will remember using an *F*-statistic to compare the variances of two independent samples before making a *t*-test. Now in this example there are more than two variances to compare so the *F*-statistic is not appropriate. Bartlett (1937) developed a procedure for comparing more than two variances that gives results equivalent to the two-tailed *F*-statistic when there are only two samples. Since Bartlett's procedure is used several places in this book, it is described in chapter 15.

Bartlett's procedure is particularly sensitive to nonnormality of the samples. In fact, nonnormality will cause a significant test statistic about as often as will differences between variances. This is not important to the data analyst in this situation since either condition invalidates the *F*-test in an analysis of variance.

The Bartlett statistic for k variances is computed as follows. Let df_i be the degrees of freedom and sd_i be the standard deviation for sample i. Also for convenience, let df be the sum of the degrees of freedom for the samples, i.e.,

$$df = df_1 + df_2 + \cdots + df_k$$

and let

$$sd^2 = \frac{\sum_{i=1}^{k} df_i sd_i^2}{df}$$

Then the test statistic M is

$$M = \frac{1}{c} \left[df \ln (sd^2) - \sum_{i=1}^{k} df_i \ln (sd_i^2) \right]$$

where ln is the natural logarithm function and

$$c = 1 + \frac{\displaystyle\sum_{i=1}^{k}(1/df_i) - 1/df}{3(k-1)}$$

M is approximately distributed as chi-square with $k - 1$ degrees of freedom. Bartlett's test statistic is somewhat complex but it can be easily computed with most scientific pocket calculators.

The sample standard deviations calculated from the summations given in table 4.5 are presented below.

	Sample				
	1	*2*	*3*	*4*	*5*
sd	6.08	4.24	7.06	2.13	4.32
n	16	13	10	15	14

The test statistic M for these five samples is 16.93. Assuming the null hypothesis that the samples came from normal populations with a common variance and the observed differences are due to chance only, the probability of this M-statistic can be determined by comparing 16.93 to percentiles of the chi-square distribution for 4 degrees of freedom. This probability, 0.0036, is too small to proceed any further with five samples. An F-test for the analysis of variance in table 4.6 would not be valid.

The original data analysis issue still persists, so the analyst can't just stop here. The test for the homogeneity of the variances has raised a new issue: Is there a subgroup of samples with equivalent variances that can be analyzed? Samples 2 and 5 have similar standard deviations that are in the middle of the group of 5. Samples 1 and 3 also have similar standard deviations, whereas sample 4 has a small standard deviation that seems to stand alone. When sample 4 is deleted, Bartlett's M-statistic drops to 4.164. The probability of a chi-square with 3 degrees of freedom this large or larger is approximately 0.25, assuming only chance differences cause the deviations from the null hypothesis assumptions.

In chapter 2, the data for two independent samples were pooled for a normality test. There is really no point in making such a test for the five samples in example 4.1, but four of the samples seem to have equivalent variances, so a direct extension of the Lilliefors procedure for two independent samples should be used here for four samples with equivalent variances. The first step is to subtract the sample averages from each of the data so that the average of each sample becomes zero. The standard deviation for completing the normalization of the data is computed from these new data, remembering that the degrees of freedom is the sum of the degrees of freedom for the four samples, i.e., 49. This, of course, is equivalent to pooling or summing the sums of squares for the samples. The Lilliefors test statistic is 0.082. If the null hypothesis that the samples are from normal populations is true and the observed deviations from normality are just chance, then the probability of a Lilliefors statistic this large or larger is much greater than 0.20.

Since the conditions for a valid *F*-test are now satisfied, a new analysis of variance is presented in table 4.7. The *F*-statistic for a test of the null hypothesis that the four samples come from populations having the same mean is 10.05. Assuming the null hypothesis is true and the observed differences are just chance variation alone, then the probability of an *F*-statistic this large or larger is approximately 0.0001. This probability is very small, so the analyst can safely conclude that there are differences among the averages of the four samples.

TABLE 4.7　　The ANOVA Table for Example 4.1

Source of variation	df	ss	ms	
Between samples	3	899.50	299.83	$F = 10.05$
Within samples	49	1461.26	29.82	
Total	52	2360.76		

In practice, an analyst would always satisfy the conditions for an *F*-test before preparing the ANOVA table. Table 4.6 was prepared for this discussion to identify more easily the conditions that must be true and to motivate examining the data for these conditions.

When samples are as large as the samples in this example, they could be examined separately for normality. The Lilliefors statistics for samples 1, 2, 3, and 5 are

$$0.143 \qquad 0.176 \qquad 0.243 \qquad 0.164$$

Now these test statistics are not really large for their respective sample sizes, but examining the samples separately can create a problem, i.e., the probability of a type I error increases with the number of tests made. The reader will remember that a type I error is the error made in rejecting a null hypothesis of no differences when in fact there are no differences. This kind of error is inescapably made with a frequency determined by the probability level chosen for statistical significance. Make enough tests and every analyst will unknowingly make these errors. Assume the analyst wants to prove normality, an impossibility, so 0.20 is chosen for the statistical significance level for the Lilliefors tests. The probability is 0.59 that one test in these four will indicate "significant" differences from normality. This is unacceptably large. Consequently, the safe approach is to combine the samples for a single test as was done in chapter 2 and for this example. Chapter 17 has a general discussion of problems in making many similar tests. The problem is also discussed below in connection with multiple paired comparisons.

Multiple Comparisons of Sample Pairs

The analysis of variance for more than two independent samples examines the homogeneity of the sample averages. If the F-statistic is small, then there is no evidence against the null hypothesis that the samples came from populations with the same mean values. The sample averages, for all practical purposes, can be considered equivalent. If, however, the F-statistic is large, then there is evidence against the null hypothesis and the sample averages cannot be considered equivalent. The analysis of variance and the F-statistic give no other useful information on the differences. Management and researchers usually need much more. One way to get additional information is by comparing each sample with every other sample, i.e., do multiple paired comparisons.

The multiple comparison of different sample pairs presents a problem that will always plague data analysts. The problem is a technical one that can be described as follows. Every time a statistical test is made, the data analyst is trying to decide if the observed differences can be attributed to chance variation or if they are caused by something else. This problem can be restated in strict statistical terminology as follows. Every time a statistical test is made, the data analyst is trying to decide if there is no evidence against the null hypothesis of no differences or if the data contain evidence contrary to the null hypothesis. If the probability of the test statistic, assuming the null hypothesis, is very low, the analyst will usually conclude that the differences are not entirely due to chance. There is a pitfall associated with every test and that is the test statistic may be very large and still due to chance. This kind of error is controlled by choosing the significance level, usually 0.10, 0.05, or 0.01, for the statistical test. Now when a series of similar statistical tests is made on samples that do differ only by chance, the significance level is the proportion of such tests that will indicate the chance difference is not due to chance. This error can be controlled, but it is always there. Make sufficient tests at a particular significance level and the error will be made.

Now unless all null hypotheses of no pairwise differences are true, and the large F-statistic indicates this is not likely for example 4.1, this probability rises rather quickly in unknown and uncontrollable ways to unacceptable levels. It may help simply to use a smaller probability level for multiple tests than for single tests but this procedure is not sufficient. The new significance level will depend on the true sample differences, which the analyst never knows.

Statisticians have developed many solutions to this problem, each with characteristics that distinguish it from the others. The objective of all multiple comparison procedures is to control the probability of erroneously concluding a pair difference exists when in fact it does not. While a somewhat general discussion of these procedures is given in chapter 15, Duncan's (1955) procedure will be illustrated here.

To begin Duncan's procedure, the sample averages are calculated and written down in either ascending or descending order. A critical range (CR) is computed for pairs of sample averages as follows.

$$CR = q \sqrt{\frac{ms(1/n_i + 1/n_j)}{2}}$$

where ms = the residual mean square from the ANOVA table,
n_i and n_j = the sample sizes, and
q = Duncan's special studentized range statistic for r samples, as explained below, df, the degrees of freedom for ms, and the significance level chosen for the tests.

The q-statistics are found in table Q of part V.

The number of samples spanned by a comparison including the two samples compared is called r. For ordered samples that are adjacent, $r = 2$. It is 4 when there are two samples between the two being compared.

Duncan's procedure, like many others in statistics, was developed for equal sample sizes. Most data analyses do not involve equal sample sizes, so to make the procedures applicable in real data analysis problems, they are modified in ways that analysts generally accept and consider reasonable. The formula for the critical range given above has been modified in accordance with procedures for the analogous t-test of two independent averages given in chapter 2.

The sample averages are compared in the following order: the largest to the smallest, the largest to the second smallest, and so on until an observed range is less than the critical range. When this happens, the comparisons with the largest cease and all averages between the two in this last test are declared not significantly different from the largest average. The process is repeated in turn for the second largest, the third largest, and so on until an average not significantly different from the smallest average is found. At this point, all comparisons cease.

The four sample averages are presented in ascending order in table 4.8.

TABLE 4.8 Multiple Comparisons
for Example 4.1 (Significance at 0.05
noted)

Sample	Mean	n
1	15.0	16
2	17.0	13
5	17.3	14
3	26.6	10

Using a significance level of 0.05, the critical range for samples 1 and 3 is computed as follows.

$$CR = 3.09 \sqrt{\frac{29.82(\frac{1}{16} + \frac{1}{10})}{2}} = 4.81$$

Samples 1 and 3 differ by more than their critical range so the difference is significant at the 0.05 level. The critical range for samples 2 and 3 is computed next.

$$CR = 3.00 \sqrt{\frac{29.82(\frac{1}{13} + \frac{1}{10})}{2}} = 4.87$$

Samples 2 and 3 differ by more than their critical range so the difference is significant at the 0.05 level. The critical range for samples 5 and 3 is 4.56 and this too is exceeded by the observed difference. Therefore, samples 1, 2, and 5 are all significantly different, at the 0.05 level, from sample 3.

The critical range for samples 1 and 5 is computed next. It is 4.33. This is greater than the actual difference so samples 1 and 5 do not differ significantly. Duncan's procedure is to stop comparisons at this point. Samples 1 and 2 are not to be tested because sample 2 falls between two samples that do not differ significantly.

A line has been drawn in table 4.8 grouping samples 1, 2, and 5 to signify that they do not differ significantly. Sample 3 is excluded from the group because it differs from the other three samples. This is a very simple structure. Often the lines connecting the samples overlap, creating indistinct associations that are difficult to interpret for management. In this simple case, samples 1, 2, and 5 have similar averages and the population means may be considered equivalent, while sample 3 indicates that its population is apparently different from the other three.

Equivalence as used here means that the sample differences are not sufficiently large to permit the analyst to reject the null hypothesis that the population means are not different. Equivalence does not, of course, imply that the population means are equal. Equivalence is not difficult to interpret for this example, but there are many situations, e.g., example 4.2, where it is most difficult to interpret.

ORDINAL MEASUREMENTS

Sometimes ratio and interval data do not satisfy the conditions for a valid *F*-test in an analysis of variance. An alternative procedure when one of the conditions is not satisfied is to transform the data to the ordinal scale, i.e., to ranks, and then to analyze the ranks. Such nonparametric methods are not without their own limitations, which will be discussed.

The Kruskal-Wallis procedure [see Kruskal (1952) and Kruskal and Wallis (1952)] is the most widely proposed and used method for comparing more than two independent ordinal data samples. One attraction of this procedure is that it reduces to the Wilcoxon procedure when there are only two samples. It is perhaps more appropriate to describe the Kruskal-Wallis procedure as a generalization of the Wilcoxon procedure for use with more than two samples.

Example 4.2

The Kruskal-Wallis procedure will be illustrated using the breaking strengths of paper in table 4.9. The manufacture of paper is a continuous process. After the paper has been dried, it is stored in large rolls until it is processed into specific products. Samples are taken between rolls to

TABLE 4.9   Breaking Strengths of
Paper for Example 4.2 (in kilograms)

	Sample		
1	*2*	*3*	*4*
9.5	8.1	8.6	9.0
8.6	8.2	8.8	7.9
8.8	7.6	9.2	8.2
6.7	7.7	8.8	7.6
7.3	8.0	9.8	8.7
6.8	7.5	9.7	7.8
7.7	7.3	9.2	8.8
7.7	8.2	9.7	7.6
8.6	8.5	10.2	8.9
9.7	8.4	9.7	8.7

ascertain that the paper meets all specifications. One specification is breaking strength. Variability in the breaking strength can be considered the result of two things: variability in the paper product and variability resulting from the measurement itself. To separate these and to minimize the measurement variability, several aliquots of each interroll sample are pulled, with the force measured in kilograms, until the paper breaks. The data in table 4.9 are the breaking strengths for 10 aliquots taken at four different times.

The principal data analysis issue concerns the differences between the four samples. The manufacturer, of course, wants small differences between the samples since this would imply a uniform product.

Bartlett's M, calculated from sample statistics presented in table 4.10, is 9.58. Assuming that the four samples are from normal populations with equal variances and the differences between the sample variances are due to chance alone, then the probability of a chi-square with 3 degrees of freedom being this large or larger is about 0.024. It would not be prudent, therefore, to assume homogeneous variances. The variance for sample 1 is obviously different. This sample could be dropped and the analysis continued using the other three. Their variances are homogeneous ($M = 0.9698$), so if the samples are normally distributed, an F-test could be used in an analysis of variance to compare their averages.

TABLE 4.10   Sample Statistics for Example 4.2: Breaking Strengths of Paper (in kilograms)

	Sample			
	1	*2*	*3*	*4*
n	10	10	10	10
Σy	81.4	79.5	93.7	83.2
Σy^2	672.70	633.49	880.47	695.04
sd^2	1.1227	0.1628	0.2779	0.3129

NOTE: Bartlett's $M = 9.58$; $p = 0.0238$.

Another approach would be to try a parametric transformation. While usually possible and sometimes useful, parametric transformations are always difficult to explain. The reader is invited to demonstrate that the usual simple transformations, i.e., logarithmic, exponential, and square root, do not produce homogeneous variances for these four samples. The log transformation of the breaking strengths plus a large constant, e.g., 50, does produce a Bartlett M with a probability in the range of 0.09, but who would want to explain this transformation to management? Instead, this discovery will be used to justify a Kruskal-Wallis procedure.

The basic conditions for a Kruskal-Wallis analysis are that the ratio and interval data be independent and that they come from populations differing at most in location. From the data analysis viewpoint, this last condition means the populations have the same shape but they may differ in average value. Nonhomogeneous variances are evidence that the shapes are not the same, at least as originally measured. The variances do appear to be homogeneous under a log transformation employing a large constant, e.g.,

$$y' = \ln (50 + y)$$

and the ranks of y' and y will be the same. This kind of argument may seem a little farfetched, but without something like it, Kruskal-Wallis cannot be justified for comparing the four samples of breaking strengths.

Parametric transformations that preserve the rank transform are monotonic transformations.

For the Kruskal-Wallis procedure, rank the data from all of the samples together as presented in table 4.11 for example 4.2. Ties are given average ranks, e.g., the third and fourth smallest breaking strengths are both 7.3 so 3.5, the average rank, is assigned to each.

TABLE 4.11   Rank-Transformed
Breaking Strengths of Paper for
Example 4.2

Sample			
1	*2*	*3*	*4*
34	15	22	31
22	17	27.5	13
27.5	7	32.5	17
1	10	27.5	7
3.5	14	39	24.5
2	5	36.5	12
10	3.5	32.5	27.5
10	17	36.5	7
22	20	40	30
36.5	19	36.5	24.5

After ranking the data, use the ranks to prepare an analysis of variance as in table 4.13. Summary statistics that will be useful in the analysis are presented in table 4.12. The Kruskal-Wallis statistic is the between sum of squares divided by the total mean square, i.e.,

$$\frac{2322.10}{136.17} = 17.05$$

TABLE 4.12 Rank Statistics for Example 4.2: Breaking Strengths of Paper

	Sample				
	1	*2*	*3*	*4*	*Total*
n	10	10	10	10	40
ΣR	168.5	127.5	330.5	193.5	820.0
ΣR^2					22,120.5

$$\frac{168.5^2}{10} + \frac{127.5^2}{10} + \frac{330.5^2}{10} + \frac{193.5^2}{10} = 19,132.10$$

$$\frac{820^2}{40} = 16,810.00$$

TABLE 4.13 The ANOVA Table for Example 4.2

Source of variation	*df*	*ss*	*ms*	
Between samples	3	2322.10	774.03	KW = 17.05
Within samples	36	2988.40	83.01	
Total	39	5310.50	136.17	

NOTE: KW = Kruskal-Wallis.

The data analyst will find many algebraic forms for the Kruskal-Wallis statistic. Every authority in nonparametric statistics has a personal favorite. The statistic is further complicated by various methods of correcting for ties in the original data. All of these complications are unnecessary since the analysis of variance procedure will produce the correct statistic for every analysis, with and without ties. Furthermore, using this analysis of variance procedure should make it easier for the analyst to remember what to do.

The exact distribution of the Kruskal-Wallis statistic depends on the number of data in each of the samples, i.e., there is a different distribution for every combination of sample size. Consequently, even a modest table of the exact distribution is physically large. It is customary in most texts to present exact tables for problems with only three samples and then for sample sizes only up to five. By arranging the tables so that the first sample has the least number of data and the last sample has the most, even then there are 34 exact distributions for three samples with five or fewer data each.

As the smallest sample size increases and the number of samples increases, the Kruskal-Wallis statistic quickly approaches the chi-square distribution with $k - 1$ degrees of freedom where k is the number of samples. The approximation is used whenever k is greater than 3. It is also used whenever k is equal to 3 and any sample has more than five data.

Example 4.2 has four samples so the Kruskal-Wallis statistic is compared with the chi-square distribution for 3 degrees of freedom. Assuming that the null hypothesis is true, i.e., the data are independent and come from populations that differ at most in location, then the probability of a chi-square with 3 degrees of freedom as large or larger than 17.05 is approximately 0.0009. This is evidence to contradict the null hypothesis of no sample differences.

A significantly large Kruskal-Wallis statistic immediately raises the question: Are any of the samples alike and if so which ones? This question is usually answered by comparing each sample with each other sample in a multiple comparison procedure.

Multiple Comparisons

The comparison of pairs of treatments in the ordinal measurement scale is based on the central limit theorem. No matter what distribution data have, the central limit theorem in its simplest form is that sums or averages of the data tend to be normally distributed and the approximation improves as the number of data summed or averaged increases. Now for rectangularly distributed data such as ranks, sums or averages of as few as 12 data are very nearly normally distributed. Furthermore, the average and the variance of the normally distributed sums or averages can be estimated from the original data. All of this means that any of the procedures available for multiple comparison of samples in the ratio and interval measurement scale can be used with sums or averages of sample ranks.

Dunn (1964) and Conover (1980), among others, have described methods for comparing sample pairs that are valid for unequal sample sizes. The calculations are remarkably similar but the properties of the results are considerably different. They are presented in parallel here to facilitate comparison. Like the objective of other multiple comparison procedures, the objective of these two procedures is to control the probability of erroneously concluding a pair difference exists when in fact it does not.

To begin either procedure, the average ranks for the k samples are written down in either ascending or descending order. Then a critical range is calculated as follows.

$$CR = q \sqrt{ms \left(\frac{1}{n_i} + \frac{1}{n_j} \right)}$$

The analysis of variance of the ranks, table 4.13, furnishes the ms for both procedures. In Dunn's procedure, it is the total mean square, while in

Conover's procedure it is the residual mean square. A significance level α is chosen for a single comparison, e.g., 0.05 or 0.01. In Dunn's procedure, q is the standard normal deviate for $\alpha/2$ divided by the total possible number of pair comparisons $k(k-1)/2$, i.e.,

$$\frac{\alpha}{[k(k-1)]}$$

For Conover's procedure, q is the value of Student's t for α and the degrees of freedom for ms. In both procedures, n_i and n_j are the sample sizes of the two average ranks compared.

Any difference between two average ranks that exceeds the critical range is significant at the α level. Dunn's procedure attempts to hold the probability of making a single erroneous decision in the group of pairwise comparisons to α while the probability is α of making an erroneous decision for each of the comparisons using Conover's procedure. The probability of a single erroneous decision in the group of pairwise Conover comparisons is something larger than α. So the data analyst must decide which critical range to compute and to present to the client. The two procedures represent the extremes in multiple comparisons. Conover's procedure is basically Fisher's least significant differences applied to ranks and will usually produce much smaller critical ranges than will Dunn's procedure. In other words, Dunn's procedure is more conservative than Conover's procedure. A good middle-ground procedure for ranks could be based on Duncan's method described earlier in this chapter for ratio and/or interval measurements.

The sample sizes in example 4.2 are all equal to 10, so the calculations to illustrate these procedures are simplified. To facilitate comparison, α will be 0.05 for all critical ranges. Dunn's critical range for these four samples is

$$CR = 2.637 \sqrt{136.17\left(\frac{1}{10} + \frac{1}{10}\right)} = 13.76$$

and Conover's critical range is

$$CR = 2.021 \sqrt{83.01\left(\frac{1}{10} + \frac{1}{10}\right)} = 8.23$$

Duncan's procedure requires three critical ranges be computed as follows for example 4.2.

$$CR_4 = 3.10 \sqrt{\frac{83.01(\frac{1}{10} + \frac{1}{10})}{2}} = 8.93$$

$$CR_3 = 3.01 \sqrt{\frac{83.01(\frac{1}{10} + \frac{1}{10})}{2}} = 8.67$$

$$CR_2 = 2.86 \sqrt{\frac{83.01(\frac{1}{10} + \frac{1}{10})}{2}} = 8.24$$

Table 4.14 presents the average ranks in ascending order for the four samples. Using Dunn's critical range, sample 3 differs significantly from samples 2 and 1 but does not differ from sample 4. Samples 2 and 1 do not differ from sample 4 and samples 2 and 1 are not different from each other. These somewhat complex relationships are illustrated in table 4.14 by the lines under "Dunn" grouping the samples from populations that do not differ significantly.

TABLE 4.14   Multiple Comparisons for Example 4.2 (Significance at 0.10 Noted)

Sample	Average rank	Dunn	Conover-Duncan
2	12.75		
1	16.85		
4	19.35		
3	33.05		

It is not easy to explain the relationships among the four populations indicated by the results of Dunn's multiple comparisons. For example, there is no evidence that populations 2, 1, and 4 do not have the same means. Likewise, there is no evidence that populations 4 and 3 do not have the same means yet there is evidence that the means for populations 2 and 1 are not the same as the mean for population 3. An interpretation problem can arise because many data analysis clients will equate "no evidence against" or the data analysis concept of "equivalence" with equality. Being aware that the results of an analysis may be misunderstood and then may be misinterpreted, the data analyst must provide very precise, simple explanations for all results that are reported.

Using Conover's critical range, sample 3 differs from the other three samples which are not different from each other. This simple relationship is illustrated by the line under "Conover" grouping the samples from populations that do not differ significantly by these criteria. Finally, Duncan's critical ranges indicate the same relationships among the populations as Conover's criteria.

If the ordinal data were obtained by ranking ratio or interval data, results such as presented in table 4.14 always suggest the need for estimates of sample pair differences in the original measurement scale. A commonly used procedure is based on the Hodges-Lehmann procedure for estimating the difference between two independent samples. First, estimate all possible Hodges-Lehmann two-sample differences. Table 4.15 gives all of the differences for making the 6 two-sample Hodges-Lehmann estimates. Table 4.16 gives these same differences after sorting. From this table, the medians are found and table 4.17 is constructed.

The Hodges-Lehmann difference for samples i and j is the negative of the difference for samples j and i, e.g., the difference is 0.2 for samples 1 and 2 while for samples 2 and 1, it is -0.2. These simple Hodges-Lehmann estimates are inconsistent in the sense that the estimate for samples i and j plus the estimate for samples j and k is probably not the estimate for samples i and k. To illustrate this, the estimate for samples 3 and 4 plus the estimate for samples 1 and 3 is

$$1.0 + (-1.1) = -0.1$$

while the estimate for 1 and 4 is -0.2. This inconsistency is difficult if not impossible to explain, so Hodges and Lehmann (1963), recognizing the problem, proposed the solution presented here.

Compute the average of each row using the sample sizes as weights. Then these row averages, included in table 4.17, are used to compute all two-sample differences, e.g.,

Sample 1 minus 2 is 0.250
Sample 1 minus 3 is -1.175
Sample 1 minus 4 is -0.175
Sample 2 minus 3 is -1.425
Sample 2 minus 4 is -0.425
Sample 3 minus 4 is 1.000

This procedure may seem a little complex, but the inconsistency described above does not exist with the adjusted sample differences. For example, the adjusted estimate for samples 3 and 4 plus the adjusted estimate for samples 1 and 3 is

$$1.000 + (-1.175) = -0.175$$

which is the adjusted estimate for samples 1 and 4.

NOMINAL MEASUREMENTS

Usually the transformation of ratio or interval data to the ordinal scale is sufficient for most analyses. Sometimes, however, ratio, interval, and even ordinal data are transformed to the nominal scale and, of course, data can originate in the nominal scale. The analysis is the same in any

TABLE 4.15 Unsorted Sample Differences for the Hodges-Lehmann Estimator for Example 4.2

Samples 1 and 2				Samples 1 and 3				Samples 1 and 4			
1.4	1.3	−1.3	0.2	0.9	−0.9	−1.8	−2.0	0.5	1.0	−2.2	−0.1
1.3	1.5	−1.4	0.4	0.7	−0.4	−2.0	−1.5	1.6	0.0	−1.1	−1.1
1.9	0.6	−0.8	−0.5	0.3	−0.9	−2.4	−2.0	1.3	1.2	−1.4	0.1
1.8	0.3	−0.9	−0.8	0.7	−1.4	−2.0	−2.5	1.9	−0.1	−0.8	−1.2
1.5	0.4	−1.2	−0.7	−0.3	−0.9	−3.0	−2.0	0.8	0.1	−1.9	−1.0
2.0	−1.4	−0.7	0.5	−0.2	−1.9	−2.9	0.0	1.7	−2.3	−1.0	−0.4
2.2	−1.5	−0.5	0.4	0.3	−2.1	−2.4	−0.2	0.7	−1.2	−2.0	0.7
1.3	−0.9	−1.4	1.0	−0.2	−2.5	−2.9	−0.6	1.9	−1.5	−0.8	0.4
1.0	−1.0	−1.7	0.9	−0.7	−2.1	−3.4	−0.2	0.6	−0.9	−2.1	1.0
1.1	−1.3	−1.6	0.6	−0.2	−3.1	−2.9	−1.2	0.8	−2.0	−1.9	−0.1
0.5	−0.8	−0.4	1.1	0.0	−3.0	−0.9	−1.1	−0.4	−1.1	−1.3	0.8
0.4	−0.6	−0.5	1.3	−0.2	−2.5	−1.1	−0.6	0.7	−2.1	−0.2	−0.2
1.0	−1.5	0.1	0.4	−0.6	−3.0	−1.5	−1.1	0.4	−0.9	−0.5	1.0
0.9	−1.8	0.0	0.1	−0.2	−3.5	−1.1	−1.6	1.0	−2.2	0.1	−0.3
0.6	−1.7	−0.3	0.2	−1.2	−3.0	−2.1	−1.1	−0.1	−2.0	−1.0	−0.1
1.1	−0.8	0.2	1.6	−1.1	−1.3	−2.0	1.1	0.8	−1.7	−0.1	0.7
1.3	−0.9	0.4	1.5	−0.6	−1.5	−1.5	0.9	−0.2	−0.6	−1.1	1.8
0.4	−0.3	−0.5	2.1	−1.1	−1.9	−2.0	0.5	1.0	−0.9	0.1	1.5
0.1	−0.4	−0.8	2.0	−1.6	−1.5	−2.5	0.9	−0.3	−0.3	−1.2	2.1
0.2	−0.7	−0.7	1.7	−1.1	−2.5	−2.0	−0.1	−0.1	−1.4	−1.0	1.0
0.7	−0.2	−0.4	2.2	0.2	−2.4	−0.9	0.0	−0.2	−0.5	−1.3	1.9
0.6	0.0	−0.5	2.4	0.0	−1.9	−1.1	0.5	0.9	−1.5	−0.2	0.9
1.2	−0.9	0.1	1.5	−0.4	−2.4	−1.5	0.0	0.6	−0.3	−0.5	2.1
1.1	−1.2	0.0	1.2	0.0	−2.9	−1.1	−0.5	1.2	−1.6	0.1	0.8
0.8	−1.1	−0.3	1.3	−1.0	−2.4	−2.1	0.0	0.1	−1.4	−1.0	1.0

Samples 2 and 3				Samples 2 and 4				Samples 3 and 4			
−0.5	−2.1	−1.1	−1.5	−0.9	−0.2	−1.5	0.4	−0.4	1.4	0.7	1.9
−0.7	−1.6	−1.3	−1.0	0.2	−1.2	−0.4	−0.6	0.7	0.4	1.8	0.9
−1.1	−2.1	−1.7	−1.5	−0.1	0.0	−0.7	0.6	0.4	1.6	1.5	2.1
−0.7	−2.6	−1.3	−2.0	0.5	−1.3	−0.1	−0.7	1.0	0.3	2.1	0.8
−1.7	−2.1	−2.3	−1.5	−0.6	−1.1	−1.2	−0.5	−0.1	0.5	1.0	1.0
−1.6	−0.9	−2.2	−0.1	0.3	−1.3	−0.3	−0.5	0.8	−0.2	1.9	1.2
−1.1	−1.1	−1.7	−0.3	−0.7	−0.2	−1.3	0.6	−0.2	0.9	0.9	2.3
−1.6	−1.5	−2.2	−0.7	0.5	−0.5	−0.1	0.3	1.0	0.6	2.1	2.0
−2.1	−1.1	−2.7	−0.3	−0.8	0.1	−1.4	0.9	−0.3	1.2	0.8	2.6
−1.6	−2.1	−2.2	−1.3	−0.6	−1.0	−1.2	−0.2	−0.1	0.1	1.0	1.5
−0.4	−2.0	−1.3	−1.2	−0.8	−0.1	−1.7	0.7	−0.2	1.0	0.2	2.4
−0.6	−1.5	−1.5	−0.7	0.3	−1.1	−0.6	−0.3	0.9	0.0	1.3	1.4
−1.0	−2.0	−1.9	−1.2	0.0	0.1	−0.9	0.9	0.6	1.2	1.0	2.6
−0.6	−2.5	−1.5	−1.7	0.6	−1.2	−0.3	−0.4	1.2	−0.1	1.6	1.3
−1.6	−2.0	−2.5	−1.2	−0.5	−1.0	−1.4	−0.2	0.1	0.1	0.5	1.5
−1.5	−0.6	−2.4	−0.2	0.4	−1.0	−0.5	−0.6	1.0	0.8	1.4	0.7
−1.0	−0.8	−1.9	−0.4	−0.6	0.1	−1.5	0.5	0.0	1.9	0.4	1.8
−1.5	−1.2	−2.4	−0.8	0.6	−0.2	−0.3	0.2	1.2	1.6	1.6	1.5
−2.0	−0.8	−2.9	−0.4	−0.7	0.4	−1.6	0.8	−0.1	2.2	0.3	2.1
−1.5	−1.8	−2.4	−1.4	−0.5	−0.7	−1.4	−0.3	0.1	1.1	0.5	1.0
−1.0	−1.7	−0.4	−1.3	−1.4	0.2	−0.8	0.6	0.2	2.0	0.7	1.9
−1.2	−1.2	−0.6	−0.8	−0.3	−0.8	0.3	−0.4	1.3	1.0	1.8	0.9
−1.6	−1.7	−1.0	−1.3	−0.6	0.4	0.0	0.8	1.0	2.2	1.5	2.1
−1.2	−2.2	−0.6	−1.8	0.0	−0.9	0.6	−0.5	1.6	0.9	2.1	0.8
−2.2	−1.7	−1.6	−1.3	−1.1	−0.7	−0.5	−0.3	0.5	1.1	1.0	1.0

TABLE 4.16 Sorted Sample Differences for the Hodges-Lehmann Estimator for Example 4.2

Samples 1 and 2				Samples 1 and 3				Samples 1 and 4			
-1.8	-0.7	0.2	1.1	-3.5	-2.0	-1.1	-0.2	-2.3	-1.1	-0.2	0.8
-1.7	-0.7	0.2	1.1	-3.4	-2.0	-1.1	-0.2	-2.2	-1.0	-0.1	0.8
-1.7	-0.7	0.3	1.2	-3.1	-2.0	-1.1	-0.2	-2.2	-1.0	-0.1	0.8
-1.6	-0.6	0.4	1.2	-3.0	-2.0	-1.1	-0.2	-2.1	-1.0	-0.1	0.8
-1.5	-0.5	0.4	1.3	-3.0	-2.0	-1.1	-0.2	-2.1	-1.0	-0.1	0.9
-1.5	-0.5	0.4	1.3	-3.0	-2.0	-1.1	-0.2	-2.0	-1.0	-0.1	0.9
-1.4	-0.5	0.4	1.3	-3.0	-2.0	-1.1	-0.1	-2.0	-0.9	-0.1	1.0
-1.4	-0.5	0.4	1.3	-2.9	-2.0	-1.1	0.0	-2.0	-0.9	-0.1	1.0
-1.4	-0.5	0.4	1.3	-2.9	-1.9	-1.1	0.0	-1.9	-0.9	0.0	1.0
-1.3	-0.4	0.4	1.3	-2.9	-1.9	-1.0	0.0	-1.9	-0.8	0.1	1.0
-1.3	-0.4	0.5	1.4	-2.9	-1.9	-0.9	0.0	-1.7	-0.8	0.1	1.0
-1.2	-0.4	0.5	1.5	-2.5	-1.8	-0.9	0.0	-1.6	-0.6	0.1	1.0
-1.2	-0.3	0.6	1.5	-2.5	-1.6	-0.9	0.0	-1.5	-0.5	0.1	1.0
-1.1	-0.3	0.6	1.5	-2.5	-1.6	-0.9	0.0	-1.5	-0.5	0.1	1.2
-1.0	-0.3	0.6	1.5	-2.5	-1.5	-0.9	0.2	-1.4	-0.5	0.1	1.2
-0.9	-0.2	0.6	1.6	-2.5	-1.5	-0.7	0.3	-1.4	-0.4	0.4	1.3
-0.9	0.0	0.7	1.7	-2.4	-1.5	-0.6	0.3	-1.4	-0.4	0.4	1.5
-0.9	0.0	0.8	1.8	-2.4	-1.5	-0.6	0.5	-1.3	-0.3	0.5	1.6
-0.9	0.0	0.9	1.9	-2.4	-1.5	-0.6	0.5	-1.3	-0.3	0.6	1.7
-0.8	0.1	0.9	2.0	-2.4	-1.5	-0.6	0.7	-1.2	-0.3	0.6	1.8
-0.8	0.1	1.0	2.0	-2.4	-1.4	-0.5	0.7	-1.2	-0.3	0.7	1.9
-0.8	0.1	1.0	2.1	-2.1	-1.3	-0.4	0.9	-1.2	-0.2	0.7	1.9
-0.8	0.1	1.0	2.2	-2.1	-1.2	-0.4	0.9	-1.1	-0.2	0.7	1.9
-0.8	0.2	1.1	2.2	-2.1	-1.2	-0.3	0.9	-1.1	-0.2	0.7	2.1
-0.7	0.2	1.1	2.4	-2.1	-1.1	-0.2	1.1	-1.1	-0.2	0.8	2.1

Samples 2 and 3				Samples 2 and 4				Samples 3 and 4			
-2.9	-1.9	-1.5	-1.0	-1.7	-0.8	-0.4	0.2	-0.4	0.5	1.0	1.6
-2.7	-1.8	-1.5	-1.0	-1.6	-0.8	-0.4	0.2	-0.3	0.6	1.0	1.6
-2.6	-1.8	-1.4	-0.9	-1.5	-0.8	-0.3	0.2	-0.2	0.6	1.0	1.6
-2.5	-1.7	-1.3	-0.8	-1.5	-0.7	-0.3	0.3	-0.2	0.7	1.0	1.6
-2.5	-1.7	-1.3	-0.8	-1.4	-0.7	-0.3	0.3	-0.2	0.7	1.0	1.8
-2.4	-1.7	-1.3	-0.8	-1.4	-0.7	-0.3	0.3	-0.1	0.7	1.0	1.8
-2.4	-1.7	-1.3	-0.8	-1.4	-0.7	-0.3	0.3	-0.1	0.7	1.1	1.8
-2.4	-1.7	-1.3	-0.7	-1.4	-0.7	-0.3	0.4	-0.1	0.8	1.1	1.9
-2.3	-1.7	-1.3	-0.7	-1.3	-0.7	-0.3	0.4	-0.1	0.8	1.2	1.9
-2.2	-1.7	-1.3	-0.7	-1.3	-0.6	-0.2	0.4	0.0	0.8	1.2	1.9
-2.2	-1.6	-1.2	-0.7	-1.3	-0.6	-0.2	0.4	0.0	0.8	1.2	1.9
-2.2	-1.6	-1.2	-0.6	-1.2	-0.6	-0.2	0.5	0.1	0.8	1.2	2.0
-2.2	-1.6	-1.2	-0.6	-1.2	-0.6	-0.2	0.5	0.1	0.9	1.2	2.0
-2.2	-1.6	-1.2	-0.6	-1.2	-0.6	-0.2	0.5	0.1	0.9	1.3	2.1
-2.1	-1.6	-1.2	-0.6	-1.2	-0.6	-0.1	0.6	0.1	0.9	1.3	2.1
-2.1	-1.6	-1.2	-0.6	-1.1	-0.6	-0.1	0.6	0.2	0.9	1.3	2.1
-2.1	-1.6	-1.2	-0.5	-1.1	-0.5	-0.1	0.6	0.2	0.9	1.4	2.1
-2.1	-1.5	-1.1	-0.4	-1.1	-0.5	-0.1	0.6	0.3	0.9	1.4	2.1
-2.1	-1.5	-1.1	-0.4	-1.0	-0.5	0.0	0.6	0.3	1.0	1.4	2.1
-2.0	-1.5	-1.1	-0.4	-1.0	-0.5	0.0	0.6	0.4	1.0	1.5	2.2
-2.0	-1.5	-1.1	-0.4	-1.0	-0.5	0.0	0.7	0.4	1.0	1.5	2.2
-2.0	-1.5	-1.1	-0.3	-0.9	-0.5	0.0	0.8	0.4	1.0	1.5	2.3
-2.0	-1.5	-1.0	-0.3	-0.9	-0.5	0.1	0.8	0.5	1.0	1.5	2.4
-2.0	-1.5	-1.0	-0.2	-0.9	-0.5	0.1	0.9	0.5	1.0	1.5	2.6
-1.9	-1.5	-1.0	-0.1	-0.8	-0.4	0.1	0.9	0.5	1.0	1.6	2.6

4.21

TABLE 4.17   Hodges-Lehmann Two-Sample
Differences for Example 4.2

First sample	Second sample 1	2	3	4	Average
1	0	0.2	−1.1	−0.2	−0.275
2	−0.2	0	−1.5	−0.4	−0.525
3	1.1	1.5	0	1.0	0.900
4	0.2	0.4	−1.0	0	−0.100
n	10	10	10	10	

case since the procedure is determined by the measurement scale, not by
the source of the data.

When data are transformed to the nominal scale, the nominal categories are ordered simply because the ratio, interval, or ordinal data are
ordered. The dividing points defining the nominal categories should have
some meaning for the investigator or for management. In example 4.2,
breaking strengths less than a specified amount might indicate an inferior
product or a product suitable for a particular purpose, while above the
specified amount indicates an acceptable product or a product suitable
for another purpose. Two definitions for nominal categories will be used
with the data in example 4.2. They are chosen to illustrate a data analysis
problem and are in no way descriptive of any actual paper product.

Consider the data in example 4.2 classified as follows: less than 8 kilograms and greater than or equal to 8 kilograms. To make the transformation, the data in each sample are classified and the number in each
nominal scale class counted. Table 4.18 presents the results.

TABLE 4.18   Breaking Strengths of Paper for
Example 4.2

	Sample 1	2	3	4	Total
	Observed counts				
<8 kilograms	5	4	0	4	13
≥8 kilograms	5	6	10	6	27
Total count	10	10	10	10	40
	Expected counts				
<8 kilograms	3.25	3.25	3.25	3.25	13
≥8 kilograms	6.75	6.75	6.75	6.75	27
Total count	10	10	10	10	40

NOTE: $\chi^2 = 6.72$, prob$(\chi^2 \geq 6.72) = 0.085$.

Now on the assumption that there are no sample differences, the data analyst would expect the 13 breaking strengths less than 8 kilograms to be divided among the samples in proportion to the sample size. This can be stated in another way as follows. The proportion of the breaking strengths that are less than 8 kilograms is 13/40. The expected number in each sample is the sample size multiplied by this proportion. In many analyses, the sample sizes are not equal but in this example they are. Therefore, the analyst expects 10 13/40 = 3.25 aliquots in each sample to have breaking strengths less than 8 kilograms. Likewise the analyst expects each sample to have 10 27/40 = 6.75 aliquots with breaking strengths equal to or greater than 8 kilograms. These expected values are also presented in table 4.18.

The test statistic for this analysis is a chi-square statistic computed as follows:

$$\chi^2 = \sum_{i=1}^{m} \frac{(O_i - E_i)^2}{E_i}$$

where O is the observed count, E is the corresponding expected count, and the summation is made over all the counts in the table. For the data in table 4.18, this calculation proceeds as follows.

$$\frac{(5 - 3.25)^2}{3.25} + \frac{(4 - 3.25)^2}{3.25} + \cdots + \frac{(10 - 6.75)^2}{6.75}$$

$$+ \frac{(6 - 6.75)^2}{6.75} = 6.72$$

The degrees of freedom for the test statistic is $(r - 1)(c - 1)$ where r and c are the number of rows and columns, respectively, in the contingency table. For example 4.2, there are 3 degrees of freedom, so 6.72 must be compared with percentiles of the chi-square distribution for 3 degrees of freedom. The probability of a chi-square this large or larger is about 0.085, assuming the null hypothesis of no differences between samples is true and the observable differences are due to chance alone. Consequently, there is probably no reason to reject the hypothesis of no sample differences.

Now consider example 4.2 data again but classify the data this time as follows: less than 8.5 kilograms and greater than or equal to 8.5 kilograms. The contingency table for this definition along with the expected counts are presented in table 4.19.

The chi-square test statistic, computed in the same way, is 16.34. The probability of a test statistic this large or larger is about 0.001, assuming the null hypothesis is true and the observable differences are due to chance alone. Consequently, there is now good evidence that the samples are not homogeneous, contradicting the results of the first nominal measurement analysis of example 4.2.

TABLE 4.19   Breaking Strengths of Paper for Example 4.2

	Sample				
	1	2	3	4	Total
Observed counts					
<8.5 kilograms	5	9	0	5	19
≥8.5 kilograms	5	1	10	5	21
Total counts	10	10	10	10	40
Expected counts					
<8.5 kilograms	4.75	4.75	4.75	4.75	19
≥8.5 kilograms	5.25	5.25	5.25	5.25	21
Total counts	10	10	10	10	40

NOTE: $\chi^2 = 16.34$, $\text{prob}(\chi^2 \geq 16.34) = 0.001$.

The two different nominal category definitions illustrate a data analysis problem and that problem, of course, is how to define nominal data categories. There are no general answers. Arbitrary definitions can lead to different results, so unless there are good reasons for a particular definition, the analysis of such transformed data is probably useless.

This problem is not limited to ratio, interval, or ordinal data that are transformed to a nominal scale. The measurement scale for many survey variables is nominal, e.g., public opinion polls. The dividing points for the categories are determined by the permitted answers to the survey questions. When the answers permitted bias the results, analysis is useless. The conclusions are predetermined and there is no research.

A requirement for contingency table analyses is that the counts be independent. That is, no individual should be counted twice. Like all rules, this one is made to be broken. An example of a study where individuals would be counted twice is any controlled experiment where baseline and treatment observations are compared. Sometimes this problem is avoided by classifying change rather than some absolute status.

This short presentation is not intended to cover in depth the subject of the analysis of nominal scale data. Two excellent books that data analysts may find useful are Fleiss (1973) and Upton (1978).

SUMMARY

Procedures for comparing more than two independent samples present more complex problems in analysis than were encountered with only two independent samples. For example, the basic procedure for ratio, interval, and ordinal data is an analysis of variance. The *F*-statistic for ratio and interval data and the Kruskal-Wallis statistic for ordinal data are calculated from quantities in this table. Furthermore, this more complex situation makes useful parametric transformations more difficult to find and nearly impossible to interpret. Unless a parametric transformation makes sense from a physical, biological, econometrical, or other theoretical basis, it should probably be avoided. A fishing expedition for a transformation just to achieve approximate normality usually creates more problems in interpretation and reproducibility than normality is worth.

Using the analysis of variance–based procedure presented here for ordinal data enables a data analyst to use a single summary procedure down to calculation of the test statistic. Furthermore, multiple comparison procedures for ordinal data are remarkably similar to those introduced for ratio and interval data.

Figure 4.1 is a flowchart that summarizes the analysis sequence suggested here.

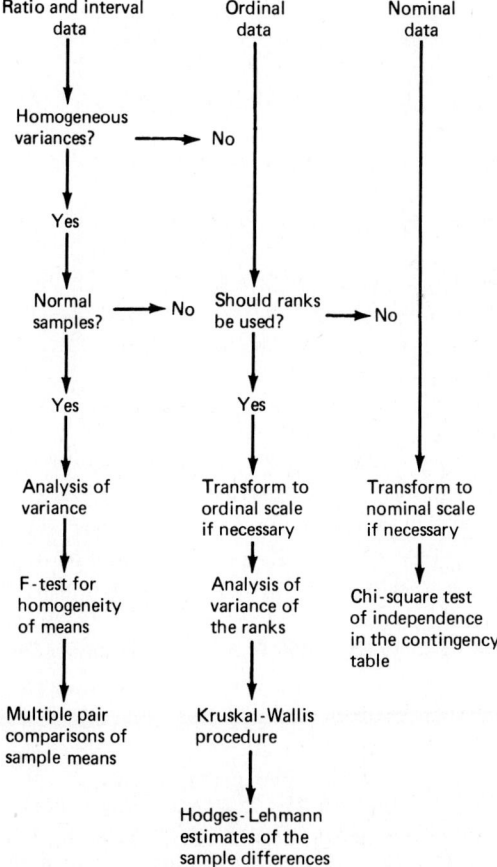

FIGURE 4.1 Analysis sequence flowchart for chapter 4.

CHAPTER 5

More Than Two Dependent Samples

Simple comparisons of more than two dependent samples are considered in this chapter. The data in more than two dependent samples can be classified in two ways, in contrast to the data in more than two independent samples which can be classified in only one way. One way for classifying the data is obvious, i.e., the several samples. The other way may not be so obvious since it associates each datum in one of the samples with data in each of the other samples. The associated data are called *blocks* in planned experiments and *strata* in surveys. These ideas parallel ideas introduced in chapter 3 for two dependent samples.

The notion of variability in data arising from more than one identifiable source will be expanded. As noted in chapter 3, variability can arise from many sources in a single problem. Two sources considered in this chapter will be associated with the two ways that the data can be classified. A third source of variability called *interaction* will be introduced in this chapter. Interaction can be identified and separated from the total only when measurements are repeated under "identical conditions."

Everyone who has analyzed data knows that measurements repeated under identical conditions are themselves seldom identical. If they are, then the measurement is gross, i.e., not very precise. Identical conditions in this chapter will mean the conditions associated with the combination of a single sample and a single block.

There are many ways that data in several samples can be associated. The most common are associations in space or time or common origin.

The differences in weight as a result of corrosion of several varieties of iron might be the research problem. The primary classification for the data would be the varieties of iron, and the secondary classification might be the conditions under which corrosion occurred, e.g., atmospheric conditions at several different geographic locations.

Another example of data that can be classified two ways is the consumer price index (CPI) for different cities throughout the United States. These data could be classified by geographic region and city size.

Sometimes in dermatology, it is possible to treat different skin lesions on a patient with different medications. Data from the different test areas can be classified by treatment and by patient.

No matter how such studies as these are planned, undesirable differences in the data often can be associated with the secondary classification. It is common practice to plan studies so that undesirable differences are identified by one or more secondary classifications. Analyses are then planned to remove the undesirable differences before making statistical tests.

In this chapter, the removal of differences that can be associated with a single secondary classification is considered. This is a simple idea but a powerful tool in the hands of a resourceful data analyst.

The analysis of data that can be classified two ways is on the borderline between problems that should have the attention of a professional consultant and problems that can be easily understood and analyzed by everyone whose job it is to interpret data. This writer therefore advises readers to seek professional help whenever situations arise that are not covered in this book.

RATIO AND INTERNAL MEASUREMENTS

Example 5.1

In the summer of 1958, the American Society for Testing and Materials (ASTM) Committee A-7 on Malleable Iron Castings initiated a project on corrosion [see Mannweiler (1972)]. Five sites representative of rural, industrial, and marine atmospheres were chosen for this study. They were State College, PA; Newark-Kearny, NJ; East Chicago-Whiting, IN; Pt. Reyes, CA; and Kure Beach, NC.

Five different pearlitic malleable irons as-cast were exposed at each of the five locations. At the end of 12 years, the specimens were cleaned in molten sodium hydroxide–sodium hydride and then weighed on a Mettler P-3 balance to the closest 0.1 gram. The weight loss was calculated.

Table 5.1 presents the data classified two ways, by type of pearlitic malleable iron and by location. Differences between the types of iron constitute the principal research interest. From this viewpoint, the locations identify a source of differences in the data that is secondary to the principal interest. This source of variability is a nuisance for the data analyst, but it is unavoidable unless the principal research interest is restricted to one kind of atmosphere.

TABLE 5.1   Example 5.1

Location	Pearlitic malleable iron types				
	A	*B*	*C*	*D*	*E*
PA	16.8	20.2	16.2	15.7	17.0
CA	35.3	36.9	34.2	33.9	34.4
NC	62.4	87.8	59.2	45.6	63.4
NJ	25.0	29.7	23.9	19.5	23.3
IN	29.4	31.9	30.7	24.8	31.6

Source: See Mannweiler (1972). Reprinted with permission from the American Society for Testing and Materials, 1916 Race Street, Philadelphia, PA 19103.

The research hypothesis is that the different kinds of pearlitic malleable iron corrode at different rates, and the purpose of this research project is to determine the significance of these differences. The null hypothesis that will be tested statistically is that no differences exist between the five irons. Use of the five locations permits comparisons that are not limited to one atmosphere. Nevertheless, any average difference between locations contributes nothing to an understanding of differences between kinds of iron and must be removed before comparing the irons. The effect of removing location differences is illustrated by first comparing the irons as independent samples, i.e., attempting to ignore the locations, and then as dependent samples.

For an analysis of variance of independent samples

1. The samples must be random.
2. The samples must be normally distributed.
3. The variances of the samples must be homogeneous.

Some summations useful for examining the data for these last two conditions are given in table 5.2.

Bartlett's procedure for comparing the sample variances was introduced in chapter 4. The test statistic for k variances is computed as follows. Let df_i be the degrees of freedom and sd_i be the standard deviation for sample i. Also for convenience, let df be the sum of the degrees of freedom for the samples, i.e.,

$$df = df_1 + df_2 + \cdots + df_k$$

and let

$$sd^2 = \frac{\sum_{i=1}^{k} df_i sd_i^2}{df}$$

TABLE 5.2   Sample Statistics for Example 5.1: Corrosion of Pearlitic Malleable Iron at Five Locations

	Sample					
	A	*B*	*C*	*D*	*E*	*Total*
Σy	168.9	206.5	164.2	139.5	169.7	848.8
Σy^2	6911.45	11,378.19	6450.42	4470.35	7033.37	36,243.78
n	5	5	5	5	5	25
sd^2	301.502	712.435	264.523	144.575	318.438	

$$\frac{168.9^2}{5} + \frac{206.5^2}{5} + \frac{164.2^2}{5} + \frac{139.5^2}{5} + \frac{169.7^2}{5} = 29{,}277.888$$

$$\frac{848.8^2}{25} = 28{,}818.4576$$

Then the test statistic M is

$$M = \frac{1}{C}\left(df \ln (sd^2) - \sum_{i=1}^{k} df_i \ln (sd_i^2)\right)$$

where ln is the natural logarithm function and

$$C = 1 + \frac{\sum_{i=1}^{k} (1/df_i) - 1/df}{3(k - 1)}$$

Bartlett's M is approximately distributed as chi-square with $k - 1$ degrees of freedom. Using the sample standard deviations given in table 5.2, M is 2.45. The null hypothesis is that the samples are from normal populations with equal variances. Assuming the null hypothesis is true and that the sample variances differ by chance only, then the probability of a chi-square with 4 degrees of freedom this large or larger is approximately 0.66 so there is no reason not to accept the variances as homogeneous. The variance of sample B may appear large compared to the other sample variances, but the small sample size of 5 makes this unimportant.

After subtracting the respective sample average from each datum, the samples are combined for a Lilliefors normality test as in table 5.3. The normalization of the data is completed on dividing these differences by the pooled standard deviation, 18.66. The Lilliefors statistic is very large, 0.228. Assuming the samples came from normal populations and that the observed differences from the normal are due to chance only, the probability of a Lilliefors statistic as large or larger is less than 0.01, so an F-test of sample differences would not be valid.

TABLE 5.3 The Lilliefors Test for Normality for Example 5.1

ID	y	$y - \bar{y}$	z	ID	z	Probabilities Normal	Probabilities Sample	da	db
A PA	16.8	−16.98	−0.910	B PA	−1.131	0.129	0.040	0.089	
A CA	35.3	1.52	0.081	A PA	−0.910	0.181	0.080	0.101	0.141
A NC	62.4	28.62	1.534	E PA	−0.908	0.182	0.120	0.062	0.102
A NJ	25.0	−8.78	−0.470	C PA	−0.892	0.186	0.160	0.026	0.066
A IN	29.4	−4.38	−0.235	D PA	−0.654	0.257	0.200	0.057	0.097
B PA	20.2	−21.10	−1.131	B NJ	−0.622	0.267	0.240	0.027	0.067
B CA	36.9	−4.40	−0.236	E NJ	−0.570	0.284	0.280	0.004	0.044
B NC	87.8	46.50	2.492	B IN	−0.504	0.307	0.320	−0.013	0.027
B NJ	29.7	−11.60	−0.622	C NJ	−0.479	0.316	0.360	−0.044	−0.004
B IN	31.9	−9.40	−0.504	A NJ	−0.470	0.319	0.400	−0.081	−0.041
C PA	16.2	−16.64	−0.892	D NJ	−0.450	0.326	0.440	−0.114	−0.074
C CA	34.2	1.36	0.073	B CA	−0.236	0.407	0.480	−0.073	−0.033
C NC	59.2	26.36	1.412	A IN	−0.235	0.407	0.520	−0.113	−0.073
C NJ	23.9	−8.94	−0.479	D IN	−0.166	0.434	0.560	−0.126	−0.086
C IN	30.7	−2.14	−0.115	E IN	−0.125	0.450	0.600	−0.150	−0.110
D PA	15.7	−12.20	−0.654	C IN	−0.115	0.454	0.640	−0.186	−0.146
D CA	33.9	6.00	0.321	E CA	0.025	0.510	0.680	−0.170	−0.130
D NC	45.6	17.70	0.948	C CA	0.073	0.529	0.720	−0.191	−0.151
D NJ	19.5	−8.40	−0.450	A CA	0.081	0.532	0.760	−0.228	−0.188
D IN	24.8	−3.10	−0.166	D CA	0.321	0.626	0.800	−0.174	−0.134
E PA	17.0	−16.94	−0.908	D NC	0.948	0.829	0.840	−0.011	0.029
E CA	34.4	0.46	0.025	C NC	1.412	0.921	0.880	0.041	0.081
E NC	63.4	29.46	1.579	A NC	1.534	0.937	0.920	0.017	0.057
E NJ	23.3	−10.64	−0.570	E NC	1.579	0.943	0.960	−0.017	0.023
E IN	31.6	−2.34	−0.125	B NC	2.492	0.994	1.000	−0.006	0.034

NOTE: The Lilliefors test statistic is 0.228. Refer to chapter 15 for details.

If these samples were truly independent, nonnormality argues for ranking the data to use a Kruskal-Wallis procedure. An alternative parametric procedure appropriate in later analysis of the data as dependent samples will be used instead so that the results of the data analysis as independent and as dependent samples may be compared.

Combining the samples for the Lilliefors normality procedure makes it easy to see that the North Carolina data are all greater than the average (zero) and account for five of the six largest deviations. This suggests that the North Carolina atmosphere produced so much corrosion that the data may be from a completely separate population. The alternative procedure followed here will be to drop North Carolina data and compare the five irons at only four sites. This of course will mean that the results will not be relevant to the North Carolina atmosphere.

Some useful summations for the independent samples after dropping the North Carolina data are given in table 5.4. Bartlett's M-statistic, calculated from the variances of samples at four locations, is 0.059. Assuming the samples are from normal populations with equal variances and that the observable differences are due to chance alone, then the proba-

TABLE 5.4 Sample Statistics for Example 5.1: Corrosion of Pearlitic Malleable Iron at Four Locations

| | Sample | | | | | |
	A	B	C	D	E	Total
Σy	106.5	118.7	105.0	93.9	106.3	530.4
Σy^2	3017.69	3669.35	2945.78	2390.99	3013.81	15,037.62
n	4	4	4	4	4	20
sd^2	60.7092	48.9758	63.1767	62.2292	62.9625	

| | Location | | | | |
	PA	CA	NJ	IN	Total
Σy	85.9	174.7	121.4	148.4	530.4
Σy^2	1488.21	6109.91	3001.44	4438.06	15,037.62
n	5	5	5	5	20

$$\frac{106.5^2}{4} + \frac{118.7^2}{4} + \frac{105.0^2}{4} + \frac{93.9^2}{4} + \frac{106.3^2}{4} = 14{,}143.46$$

$$\frac{85.9^2}{5} + \frac{174.7^2}{5} + \frac{121.4^2}{5} + \frac{148.4^2}{5} = 14{,}931.884$$

$$\frac{530.4^2}{20} = 14{,}066.208$$

bility of a chi-square with 4 degrees of freedom this large or larger is greater than 0.99. Thus there is no reason, based on Bartlett's procedure, not to accept equivalent variances. The Lilliefors statistic for the combined samples is 0.093. Again assuming the samples are from normal populations and that the observable differences are due to chance, then the probability of a Lilliefors statistic this large or larger is greater than 0.20, so there is no reason not to accept normality also.

The conditions for a valid F-test are satisfied. The analysis of variance, abbreviated ANOVA for tables, presented in table 5.5, indicates *no* significant differences among the averages of the five independent samples after removing the North Carolina data from each sample. The analyst can report that the five kinds of malleable iron castings corroded about the same at the four remaining locations. The reader should verify the

TABLE 5.5 The ANOVA Table for Example 5.1 (Assuming independent samples)

Source of variation	df	ss	ms	
Between samples	4	77.252	19.313	$F = 0.32$
Within samples	15	894.160	59.611	
Total	19	971.412		

Lilliefors and Bartlett statistics as well as the calculations for the F-statistic in the analysis of variance.

The analysis of the data in table 5.1 as dependent samples is considered next.

The atmospheres under which the corrosion occurred identify a source of variability that may interfere with a comparison of the five kinds of pearlitic malleable iron. Consequently, differences that were the result of the atmospheres must be removed first. Since there is only one datum for each kind of iron at each location, this is easily done by subtracting the average for each location, i.e., row, from each observation at that location. The results, presented in table 5.6, are then examined for normality and equal variances.

TABLE 5.6 Location Averages Subtracted for Example 5.1

| Location | *Pearlitic malleable iron types* | | | | |
	A	B	C	D	E
PA	−0.38	3.02	−0.98	−1.48	−0.18
CA	0.36	1.96	−0.74	−1.04	−0.54
NC	−1.28	24.12	−4.48	−18.08	−0.28
NJ	0.72	5.42	−0.38	−4.78	−0.98
IN	−0.28	2.22	1.02	−4.88	1.92

The sample standard deviations, using the data in table 5.6, are given below.

| | *Sample* | | | | |
	A	B	C	D	E
sd	0.768	9.475	2.036	6.959	1.123

Using these standard deviations and 4 degrees of freedom for each (five geographic locations), Bartlett's M-statistic is 26.565. Assuming the samples are from normal populations and that the observable differences are just chance variation, then the probability of a chi-square with 4 degrees of freedom (five malleable irons) as large or larger than this is approximately 0.00004. This is a very small probability, so an F-test of the differences between the five irons would not be valid. Having no other information, an analyst might try omitting type B iron and analyzing the data for the other four samples. The reader can verify that the M-statistic for the four samples will be 19.69 and the probability of a chi-square with 3 degrees of freedom this large or larger is approximately 0.00036. Further investigation along this line will show that the five types of pearlitic iron can be divided into two subgroups. Samples A, C, and E have homogeneous variances and samples B and D also have homogeneous variances, but separate analyses of these two subgroups is not a good strategy. At this point, the analyst might wish for a better alternative, and that is precisely what a test for normality will suggest.

In the above paragraph, "4 degrees of freedom" is used twice. The first use results from there being five geographic locations from which to calculate each sample standard deviation while the second results from there being five different malleable irons.

The Lilliefors test for normality is a modification of the procedure explained in chapter 4 for more than two independent samples. In chapter 4, the degrees of freedom for the standard deviation used to normalize the data was the sum of the degrees of freedom for the individual samples. The degrees of freedom for dependent samples is this same sum decreased by the degrees of freedom for the secondary classification since these differences have been removed, e.g., 16 is the degrees of freedom for a Lilliefors test in example 5.1. The Lilliefors statistic for example 5.1 is 0.241 and the probability of one this large or larger for $n = 25$ is less than 0.01, assuming the samples are from normal populations and the observable sample departures from normality are just chance variation.

Bartlett's procedure identifies a problem in data analysis. In calculating the sample standard deviations, 4 was accepted as the degrees of freedom for each sample to give a total of 20. Yet the degrees of freedom for the same data used in the Lilliefors procedure was 16. This problem illustrates how quickly simple data analysis can arrive at anomalies statisticians have not yet considered.

All is not lost, however. Any adjustment in Bartlett's procedure for the locations would reduce the degrees of freedom for each sample. Use of 4 in example 5.1 is the conservative approach since it produces the largest value of Bartlett's M-statistic. Any value less than 4 will produce a smaller M-statistic resulting in a higher acceptance rate. Consequently, the procedure illustrated here will result in the analyst using nonparametric methods occasionally when parametric methods would be acceptable.

At this point, the analyst must choose between

1. Trying to find a subgroup of iron types that meet the normality and homogeneous variance criteria
2. Considering a subgroup of the locations

The standard deviations given above for the five kinds of iron fall naturally into two groups: A, C, and E in one group, and B and D in the other. The analysis can indeed proceed this way, but when completed, the results will appear fragmented. The second alternative proves to be the better choice.

Even a cursory examination of the data presented in table 5.6 indicates the North Carolina atmosphere produced abnormal differences. If North Carolina data are dropped, Bartlett's M-statistic is 4.906 and the Lilliefors statistic is 0.178. The probabilities of statistics as large or larger than these are approximately 0.18 and 0.0875, respectively, when the null hypotheses are true and the observable differences are just chance variation. The Lilliefors probability is not really large, but it is about as large as can be achieved without fragmenting the analysis as rejected in the preceding paragraph.

It is interesting from the data analysis viewpoint to note that statistics books never suggest that the data in a block can invalidate comparison of the primary classification. Quite the contrary is true. Removal of the differences in the data associated with the secondary classification is always considered sufficient for any analysis. While it may be sufficient for some analyses, it is not appropriate for the five locations in example 5.1.

The data in table 5.1, omitting North Carolina, were used for the sums in table 5.4 which are needed for the analysis of variance.

The analysis of variance begins with the total sum of squares for m samples in n blocks

$$\sum_{j=1}^{m} \sum_{i=1}^{n} (y_{ij} - \bar{\bar{y}})^2$$

where $\bar{\bar{y}}$ is the overall average of the data, i identifies the datum in a sample, and j identifies the sample.

Since these are dependent data, i.e., they can be classified in two ways, the index i identifies the block or secondary classification. This overall sum of squares is divided into three components that can be associated with samples which are types of iron, blocks which are geographic locations, and a remainder or residual portion. Using procedures similar to those described in chapter 20, the results are the formulas below. For samples,

$$\sum_{j=1}^{m} n(\bar{y}_j - \bar{\bar{y}})^2 = \frac{\sum_{j=1}^{m} \left(\sum_{i=1}^{n} y_{ij} \right)^2}{n} - \frac{\left(\sum_{j=1}^{m} \sum_{i=1}^{n} y_{ij} \right)^2}{mn}$$

and for blocks,

$$\sum_{i=1}^{n} m(\bar{y}_i - \bar{\bar{y}})^2 = \frac{\sum_{i=1}^{n} \left(\sum_{j=1}^{m} y_{ij} \right)^2}{m} - \frac{\left(\sum_{j=1}^{m} \sum_{i=1}^{n} y_{ij} \right)^2}{mn}$$

and for residual,

$$\sum_{j=1}^{m} \sum_{i=1}^{n} y_{ij}^2 - \frac{\sum_{j=1}^{m} \left(\sum_{i=1}^{n} y_{ij} \right)^2}{n} - \frac{\sum_{i=1}^{n} \left(\sum_{j=1}^{m} y_{ij} \right)^2}{m} + \frac{\left(\sum_{j=1}^{m} \sum_{i=1}^{n} y_{ij} \right)^2}{mn}$$

These formulas were used to prepare the analysis of variance in table 5.7. On removing the differences associated with the secondary classification, the denominator for the F-statistic was reduced from 59.611 in table 5.5 to 2.374. This is a large reduction that completely changes the results of the analysis. The null hypothesis is that the samples, after removing the geographic location differences, are from normal populations with equal means. Assuming this is true and that the observable differences are just chance variation, then the probability of an F-statistic with 4 and 12 degrees of freedom as large or larger than 8.14 is much less than 0.005, the smallest probability for which values are provided in table D of part V. As independent samples, the corrosion was essentially the same for the five irons. As dependent samples, the corrosion was *not* the same for the five irons. The next data analysis question is this: Which irons corroded about the same and which corroded at higher or lower rates?

TABLE 5.7   The ANOVA Table for Example 5.1 (Assuming dependent samples)

Source of variation	df	ss	ms	
Between samples	4	77.252	19.313	$F = 8.14$
Between blocks	3	865.676	288.559	
Residual	12	28.484	2.374	
Total	19	971.412		

Multiple Comparisons of Sample Pairs, Newman-Keuls

The analysis of variance for more than two dependent samples examines the homogeneity of the sample averages. A small F-statistic implies, for all practical purposes, that the population means can be considered equivalent. If the F-statistic is large, the population means cannot be considered equivalent but the analysis of variance and the F-statistic do not show how the means differ. A report on the analysis will be pretty sterile without something more. Multiple paired comparisons of the samples provide additional information that management often finds very useful.

The multiple comparison of different sample pairs presents a technical problem for data analysts. Every null hypothesis is a statement that no differences exist. Assuming that this is true, the purpose of every statistical test is to decide if observed differences can be attributed to chance variation in the data or if they are caused by something else. If the probability of the test statistic is very low, the analyst will usually conclude that the difference is not entirely due to chance. There is a pitfall associated with every test made, and that is the probability associated with the test statistic, assuming only chance variation in the data, may be very small and the difference still due to chance. It would be an error under these conditions to reject the null hypothesis, i.e., to conclude that important differences probably do exist. This kind of error is controlled by choosing the significance level for the statistical test, usually 0.10, 0.05, or 0.01. Now when a series of similar statistical tests are made on samples that differ only by chance, the significance level chosen for the tests is the proportion of such tests that will indicate the chance difference is *not* due to chance. This kind of error can be controlled but it cannot be eliminated. Make sufficient tests at a particular significance level and errors like this will be made. Statistics texts refer to this kind of error as a *type I error*.

When more than two samples with nonequivalent averages are compared to each other in pairs, the probability of this kind of error rises rather quickly in unknown and uncontrollable ways to unacceptable levels. An analyst may be tempted simply to use a smaller significance level for multiple tests than is used for single tests but this is not sufficient. The new significance level depends on the true sample differences which the analyst never knows but which the *F*-test indicates are not zero.

In chapter 4, Duncan's (1955) procedure for multiple comparisons was introduced. Duncan's procedure could be used for dependent samples but another procedure, the Newman-Keuls [see Newman (1939) and Keuls (1952)], will be introduced instead. The Newman-Keuls procedure differs from the Duncan procedure in only one aspect, the statistic used to calculate the critical ranges.

Among all the procedures developed for multiple comparisons, the Duncan and Newman-Keuls are considered middle-of-the-road procedures. They are more conservative, in the sense that they reject null hypotheses less, than Fisher's least significant difference (LSD) method, but they are less conservative than procedures proposed by Tukey (1953) and Scheffè (1953). A comparison of the tables for the Duncan and Newman-Keuls procedures shows that the Newman-Keuls is slightly more conservative than the Duncan. Keppel (1973) presents a good discussion of multiple comparison procedures including these four.

A critical range for the Newman-Keuls procedure is calculated for each pair of samples as follows:

$$CR = q \sqrt{\frac{ms}{n}}$$

where ms = the residual mean square from the ANOVA table,
      n = the sample size, and
      q = the studentized range statistic for r samples, as explained
           below; df, the degrees of freedom for ms; and the
           significance level chosen for the tests.

The q-statistics are found in table R of part V.

The number of samples spanned by a comparison including the two samples compared is called r. For ordered samples that are adjacent $r = 2$. r is 4 when there are two samples with averages between the two being compared.

The F-test from the analysis of variance for example 5.1 indicates that the five sample averages should not be considered homogeneous. To determine which samples should be considered different from which other samples, the sample averages are arranged in ascending order as in table 5.8. Then the Newman-Keuls critical range is calculated for each pair.

The quantity $\sqrt{ms/n}$ can be calculated once and used for all comparisons. It is 0.7704. The critical ranges for $\alpha = 0.10$ are computed in table 5.9 by multiplying each q-statistic by this quantity. The actual ranges in table 5.8 are then compared to the critical ranges. All averages in a nonsignificant range are considered equivalent.

TABLE 5.8 Multiple Comparisons for Example 5.1

Sample	Average	n
D	23.475	4
C	26.25	4
E	26.575	4
A	26.625	4
B	29.675	4

TABLE 5.9 Critical Ranges (CR) for Example 5.1

r	q	CR
2	2.52	1.94
3	3.20	2.47
4	3.62	2.79
5	3.92	3.02

NOTE: $\alpha = 0.10$, df = 12, ms = 2.374.

The comparisons yield a structure which is indicated by the vertical line in table 5.8 connecting samples C, E, and A. The multiple comparisons indicate that the population means from which these samples came are essentially the same, while populations D and B have means different from the middle three at a significance level of 0.10.

The analysis can be summarized as follows. Five different pearlitic malleable irons as-cast were allowed to corrode under five different atmospheric conditions. The corrosion at one location was so large that it invalidated a comparison of the five irons. The data from this location were removed and only the remaining four locations were used in the analysis to compare the irons. Types A, C, and E irons all corroded about the same. Type D iron corroded significantly less, while type B iron corroded significantly more. Significance was set at $p = 0.10$ for these comparisons.

Example 5.2

Tests were made by the Bureau of Public Roads in 1956 to determine the difference in compression strength between two methods of applying sulfur cement caps to high-strength concrete cylinders. In these tests, caps were applied vertically and horizontally, and the strengths obtained were compared with strengths of similar cylinders capped with high-alumina cement.

Nine cylinders were cast from each batch of high-strength concrete produced. Three cylinders chosen at random were capped with high-alumina cement and another three were capped with a sulfur cement while the cylinders were standing on their ends. The remaining three cylinders were capped with the same sulfur cement while lying horizontally on their sides. The compression strengths obtained can be classified in two ways, by capping method and material and by concrete batch. The principal research problem concerns capping method and material differences. The concrete batches identify a possible source of differences that can interfere with an analysis of this problem. This study, reported by Werner (1958), was large and experienced some cylinder failures not related to compression strength. For these reasons, a subset of data was selected for example 5.2. These data are presented in table 5.10.

In example 5.1, there was one observation for each combination of the two ways the data could be classified, whereas in this example there are

TABLE 5.10 Compression Strengths ÷ 10 for Example 5.2

Batch	High alumina	Type A sulfur Horizontal	Vertical
1	656	618	614
	637	613	617
	649	619	634
2	648	575	641
	638	608	634
	649	612	614
3	602	545	597
	585	534	566
	608	547	593
4	599	594	615
	634	507	588
	587	505	591
5	615	579	591
	610	542	595
	634	596	596

three observations for each combination. Multiple observations for each combination present several new problems in the analysis of more than two dependent samples.

The research hypothesis is that the different methods and materials of capping the concrete cylinders result in different compression strengths, and the purpose for this research project was to determine the significances of these differences. The null hypothesis that will be tested statistically is that no differences exist between the three methods and materials classifications. Concrete batches differ, so to make the results of the study more widely applicable, several smaller, more manageable batches were used. Nevertheless, any average difference between concrete batches contributes nothing to an understanding of method and material differences and so must be removed in the analysis. This is easily done before examining the data for normality and equal variances by subtracting the average for a batch from the compression strength for each cylinder in the batch. These differences are presented in table 5.11.

The three sample standard deviations for computing Bartlett's M-statistic are given below.

	High alumina	Type A sulfur	
		Horizontal	Vertical
sd	12.453	25.929	14.082

TABLE 5.11   Compression Strengths $\div$ 10 for Example 5.2 (Batch averages subtracted)

Batch	High alumina	Type A sulfur	
		Horizontal	Vertical
1	27.444	-10.556	-14.556
	8.444	-15.556	-11.556
	20.444	-9.556	5.444
2	23.667	-49.333	16.667
	13.667	-16.333	9.667
	24.667	-12.333	-10.333
3	26.778	-30.222	21.778
	9.778	-41.222	-9.222
	32.778	-28.222	17.778
4	19.000	14.000	35.000
	54.000	-73.000	8.000
	7.000	-75.000	11.000
5	19.667	-16.333	-4.333
	14.667	-53.333	-0.333
	38.667	0.667	0.667

Using 14 degrees of freedom for each of the standard deviations, Bartlett's M-statistic is 8.94. The null hypothesis is that the samples came from normal populations with equal variances. Assuming this is true and that the observable differences in the sample variances are due to chance variation, then the probability of a chi-square with 2 degrees of freedom this large or larger is approximately 0.012. This is far too small a probability to proceed with an analysis of variance. The data must be analyzed, so the problem is: What next?

At this point in the analysis, there are two alternative courses of action.

1. The three method and material samples can be compared in pairs according to procedures in chapter 3.
2. Batch 4, which has a very large standard deviation compared to the other four batches, can be dropped from further consideration.

Since there are only three method and material samples, the first of these alternatives is a viable course of action but the second is adopted in keeping with the subject of this chapter. Furthermore, the first of the two alternatives would be quite unattractive because, for more than three method and material samples, the number of comparisons would quickly become very large.

The three sample standard deviations, using batches 1, 2, 3, and 5 in table 5.11, are given below.

	Sample		
	High alumina	*Type A sulfur*	
		Horizontal	*Vertical*
sd	9.143	16.991	12.463

Bartlett's M-statistic is 3.94, assuming 11 degrees of freedom for each of the standard deviations. Assuming the same null hypothesis is true and that the observable differences are just chance variation, then the probability of a chi-square with 2 degrees of freedom this large or larger is approximately 0.148. This is not a very large probability, but further reduction in the M-statistic while retaining three method and material samples does not seem feasible.

The Lilliefors statistic, using these same batches, is 0.059. The null hypothesis for a Lilliefors test is that the sample or samples came from normal populations. Assuming this is true and that the observable sample deviations from normality are chance alone, then the probability of a Lilliefors statistic this large or larger for $n = 36$ is much, much greater than 0.20. It now seems reasonable to proceed with computation of the analysis of variance.

The analysis of variance begins with the total sum of squares for m samples in n blocks with r replicates in each sample-block combination.

$$\sum_{k=1}^{m}\sum_{j=1}^{n}\sum_{i=1}^{r}(y_{ijk}-\bar{\bar{y}})^2$$

where $\bar{\bar{y}}$ is the overall average of the data, k identifies the sample, j identifies the block, and i identifies the datum in a sample-block combination.

This overall sum of squares is divided into four components using procedures similar to those described in chapter 20. These components and the formulas for computing the sums of squares are, for samples,

$$\sum_{k=1}^{m} nr\,(\bar{y}_k-\bar{\bar{y}})^2 = \frac{\displaystyle\sum_{k=1}^{m}\left(\sum_{j=1}^{n}\sum_{i=1}^{r}y_{ijk}\right)^2}{nr} - \frac{\left(\displaystyle\sum_{k=1}^{m}\sum_{j=1}^{n}\sum_{i=1}^{r}y_{ijk}\right)^2}{mnr}$$

and for blocks

$$\sum_{j=1}^{n} mr\,(\bar{y}_j-\bar{\bar{y}})^2 = \frac{\displaystyle\sum_{j=1}^{n}\left(\sum_{k=1}^{m}\sum_{i=1}^{r}y_{ijk}\right)^2}{mr} - \frac{\left(\displaystyle\sum_{k=1}^{m}\sum_{j=1}^{n}\sum_{i=1}^{r}y_{ijk}\right)^2}{mnr}$$

and for interaction

$$\frac{\displaystyle\sum_{k=1}^{m}\sum_{j=1}^{n}\left(\sum_{i=1}^{r}y_{ijk}\right)^2}{r} - \frac{\displaystyle\sum_{k=1}^{m}\left(\sum_{j=1}^{n}\sum_{i=1}^{r}y_{ijk}\right)^2}{nr} - \frac{\displaystyle\sum_{j=1}^{n}\left(\sum_{k=1}^{m}\sum_{i=1}^{r}y_{ijk}\right)^2}{mr}$$

$$+ \frac{\left(\displaystyle\sum_{k=1}^{m}\sum_{j=1}^{n}\sum_{i=1}^{r}y_{ijk}\right)^2}{mnr}$$

and for residual

$$\sum_{k=1}^{m}\sum_{j=1}^{n}\sum_{i=1}^{r}y_{ijk}^2 - \frac{\displaystyle\sum_{k=1}^{m}\sum_{j=1}^{n}\left(\sum_{i=1}^{r}y_{ijk}\right)^2}{r}$$

Summations of the data needed in these formulas to prepare the analysis of variance table are given in table 5.12 while the analysis of variance is given in table 5.13.

In contrast with example 5.1, there are two F-statistics in this problem. The first of these, the ratio of the mean square for interaction to the mean

TABLE 5.12 Summations for Preparing the ANOVA Table for Example 5.2

		Cell sums		
			Type A sulfur	
Batch	*High alumina*	*Horizontal*	*Vertical*	*Batch totals*
1	1942	1850	1865	5657
2	1935	1795	1889	5619
3	1795	1626	1756	5177
5	1859	1717	1782	5358
Sample totals	7531	6988	7292	21,811

$$\frac{7531^2}{12} + \frac{6988^2}{12} + \frac{7292^2}{12} = 13{,}226{,}780.7500$$

$$\frac{5657^2}{9} + \frac{5619^2}{9} + \frac{5177^2}{9} + \frac{5358^2}{9} = 13{,}231{,}589.2222$$

$$\frac{1942^2}{3} + \frac{1935^2}{3} + \cdots + \frac{1756^2}{3} + \frac{1782^2}{3} = 13{,}245{,}197.0000$$

$$\frac{21{,}811^2}{36} = 13{,}214{,}436.6944$$

TABLE 5.13 The ANOVA Table for Example 5.2

Source of variation	*df*	*ss*	*ms*	
Between samples	2	12,344.06	6172.03	$F = 29.3$
Between batches	3	17,152.53	5717.51	
Interaction	6	1263.72	210.62	$F = 1.11$
Residual	24	4540.00	189.17	
Total	35	35,300.31		

square for residual, equals 1.11 and has 6 and 24 degrees of freedom. The null hypothesis tested by this F-statistic is that the differences between populations are independent of the blocks, i.e., the concrete batches. The null hypothesis can be explained this way. The differences between any two populations do not depend on which block or blocks one might consider. Assuming this is true and that all observable inconsistencies are just chance variation, then the probability of an F-statistic this large or larger is approximately 0.40.

This is a large probability, so no significance should be attached to the interaction. It is important, however, for data analysts to understand how to interpret interactions and what to do about them. Interaction, as its name implies, measures the interdependence of the two ways the data can be classified. If the interaction F-statistic is large, then the population differences depend on the secondary classification. If the interaction F-statistic is small, then whatever population differences exist may be considered independent of the secondary classification.

Interaction has the following meaning for example 5.2. Since the F-statistic is not significant, differences that exist between the three method and material populations are not dependent on the concrete batches. This means that the relationships among the method and material populations for one batch are essentially the same for all batches. Consider the cell sums in table 5.12. The sum for high-alumina content is the largest sum for each batch. The sum for horizontal capping is the smallest sum for each batch. Thus the sample relationships are the same except for magnitude.

Table 5.14 presents the cell sums from table 5.12 except that the sums for samples B and C in batch 5 have been interchanged. The relationships among the samples are essentially the same for batches 1, 2, and 3, but they are different for batch 5. The reader should verify that the F-statistic for interaction is now 3.395 and the probability of an F-statistic this large or larger is approximately 0.016, assuming the null hypothesis is true and the observable differences are chance variation.

TABLE 5.14 Arbitrary
Rearrangement for Example 5.2 of
Cell Sums to Produce a Significant
Interaction

	Sample			
Batch	A	B	C	Total
1	1942	1850	1865	5657
2	1935	1795	1889	5619
3	1795	1626	1756	5177
5	1859	1782	1717	5358
Sample totals	7531	7053	7227	21,811

In general, a data analyst would not make an F-test of the primary classifications in the presence of significant interaction. A significant interaction defines a new principal research problem and that is: Why do the differences among the primary classes depend on the secondary classes? If the secondary classes are as homogeneous as batches of concrete produced with the same formula should be, the researcher and the analyst have very big problems. The analyst must explain the interaction, perhaps using tables and graphics, to the researcher. The researcher must then determine why the interaction occurred. Until this problem is resolved, comparison of the primary classes is useless or, at best, dependent on the secondary classification. A significant interaction is a dilemma and there are no easy answers to the puzzle it creates, especially none general enough to be included in a book.

The second F-statistic in this problem, the ratio of the mean square for the primary classification to the mean square for interaction, equals 29.3 and has 2 and 6 degrees of freedom. The null hypothesis tested by this F-statistic is that the method and materials populations have equal means after removing the batch differences, i.e., there are no differences between the means of the populations represented by the primary classification after accounting for the differences that can be associated with the secondary classification. Assuming the null hypothesis is true and that the observable differences in the data are only chance variation, then the probability of an F-statistic this large or larger is less than 0.005. This is a very small probability, so much significance should be attached to the differences between the three methods and materials used for capping the cylinders.

Some statisticians recommend pooling the residual with the interaction before testing the primary classes whenever the interaction F-statistic is not significant. There is a reasonable theoretical basis for pooling, so the procedure should be considered. To pool the residual and interaction, sum the sums of squares for a new residual sum of squares and sum the degrees of freedom for a new residual degrees of freedom. Then compute a mean square from these new values for the F-statistic measuring the homogeneity of the primary classes. For example 5.2, the pooled mean square is 193.46, and the new F-statistic is 31.90. This is a larger F-statistic with more degrees of freedom, so the probability of an F-statistic as large or larger will be even smaller, assuming only chance variation.

Multiple Comparisons of Sample Pairs, Dunnett

The analysis for example 5.2 has examined the methods and materials averages and found that the population means, after removing common differences associated with the concrete batches, are not homogeneous. The general relationship, however, among the methods and materials is the same for four different batches of concrete. The analysis, thus far, has not established how the methods and materials differ from each other. The high-alumina caps were added as a kind of control against which the two type A sulfur cappings could be compared. This viewpoint permits use of a special multiple comparison procedure proposed by Dunnett (1955).

Dunnett's procedure uses a critical difference (CD) to compare several samples to a "single" control. This critical difference is calculated as follows:

$$CD = q \sqrt{\frac{2ms}{n}}$$

where ms = the residual mean square from the ANOVA table, assuming that the interaction can be and has been pooled with it

n = the sample size

q = a factor from Dunnett's table for r, the number of samples in the study; df, the degrees of freedom for ms; and the significance level chosen for the tests.

The q-statistics are found in table S of part V.

The value of q for r = 3, df = 30, and a significance level of 0.05 is 1.99, hence Dunnett's critical difference is

$$CD = 1.99 \cdot 5.678 = 11.30$$

Table 5.15 presents the sample averages. Comparing the sample differences to Dunnett's critical range indicates that the populations for both vertical and horizontal type A sulfur caps have less compression strengths than does the control population of high-alumina caps. Dunnett's critical range should be used only to compare experimental samples to a control. It should not be used to compare experimental samples with each other.

TABLE 5.15 Multiple Comparisons for Example 5.2

Sample	Average	n
Control	627.6	12
Vertical	607.7	12
Horizontal	582.3	12

NOTE: Dunnett's critical difference is 8.08.

The analysis can be summarized now. Nine cylinders were prepared from each of five batches of concrete for comparing vertically and horizontally applied type A sulfur caps with high-alumina cement caps. The data from one concrete batch caused an abnormally large variance for the horizontal sample data and were dropped from the analysis. The analysis using the remaining four batches indicated that differences between the capping methods and materials were independent of the concrete batches and that both vertically and horizontally applied sulfur caps had less compression strengths than did the control high-alumina caps. Normally an aberrant concrete batch such as batch 4 would be of considerable concern. The entire research program might be refocused on that one concrete batch until an explanation was found. Such is not the case in this investigation since other cylinder failures unrelated to compression strength were identified.

Random and Fixed Classifications

At this point, random and fixed classifications must be differentiated. All of the primary classifications in this chapter and chapter 3 are fixed classifications, while the secondary classifications are random. Random classes are classes selected from a larger group or population of possible classes. For instance, the locations in example 5.1 were really five places in the United States from the thousands that could have been chosen. If the five locations were chosen at random from some large delineation of possible locations, then the locations would be random representatives of this large group. A similar situation exists for the batches of concrete in example 5.2. Many, many batches of concrete can be created, so in a sense, the five batches created for the data in table 5.10 are a random sample of the many batches that could be created.

In contrast to the locations of example 5.1, the five kinds of pearlitic iron are fixed, at least fixed for the study in the sense that they were not chosen at random from some larger group of irons. The principal research interest focuses on differences among these five irons and the results, when the analysis is completed, apply to these kinds of iron only. Similarly for example 5.2, the principal research interest focuses on differences among three methods and materials for capping concrete cylinders and the results of the analysis apply only to the three methods studied.

The differences between random and fixed classifications should now be clear. Random classifications are associated with classes that are somehow chosen from a larger group or population of classes while fixed classifications are associated with specific things that do not represent some larger group or population.

It has been my experience that in most practical problems, the primary classes are fixed and the secondary classes are random. This is called a type III, or mixed, analysis of variance model. In a type I model, both classifications are fixed, while in a type II model, both classifications are random. The computations to prepare an analysis of variance are the same for all three models, but the computations for the F-statistics and the interpretation of the results depend on the model type. The procedures presented here are for type III models and should not be used for the other model types.

ORDINAL MEASUREMENTS

For an *F*-test in an analysis of variance to be valid,

1. The samples must be random.
2. The data must be normally distributed.
3. The variances of the samples must be homogeneous.

Sometimes ratio and interval data do not satisfy these important conditions and an alternative to the analysis of variance must be used. The most common technique is a transformation of the data to the ordinal scale, i.e., to ranks, and then an analysis made of the ranks. As in the case of more than two independent samples, this procedure is not without its limitations and these will be discussed.

The most widely used procedure for transforming ratio and interval data in more than two dependent samples to the ordinal measurement scale is one proposed by Friedman (1937). The data in each secondary class, i.e., each block, are ranked independently of the other blocks to remove block differences. Friedman then proposed a statistic for testing differences between the primary classes that is distributed asymptotically as a chi-square statistic. Iman and Davenport (1980) have proposed a superior testing procedure based on a statistic that is distributed asymptotically as an *F*-statistic, so their procedure is used in this chapter. In this context, asymptotically means that the distribution of the test statistic is approximated by the chi-square or the *F* distribution, and this approximation improves as the number of secondary classes increases.

Sometimes data originate in the ordinal measurement scale. Intelligence quotients (IQs), aptitude and/or ability scores, and many other psychological measures are generally considered to be ordinal measurements. Political and consumer polls where participants are asked to rate something on a scale, for example, from 1 to 10, sometimes achieve ordinal measurement. Whenever such data can be classified two ways, the procedures in this section of chapter 5 may be appropriate to answer management and research questions.

Example 5.3

Psoriasis is a skin disease characterized by itchy, red patches that become covered with loose scales. The condition can be acute, but in most cases it is chronic and mild. The eruptions may occur anywhere on the human body but are most common on the scalp, forearms, elbows, knees, and legs. The cause of psoriasis is unknown, but it seems to run in families and is not communicable.

There is no known cure for psoriasis, but the itching can be relieved and the red patches cleared with ointments and lotions. In developing new products or in retesting old products for the treatment of psoriasis, two or more products are applied to different lesions on the same patient. The assigned treatment will be continued for perhaps 2 months and then the results assessed. This procedure is, of course, repeated for several patients in each study.

Data from studies such as described above can be classified in two ways. The primary classification is treatment; the secondary classification is patient. Patients do not all respond in the same way or to the same degree. Any differences in response between patients that affects all treatments alike will increase the residual and thus obscure the all-important treatment differences. Consequently, such differences are a nuisance and must be removed during the course of the analysis. Friedman ranks accomplish this for the analyst.

Human skin is a complex barrier that protects the rest of the body from germs and a tough, resilient cushion that protects the softer tissues beneath. The two main layers of the skin are the epidermis and the dermis beneath it. The border between the two layers is very irregular due to numerous cone-shaped dermal papillae that extend upward into the epidermis. The epidermis has a network of ridges around these papillae that appear as pegs in microscopic examination of histological cross sections.

Cells in the epidermis are formed in the single cell layer next to the dermis and then move outward through the epidermis to be sloughed off as "dead skin." The normal transit time for this process has been estimated to be about 75 days.

The dermal papillae beneath psoriatic lesions may reach heights of 0.5 to 0.8 millimeters in contrast with 0.1 millimeters in normal skin. They are long and thin, irregular in size and often club shaped. Beneath active psoriatic lesions, the rate of epidermal cell replication is greatly accelerated. The result is a thickening of the epidermis. Treatment tends to reduce the heights of the dermal papillae and the thickness of the epidermal layer.

Most variables observed in these studies are subjective opinions, e.g., redness, itching, and scaliness. The principal objective measure is the thickness of the epidermis, in spite of its irregularity because of the dermal papillae. Consequently, skin biopsies are taken before and after treatment for measuring the thickness in millimeters. Table 5.16 presents such measurements for the comparison of three products in a study that used 10 patients.

Coal tar, a frequent if sometimes controversial treatment for psoriasis, is a complex of organic compounds obtained by the distillation of bituminous coal during the production of coke. Like crude oil, it can be fractionated into many different organic compounds characterized chiefly by their boiling temperatures. Fractions A and B for this study were obtained in this manner from crude coal tar.

Using procedures presented earlier in this chapter, the reader should

TABLE 5.16 Changes in Epidermis Thickness for Example 5.3

Subject	Crude coal tar	Fraction A	Fraction B
1	0.42	0.33	0.39
2	0.43	0.55	0.51
3	0.40	0.52	0.70
4	0.15	0.24	0.41
5	0.31	0.35	0.41
6	0.25	0.23	0.31
7	0.41	0.41	0.48
8	0.29	0.21	0.27
9	0.58	0.68	0.68
10	0.19	0.31	0.25

verify for the data in table 5.16 that Bartlett's M-statistic is 1.23 and that Lilliefors statistic is 0.192. The probability of a chi-square with df = 2 being larger than 1.23 is large at 0.55, while the probability of a Lilliefors statistic with n = 30 being larger than 0.192 is less than 0.01, assuming for both tests that the observable differences are just chance variation. This latter probability is so small that a transformation to the ordinal measurement scale seems the only safe procedure.

The data for each subject, i.e., block, are transformed to ranks without regard for the other subjects. Table 5.17 presents the data so transformed. Note that tied data for subjects 7 and 9 are given average ranks and that average differences between subjects no longer exist, i.e., the average rank is 2 for all subjects. Removing the differences between subjects requires

TABLE 5.17 Changes in Epidermis Thickness after Transforming to the Ordinal Scale for Example 5.3

Subject	Crude coal tar	Fraction A	Fraction B
1	3	1	2
2	1	3	2
3	1	2	3
4	1	2	3
5	1	2	3
6	2	1	3
7	1.5	1.5	3
8	3	1	2
9	1	2.5	2.5
10	1	3	2

that the total degrees of freedom be reduced by the degrees of freedom for subjects, i.e.,

$$\text{Total df} = N - b$$

where $N =$ the total number of data in the analysis
$b =$ the number of blocks

The Iman-Davenport procedure is simply to use these ranks in an analysis of variance. Since the ranking procedure removed the secondary or block differences, the total sum of squares of the ranks can be divided into only two components. One component is identified with the samples or primary classification and the other with the residual. These components are computed using the following formulas. For samples,

$$\sum_{j=1}^{m} (\bar{y}_j - \bar{\bar{y}})^2 = \frac{\sum_{j=1}^{m} \left(\sum_{i=1}^{n} y_{ij} \right)^2}{n} - \frac{\left(\sum_{j=1}^{m} \sum_{i=1}^{n} y_{ij} \right)^2}{mn}$$

and for residual,

$$\sum_{j=1}^{m} \sum_{i=1}^{n} y_{ij}^2 - \frac{\sum_{j=1}^{m} \left(\sum_{i=1}^{n} y_{ij} \right)^2}{n}$$

Some useful summations of the ranks are given in table 5.18 and then the analysis of variance is presented in table 5.19. Note that subjects are not a source of differences and that the degrees of freedom for total have

TABLE 5.18 Sample Statistics for Example 5.3: Thickness of the Epidermis

| | Sample | | | |
	Crude coal tar	Fraction A	Fraction B	Total
Σy	15.5	19.0	25.5	60.0
Σy^2	30.25	41.50	67.25	139.0

n, the number of subjects $= 10$
k, the number of products $= 3$

$$\frac{15.5^2}{10} + \frac{19.0^2}{10} + \frac{25.5^2}{10} = 125.15$$

$$\frac{60.0^2}{30} = 120.00$$

TABLE 5.19   The ANOVA Table for Example 5.3
Using Friedman Ranks

Source of variation	df	ss	ms	
Between samples	2	5.15	2.575	$F = 3.35$
Residual	18	13.85	0.7694	
Total	20	19.00	0.950	

NOTE: Patients were removed as a source of variation by the Friedman ranking procedure.

been reduced by 9 for this fact. The null hypothesis tested here is that there are no differences between the populations represented by the primary classification after removing differences associated with the secondary classification. This means that all arrangements, i.e., permutations, of the ranks within each secondary class are equally likely and the sum of squares for samples will be very near zero. Assuming the null hypothesis is true and the observable differences are just chance variation, then the probability of an F-statistic with 2 and 18 degrees of freedom as large or larger than 3.35 is approximately 0.061. While this probability is small, the analyst should consider the result inconclusive and recommend that the study be continued with additional patients.

Friedman's chi-square statistic can be obtained from the ANOVA table too. It is

$$\frac{\text{ss for samples}}{\text{ms for total}}$$

The Friedman statistic has $m - 1$ degrees of freedom where m is the number of primary classes. Note that this definition of the statistic is exactly the definition for the Kruskal-Wallis statistic given in chapter 4.

Even though the F-test of the ranks is inconclusive, management might like to base the decision for continuing or terminating the study on the results of the individual paired comparisons of the three treatments. For example, if one of the coal tar fractions held promise of being superior to crude coal tar, management might have more enthusiasm for extending the study.

Multiple Comparisons

Statisticians have proposed several methods for doing multiple paired comparisons of more than two dependent samples in the ordinal scale. These methods seem to have one thing in common, they make use of the central limit theorem to justify adapting ratio and interval procedures that require normality. This is really not a bad thing to do because

1. Ranks qualify as a rectangular distribution.

2. Sums or averages of about 12 from a rectangular distribution are themselves distributed very nearly normal.

Always a big problem in data analysis is the calculation of the residual mean square. Hollander and Wolfe (1973) simply use the pooled mean square of the ranks within secondary classifications. They treat this as a population statistic rather than as a sample statistic, which is all right, and then base critical ranges on the expected range of m normal variables, i.e., m primary class averages. In contrast, Conover (1980) uses the residual mean square from the analysis of variance of the Friedman ranks and bases critical ranges on Fisher's least-significant-difference procedure. Hollander and Wolfe's procedure is very conservative, i.e., produces larger critical ranges, when compared to Conover's procedure.

Accepting that the central limit theorem justifies using ratio and interval analogues that require normality, then the position taken here is that the Duncan or the Newman-Keuls procedure is more appropriate. None of these procedures seem particularly justified unless the number of secondary classes is at least 10 or 12. The Newman-Keuls critical ranges are calculated from the following.

$$\text{CR} = q \sqrt{\frac{\text{ms}}{n}}$$

Everything in this equation has been defined earlier in this chapter.

Since there are only three primary classes in example 5.3, only two values of q are required. For a significance level of 0.05, they are

$$q_{2.18} = 2.97 \qquad q_{3.18} = 3.61$$

There are 10 secondary classes, i.e., patients, in example 5.3, so

$$\sqrt{\frac{\text{ms}}{n}} = 0.277$$

The two critical ranges are then

$$\text{CR}_2 = 2.97 \cdot 0.277 = 0.82$$

$$\text{CR}_3 = 3.61 \cdot 0.277 = 1.00$$

Table 5.20 presents the averages for the three primary classes. As anticipated from the results of the analysis of variance, the observed ranges are small. The difference between crude coal tar and fraction B is the largest and just equal to the critical range. This is a promising result on which to base further research. The vertical bar indicates that the sample differ-

TABLE 5.20 Multiple Comparisons for Example 5.3

Sample	Average	n
Crude coal tar	1.55	10
Fraction A	1.90	10
Fraction B	2.55	10

ence for crude coal tar and fraction A is less than the Newman-Keuls critical range.

Example 5.4

Batch 4 in example 5.2 was found to be different from the other four batches. This presented a problem in the analysis that was solved by dropping batch 4 and using the remaining four batches in the analysis. Another way of solving this problem is to transform all of the data to the ordinal scale before analysis. Example 5.4 illustrates analysis of these data in the ordinal scale.

The data in each batch are transformed to ranks without regard for the other batches. This is an extension or generalization of Friedman ranking for one observation per primary-secondary classification to several observations each. It removes the differences associated with the secondary classification without removing the interaction.

The ordinal data following the transformation are presented in table 5.21. These ranks are now analyzed just as the ratio and interval data were analyzed for example 5.2, except that the Bartlett and the Lilliefors procedures are not required. Summations of the data needed for the analysis of variance are given in table 5.22, and the analysis of variance, calculated from formulas used for example 5.2, is presented in table 5.23.

As with example 5.2, there are two F-statistics in this problem that must be examined. The first tests interaction while the second tests sample differences. The F-statistic for interaction, 0.737, is very small, indicating that the two classifications, capping and batch, are independent of each other. The F-statistic for the samples, 34.4, is very large, indicating that the capping methods do differ from each other. This F-statistic far exceeds the tabled values for an F-statistic with 2 and 8 degrees of freedom, so the probability is much less than 0.005, assuming the null hypothesis of no method or material differences is true and that the observable differences are just chance variation.

The results of this analysis are essentially the same as the results for example 5.2. What then has been gained by transforming the data to the ordinal measurement scale? The exclusion of batch 4 in the analysis of

TABLE 5.21 Compression Strengths for Example 5.4 after Transforming to the Ordinal Scale

Batch	High alumina	Type A sulfur Horizontal	Type A sulfur Vertical
1	9	4	2
	7	1	3
	8	5	6
2	8	1	7
	6	2	5
	9	3	4
3	8	2	7
	5	1	4
	9	3	6
4	7	6	8
	9	2	4
	3	1	5
5	8	2	3
	7	1	4
	9	5.5	5.5

TABLE 5.22 Summations for Preparing the ANOVA Table for Example 5.4

	Cell sums			
	High	Type A sulfur		Batch
Batch	alumina	Horizontal	Vertical	totals
1	24	10	11	45
2	23	6	16	45
3	22	6	17	45
4	19	9	17	45
5	24	8.5	12.5	45
Sample totals	112	39.5	73.5	225

$$\frac{112^2}{15} + \frac{39.5^2}{15} + \frac{73.5^2}{15} = 1300.43$$

$$\frac{24^2}{3} + \frac{23^2}{3} + \cdots + \frac{17^2}{3} + \frac{12.5^2}{3} = 1320.83$$

$$\frac{225^2}{45} = 1125.0$$

TABLE 5.23 The ANOVA Table for Example 5.4

Source of variation	df	ss	ms	
Between samples	2	175.43	87.72	F = 34.4
Interaction	8	20.40	2.55	F = 0.738
Residual	30	103.67	3.46	
Total	40	299.50	7.4875	

NOTE: Batches were removed as a source of variation by the Friedman ranking procedure.

example 5.2 will forever raise the question, why was it necessary? Batch 4 looks like all the other batches, at least to management, so discarding it is hard to explain. In example 5.4, batch 4 is not discarded so no explanation is needed. Results are now based on all batches used in the study. This is much, much easier to explain to management.

Since the F-statistic for interaction is very small, the interaction and residual may be pooled before computing the F-statistic for the samples. This is done by dividing the sum of the two sums of squares by the sum of the two degrees of freedom. This new mean square for residual is 3.265 and the F-statistic is 26.9. While somewhat less than the F-statistic in table 5.23, i.e., 34.4, this F-statistic has a considerably larger degrees of freedom for the denominator so the probability may even be smaller than the probability for 34.4.

Multiple Comparisons of Sample Pairs, Dunnett

The analysis, thus far, indicates that the methods and materials populations do not have equivalent means and that the differences between the populations are independent of the batches of concrete. Just how the populations differ has not been established but it can be by comparing each sample with each other sample. Such paired comparisons should not be made unless the population means are not equivalent.

In the parallel example 5.2, the high-alumina cappings were considered a control and this permitted Dunnett's procedure to be demonstrated. This will also be assumed for example 5.4 so that a comparison can be made of the two results.

Dunnett's procedure is specifically to compare several samples to a control. A critical difference (CD) is computed for the differences between the control sample and any of the other samples as follows:

$$CD = q \sqrt{\frac{2ms}{n}}$$

where ms = the residual mean square from the ANOVA table of the ranks, assuming that the interaction can be and has been pooled with it

n = the sample size

q = a factor from Dunnett's table for r, the number of samples in the study; df, the degrees of freedom for ms; and the significance level chosen for the tests

The q-statistics are found in table S of part V.

For this example, $r = 3$ and $df_{ms} = 38$ so that for a significance level of 0.05, $q = 1.97$. The sample size is 15, so

$$\sqrt{\frac{2ms}{n}} = 0.679$$

and the critical difference is

$$CD = 1.97 \cdot 0.679 = 1.338$$

Table 5.24 presents the average rank for the samples with the control sample first. Both vertical and horizontal type A sulfur caps had less compression strengths, i.e., smaller average ranks, than did the control high-alumina caps. The differences from the control were greater in both cases than Dunnett's critical difference.

TABLE 5.24 Comparison of Average Ranks for Example 5.4

Sample	Average	n
Control	7.467	15
Vertical	4.900	15
Horizontal	2.633	15

NOTE: Dunnett's critical difference is 0.92.

Analysts must be careful in reporting such things as average ranks, so that management won't attach a ratio or interval interpretation to the differences. If sample pair differences are required, as they often are, then a procedure for making estimates can be developed by combining Hodges-Lehmann differences and Walsh averages. Determine the median Hodges-Lehmann differences between sample pairs for each batch separate from the other batches. Next compute Walsh averages for these median Hodges-Lehmann differences. Sort these averages for each pair of samples and use the median as a raw estimate of the difference. The Hodges-Lehmann procedure presented for example 4.2 to establish consistency among the two-sample estimates should be used to obtain the final estimates. This process is not demonstrated here since all of the individual steps have been demonstrated in chapters 2, 3, and 4.

NOMINAL MEASUREMENTS

When ratio or interval data for more than two independent samples were transformed to the nominal scale, the measurement scale was divided into two or more ordered categories and the data classified into a two-way contingency table. The same procedure can be followed for ratio or interval data for more than two dependent samples. Three-way contingency tables result with the classifications corresponding to the original primary and secondary classifications and to the new measurement scale categories. Unless there are several data in each primary-secondary combination, many cells in the three-way contingency table will be empty, i.e., they will contain no data.

The analysis of three-way contingency tables can be very complex and is beyond the scope of this book. The interested reader is referred to Upton (1978) and to Bishop, Fienberg, and Holland (1975).

Throughout chapters 3 and 5, the secondary classification has been associated with nuisance differences, at least insofar as comparison of the primary classes is concerned. From this viewpoint, differences associated with the secondary classification can be removed when the data are transformed to the nominal scale. The data in example 5.2 will be used to illustrate the procedure.

Example 5.5

This example begins with the data in table 5.10 and quickly moves to the data in table 5.11. The difference between the data in these two tables is that batch averages, i.e., block averages, have been subtracted from the data in table 5.10 to get the data in table 5.11. Since this causes the batch averages to be zero, there are no secondary classification differences in table 5.11. The batch classifications are ignored and the data analyzed just as nominal data were analyzed in chapter 4.

The problem, as it has been in previous chapters, is to define meaningful nominal scale categories in the ratio or interval measurement scale. Since analyzing these data in this fashion is kind of artificial, the nominal classes are also kind of artificial. The dividing point chosen for this example is 5.0. Statisticians suggest that the simple chi-square procedure not be used to analyze contingency tables which have expected values less than 5 for any cell. In a problem with a minimum of cells, dividing points must be chosen carefully to satisfy this suggestion, and that has been done here. The expected values are presented in the bottom half of table 5.25. Chi-square for this table is

$$\frac{(15 - 8)^2}{8} + \frac{(1 - 8)^2}{8} + \cdots + \frac{(7 - 7)^2}{7} = 26.25$$

Example 5.5 shows how an analyst can take ratio, interval, or even ordinal data, such as the data in examples 5.1, 5.2, and 5.3, and transform them to the nominal scale for analysis. Sometimes data that can be classified two ways originate in the nominal scale. When this happens, the principal analysis issue is usually not the differences between the primary classes but the independence of the two classifications. It is still possible to interpret the results relative to the differences between the primary classes as will be demonstrated.

TABLE 5.25 Example 5.2 Data Transformed to the Nominal Scale for Example 5.5

	High alumina	Type A sulfur	
		Horizontal	Vertical
Greater than 5.0	15	1	8
Less than 5.0	0	14	7
Expected values based on marginal probabilities			
Greater than 5.0	8	8	8
Less than 5.0	7	7	7

NOTE: $\chi^2 = 26.25$, prob$(\chi^2 > 26.25) \leq 0.00001$.

Example 5.6

The data for this example are taken from Hunt (1948). Metal-casting rejects were classified by the reason for rejection. The number of rejects in each class is presented in table 5.26 for three different weeks. The principal research issue for this kind of data is the independence of the two classifications, reason for rejection and weeks. The null hypothesis generally tested is that they are independent. If the classifications are independent, then the rejection problem is stable, whatever its causes may be. If they are not independent, then the rejection problem is not stable and the reasons for the instability could be investigated. Table 5.27 presents the expected counts computed from the marginal sums.

The chi-square statistic is computed from these two tables as follows.

$$\frac{(97 - 93.9)^2}{93.9} + \frac{(8 - 8.5)^2}{8.5} + \cdots + \frac{(39 - 19.8)^2}{19.8} = 45.572$$

TABLE 5.26 Casting Rejects for Example 5.6

	Reason for rejection							
	Sand	Misrun	Shift	Drop	Core	Broken	Other	Total
Week 1	97	8	18	8	23	21	5	180
Week 2	120	15	12	13	21	17	15	213
Week 3	82	4	0	12	38	25	19	180
Total	299	27	30	33	82	63	39	573

SOURCE: See Hunt (1948). Copyright American Society for Quality Control, Inc. Reprinted by permission.

TABLE 5.27 Expected Casting Rejects for Example 5.6

	Reason for rejection						
	Sand	Misrun	Shift	Drop	Core	Broken	Other
Week 1	93.9	8.5	9.4	10.4	25.8	19.8	12.3
Week 2	111.2	10.0	11.2	12.2	30.4	23.4	14.4
Week 3	93.9	8.5	9.4	10.4	25.8	19.8	12.3

NOTE: $\chi^2 = 45.6$, prob($\chi^2 > 45.6$) ≤ 0.00001. (Expected values for week 2 have been adjusted to preserve the column counts.)

Assuming the null hypothesis is true and that the observable differences are simply chance variations, then the probability of a chi-square statistic with 12 degrees of freedom exceeding 45.572 is approximately 0.0000092. Such a small probability invites the analyst to try to determine from the data the reason why the two classifications are not independent. Most likely the nominal data do not contain sufficient information to determine why the classifications are not independent and additional research would be required. A reader faced with this kind of situation is advised to find a consultant.

SUMMARY

There is no procedure for comparing more than two dependent samples that is correct to use all of the time. The analysis sequence presented here is one approach. While the analysis of two dependent samples was simpler than the analysis of two independent samples, the opposite is true for more than two samples. Analysis of more than two dependent samples is usually much more complex than the analysis of more than two independent samples.

Statistics texts often present procedures for data that can be classified in three or more ways. It has been the experience of this writer that there is still only one primary classification. Consequently, all cross classifications resulting from the second, third, fourth, etc. ways of classifying the data can be treated as a single large group of secondary classes and differences associated with these nonprimary classifications removed by the procedures presented in this chapter. For example, suppose two secondary classifications are sex and age with two classes each. The combinations could be

male, under 40 years

female, under 40 years

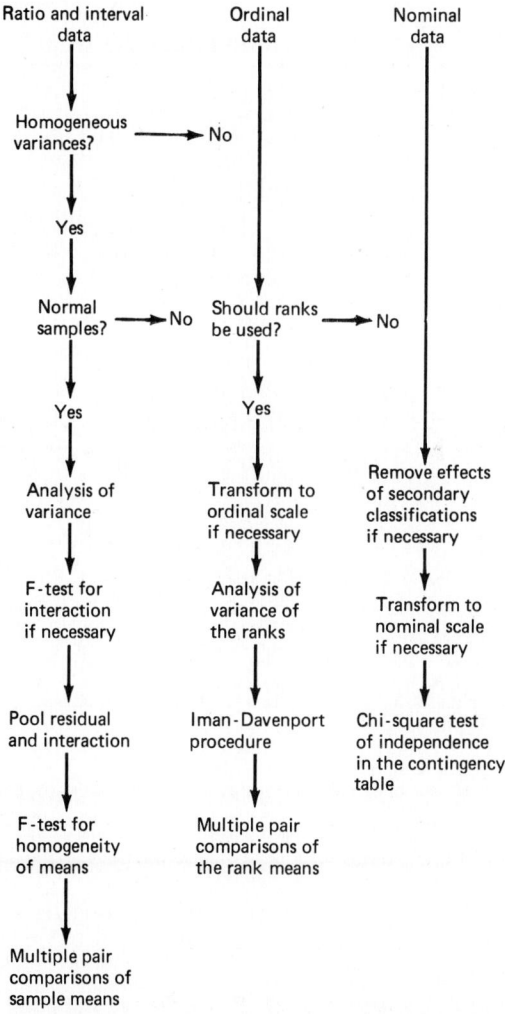

FIGURE 5.1   Analysis sequence flowchart for chapter 5.

male, over 40 years

female, over 40 years

and these combinations define four secondary classes that can be used in the analysis.

Formulas have been given for most situations described in this chapter as if the primary classes had different numbers of data. Such situations often occur in data analysis, but this has been done here more to help readers understand the formulas than to equip them for analyzing such data. This writer recommends a consultant whenever classes do not have equal numbers of data, because this creates problems that should not be taken lightly.

Figure 5.1 is a flowchart that summarizes the sequence suggested in this chapter.

PART II

Linear Relationships

Dependent samples were defined in part I, in contrast to independent samples, as samples that could be classified in two ways. One way was by sample; the other way associated each datum in a sample with data in the other samples. Associated data, one from each sample, were called data pairs in chapter 3 and blocks in chapter 5. The principal data analysis issue was always comparison of the samples.

It was implied in part I, but not formally stated, that associated data measured the same thing. Some examples were the corrosion of several kinds of iron, right and left brachial systolic blood pressure, and crop yields for competing varieties of apples. In all of these examples, the samples were compared or contrasted with each other. The principal issue was the difference between the samples. Whether this was a naturally occurring difference or an experimentally induced difference was inconsequential.

Samples that measure totally different things can still be associated. Some examples are the following:

Year and consumer price index

Heights of men and heights of their sons

Temperature and energy consumption

Crop yield, preplanting precipitation, and growing season precipitation

Other examples are easy to find, but in all such cases sample comparisons would be meaningless. For example, what would it mean to compare temperature with energy consumption? An average temperature in, say, degrees Celsius is simply not comparable to energy consumption measured on any scale. It makes a lot more sense to ask how temperature influences energy consumption or, given a specific energy consumption for a 24-hour period, what was the most likely temperature during the period? So instead of comparison or contrast, related samples that measure totally different things invite data analysts to investigate relationships. Given preplanting and growing season precipitation, what is the most likely corn yield? Given a preplanting precipitation, what is the necessary growing season precipitation to achieve a target corn yield? Given the height of a man, what is the most likely height of his son? Given the height of a son, what is the most likely height of his father?

Related samples that measure different things present a different set of issues from those studied in part I. Part II will be concerned with defining and measuring the relationship between the samples, and part III will present procedures for comparing these relationships. As in part I, ratio and interval measurements will be considered first, followed by ordinal and nominal data.

VARIABLES AND EQUATIONS

It will be convenient now to introduce the notion of variables. Generally, any quantity that can have different values is a variable. For example, temperature is a variable. Growing season precipitation is a variable. Energy consumption is a variable. In the more precise mathematical sense, a variable is a quantity which may have any one of a specified set of values. It is convenient to use the term variable to denote nonmeasurable but classifiable characteristics, as well as measurable characteristics. Sex is a variable in this sense for humans and many other species since each individual can be classified either male or female. Samples are then specific groups of values that the variable has assumed.

Algebraic equations are used to express relationships between the variables represented by the samples. In these equations, it is customary to use letters near the end of the Roman alphabet—x, y, z—to stand for variables. Greek letters, usually near the beginning of the alphabet, are used for parameters that make the algebraic relationship specific. Roman letters, often corresponding to the Greek equivalents, are used to represent numeric estimates of the parameters calculated from the data.

For example, a simple relationship between x and y might be

$$y = \alpha + \beta x$$

Using a specific set of x, y data, estimates a and b of α and β can be calculated. Then an estimate of the relationship, based on the specific set of data, would be

$$y = a + bx$$

Parameters are considered to be constants, but estimates of the parameters are variables in the sense that the estimates based on different sets of data will be different. Furthermore, in procedures for estimating parameters from data, the estimates are treated as variables, while the data, x, y, and z, are treated as constants.

It is usually unnecessary in data analysis to distinguish between parameters and their estimates. For example, the explanation of linear as opposed to nonlinear relationships that follows does not require parameters and their estimates be distinguished. There are, however, a few situations where the distinction facilitates explanations, e.g., when contrasting the deterministic and multivariate viewpoints.

Linear vs. Nonlinear

An algebraic equation or expression is linear in a particular variable if that variable appears in the equation only in the first degree and is not an argument for a trigonometric, logarithmic, exponential, or other transcendental function. Only equations linear in the quantities that make the expression specific for the data, i.e., the a, b, c, are considered in this book. No restrictions, however, are placed on the variables, i.e., x, y, z, that represent the data to be analyzed.

The easiest way to explain the concept is probably through examples,

$$y = a + bx$$

is a linear equation relating x and y, since it is linear in a and b.

$$y = a + bx + cx^2$$

and

$$y = a + bx + c \log x^2$$

are linear equations for the same reason, but

$$y = a + b \log (x + c)$$

is not a linear equation since the quantity c is within the argument of the logarithmic function of x. The fact that x is an argument of the logarithmic function in the last two examples is of no consequence.

Some equations are easy to make linear. For example, an exponential transformation of the y variable in

$$y = \log (a + bx)$$

will produce the following:

$$e^y = a + bx$$

The equation is now linear in a and b.

Deterministic vs. Multivariate

Linear relationships are discussed in two different contexts in statistical texts. These are not ordinarily given names, but they can be associated with the words *deterministic* and *multivariate,* which are reasonable names to use here. Linear relationships in both of these contexts are described below and then the data analysis implications are discussed. For simplicity, the descriptions are presented in terms of two variables x and y, but the reader will have no difficulty generalizing to more than two variables.

From the deterministic viewpoint, values of x are selected by the researcher and then y is measured for these values of x. The variable x is said to be the independent variable, i.e., independent of y, which is said to be the dependent variable, i.e., dependent on x. The variable x is not a random variable since the values of x are selected by the researcher and it is assumed that x can be measured without appreciable error. Serious reflection on statistical discussions of this leads one to believe x is more "determined" than it is measured. The variable y is assumed to be distributed about an expected value μ with variance σ^2 and that all observations of y are independent of each other. It is also assumed that μ is a linear function of x,

$$\mu = \alpha + \beta(x - \bar{x})$$

The implication is that x and y are cause and effect. The problem then is to obtain from the data estimates a, b, and sd^2 of the parameters α, β, and σ^2.

From the multivariate viewpoint, both x and y are random variables drawn from some hypothetical population of x and y pairs. The variables may be related but neither is considered determined by the other. It is quite appropriate to seek the best estimate of y for a given x, $E(y|x)$, or to seek the best estimate of x for a given y, $E(x|y)$. If the relationship is believed to be linear in the coefficients, then

$$E(y|x) = \mu + \beta(x - \nu)$$

and

$$E(x|y) = \nu + \beta'(y - \mu)$$

The parameters μ and ν are, respectively, the means of the hypothetical populations of y and x. The problem here is to obtain from the data estimates of the parameters ν, μ, β, β', and appropriate variances.

From the multivariate viewpoint, the x and y pairs are assumed to be a random sample, but there is no reason why the sample has to be a simple random sample. The x and y pairs can just as well be a stratified random sample or even, as most often occurs, a random sample from selected strata. In this last situation, estimates of the simple x and y variances, i.e., without regard to the strata, will be meaningless, but this happens not to be important in data analysis. The important estimates from the data analysis viewpoint are the variance of y for a given x or the variance of x for a given y and these can be estimated.

The fundamental difference in these two viewpoints lies in the assumption made for the deterministic viewpoint, i.e., x and y are cause and effect. No such assumption is made for the multivariate viewpoint. The data analysis implications are enormous. On adopting the multivariate viewpoint, the data analyst is free to calculate parameters for an equation to estimate y from x or to estimate x from y. The deterministic viewpoint restricts the analyst to estimates of y from x. Should x need to be estimated from y, the regression equation must be solved for x rather than developing an equation to estimate x specifically.

A calibration problem may help to explain the data analysis viewpoint. Consider an analytical chemical laboratory doing analyses for a particular element. A researcher who wants to know the precision of laboratory results and any bias that might exist prepares a series of samples at different concentrations of the element for the laboratory to analyze. The concentrations can be called x and the laboratory results called y. So far, this looks like a typical deterministic problem, but it really isn't from the data analysis viewpoint.

The samples may have been prepared very carefully and the concentrations calculated without arithmetic error; but the procedures for preparing the samples can never be duplicated exactly from one sample to another. Therefore, the calculated concentrations are still only estimates of unknown concentrations, even though they may be far more precise than the laboratory results. A calculated concentration and a laboratory result are both estimates of some unknown concentration.

This calibration problem now has the attributes of the multivariate viewpoint and it is perfectly reasonable to ask:

What is the most likely prepared concentration for a particular laboratory result?

The answer to this question is the purpose for the calibration exercise. The answer to the question:

What is the most likely laboratory result for a particular prepared sample?

is usually of only passing interest. The first question requires that the prepared concentration play the role of dependent variable in the analysis, whereas the second question requires that the laboratory result play this role.

It is rare in data analysis to find a variable x that can be determined or measured without appreciable error as required for deterministic problems. Classifications and counts may be the only kinds of measurements that meet the requirements, and even counts may be inexact under some conditions, e.g., the U.S. census.

For data analysis, the deterministic viewpoint is artificial. The true relationship between variables is often very complex, involving more variables than the analyst realizes or cares to consider. The multivariate viewpoint, wherein the purpose of the analysis defines the formulation used, seems much more reasonable for data analysis. In practice, the function that the data analyst defines and studies is always a gross simplification of true nature since only simple formulations are tractable.

Francis Galton (1889) may have been the first to use the term *regression,* and his use of it was much nearer the multivariate viewpoint than the deterministic viewpoint. In fact, the first example chosen for chapter 6 uses one of Galton's original problems, but not his data.

Variables are related in the multivariate viewpoint, but it is not strictly correct to view any one variable as being dependent on the other variables. It is, nevertheless, convenient to think of the estimated quantity as dependent on the variables used for its estimation. Data analysis nomenclature is then made completely compatible with statistical discussions of regression analysis, but remember, cause and effect is not implied.

Regression analysis, as presented in statistical texts, is more concerned with estimation than with helping the analyst to choose the best functional form, i.e., the most appropriate simplification, for the relationship. I hope the chapters in part II will partially remedy this situation.

Getting a Picture

The first step in the study of relationships between variables is to plot the data. It is customary to measure along the horizontal or x-axis to plot the data designated independent and along the vertical or y-axis to plot the data designated dependent. Figure II.1 illustrates this for the consumer price index (CPI) for years 1950 through 1980. Obviously, the CPI is related to the year, but the year is not deterministic. The plot provides a picture of the relationship between the variables that data analysts find

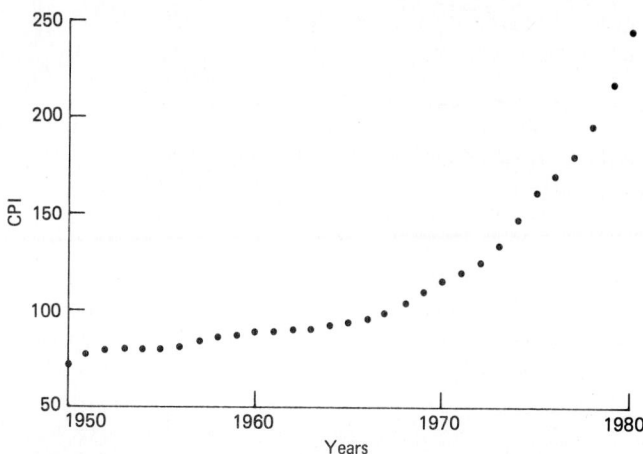

FIGURE II.1 Consumer price index (CPI) for 1950 to 1980 (1967 CPI = 100).

helpful. When there are more than two variables, it is difficult to plot the data in a meaningful way for data analysis, but sometimes plots of all or most of the possible pairs are useful. Chapter 9 presents a problem involving more than two variables plotted on a single graph.

SUMMARY

To summarize, part II introduces regression analysis wherein

1. Simple and usable algebraic relationships for the variables are defined.
2. Procedures are developed for determining the numerical quantities a, b, and c to make the chosen algebraic relation specific for the data.
3. Significance tests of these numerical quantities are made.

The multivariate viewpoint avoids the artificial assumptions associated with cause and effect and permits the data analyst to focus attention on the client's real issue of obtaining the best estimate of the unknown variable in each situation.

CHAPTER 6

Simple Two-Variable Relationships

Estimation and testing of the relationship between two variables expressed in a linear equation are presented in this chapter.

RATIO AND INTERVAL MEASUREMENTS

Example 6.1

Consider the heights of 30 fathers and sons presented in table 6.1. There are several research issues here.

1. Is there a relationship between the heights of fathers and sons?

If there is, then

2. Given the height of a father, what is the most likely height of his son?
3. Given the height of a son, what is the most likely height of his father?

The data in table 6.1 should be plotted first, as in figure 6.1. The data form an elliptical cluster indicating that there is a relationship between father and son heights. A circular cluster would indicate no relationship existed between the two variables. However, the data analyst must determine more about the possible relationship than this. What is the best algebraic approximation to use? Could the apparent relationship, i.e., the elliptical clustering of the data, be due to chance?

TABLE 6.1 Father and Son Heights for Example 6.1

Son	Father	Son	Father	Son	Father
67.7	68.0	68.6	68.0	68.3	70.1
70.1	68.0	68.9	69.1	70.5	69.3
69.7	71.9	67.1	69.1	69.3	68.0
70.3	70.3	69.0	67.6	70.5	69.6
68.7	66.2	69.5	68.8	67.7	67.5
68.6	67.9	69.7	69.1	67.5	67.1
68.4	65.9	67.4	67.5	70.2	70.6
67.3	68.5	71.3	72.4	69.3	67.9
64.8	64.4	65.3	66.3	71.6	68.5
67.6	67.3	67.1	65.8	72.7	70.2

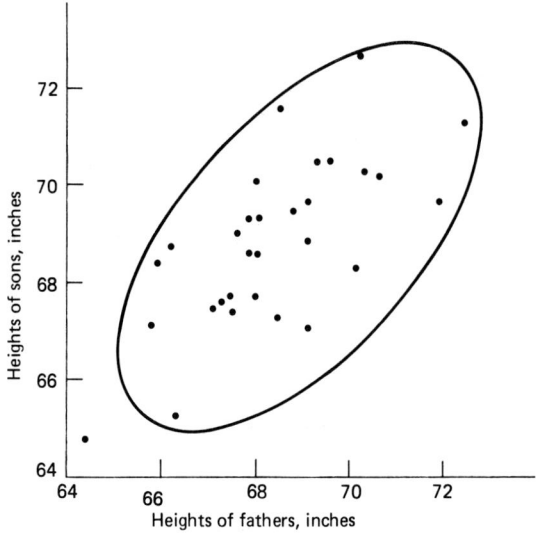

FIGURE 6.1   Plot of data for example 6.1.

Data analysis is sometimes an art, and the first of these two questions illustrates this. The simple elliptical cluster these data exhibit indicates an equation of the form

$$y = \alpha + \beta x$$

can be used to approximate the relationship. But things are not always this simple.

Data often exhibit much more complex patterns, and therein lies the art. Two data analysts, based on their individual experience, may choose widely differing algebraic approximations, i.e., models, for the same data set. Add to this the use or lack of use of other variables that might be related, and the art of data analysis becomes even more evident. For example, the mother's height probably has as much influence on the son's height as does the father's, but this is not considered in example 6.1. More will be said about the art of model selection, sometimes called curve fitting.

The analyst can expect to answer two questions with these data. These questions are

1. Given a specific height for a father, what is the most likely height for his son?
2. Given a specific height for a son, what is the most likely height for his father?

These questions presuppose no cause-and-effect relationship. Heredity is not an issue. The data analysis problem is simply:

> How is the best estimate of either of these variables obtained from the other?

Sons' heights are dependent for estimation on their fathers' heights in the first question, while in the second question, the fathers' heights are dependent for estimation on their sons' heights.

The next problem is to estimate α and β in the algebraic equation

$$y = \alpha + \beta x$$

The standard method for obtaining estimates of α and β is the method of least squares discussed in chapter 18. This method determines estimates a and b which minimize D, the sum of squares of the deviations between the observed values of y and the values of y determined by substituting the associated x into the equation. These calculated values of y are called Y. Thus

$$D = \Sigma d^2 = \Sigma(y - Y)^2 = \Sigma(y - \alpha - \beta x)^2$$

is minimized. One of the d has been noted by a dashed line on figure 6.2.

To obtain the least-squares estimates, the derivatives (reference the calculus) of D with respect to α and β are set equal to zero at the same time the roman letters, a and b, indicating estimates are substituted for α and β. The equations are rearranged slightly to give

$$na + b\Sigma x = \Sigma y$$
$$a\Sigma x + b\Sigma x^2 = \Sigma xy$$

Then these equations are solved for a and b.

$$b = \frac{\Sigma xy - \Sigma x \, \Sigma y/n}{\Sigma x^2 - (\Sigma x)^2/n}$$

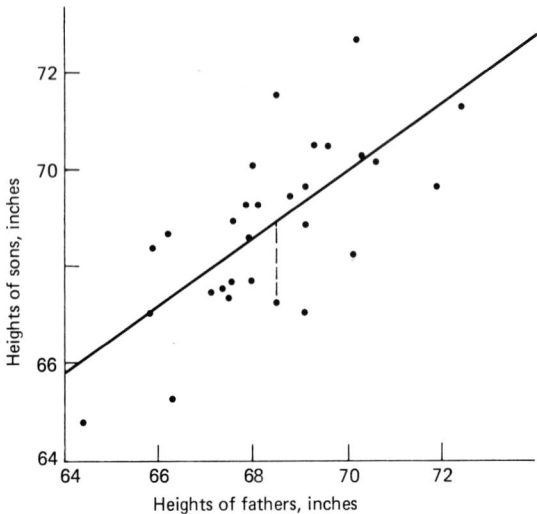

FIGURE 6.2   Plot of data for example 6.1. Approximate regression line for sons' heights as a function of fathers' heights.

$$a = \frac{\Sigma y - b\Sigma x}{n}$$

The quantity b is the regression coefficient. It determines the orientation of the graph of the regression equation. The quantity a is the y-intercept or more simply, just the intercept. This name is given to a because that is the point where the graph of the regression equation crosses the y-axis, i.e., the value of the regression function when x is equal to zero.

In preparation for estimating these coefficients, the data in example 6.1 are summarized.

Sum of fathers' heights = 2050.9

Sum of fathers' heights squared = 140,297.61

Sum of sons' heights = 2064.7

Sum of sons' heights squared = 142,187.45

Sum of fathers' heights multiplied

by the sons' heights = 141,212.28

Count of the data pairs = 30

Consider the problem: Given a specific height for a father, what is the most likely height for his son? This question implies that a father's height is to be the basis for estimating his son's height. Consequently, estimation of son heights y is dependent on the father heights x.

The regression equation is determined from the data of example 6.1 and the equations for a and b as follows:

$$\Sigma x = 2050.9$$
$$\Sigma x^2 = 140{,}297.61$$
$$\Sigma y = 2064.7$$
$$\Sigma y^2 = 142{,}187.45$$
$$\Sigma xy = 141{,}212.28$$
$$n = 30$$
$$b = 0.684996$$
$$a = 21.994730$$

Therefore

$$\text{Sons' heights} = 21.99 + 0.6850 \cdot \text{fathers' heights}$$

The line that represents this relationship is plotted in figure 6.3.

Now consider this problem: Given a specific height for a son, what is the most likely height for his father? This question implies that a father's height is to be estimated from his son's height. Consequently, for this problem, son heights assume the role of independent variable and are designated x, while the father heights are considered dependent and are designated y.

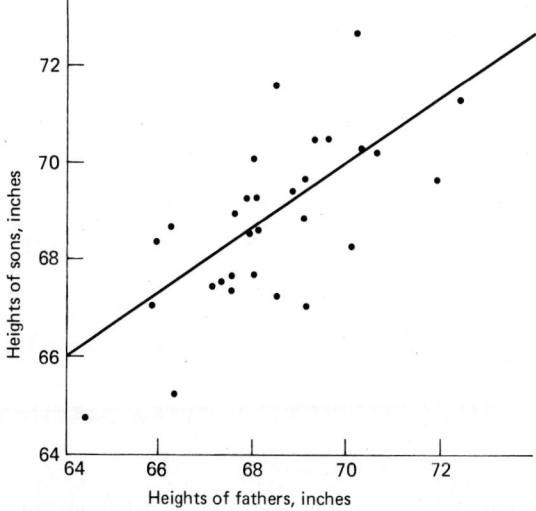

FIGURE 6.3 Plot of data for example 6.1. Regression equation plotted for sons' heights as a function of fathers' heights.

The regression equation for this relationship is determined from the data of example 6.1 as follows:

$$\Sigma x = 2064.7$$
$$\Sigma x^2 = 142,187.45$$
$$\Sigma y = 2050.9$$
$$\Sigma y^2 = 140,297.61$$
$$\Sigma xy = 141,212.28$$
$$n = 30$$
$$b = 0.710989$$
$$a = 19.430713$$

Hence

$$\text{Fathers' heights} = 19.43 + 0.7110 \cdot \text{sons' heights}$$

The line that represents this relationship is plotted in figure 6.4.

Another way of writing the formula for the coefficient a is the following:

$$a = \bar{y} - b\bar{x}$$

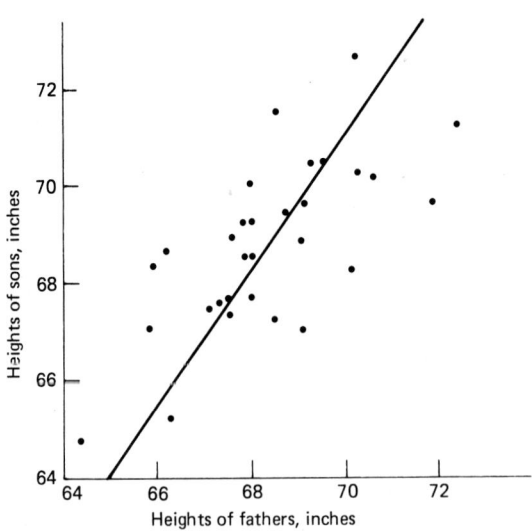

FIGURE 6.4   Plot of data for example 6.1. Regression equation plotted for fathers' heights as a function of sons' heights.

This can be solved for $\bar{y}$ to get

$$\bar{y} = a + b\bar{x}$$

which demonstrates that the average height of the sons can be obtained from the average height of the fathers. The converse is also true. The average height of the fathers can be obtained from the average height of the sons.

A consequence of the above is that the lines in both figure 6.3 and figure 6.4 pass through the point (father average, son average), i.e., they cross at that point if plotted on the same figure.

The next data analysis issue concerns the importance of the regression relationship. An ellipse that encloses all but 1 of the 30 data points has been drawn on figure 6.2. A circle could have been drawn to enclose all or nearly all of the points but a circle would have included more area where there were no points. A square or a fishhook-shaped enclosure could have been used, but these would have also enclosed more unpopulated area of the graph. The fact that an ellipse can enclose the points more tightly than can some other figure is evidence that a linear relationship exists. The orientation of the ellipse even suggests the regression lines plotted on figures 6.3 and 6.4.

A regression equation, like the average, is a summary of the data. If a father's height is unknown, the average height of the sample of sons would be the best estimate of the height of any one of the sons. Consider again the equation:

$$\text{Sons' height} = 21.99 + 0.6850 \cdot \text{fathers' height}$$

The question is: Does knowledge of a father's height help in estimating his son's height? If it does, then the regression relationship is important. It is that simple. What then can the data analyst measure to answer this question?

In figure 6.5, the sons' heights have been plotted parallel to the vertical axis above a different point for each on the horizontal axis. The fathers' heights are not considered in this plot of the sons' heights. The average sons' height is plotted as a straight line on this figure and the deviation of son 21 from this average is identified with a dashed line. In computing the standard deviation of the sons' heights, the 30 deviations from the average are computed, squared, and summed, i.e.,

$$\Sigma(y - \bar{y})^2$$

A similar quantity,

$$\Sigma(y - a - bx)^2$$

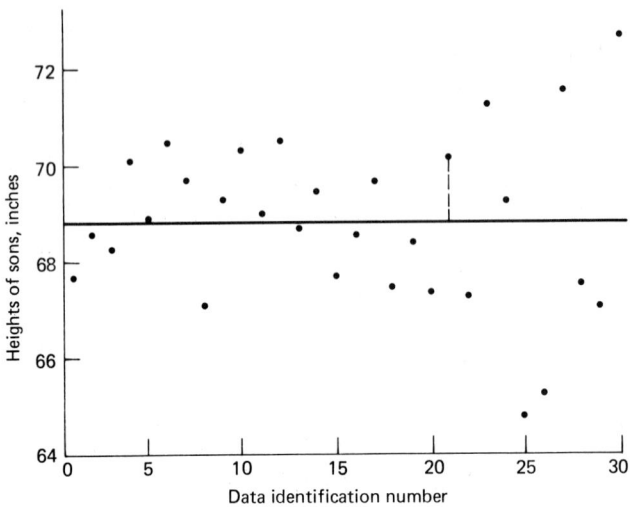

FIGURE 6.5 Plot of data for example 6.1 to show scatter of sons'
heights about the average height of sons.

measures the deviations of the data from the estimated regression equation. The regression equation is plotted in figure 6.2 where the deviation,

$$d = y - a - bx$$

for one son, 7, is identified with a dashed line.

   Just as the standard deviation, i.e.,

$$sd = \sqrt{\frac{\Sigma(y - \bar{y})^2}{(n - 1)}}$$

is a measure of the dispersion of the data about the average as shown in
figure 6.5,

$$sd_{y|x} = \sqrt{\frac{\Sigma(y - a - bx)^2}{(n - 2)}}$$

is a measure of the dispersion of the data about the regression equation
as illustrated in figure 6.3. This new measure of dispersion is the standard
deviation of y given x. Now if the fathers' heights help in estimating sons'
heights, then

$$\Sigma(y - a - bx)^2$$

should be smaller than

$$\Sigma(y - \bar{y})^2$$

This is the basis for a statistical test of the importance of the regression relationship.

In chapter 18, procedures for calculating

$$D = \Sigma(y - a - bx)^2$$

are given. The calculation formula is

$$D = \Sigma y^2 - a\,\Sigma y - b\,\Sigma xy$$

An analysis of variance, like table 6.2 is prepared from these sums. The last entry in the sum of squares column is the sum of squared deviations from the average y,

$$\Sigma(y - \bar{y})^2$$

The middle entry in the sum of squares column is the sum of squared deviations about the regression equation,

$$\Sigma(y - a - bx)^2$$

The first entry in this column is the last entry minus the middle entry. Working from the bottom up may seem an awkward procedure, but it has advantages and analysts get used to it. The df column is as indicated.

TABLE 6.2 The ANOVA Table

Source of variation	df	ss	ms	F
Due to regression	1	Difference		
About regression	$n - 2$	$\Sigma(y - a - bx)^2$		
Total	$n - 1$	$\Sigma(y - \bar{y})^2$		

The two entries in the mean square column are the corresponding sums of squares divided by the degrees of freedom. Finally, the ratio of these mean squares is an F-statistic with the associated degrees of freedom. Incidentally, the first of these two mean squares should always be larger than the second so that the F-statistic must be compared to a standard one-tailed F-table.

Consider the first of the two regression equations,

Sons' height $= 21.99 + 0.6850 \cdot$ fathers' height

The completed analysis of variance is given in table 6.3. The research hypothesis is that there is a relationship between a father's height and the height of his son. In fact, tall fathers are expected to have tall sons and short fathers are expected to have short sons. So the research hypothesis can even incorporate this aspect of the expected relationship. The corre-

TABLE 6.3 The ANOVA Table for Example 6.1

Source of variation	df	ss	ms	
Due to regression	1	42.82	42.82	F = 26.58
About regression	28	45.10	1.61	
Total	29	87.91		

sponding null hypothesis tested by the *F*-statistic is that no relationship exists. Assuming that the null hypothesis is true and that the apparent relationship in the *x, y* data pairs is just due to chance, then the probability of an *F*-statistic with 1 and 28 degrees of freedom larger than 26.58 is something less than 0.0001. Therefore, the regression does improve an estimate of a son's height. In other words, the standard deviation of *y* given *x* is significantly less than the standard deviation of *y* with *x* not given.

Now consider the analysis of variance for the second of the two regression equations,

$$\text{Fathers' height} = 19.43 + 0.7110 \cdot \text{sons' height}$$

The sums of squares and mean squares will not be the same, but the *F*-statistic will be identical, provided sufficient decimal places are used in computing the slopes and intercepts. Consequently, the significance, or importance, of the two relationships will be the same. Readers may want to verify all this for one problem by computing the ANOVA table for the second relationship.

The *F*-test here is valid only under the conditions of normality, independent observations, and equal variance. Statistics books often fail to make it clear that normality and equal variance are conditional on the regression equation. It is really the deviations about the regression equation, i.e.,

$$d = y - a - bx$$

that must be normally distributed, independent, and have equal variance. Consequently, tests of normality and equal variance to validate the *F*-test must be tests of these deviations. There are tests for independence, but these are seldom used unless the data pairs were observed in some natural order, e.g., *x* equal to time.

The equal variance condition is a most complex issue. It literally means that if many more observations could be made for each *x*, the *y* data at each *x* would exhibit the same variance as the data for any other *x*. In general, this issue cannot be examined by observing additional data, e.g., the father-son data.

The most common form of unequal variance is probably one in which the *d* are related, either directly or inversely, to the *x*. To examine this form of unequal variance, divide the x-axis into three or more regions such that the regions contain approximately equal numbers of data pairs. Now compute the variance of the d's within each region and compare them using Bartlett's procedure as discussed in chapter 15.

Using the equation

$$\text{Sons' height} = 21.99 + 0.6850 \cdot \text{fathers' height}$$

table 6.4 presents d and father's heights sorted on fathers' heights. Each column in the table contains approximately a contiguous one-third of the data pairs and the variance of the respective column. The null hypothesis to be tested is that the three subsamples have equal variances. Assuming this is true and that any observable differences are just chance, Bartlett's M-statistic, 0.46, with 2 degrees of freedom is not significant.

Since the d's are differences between observed y and y dependent on x, i.e.,

$$d = y - a - bx$$

it can be argued that the average has already been subtracted from the observations and so Bartlett's M-statistic should be calculated from

$$\Sigma \frac{d^2}{\text{df}}$$

rather than the variances

$$\Sigma \frac{(d - \bar{d})^2}{\text{df}}$$

TABLE 6.4   Groups for Bartlett's Procedure for Example 6.1

Lower portion		Middle portion		Upper portion	
d	x	d	x	d	x
−1.3085	64.4	0.0940	67.9	−0.4279	69.1
0.0325	65.8	0.7940	67.9	−2.2279	69.1
1.2640	65.9	−0.8744	68.0	0.3721	69.1
1.3585	66.2	1.5255	68.0	1.0351	69.3
−2.1100	66.3	0.0255	68.0	0.8296	69.6
−0.4580	67.1	0.7255	68.0	−1.7129	70.1
−0.4950	67.3	−1.6169	68.5	2.6186	70.2
−0.8320	67.5	2.6831	68.5	0.1501	70.3
−0.5320	67.5	0.3776	68.8	−0.1554	70.6
0.6995	67.6			−1.5459	71.9
				−0.2884	72.4
Variance = 1.217885		Variance = 1.583107		Variance = 1.918912	
$n = 10$		$n = 9$		$n = 11$	

Such a refinement is probably unnecessary since the entire procedure is just to take a gross look at the variance along the regression equation. Furthermore, M calculated this way could be made large by inappropriately using simple linear regression when a more complex formulation would be better.

Normality can be investigated with the Lilliefors or the Shapiro-Wilk procedures. The reader should verify that the Shapiro-Wilk statistic is 0.9745 and that the probability of a statistic this large or larger is approximately 0.688, assuming the deviations are normally distributed and that any apparent deviation from normality is due to chance variation in the data. This is much too large to reject normality of the deviations from the regression line.

There are a number of important issues to consider at this point.

1. Statistical tests for a and b compared to arbitrary constants
2. Confidence limits and intervals for a and b and the regression equation
3. Prediction intervals for the regression equation
4. Linearity of the regression relationship
5. Correlation

Each of these will be discussed.

Occasionally data analysts need to compare the regression coefficients with some specified constant. For example, 45° might be the expected slope of the regression equation for estimating sons' heights from fathers' heights, i.e., $b = 1.0$. A nonzero intercept could mean that sons were growing taller or shorter, in general, than their fathers, so zero might be the expected intercept α.

A t-test can be used to compare a regression coefficient b with any arbitrary constant. The form of this test statistic is

$$t = \frac{b - c}{se_b}$$

where b = the regression coefficient
c = the arbitrary constant
se_b = the standard error of b

This t-statistic has $n - 2$ degrees of freedom and the standard error of b is

$$se_b = \frac{sd_{y|x}}{\sqrt{\Sigma(x - \bar{x})^2}}$$

Without trying to define a research hypothesis, suppose the null hypothesis is that the slope in

$$\text{Sons' height} = \alpha + \beta \cdot \text{fathers' height}$$

is equal to 1.0. This null hypothesis does not exclude the slope being greater than 1.0 or being less than 1.0, so this will be a two-tailed test. The standard error of b calculated from the above formula is 0.1329. The t-statistic is then -2.37 with 28 degrees of freedom. Assuming the null hypothesis is true and the observable slope is just due to chance, then the probability of a t-statistic smaller than -2.37 in a two-tailed test is about 0.027. This is fairly small, so it might be concluded that the slope is not 1.0.

A t-test can also be used to compare an intercept a with any arbitrary constant. The form of this test statistic is:

$$t = \frac{a - c}{se_a}$$

where a = the intercept
c = the arbitrary constant
se_a = the standard error of a

This t-statistic also has $n - 2$ degrees of freedom and the standard error of a is

$$se_a = sd_{y|x} \sqrt{\frac{1}{n} + \frac{\bar{x}^2}{\Sigma(x - \bar{x})^2}}$$

It is not unreasonable to expect the intercept in

$$\text{Son's height} = \alpha + \beta \cdot \text{father's height}$$

to be zero, i.e., $\alpha = 0$, because such an intercept would indicate that the sons were neither taller nor shorter than their fathers, i.e., there was no difference between generations. This hypothesis does not exclude the intercept being greater than 0.0 or being less than 0.0, so this will also be a two-tailed test. The standard error of the intercept a calculated from the above formula is

$$1.27 \sqrt{\frac{1}{30} + \frac{4673.09}{91.22}} = 9.09$$

The t-statistic is then 2.42 with 28 degrees of freedom. The probability of a t-statistic larger than 2.42 in a two-tailed test is about 0.023, assuming the null hypothesis is true and that the observable difference from zero is chance variation. This is fairly small, so it might be concluded that the intercept is not zero, that sons are in general growing taller than their fathers. Before concluding this, however, the reader should make a similar test of the intercept in the equation for estimating fathers' height from sons' height and then resolve the two conclusions.

Confidence intervals can be estimated for the coefficients of the regression equation. For the regression coefficient b, confidence limits can be determined from

$$\text{cl} = b \pm t \cdot \text{se}_b$$

and for the intercept a, confidence limits can be determined from

$$\text{cl} = a \pm t \cdot \text{se}_a$$

Suppose 95% confidence limits are needed for b. Referring to a table of the t-statistic for 28 degrees of freedom, 95% of the t-distribution is between -2.048 and $+2.048$. Then the confidence limits for b are

$$\text{cl}_{\text{lower}} = 0.6850 - 2.048 \cdot 0.1329 = 0.4128$$

$$\text{cl}_{\text{upper}} = 0.6850 + 2.048 \cdot 0.1329 = 0.9572$$

The analyst can be reasonably (95%) confident that this interval includes the "true" regression coefficient.

It is sometimes desirable to display confidence limits on either side of the regression equation plotted on a graph. Let X be an arbitrary value of x. Then the following formula can be used to calculate values of y for plotting these confidence limits.

$$y = a + bX \pm t \cdot \text{sd}_{y|x} \sqrt{\frac{1}{n} + \frac{(X - \bar{x})^2}{\Sigma(x - \bar{x})^2}}$$

Table 6.5 demonstrates how this is done for example 6.1 and the plot in figure 6.3 has been reproduced with the limits in figure 6.6. These limits will include the regression line with the confidence chosen for t.

More often than the above, the analyst wants limits for the value of y predicted for a specific value of x. These limits can be computed with the following formula and plotted on either side of the graph of the regression equation.

$$y = a + bX \pm t \cdot \text{sd}_{y|x} \sqrt{1 + \frac{1}{n} + \frac{(X - \bar{x})^2}{\Sigma(x - \bar{x})^2}}$$

Table 6.6 demonstrates how this is done for example 6.1 and these limits have also been plotted in figure 6.6. For any given value of x, these are limits for the predicted value of y.

TABLE 6.5 Confidence Limits for the Regression Equation in Example 6.1

	(2)	(3)	(4)	(5)	(6)	(7)
X	$X - \bar{x}^2$	$\dfrac{col\ (2)}{\Sigma(x - \bar{x})^2}$	$\sqrt{\dfrac{1}{n} + col\ (3)}$	$t \cdot sd_{y\mid x} \cdot$ $col\ (4)$	$a + bX +$ $col\ (5)$	$a + bX -$ $col\ (5)$
65	11.29	0.1237	0.3963	0.86	67.4	65.7
66	5.57	0.0610	0.3072	0.66	67.9	66.5
67	1.85	0.0203	0.2315	0.50	68.4	67.4
68	0.13	0.0014	0.1864	0.40	69.0	68.2
69	0.41	0.0045	0.1945	0.42	69.7	68.8
70	2.69	0.0295	0.2506	0.54	70.5	69.4
71	6.97	0.0764	0.3312	0.72	71.3	69.9
72	13.25	0.1452	0.4225	0.91	72.7	70.4

$\bar{x} = 68.36$ $\Sigma(x - \bar{x})^2 = 91.2497$ $sd_{y\mid x} = 1.2691$ $a = 21.99$ $b = 0.6850$
confidence level $= 0.90$ $t = 1.701$ $n = 30$

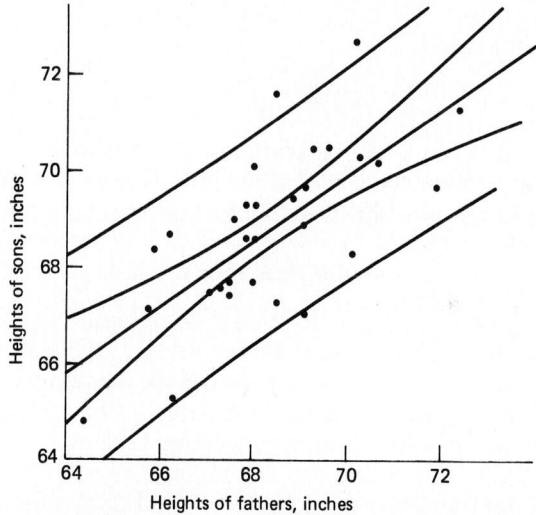

FIGURE 6.6 Plot of data for example 6.1. The center-line is the regression line,

$$\text{Sons' height} = 21.99 + 0.6850 \cdot \text{fathers' height}$$

The two lines next to the regression line are the 90% confidence limits for the regression line. The outermost two lines are the 90% confidence limits for predicted values of sons' heights.

All of the tools for a simple test of linearity, i.e., that the regression equation does not need terms involving x^2, x^3, etc., were presented in chapter 2. This particular test probably cannot be found in any statistics text. It is presented here as an example of how data analysts can adapt their tools to new situations. The regression equation of sons' heights as a function of fathers' heights will be used for illustration.

TABLE 6.6   Confidence Limits for Predicted y Values for Example 6.1

(2)	(3)	(4)	(5)	(6)	(7)
X	$(X - \bar{x})^2$	$\dfrac{col\ (2)}{\Sigma(x - \bar{x})^2}$	$\sqrt{1 + \dfrac{1}{n} + col\ (3)}$	$t \cdot sd_{y\mid x} \cdot$ $col\ (4)$	$a + bX +$ $col\ (5)$	$a + bX -$ $col\ (5)$
65	11.29	0.1237	1.0757	2.32	68.8	64.2
66	5.57	0.0610	1.0461	2.26	69.5	64.9
67	1.85	0.0203	1.0264	2.22	70.1	65.7
68	0.13	0.0014	1.0172	2.20	70.8	66.4
69	0.41	0.0045	1.0187	2.20	71.5	67.1
70	2.69	0.0295	1.0309	2.23	72.2	67.7
71	6.97	0.0764	1.0534	2.27	72.9	68.4
72	13.25	0.1452	1.0856	2.34	73.7	69.0

$\bar{x} = 68.36$ $\Sigma(x - \bar{x})^2 = 91.2497$ $sd_{y\mid x} = 1.2691$ $a = 21.99$ $b = 0.6850$
confidence level = 0.90 $t = 1.701$ $n = 30$

Now if the function

$$y = a + bx$$

is satisfactory, the points will be scattered about it randomly. If it isn't, then an analyst should expect the points in the center to be on the opposite side of the function from those on the end implying that

$$y = a + bx + cx^2$$

or some other more complex relationship is required. So divide the fathers' heights into three groups: the shortest 7, the tallest 8, and the middle 15. Now count the number above the regression line in each group. Referring to figure 6.3, these counts are 3, 9, and 4 and can be used to prepare the two-by-two contingency table, table 6.7. The chi-square statistic with Yates' correction for this table, 0.13, is very small, so the proportion of points above the middle of the line is not very different from the proportion near the ends. Linearity appears to be an acceptable model for these data. More sophisticated tests for linearity are presented in chapter 7.

TABLE 6.7   Counts for Test of
Linearity for Example 6.1

	Center	Ends	Total
Above	9	7	16
Below	6	8	14
Total	15	15	30

The correlation coefficient is an often used measure of association between two variables. Without discussing the properties or the merits of this statistic, its relationship to regression should be noted. The correlation coefficient is the square root of the product of the linear regression coefficients for the two regression equations.

$$r = \frac{\Sigma xy - \Sigma x \Sigma y/n}{\sqrt{[\Sigma x^2 - (\Sigma x)^2/n]\,[\Sigma y^2 - (\Sigma y)^2/n]}}$$

For example, the correlation coefficient for the heights of fathers and the heights of sons is

$$r = \sqrt{0.6850 \cdot 0.7110} = 0.6979$$

The regression coefficients will always have the same sign and this sign should be given to the correlation coefficient. The significance of a correlation coefficient, i.e., compared to zero, can be determined by comparing

$$t = \frac{r}{\sqrt{(1 - r^2)/(n - 2)}}$$

to the t-distribution for $n - 2$ degrees of freedom. This t-test for the correlation coefficient is dependent on the x and y being bivariate normally distributed.

A summary or overview will help put the regression problem in perspective. A regression analysis can be divided into three parts.

1. A procedure for estimating the coefficients of the regression equation
2. A procedure for comparing these coefficients with any specified constants, e.g., zero or one
3. A procedure for estimating confidence limits

In fact, these are the essentials for any analysis. Simple, monomial regression, i.e., regression with only one slope coefficient in the equation, is a bit more complex than comparing two samples in the sense that procedures must be provided for estimating and evaluating the significance of both a slope and an intercept. Confidence limits pose a very special problem in that procedures may be needed for the slope, the intercept, the regression line itself, and finally limits for values of y predicted from a given x. Furthermore, there is always a complementary correlation problem. Correlation is an attempt to measure association or degree of relationship, often without the analyst recognizing that simple correlation assumes a linear relationship between x and y. The correlation problem has the same three parts: estimation, testing, and limits.

Example 6.2

Before introducing nonparametric approaches to the regression problem, another complete example will help to clarify all of the above ideas. A new drug, a beta-blocker, is expected to lower blood pressure. Fifteen patients are given the new drug for 9 weeks to test this research hypothesis. Before the test begins and at the end of each week, the physician measures their blood pressures. Table 6.8 presents the average systolic pressure at each measurement time.

TABLE 6.8   Data for Example 6.2

Week	Average systolic blood pressure
0	142.4
1	139.5
2	138.6
3	134.7
4	136.0
5	133.8
6	133.4
7	134.0
8	129.4
9	131.8

This study is not a clinical trial. It is one kind of initial response study that might be done very soon after the human dose has been established. When repeated several times at different doses, it might be part of a program to determine an effective human dose.

Now the first thing is to summarize the data as follows where x is time in weeks and y stands for the weekly blood pressures.

$$\Sigma x = 45.0$$

$$\Sigma x^2 = 285.0$$

$$\Sigma y = 1353.6$$

$$\Sigma y^2 = 183,357.66$$

$$\Sigma xy = 5993.6$$

$$n = 10$$

Several useful functions of these sums are

$$\Sigma(x - \bar{x})^2 = \Sigma x^2 - \frac{(\Sigma x)^2}{n} = 82.5$$

$$\Sigma(y - \bar{y})^2 = \Sigma y^2 - \frac{(\Sigma y)^2}{n} = 134.364$$

$$\Sigma(x - \bar{x})(y - \bar{y}) = \Sigma xy - \frac{\Sigma x\, \Sigma y}{n} = -97.6$$

Next the slope b and the intercept a are calculated.

$$b = \frac{5993.6 - 45.0 \cdot 1353.6/10}{285.0 - 45.0 \cdot 45.0/10} = -1.183$$

$$a = \frac{1353.6 - (-1.183) \cdot 45.0}{10} = 140.68$$

The three sums of squares for the ANOVA table are calculated as follows.

$$\text{Total} = 183{,}357.66 - \frac{1353.6 \cdot 1353.6}{10} = 134.364$$

$$\text{About regression} = 183{,}357.66 - 140.68 \cdot 1353.6$$
$$- (-1.183) \cdot 5993.6 = 18.903$$

$$\text{Due to regression} = 134.364 - 18.899 = 115.464$$

The analysis of variance is presented in table 6.9. The F-statistic is very large, indicating a significant relationship between time and systolic blood pressure. Assuming the null hypothesis that no relationship exists is true and that the observable relationship is just due to chance, then the probability of an F-statistic this large or larger is beyond the limit of table D in part V.

TABLE 6.9   The ANOVA Table for Example 6.2

Source of variation	df	ss	ms	
Due to regression	1	115.465	115.465	$F = 48.88$
About regression	8	18.899	2.362	
Total	9	134.364		

The standard deviation of y given x is obtained from the analysis of variance,

$$\sqrt{2.363} = 1.537$$

The standard error of b is

$$se_b = \frac{1.537}{\sqrt{82.5}} = 0.1692$$

The t-statistic for comparing b with zero is

$$t = \frac{1.183 - 0}{0.1692} = -6.990$$

which has $n - 2$ degrees of freedom. Note that the square of this t-statistic is the F-statistic in the analysis of variance, table 6.9.

The standard error of a is

$$se_a = 1.537 \sqrt{\frac{1}{10} + \frac{20.25}{82.5}} = 0.9034$$

The t-statistic for comparing a with zero is

$$t = \frac{140.68 - 0}{0.9034} = 155.73$$

which has $n - 2$ degrees of freedom.

Table 6.10 illustrates the calculation of confidence limits for the regression line while table 6.11 illustrates the calculation of limits for y-values predicted for specific values of x.

Finally, the correlation coefficient is

$$r = \frac{-97.6}{\sqrt{82.5 \cdot 134.364}} = -0.927$$

TABLE 6.10 Confidence Limits for the Regression Equation in Example 6.2

X	(2) $(X - \bar{X})^2$	(3) $\dfrac{col\ (2)}{\Sigma(x - \bar{x})^2}$	(4) $\dfrac{1}{n} + col\ (3)$	(5) $t \cdot sd_{y\mid x} \cdot$ $col\ (4)$	(6) $a + bX +$ $col\ (5)$	(7) $a + bX -$ $col\ (5)$
0	20.25	0.245	0.587	1.53	142.2	139.2
1	12.25	0.148	0.498	1.30	140.8	138.2
2	6.25	0.076	0.420	1.10	139.4	137.2
3	2.25	0.027	0.356	0.931	138.1	136.2
4	0.25	0.003	0.321	0.839	136.8	135.1
5	0.25	0.003	0.321	0.839	135.6	133.9
6	2.25	0.027	0.356	0.931	134.5	132.6
7	6.25	0.076	0.420	1.10	133.5	131.3
8	12.25	0.148	0.498	1.30	132.5	129.9
9	20.25	0.245	0.587	1.53	131.5	128.5

$\bar{x} = 4.5$ $\Sigma(x - \bar{x})^2 = 82.5$ $sd_{y\mid x} = 1.537$ $a = 140.68$ $b = -1.183$
confidence level $= 0.90$ $t = 1.701$ $n = 10$

TABLE 6.11 Confidence Limits for Predicted y Values for Example 6.2

	(2)	(3)	(4)	(5)	(6)	(7)
X	$(X - \bar{x})^2$	$\dfrac{\text{col 2}}{\Sigma\,(x - \bar{x})^2}$	$\sqrt{1 + \dfrac{1}{n} + \text{col (3)}}$	$t \cdot sd_{y\mid x} \cdot$ col (4)	$a + bX +$ col (5)	$a + bX -$ col (5)
0	20.25	0.245	1.160	3.03	143.7	137.6
1	12.25	0.148	1.117	2.92	142.4	136.6
2	6.25	0.076	1.084	2.83	141.1	135.5
3	2.25	0.027	1.062	2.78	139.9	134.3
4	0.25	0.003	1.050	2.75	138.7	133.2
5	0.25	0.003	1.050	2.75	137.5	132.0
6	2.25	0.027	1.062	2.78	136.3	130.8
7	6.25	0.076	1.084	2.83	135.2	129.5
8	12.25	0.148	1.117	2.92	134.1	128.3
9	20.25	0.245	1.160	3.03	133.0	127.0

$\bar{x} = 4.5$ $\Sigma(x - \bar{x})^2 = 82.5$ $sd_y\mid x = 1.537$ $a = 140.68$ $b = -1.183$
confidence level $= 0.90$ $t = 1.701$ $n = 10$

The t-statistic for comparing r with zero is

$$t = \frac{-0.927}{\sqrt{(1.0 - 0.859)/8}} = -6.991$$

Note that this t-statistic is identical to the t-statistic calculated to compare the slope b with zero and that this t-statistic squared is the F-statistic in the analysis of variance, table 6.9. When the slope and correlation coefficient are compared to zero, these three tests are algebraically equivalent for simple two-variable problems. (This t-statistic compares r to zero but it cannot be used to compare r to any other constant.)

To complete this example, figure 6.7 shows the limits on the regression line and the limits for y predicted for a specified value of x. The reader is well advised to work through all of the arithmetic in this example.

The discussion for ratio and interval measurements would not be complete without mentioning parametric transformations. In the context of this chapter, parametric transformations are made for two reasons:

1. To permit a linear equation to be used to approximate a relationship that is otherwise not linear
2. To stabilize the variance, i.e., to make the variance of the d's equivalent throughout the range of x

Sometimes a transformation will accomplish both of these objectives at the same time. More often one is accomplished while the other is exac-

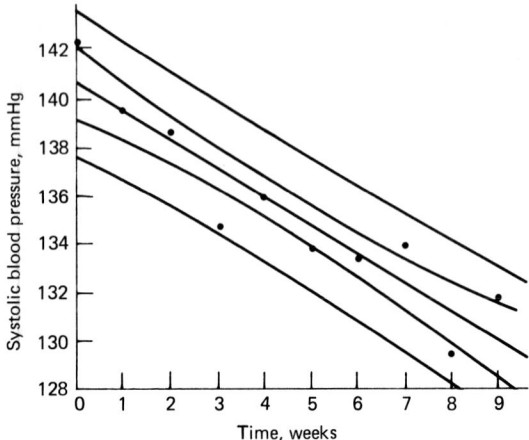

FIGURE 6.7 Plot of data for example 6.2. The center-line is the regression line

Systolic blood pressure = 140.68 − 1.183 · week

The two lines next to the regression line are the 90% confidence limits for the regression line. The outermost two lines are the 90% confidence limits for predicted values of systolic blood pressure.

erbated. There are no rules for the analyst to follow except this: If a transformation works and makes sense, use it.

A parametric transformation should always be viewed as something necessary arising from the nature of the data or the physical situation that generated the data. To be consistent with this viewpoint, a parametric transformation found to be necessary for a particular data set should be necessary for other equivalent data sets. It would be philosophically wrong to use different transformations on equivalent data sets.

ORDINAL MEASUREMENTS

There are several nonparametric approaches to this problem. Two will be presented. They are completely different and yet they complement each other.

As the reader will remember, the least-squares estimation procedure does not require normality. It is only when tests of the slope and intercept are to be made that the deviations of the points from the regression line must be normally distributed. Thus least squares is in a sense a nonparametric procedure for estimating regression parameters. So the focus of a nonparametric regression procedure may begin with the testing of the slope and intercept.

The reader will also remember that the parametric tests for

The reduction in variance in the ANOVA table

The slope compared to zero

The correlation coefficient compared to zero

are all equivalent. Since they differ only in viewpoint, just one of the tests needs to be made in order to explain the results from any of the three viewpoints. This leads the analyst to substitute a nonparametric analogue for one of the parametric procedures and then to ignore the others. There are several candidates for this but the one that seems to have the best properties is Spearman's rho, a nonparametric measure of correlation. Rho is the ordinary correlation coefficient calculated from the ranks of the observations. Hence it can be used with ratio and interval data that have been transformed to ranks or with data that originate as ranks, i.e., ordinal data. An example will make the procedure clear.

Example 6.3

The data in table 6.8 for example 6.2 are reproduced in table 6.12 along with the rank transformations of the x and y data. Note that the data are transformed by first sorting the pairs on the y values, substituting ranks for the y values, then sorting on the x values, and finally substituting ranks for the x values. While the data pairs were sorted and ranked on y before x for table 6.12, the final results are the same no matter which variable is sorted and ranked first.

TABLE 6.12 Data for Example 6.3

Week	Rank of week	Average systolic blood pressure	Rank of blood pressure
0	1	142.4	10
1	2	139.5	9
2	3	138.6	8
3	4	134.7	6
4	5	136.0	7
5	6	133.8	4
6	7	133.4	3
7	8	134.0	5
8	9	129.4	1
9	10	131.8	2

Letting $R(x)$ mean the rank of x and $R(y)$ the rank of y, the ranks are summarized just as the original data were summarized above.

$$\Sigma R(x) = 55$$
$$\Sigma R(x)^2 = 385$$
$$\Sigma R(y) = 55$$
$$\Sigma R(y)^2 = 385$$
$$\Sigma R(x)R(y) = 225$$
$$n = 10$$

The correlation coefficient computed from the ranks is

$$\text{Rho} = \frac{225 - 55 \cdot 55/10}{385 - 55 \cdot 55/10} = -0.9394$$

The null hypothesis is that there is no relationship between the x and y data. Referring to table L in part V, the probability of rho smaller than -0.9394 is something less than 0.0001, assuming the null hypothesis is true and that the apparent relationship is due to chance. So the analyst may safely conclude that there is a relationship between systolic blood pressure and the number of weeks on medication. The least-squares coefficients, which have already been calculated, are the nonparametric coefficients that define the relationship.

The above is essentially a test of b compared to zero. Suppose the issue is not b compared to zero but b compared to some other constant, e.g., one. How can this be done using Spearman's rho? To motivate the answer to this question, consider the following.

Table 6.13 presents

$$d = y - a - bx$$

TABLE 6.13 Data for Example 6.3

Week	Rank of week	$y - a - bx$	Rank of $y - a - bx$
0	1	1.7	9
1	2	0.0	5
2	3	0.3	7
3	4	−2.4	1
4	5	0.1	6
5	6	−1.0	3
6	7	−0.2	4
7	8	1.6	8
8	9	−1.8	2
9	10	1.8	10

for each week. Since $a + bx$ for any week x is a position on the regression line, the d are vertical distances between the data points and the regression line. The sum of their squares is, of course, the sum of squares about the regression line and the basis for computing the standard error of estimate. Table 6.13 also presents the ranks of these quantities.

Now compute the correlation of week and $d = y - a - bx$ and also Spearman's rho for the ranks.

$$r = 0.0038 \qquad rho = -0.0061$$

When the correlation and Spearman's rho were computed for weeks and the average systolic pressure, b was assumed to be zero and a very high correlation was obtained, viz.,

$$r = -0.9271 \qquad rho = -0.9394$$

but when b is the least-squares estimate of the slope, correlation virtually disappears. This can be the basis for comparing b to any constant.

So compute

$$y - \text{constant} \cdot x$$

for each data pair and determine the Spearman correlation coefficient, rho, for these new quantities and the x. If the constant is near b, then rho will be small and not significant. Rho will become large, and finally significant, as constants more remote from b are considered.

Incidentally, the value used for the intercept a is immaterial. No matter what value is used, the ranks are unchanged. In table 6.13, the intercept a was 140.68. In

$$y - \text{constant} \cdot x$$

it is zero.

Spearman's rho doesn't lead to ratio or interval confidence limits for the slope of the regression equation. Theil (1950) proposed a method for finding a slope and confidence limits for the slope that is analogous to the nonparametric Hodges-Lehmann procedure for finding the difference between two dependent samples and the confidence limit for the difference. A natural consequence of the procedure is an alternative test of the significance of any specified slope based on Kendall's rank correlation statistic.

Spearman's rho is a measure of correlation in the ordinal measurement scale, while Kendall's so-called rank correlation, tau, is really a measure of correlation in the nominal measurement scale. Tau is computed by comparing the x_i, y_i with every x_j, y_j such that $i < j$. Create a new set of data pairs which can be called a, b such that a is $+1$ whenever $x_i < x_j$, a is -1 whenever $x_i > x_j$, and a is zero whenever $x_i = x_j$. The new variable

b is defined in the same way using the y data. Kendall's tau is then given by

$$\text{Tau} = \frac{\Sigma ab}{\sqrt{\Sigma a^2 \, \Sigma b^2}}$$

Table K in part V gives probabilities for values of Σab, not for values of tau. This makes the table succinct and facilitates the estimation of confidence limits. Plus and minus ones have been used because they make the explanation for computing tau a little simpler, but nominal scale symbols could have been defined and counted just as easily.

Theil's procedure will be illustrated using the data of example 6.3 where week is the x-variable and the average systolic blood pressures are designated y. The test is most easily applied if the data are sorted on x as already done in table 6.12.

The procedure is this. Compute the slope for each pair of data

$$s = \frac{y_i - y_j}{x_i - x_j}$$

such that the subscript j is greater than i. The first such slope would be

$$\frac{139.5 - 142.4}{1 - 0} = -2.9$$

Several more slopes are

$$\frac{138.6 - 142.4}{2 - 0} = -1.9$$

$$\frac{134.7 - 142.4}{3 - 0} = -2.5667$$

After each datum has been paired with the first, each with subscript greater than 2 is paired with the second.

$$\frac{138.6 - 139.5}{2 - 1} = -0.9$$

$$\frac{134.7 - 139.5}{3 - 1} = -2.4$$

and so on. The only restrictions are that the subscript j must always be greater than i and, to keep matters simple, slopes are not computed when $x_i = x_j$ since dividing by zero is not defined in arithmetic. The a's in the formula for Kendall's tau are simply 1.0 given the sign of these numerators, the b's are 1.0 given the sign of the denominators, and the ab's are 1.0 given the sign of the slopes.

Table 6.14 presents the 45 slopes that can be calculated in this manner from 10 points and the subscripts to identify the points used for each. The next step is to sort these slopes from smallest to largest. Table 6.15 presents the slopes sorted along with the rank of each.

The nonparametric estimate based on Theil's procedure is the median. Since there is an odd number of slopes for 10 data points, the median is uniquely determined by the middle value, i.e., it is -1.2. The median is, of course, the average of the two middle values whenever there is an even number of slopes.

It is not uncommon to compare nonparametric estimates to least-squares estimates in examples. The reader will remember that the least-squares estimate for this slope was -1.183, indicating reasonable agreement of the two methods for this problem.

Confidence limits for the Theil estimate are determined as follows.

1. Choose the confidence level $1 - \alpha$.
2. If there is to be an upper and a lower limit, divide α by 2 before doing step 3.

TABLE 6.14   Theil Slopes for Example 6.3

i	j	Slope	i	j	Slope
1	2	−2.9	3	10	−0.9714
1	3	−1.9	4	5	1.3
1	4	−2.5667	4	6	−0.45
1	5	−1.6	4	7	−0.4333
1	6	−1.72	4	8	−0.175
1	7	−1.5	4	9	−1.06
1	8	−1.2	4	10	−0.4833
1	9	−1.625	5	6	−2.2
1	10	−1.1778	5	7	−1.3
2	3	−0.9	5	8	−0.6667
2	4	−2.4	5	9	−1.65
2	5	−1.1667	5	10	−0.84
2	6	−1.425	6	7	−0.4
2	7	−1.22	6	8	0.1
2	8	−0.9167	6	9	−1.4667
2	9	−1.4429	6	10	−0.5
2	10	−0.9625	7	8	0.6
3	4	−3.9	7	9	−2.0
3	5	−1.3	7	10	−0.5333
3	6	−1.6	8	9	−4.6
3	7	−1.3	8	10	1.1
3	8	−0.92	9	10	2.4
3	9	−1.5333			

TABLE 6.15 Sorted Theil Slopes for Example 6.3

Rank	Slope	Rank	Slope	Rank	Slope
1	−4.6	16	−1.4667	31	−0.9167
2	−3.9	17	−1.4429	32	−0.9
3	−2.9	18	−1.425	33	−0.84
4	−2.5667	19	−1.3	34	−0.6667
5	−2.4	20	−1.3	35	−0.5333
6	−2.2	21	−1.3	36	−0.5
7	−2.0	22	−1.22	37	−0.4833
8	−1.9	23	−1.2	38	−0.45
9	−1.72	24	−1.1778	39	−0.4333
10	−1.65	25	−1.1667	40	−0.4
11	−1.625	26	−1.1	41	−0.175
12	−1.6	27	−1.06	42	0.1
13	−1.6	28	−0.9714	43	0.6
14	−1.5333	29	−0.9625	44	1.3
15	−1.5	30	−0.92	45	2.4

3. For the appropriate number of data n, find in the table of Kendall's tau (table K, part V) the first probability less than α.

4. Determine t for this probability.

5. The upper confidence limit is the slope with rank

$$m_{upper} = \frac{(k + t)}{2} + 1$$

where k = the number of slopes.

6. The lower confidence limit is the slope with rank

$$m_{lower} = \frac{(k - t)}{2}$$

where k = the number of slopes.

If some of the x data were tied, the number of slopes k will be less than $n(n - 1)/2$ and the confidence for the limits determined by this procedure will tend to be slightly larger than the desired level.

To illustrate the procedure, choose $\alpha = 0.10$ for the confidence level. Then in the column headed 10 of table K, find the first probability less than $0.10/2 = 0.05$. This is 0.036. Now t for this probability is 21 and

$$m_{lower} = \frac{45 - 21}{2} = 12$$

$$m_{upper} = \frac{45 + 21}{2} + 1 = 34$$

The upper and lower 0.90 confidence limits are the Theil slopes with ranks 12 and 34, i.e., -0.6667 and -1.6.

There is a normal approximation to use when n is greater than the largest n in table K. Compute t as follows.

$$t = z \sqrt{n(n-1)\frac{(2n+5)}{18}}$$

where z is the standard normal deviate for the desired confidence level. For example 6.3,

$$t = 1.645 \sqrt{10 \cdot 9 \cdot \frac{25}{18}} = 18.39$$

Rounding up,

$$m_{\text{lower}} = \frac{45-19}{2} = 13$$

and

$$m_{\text{upper}} = \frac{45+19}{2} + 1 = 33$$

Since the upper limit is between the median slope and zero and since these are 90% confidence limits, the null hypothesis that the true slope is greater than or equal to zero is rejected at the 0.05 probability level. Sometimes, however, the data analyst needs a more precise estimate of the p-value. This can be obtained as follows using Kendall's rank correlation procedure to test the null hypothesis of no relationship.

Compute

$$e_i = y_i - bx_i$$

for each data point in the sample where b is the hypothesized slope, usually zero or one. Now for each pair of e's such that the j subscript is greater than the i subscript, create a new set of data which can be called c_k such that

$$
\begin{aligned}
c_k &= 1 && \text{if } (e_i - e_j)(x_i - x_j) > 0 \\
c_k &= -1 && \text{if } (e_i - e_j)(x_i - x_j) < 0 \\
c_k &= 0 && \text{if } (e_i - e_j)(x_i - x_j) = 0
\end{aligned}
$$

The test statistic is the sum of the c_k where k goes from 1 to $n(n-1)/2$. The probability of the test statistic is determined from table K in part V.

To illustrate this procedure for example 6.3, let the hypothesized slope be zero. Then the number of positive c_k should be about the same as the number of negative c_k. Table 6.16 presents the observed c_k. There are 4 positive ones and 41 negative ones, resulting in a test statistic equal to -37. Now table K does not give probabilities for negative test statistics. The distributions are symmetrical, however, so the table can still be used by ignoring the sign. Assuming the null hypothesis is true and that the differences between the number of positive and negative c_k is just chance variation, then the probability of a test statistic this small or smaller is less than 0.0002.

TABLE 6.16 c_k for Example 6.3

-1	-1	-1	-1	$+1$
-1	-1	-1	-1	-1
-1	-1	-1	-1	-1
-1	-1	-1	-1	$+1$
-1	-1	-1	-1	-1
-1	-1	-1	-1	-1
-1	-1	$+1$	-1	-1
-1	-1	-1	-1	-1
-1	-1	-1	-1	$+1$

SUMMARY

Part II is about relationships between variables and chapter 6 has introduced the simplest model for these relationships. "True" relationships between variables are probably very complex. Furthermore, they are made obscure by measurement errors and our own inadequacies as analysts and investigators. Analysts must not get caught up in the "cause-and-effect" syndrome. The procedures described in chapter 6 produce gross simplifications of real-world or true relationships. These simplifications are sufficient for many problems, but they can be useless for other problems. They are often all that we know about a true relationship so there is a tendency to use the results of regression analyses without critical evaluation of their usefulness. Analysts must recognize and interpret the limitations of their analyses for client investigators and managers lest too much confidence be placed in the results.

Ordinal and nominal scale procedures have been presented for the simple models discussed in chapter 6. These procedures are not easily adapted to the problems discussed in the remainder of part II. Readers interested in using ordinal or nominal scale procedures for more complex problems should have the advice of an experienced consultant.

CHAPTER 7

Simple More-Than-Two-Variable Relationships

INTRODUCTION

Chapter 6 provides the basic structure for estimating and testing the relationship between two variables expressed in a linear equation. Chapter 7 extends this basic structure to relationships involving more than two variables and sets the stage for chapters 8 and 9 on more complex relationships.

To parallel the ideas presented in chapter 6, the focus of an analysis will be to examine the dependence of one variable on the other variables. This variable will be considered dependent in the sense that the principal result of the analysis will be an equation for estimating it from the so-called independent variables. The same data can often be used for other problems that focus on a different dependent variable.

In chapter 5, the notion of identifying and measuring variability from more than one source was introduced. That idea will be extended here with the sources being the different independent variables.

The estimation procedure will be least-squares just as it was in chapter 6 but extended to two or more independent variables. The choice of independent variables to be studied for a particular dependent variable is a problem more for the researcher or for management than for the data analyst. Nevertheless, data analysts often can and should provide help, especially if the analyst is familiar with the particular field of research. The selection of variables for the regression relationships from those studied is a data analysis problem and several ways for doing this will be illustrated.

RATIO AND INTERVAL MEASUREMENTS

Example 7.1

Prater (1956) used the data given in table 7.1 to estimate gasoline yield from characteristics of crude oil input to a refinery and the gasoline end-point. Other questions could have been asked of the same data, thereby defining different dependent variables. For example, the problem could have been to estimate the gasoline endpoint necessary for a given percent yield from available crude.

TABLE 7.1   Gasoline Yield from Crude Oil Data for Example 7.1

gr	EP	VP	10%	y	gr	EP	VP	10%	y
38.4	235	6.1	220	6.9	32.2	360	5.2	236	24.8
40.3	307	4.8	231	14.4	38.4	365	6.1	220	26.0
40.0	212	6.1	217	7.4	40.3	395	4.8	231	34.9
31.8	365	0.2	316	8.5	40.0	272	6.1	217	18.2
40.8	218	3.5	210	8.0	32.2	424	2.4	284	23.2
41.3	235	1.8	267	2.8	31.8	428	0.2	316	18.0
38.1	285	1.2	274	5.0	40.8	273	3.5	210	13.1
50.8	205	8.6	190	12.2	41.3	358	1.8	267	16.1
32.2	267	5.2	236	10.0	38.1	444	1.2	274	32.1
38.4	300	6.1	220	15.2	50.8	345	8.6	190	34.7
40.3	367	4.8	231	26.8	32.2	402	5.2	236	31.7
32.2	351	2.4	284	14.0	38.4	410	6.1	220	33.6
31.8	379	0.2	316	14.7	40.0	340	6.1	217	30.4
41.3	275	1.8	267	6.4	40.8	347	3.5	210	26.6
38.1	365	1.2	274	17.6	41.3	416	1.8	267	27.8
50.8	275	8.6	190	22.3	50.8	407	8.6	190	45.7

NOTE: gr = gravity, EP = gasoline endpoint, VP = vapor pressure, 10% = ASTM 10% point, y = percent yield.
SOURCE: See Prater (1956). Reprinted with permission from Petroleum Refiner, Gulf Publishing Company.

Values for the four independent variables were not selected, indeed could not be selected, in the sense required for regression analysis in statistics texts. Instead, the data resulted from 32 diverse batches of crude oil and operating conditions (gasoline endpoints) chosen to give a broad range of percent yields.

Just as with two variables, the dependent variable should be plotted with each independent variable as in figure 7.1. The usefulness of such plots may be limited since relationships can be obscured by variability associated with a different independent variable. The elliptical clusters on the plots in figure 7.1 imply linear relationships, and the greater the difference between the lengths of the two axes of the ellipse, the stronger the relationship. In fact, an analyst can almost order the importance of the independent variables from the ellipses in these plots.

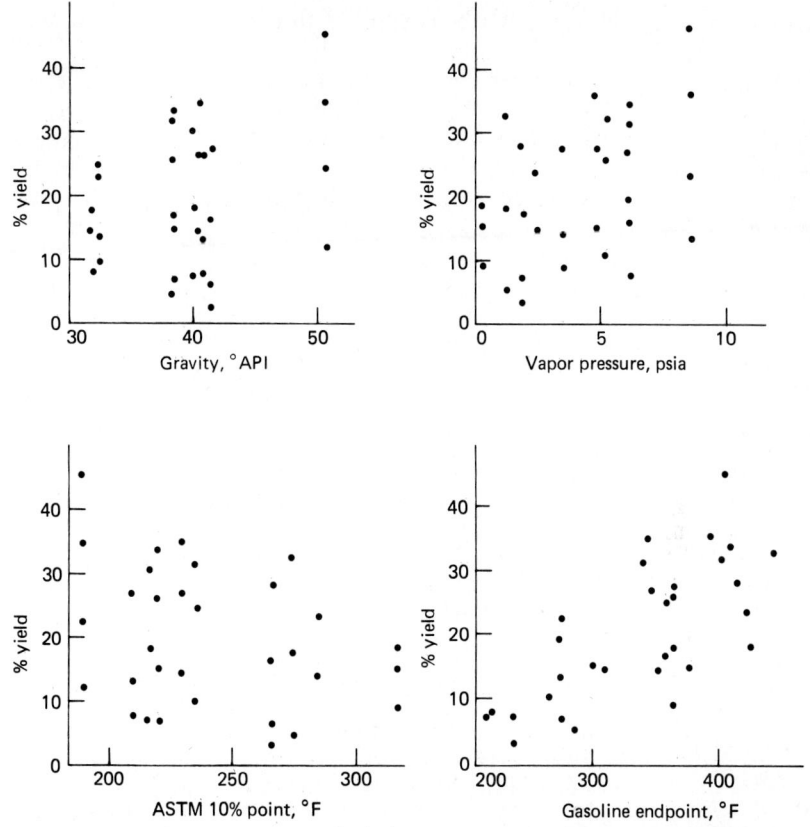

FIGURE 7.1 Percent gasoline yield plotted with each of the independent variables.

An analyst always wants the simplest usable formulation for a regression equation. The goal and the scatter displayed in figure 7.1 preclude using any formulation other than linear in the independent variables. So the regression equation to examine is

$$y = b_0 + b_1 x_1 + b_2 x_2 + b_3 x_3 + b_4 x_4$$

where y = yield as a percent of the crude oil input
$\quad x_1$ = the crude oil gravity
$\quad x_2$ = the vapor pressure of the crude oil
$\quad x_3$ = the ASTM 10% distillation point for the crude oil
$\quad x_4$ = the gasoline endpoint
$\quad b$'s = constants to be determined from the data

The problem now is to determine which of the four independent variables are important for estimating gasoline yield as a percent of the crude oil input. The procedure described parallels the procedures presented in chapter 6.

If the four independent variables were not known for a batch of crude, the best estimate for the percent gasoline yield would be the simple average of previous y data. But the four independent variables are all known for the 32 gasoline yields in table 7.1, so the procedures described in chapter 6 can be used to choose the independent variable that reduces the total sum of squares, i.e.,

$$\Sigma(y - \bar{y})^2$$

the most. The four equations, one for each of the independent variables, and the sum of squared deviations about each regression equation are presented in table 7.2. The first three of these equations do not reduce the total sum of squares, 3564.08, very much, but gasoline endpoint does, so it should obviously be considered for any formulation to estimate percent gasoline yield.

TABLE 7.2 Percent Yield as a Function of Each Independent Variable and the Residual Sums of Squares for Example 7.1

Equation	Residual ss
$y = \quad 1.264 + 0.4687 \cdot$ gravity	3347.82
$y = \quad 13.087 + 1.5719 \cdot$ vapor pressure	3038.34
$y = \quad 41.389 - 0.0900 \cdot$ ASTM 10% point	3210.38
$y = -16.662 + 0.1094 \cdot$ gasoline endpoint	1759.69

Table 7.3 presents an analysis of variance like those in chapter 6 to examine the significance of this reduction. The null hypothesis is that there is no relationship. Assuming this is true and that the observed relationship occurred just by chance, then the probability of an F-statistic with 1 and 30 degrees of freedom as large or larger than 30.76 is much, much less than 0.005, the limit of table D in part V. So any formulation to estimate percent gasoline yield must include the gasoline endpoint. Incidentally, these separate relationships indicate that vapor pressure is the only characteristic of crude oil that is even marginally related to the

TABLE 7.3 The ANOVA Table for Gasoline Endpoint for Example 7.1

Source of variation	df	ss	ms	
Gasoline endpoint	1	1804.38	1804.38	$F = 30.76$
Residual	30	1759.69	58.66	
Total	31	3564.08		

percent gasoline yield ($p = 0.03$). The other crude oil characteristics look rather useless. Further analysis, however, will show that the crude oil gravity and the ASTM 10% distillation point are the important crude oil characteristics and vapor pressure is relatively useless.

The procedure is repeated now to select the variable from the remaining three that most reduces the sum of squared deviations about

$$y = -16.662 + 0.1094 \cdot \text{gasoline endpoint}$$

The new equation will be

$$y = b_0 + b_1 x_1 + b_2 x_2$$

where y = the percent yield
x_1 = the gasoline endpoint
x_2 = one of the remaining three variables
b's = constants to be determined from the data

The deviations of the y data about the regression equation are given by

$$d = y - b_0 - b_1 x_1 - b_2 x_2$$

The least-squares procedure is to minimize D, the sum of the squares of these deviations. Chapter 18 presents general procedures for doing this, but this specific situation is not discussed. Consequently, it will be presented here.

To minimize D, the partial derivatives of D with respect to the b's are set equal to zero and the system of three equations in three unknowns, i.e., in the b's, is solved.

$$D = \Sigma d^2 = \Sigma(y - b_0 - b_1 x_1 - b_2 x_2)^2$$

$$\frac{\partial D}{\partial b_0} = -2\Sigma(y - b_0 - b_1 x_1 - b_2 x_2) \qquad = 0$$

$$\frac{\partial D}{\partial b_1} = -2\Sigma(y - b_0 - b_1 x_1 - b_2 x_2)x_1 \qquad = 0$$

$$\frac{\partial D}{\partial b_2} = -2\Sigma(y - b_0 - b_1 x_1 - b_2 x_2)x_2 \qquad = 0$$

Rearranging these last three equations gives

$$b_0 n + b_1 \Sigma x_1 + b_2 \Sigma x_2 = \Sigma y$$

$$b_0 \Sigma x_1 + b_1 \Sigma x_1^2 + b_2 \Sigma x_1 x_2 = \Sigma x_1 y$$

$$b_0 \Sigma x_2 + b_1 \Sigma x_1 x_2 + b_2 \Sigma x_2^2 = \Sigma x_2 y$$

The appendix to chapter 18 presents the procedure for solving four equations in four unknowns and this can be easily generalized to any system of equations. The results of setting up and solving this system of equations for the three remaining independent variables are presented in table 7.4. The sum of the squared deviations about each regression equation can be calculated from

$$D = \Sigma y^2 - b_0 \Sigma y - b_1 \Sigma x_1 y - b_2 \Sigma x_2 y$$

where y = the percent yield
$\quad x_1$ = the gasoline endpoint
$\quad x_2$ = one of the characteristics of the crude oil

D for each equation is also presented in table 7.4 in the column labeled residual ss.

TABLE 7.4   Percent Yield as a Function of Gasoline Endpoint and Each of the Other Independent Variables and the Residual Sums of Squares for Example 7.1

Equation	Residual ss
$y = -64.951 + 0.1356 \cdot$ endpoint $+ 1.0085 \cdot$ gravity	861.95
$y = -37.808 + 0.1393 \cdot$ endpoint $+ 2.6774 \cdot$ vapor pressure	369.87
$y = 18.468 + 0.1558 \cdot$ endpoint $- 0.2093 \cdot$ ASTM 10%	170.61

While by themselves none of the crude characteristics reduced the total sums of squares very much, they all reduce the sum of squared deviations about the gasoline endpoint equation rather substantially. The ASTM 10% point causes the largest reduction, so it should be considered for use in estimating percent gasoline yield. Table 7.5 extends the analysis of variance in table 7.3 to examine the significance of this reduction. The null hypothesis is that after removing the effect of gasoline endpoint, the ASTM 10% point is not related to the percent gasoline yields. Assuming the null hypothesis is true and that the apparent relationship is just due to chance, then the probability of an F-statistic with 1 and 29 degrees of

TABLE 7.5   The ANOVA Table for Example 7.1 after Adding the ASTM 10% Point

Source of variation	df	ss	ms	
Gasoline endpoint	1	1804.38	1804.38	
ASTM 10% point	1	1589.08	1589.08	$F = 270.11$
Residual	29	170.61	5.883	
Total	31	3564.08		

freedom as large or larger than 270.11 is very much less than 0.005, the limit of table D in part V. So this characteristic of the crude should be included in any equation to estimate the percent gasoline yield. It is interesting to note that the ASTM 10% point alone did not reduce the total sum of squares very much but, with the gasoline endpoint, it causes a very large reduction.

As might be anticipated, the procedure is repeated again to select the next variable for the percent gasoline yield estimation equation. The sum of squared deviations D is defined with one more term, partial derivatives are set equal to zero, and the resulting system of four equations in four unknowns is solved for the b's.

$$D = \Sigma d^2 = \Sigma(y - b_0 - b_1 x_1 - b_2 x_2 - b_3 x_3)^2$$

$$\frac{\partial D}{\partial b_0} = -2\Sigma(y - b_0 - b_1 x_1 - b_2 x_2 - b_3 x_3) \quad = 0$$

$$\frac{\partial D}{\partial b_1} = -2\Sigma(y - b_0 - b_1 x_1 - b_2 x_2 - b_3 x_3)x_1 \quad = 0$$

$$\frac{\partial D}{\partial b_2} = -2\Sigma(y - b_0 - b_1 x_1 - b_2 x_2 - b_3 x_3)x_2 \quad = 0$$

$$\frac{\partial D}{\partial b_3} = -2\Sigma(y - b_0 - b_1 x_1 - b_2 x_2 - b_3 x_3)x_3 \quad = 0$$

Rearranging these last four equations gives

$$b_0 n + b_1 \Sigma x_1 + b_2 \Sigma x_2 + b_3 \Sigma x_3 = \Sigma y$$
$$b_0 \Sigma x_1 + b_1 \Sigma x_1^2 + b_2 \Sigma x_1 x_2 + b_3 \Sigma x_1 x_3 = \Sigma x_1 y$$
$$b_0 \Sigma x_2 + b_1 \Sigma x_1 x_2 + b_2 \Sigma x_2^2 + b_3 \Sigma x_2 x_3 = \Sigma x_2 y$$
$$b_0 \Sigma x_3 + b_1 \Sigma x_1 x_3 + b_2 \Sigma x_2 x_3 + b_3 \Sigma x_3^2 = \Sigma x_3 y$$

Table 7.6 shows that the gravity of the crude oil reduces the residual the most. The analysis of variance is presented in table 7.7. The null hypothesis is that after removing the effect of the gasoline endpoint and the ASTM 10% distillation point, the gravity of the crude oil is not related

TABLE 7.6 Percent Yield as a Function of Gasoline Endpoint, ASTM 10%, and Each of the Other Independent Variables and the Residual Sums of Squares for Example 7.1

Equation	Residual ss
$y = 4.032 + 0.1565 \cdot$ endpoint $- 0.1866 \cdot$ ASTM $+ 0.2217 \cdot$ gravity	146.00
$y = 8.562 + 0.1540 \cdot$ endpoint $- 0.1749 \cdot$ ASTM $+ 0.5227 \cdot$ vapor pressure	160.62

TABLE 7.7   The ANOVA Table for Example 7.1
after Adding the Gravity of Crude

Source of variation	df	ss	ms	
Gasoline endpoint	1	1804.38	1804.38	
ASTM 10% point	1	1589.08	1589.08	
Gravity of crude	1	24.61	24.61	$F = 4.72$
Residual	28	146.00	5.214	
Total	31	3564.08		

to the percent gasoline yields. Assuming the null hypothesis is true and that the apparent relationship is just due to chance, then the probability of an F-statistic with 1 and 28 degrees of freedom as large or larger than 4.72 is 0.041. This probability is not very small, indicating that the gravity of the crude is not very important in estimating the percent gasoline yield.

The procedure is repeated one more time to add vapor pressure. The resulting estimation equation is

$$y = -6.821 + 0.1547 \cdot \text{endpoint} - 0.1495 \cdot \text{ASTM } 10\%$$
$$+ 0.2272 \cdot \text{gravity} + 0.5537 \cdot \text{vapor pressure}$$

and the analysis of variance is presented in table 7.8. The null hypothesis is that after removing the effect of the gasoline endpoint, the ASTM 10% distillation point, and the gravity of the crude oil, the vapor pressure is not related to the percent gasoline yields. Assuming the null hypothesis is true and that the apparent relationship is just due to chance, then the probability of an F-statistic with 1 and 27 degrees of freedom as large or larger than 2.24 is 0.17. This large a probability indicates that vapor pressure contributes little in estimating percent gasoline yield so there is no reason to include it in the equation.

This is an interesting result since it seemingly contradicts the results of the first step of the analysis. When considered separately, the relationship of vapor pressure with percent yield was second only to the relation-

TABLE 7.8   The ANOVA Table for Example 7.1
after Adding the Vapor Pressure

Source of variation	df	ss	ms	
Gasoline endpoint	1	1804.38	1804.38	
ASTM 10% point	1	1589.08	1589.08	
Gravity of crude	1	24.61	24.61	
Vapor pressure	1	11.20	11.20	$F = 2.24$
Residual	27	134.80	4.993	
Total	31	3564.08		

ship of gasoline endpoint to percent yield. Yet now it appears that vapor pressure is not useful at all. The explanation is this. Independent variables are seldom independent of each other. Some of the information that one independent variable contains will be available in other variables. Gasoline endpoint contained some of the same information that vapor pressure contained so vapor pressure dropped from being the second most important variable when considered separate from the others to being third most important when considered with gasoline endpoint.

The ASTM 10% point contained the most new information about percent gasoline yields so it was included in the estimation equation next. It also contained some of the same information that vapor pressure contained, so vapor pressure fell to fourth most important variable. The final step in the analysis discovered that most of the information about percent gasoline yields was contained in the gasoline endpoint, the ASTM 10% point, and the gravity of the crude. Vapor pressure contributed so little new information that there was no need to include it in the final equation. In fact, the gravity of the crude may not add enough new information, i.e., reduce the residual standard deviation sufficiently, to justify leaving it in the final equation either.

This procedure is a *step-forward* procedure. It began by picking the single best variable to predict percent gasoline yield and added other variables one at a time. At each stage, a test was made to determine if the best variable reduced the sum of squares by a significant amount. When the best failed to do that, the process was terminated and variables not entered in the estimation equation were discarded.

It is interesting to note that Prater did not use a stepwise procedure in 1956, probably because such procedures were not generally available until the advent of electronic computers. It is also of interest to note that the results of this analysis do not agree with Prater's. The most obvious possibility for the differences is typographical errors in the raw data presented in his paper. Finally, the results of this analysis are dependent upon the particular independent variables chosen for study. Given a different set of independent variables including none, some, or all of these four and the results might be quite different.

At each stage in this procedure, an F-test has been used to determine if the new variable significantly reduced the residual sum of squared deviations about the regression equation. A test of normality of the deviations should precede each of these F-tests, but in practice such tests are never made, at least not by the numerous computer programs available for step-forward regression analysis.

The interested reader can verify that the Lilliefors statistics at the end of each step in the step-forward procedure are 0.082, 0.090, 0.089, and 0.061. The null hypothesis for each is that the distribution of the deviations from the regression equation are not different from the normal distribution. All of these sample statistics are much smaller than the Lilliefors statistic for $p = 0.20$, hence the p-values are much greater than 0.20, assuming the null hypothesis is true in each case and that the observable deviations from the normal are just chance variations. Thus normality of the residual errors is an acceptable thesis. Table 7.9 illustrates the calculation of the third Lilliefors statistic, which is for the three-term model.

It is not unusual for the maximum deviation from the normal distribution to trend downward as the number of terms in a regression equation increases, and this is precisely what happens in example 7.1. This writer has never found this discussed in statistical literature so has no satisfactory explanation. It may be a practical result of the central limit theorem, which in one of its least-restrictive forms simply states that the

TABLE 7.9 The Lilliefors Test for Normality for Example 7.1

			Probabilities			
d	z	z sorted	Normal	Sample	da	db
−1.37	−0.60	−1.54	0.062	0.031	0.030	0.000
−3.51	−1.54	−1.54	0.062	0.062	0.000	0.031
1.81	0.79	−1.44	0.075	0.094	−0.019	0.012
−0.74	−0.32	−1.42	0.078	0.125	−0.047	−0.016
−0.01	0.00	−1.15	0.126	0.156	−0.030	0.001
2.66	1.16	−0.90	0.184	0.187	−0.004	0.028
−0.95	−0.42	−0.70	0.242	0.219	0.023	0.055
0.28	0.12	−0.60	0.274	0.250	0.024	0.055
1.08	0.47	−0.58	0.280	0.281	−0.002	0.030
−3.24	−1.42	−0.48	0.315	0.312	0.003	0.034
−0.50	−0.22	−0.42	0.338	0.344	−0.006	0.026
0.89	0.39	−0.38	0.351	0.375	−0.024	0.007
3.27	1.43	−0.32	0.373	0.406	−0.033	−0.002
0.00	0.00	−0.29	0.385	0.437	−0.053	−0.022
−0.87	−0.38	−0.25	0.400	0.469	−0.069	−0.037
−0.58	−0.25	−0.22	0.414	0.500	−0.086	−0.055
1.33	0.58	0.00	0.499	0.531	−0.033	−0.001
−2.62	−1.15	0.00	0.499	0.562	−0.063	−0.032
3.22	1.41	0.12	0.548	0.594	−0.046	−0.014
3.22	1.41	0.38	0.648	0.625	0.023	0.054
−1.33	−0.58	0.39	0.652	0.656	−0.004	0.027
−1.10	−0.48	0.47	0.682	0.687	−0.005	0.026
−3.52	−1.54	0.55	0.710	0.719	−0.009	0.023
−3.29	−1.44	0.58	0.720	0.750	−0.030	0.001
1.26	0.55	0.72	0.766	0.781	−0.016	0.016
0.87	0.38	0.79	0.786	0.812	−0.026	0.005
1.65	0.72	0.95	0.828	0.844	−0.016	0.016
−2.06	−0.90	1.16	0.877	0.875	0.002	0.034
4.78	2.09	1.41	0.921	0.906	0.015	0.046
−1.60	−0.70	1.41	0.921	0.937	−0.016	0.015
−0.67	−0.29	1.43	0.924	0.969	−0.045	−0.014
2.16	0.95	2.09	0.982	1.000	−0.018	0.013

NOTE: The Lilliefors statistic is 0.086. Refer to Chapter 15 for details.

sum of a number of variables from distributions with finite variances tends to the normal. In most practical problems, the variables used in a regression equation certainly have finite variances, so the residuals which are sums of regression variables should tend toward the normal.

Step-forward is just one of four stepwise regression procedures in general use. A step-backward procedure begins with a regression equation containing all of the independent variables. At each stage, the variable that increases the residual the least is taken out if the increase is not significant. The procedure is terminated when the removal of any one of the variables remaining significantly increases the residual. When the number of independent variables is small, these two procedures often lead to the same result, as indeed they do for example 7.1.

Another popular procedure considers all possible combinations of independent variables at each stage of the process. Stage one of the step-forward version of this technique is identical to the procedure followed for example 7.1. Once gasoline endpoint was chosen as the first independent variable in the estimation equation, only variable combinations

Gasoline endpoint and gravity

Gasoline endpoint and vapor pressure

Gasoline endpoint and ASTM 10% point

were considered for stage two in the step-forward procedure. Three other combinations,

Gravity and vapor pressure

Gravity and ASTM 10% point

Vapor pressure and ASTM 10% point

were not considered. This third procedure considers all six of these two independent variable combinations and chooses the one that most reduces the residual sum of squares. The process is repeated until the reduction caused by the addition of a single new independent variable does not significantly reduce the residual sum of squares. Table 7.10 presents all of the residual sums of squares for selecting the best estimation equation by this procedure. The result for example 7.1 is the same just as it often is for problems with a small number of independent variables.

The fourth regression procedure steps backward while using all combinations, just as the third procedure does.

Each of these procedures has strengths and weaknesses. The strengths of one are often the weaknesses of another. For example, the step-forward procedure used for example 7.1 can step right past a better formulation simply because that formulation didn't lie in the path determined by the first and subsequent variables chosen. The all-combinations procedure can be a lot of work even for large-scale computers if the number of inde-

TABLE 7.10 Residuals for Using All Combinations of Regression Procedure for Example 7.1

Independent variables	Residual
Gravity	3347.82
Vapor pressure	3038.34
ASTM 10% point	3210.38
Gasoline endpoint	1759.69
Gravity and gasoline endpoint	861.95
Gravity and vapor pressure	3037.97
Gravity and ASTM 10% point	3205.74
Gasoline endpoint and vapor pressure	369.87
Gasoline endpoint and ASTM 10% point	170.61
Vapor pressure and ASTM 10% point	3016.59
Gravity, gasoline endpoint, and vapor pressure	265.48
Gravity, gasoline endpoint, and ASTM 10% point	146.00
Gravity, vapor pressure, and ASTM 10% point	3008.76
Gasoline endpoint, vapor pressure, and ASTM 10% point	160.62
Gasoline endpoint, ASTM 10% point, gravity, and vapor pressure	134.80

pendent variables and the number of data are large. If there are 10 independent variables, for example, there could be 1023 formulations to examine.

Much research in statistics seeks to speed up and improve these basic procedures. For example, the stepwise procedures are improved by defining a mechanism for removing independent variables if new variables entering at a later stage can take their place. A stepwise procedure that incorporates this feature may find formulations outside paths dictated by earlier choices of independent variables for the regression equation. The all-combinations procedures can be speeded up considerably by examining the variances and covariances along the way rather than the individual equations. All of this is beyond the scope and purpose of this book but is mentioned so that the reader will know other procedures exist when using the statistical packages available for both large and small computers. An important problem for data analysts is to understand adequately the procedures employed and the results reported by the statistical analysis packages used.

Example 7.2

Data were collected for each of the 50 states following the 1980 census on nine variables that might be related to crime (Table 7.11). The variables examined were

1. Age: the number of 14 to 24 year olds per 1000 total population
2. Sex: the number of males per 1000 females

TABLE 7.11 Crime Data for Example 7.2

st	Age	Sex	nw	%u	pw	pnw	pr	%t	%a	Crime
AL	202	926	262	62.0	112	354	0.89	24.0	4.3	48.99
AK	214	1127	229	43.4	68	196	1.17	13.3	8.5	65.95
AZ	199	970	176	75.1	100	226	2.06	12.1	3.9	76.14
AR	191	935	173	39.2	140	386	1.73	18.6	3.5	37.96
CA	199	972	238	94.9	87	185	1.51	17.4	4.5	75.90
CO	207	985	110	80.9	85	223	3.07	13.8	2.9	73.53
CT	197	930	99	88.3	61	294	1.88	14.9	3.3	58.37
DE	212	932	178	67.0	80	283	1.30	20.7	4.8	66.89
FL	177	922	160	87.9	92	315	1.53	18.2	3.9	80.32
GA	206	935	278	60.0	99	317	0.81	14.6	2.7	56.28
HI	208	1053	669	79.0	94	96	0.48	25.0	5.0	65.43
ID	194	996	44	18.3	115	333	7.43	14.6	3.9	45.31
IL	199	940	192	81.0	68	299	0.96	14.4	3.9	49.50
IN	203	944	89	69.8	81	247	3.36	18.4	4.4	45.40
IA	199	946	26	40.1	87	253	13.00	11.5	2.9	47.16
KS	199	959	83	46.8	86	235	4.04	7.6	2.7	54.04
KY	203	956	77	44.5	164	365	5.38	18.6	3.4	35.32
LA	214	942	308	63.4	109	356	0.69	22.5	3.9	52.68
ME	198	945	13	33.0	124	133	69.00	14.0	5.9	42.43
MD	205	940	251	88.8	65	194	1.00	16.1	3.7	65.58
MA	206	909	65	85.3	84	265	4.53	14.7	4.3	58.35
MI	207	952	150	82.8	84	249	1.90	15.8	6.2	68.54
MN	206	962	34	64.6	85	250	9.57	12.7	3.2	47.37
MS	209	929	359	27.1	123	443	0.50	18.4	3.3	35.37
MO	196	927	116	65.3	102	259	3.00	14.6	3.2	53.51
MT	197	997	60	24.0	107	340	4.94	8.8	3.5	50.19
NB	197	953	51	44.1	95	225	7.83	9.1	2.0	41.78
NV	194	1025	125	82.0	71	170	2.94	11.8	4.1	85.92
NH	200	949	12	50.7	85	91	77.00	7.5	2.1	43.22
NJ	190	922	168	91.4	66	236	1.39	18.9	5.3	61.80
NM	209	971	249	42.3	129	295	1.31	19.3	4.5	62.01
NY	193	905	205	90.1	94	289	1.26	21.2	5.4	69.05
NC	207	943	242	52.7	94	287	1.02	13.0	3.1	45.20
ND	210	1012	41	35.9	115	296	9.00	8.6	3.1	29.91
OH	199	935	111	80.3	83	259	2.56	16.6	4.0	54.47
OK	195	954	141	58.5	108	258	2.55	8.2	2.2	48.37
OR	183	970	54	64.9	104	225	8.09	15.8	4.8	70.37
PA	193	919	102	81.9	82	282	2.55	20.0	5.2	36.83
RI	204	909	53	92.2	89	280	5.71	17.9	5.6	58.52
SC	215	946	312	59.8	82	312	0.58	14.5	2.7	53.19
SD	203	974	74	15.8	133	431	3.86	7.9	2.3	30.13
TN	198	933	165	62.8	131	339	1.97	17.9	3.8	43.11
TX	205	968	213	80.0	112	265	1.56	13.6	2.5	60.50
UT	208	984	53	79.0	95	282	6.00	11.0	3.2	57.50
VT	209	950	8	22.3	108	250	55.00	8.7	3.2	50.61
VA	210	959	209	69.6	72	261	1.04	16.0	3.0	46.71
WA	195	987	85	80.4	91	184	5.31	13.0	5.4	67.42
WV	189	941	38	37.1	137	267	12.80	23.5	5.1	26.19
WI	208	960	56	66.8	71	281	4.24	12.1	3.1	47.67
WY	201	1052	51	15.3	72	208	6.40	10.0	2.4	51.32

NOTE: st = state; age = 14 − 24 year olds per 1000 population; sex = males per 1000 females; nw = nonwhites per 1000 population; %u = percent urban; pw = whites below poverty level per 1000 white population; pnw = nonwhites below poverty level per 1000 nonwhite population; pr = white/ nonwhite poverty ratio; %t = percent unemployed teens; %a = percent unemployed adult; crime = total reported crimes per 1000 population.

7.13

3. Nonwhite: the number of nonwhites per 1000 total population
4. Urban: the percentage of the population living in urban areas
5. Poverty whites: the number of whites below the established poverty level per 1000 whites
6. Poverty nonwhites: the number of nonwhites below the established poverty level per 1000 nonwhites
7. Poverty ratio: the ratio of the whites below the poverty level to the nonwhites below the poverty level
8. Teens: the percentage of 14 to 24 year olds that are unemployed
9. Adults: the percentage of adult males that are unemployed.

The dependent variable for the study was the total reported crimes per 1000 population.

This study has several characteristics that are worth noting. The independent variables were selected because someone thought they might be related to crime. The values of the independent variables were in no way selected for this study. Each was measured about as accurately and precisely as the dependent variable but probably no more so.

The research problem is to determine which of these nine independent variables is related to crime and something about the strength of that relationship. It is a kind of fishing expedition that occurs routinely in the social sciences and is not unknown in the other sciences. The success of the program depends on the researcher's knowledge and good judgment in choosing the independent variables. It can also depend on the researcher's prejudices through omission of variables. This example was chosen for illustrating data analysis rather than any sociological issue.

The total sum of squares, 8979.19, for the response or dependent variable is calculated from the following summations.

$$\Sigma y = 2698.31$$
$$\Sigma y^2 = 154,596.7259$$
$$n = 50$$

Procedures presented in chapter 6 are used to determine the reductions in the total sum of squares that can be made by each of the nine independent variables considered separately. Table 7.12 gives these reductions and table 7.13 presents the analysis of variance for the independent variable producing the largest reduction, i.e., percent urban. As usual, the null hypothesis is that crime is not related to the percentage of the population living in urban areas. Assuming that this is true and that the apparent relationship is just due to chance, then the probability of an F-statistic as large or larger than 31.52 is much less than 0.005, the limit of table D in part V. So any regression equation relating crime to population characteristics must include percent urban as an independent variable.

TABLE 7.12 Reductions in Total
Sum of Squares for Example 7.2

Variable	Reduction
Age	111.37
Sex	285.54
Nonwhite	608.49
Percent urban	3559.12
Poverty whites	2170.06
Poverty nonwhites	1621.66
Poverty ratio	492.45
Teens	93.41
Adults	944.66

TABLE 7.13 The ANOVA Table for Example 7.2
for Percent Urban

Source of variation	df	ss	ms	
Percent urban	1	3559.12	3559.12	$F = 31.52$
Residual	48	5420.07	112.92	
Total	49	8979.19		

A perusal of the reductions in table 7.12 might be expected to provide clues for the next variable most likely to reduce the total sum of squares by an important amount and, hence, be useful in the regression equation for crime rate. Age, sex, and teenage unemployment all appear to be useless. The usefulness of the nonwhite population per 1000 total population and the poverty ratio is certainly questionable. Based on the sums of squares in table 7.12, the most likely candidate for the second most useful variable in the regression equation for crime is either the number of whites per 1000 below the established poverty level or the number of nonwhites per 1000 below the poverty level. Table 7.14 presents the fur-

TABLE 7.14 Reductions in Total
Sum of Squares with Percent Urban
Included for Example 7.2

Variable	Reduction
Age	12.12
Sex	1348.06
Nonwhite	99.10
Poverty whites	314.06
Poverty nonwhites	577.83
Poverty ratio	5.27
Teens	163.46
Adults	223.52

ther reductions in the total sum of squares for each of the remaining eight independent variables and, surprisingly, the sex ratio is the next most important. Obviously, the percent urban variable contains much the same information as either of the two poverty measures. Table 7.15 gives the analysis of variance after this new variable is added to the model. The null hypothesis is that after removing the effect of the percentage of the population living in urban areas, the crime rate is not related to the sex ratio. Assuming this null hypothesis is true and that the apparent relationship is just due to chance, then the probability of an F-statistic as large or larger than 15.56 is also much less than 0.005. This is very small, so any regression equation to estimate crime must include the sex ratio in addition to percent urban.

TABLE 7.15 The ANOVA Table for Example 7.2 after Adding Sex Ratio

Source of variation	df	ss	ms	
Percent urban	1	3559.12	3559.12	
Sex ratio	1	1348.06	1348.06	$F = 15.56$
Residual	47	4072.02	86.64	
Total	49	8979.20		

Based on the reductions in table 7.12, one might have expected the number of whites below the poverty level to have been the second most important variable related to crime. This is obviously not so and an explanation is in order. The variables in table 7.12 are not independent of each other. Apparently the percent living in urban areas provides much the same information on the crime rate as does the number of whites or the number of nonwhites below the established poverty level. This duplication of information reorders the importance of the remaining eight independent variables so that the sex ratio provides the most new information on crime.

The residual sum of squares was computed for seven equations like

$$y = b_0 + b_1 \cdot \text{percent urban} + b_2 \cdot \text{sex ratio} + b_3 x_3$$

where x_3 was one of the remaining independent variables. The further reductions are presented in table 7.16 where age, which seemed almost useless before the sex ratio was added to the equation, reduces the residual the most. The analysis of variance is presented in table 7.17. The null hypothesis is that after removing the effect of the percent living in urban areas and the sex ratio, age is not related to the crime rate. This can be stated another way as follows: A regression equation with percent urban, the sex ratio, and age provides no better explanation for differences in the

TABLE 7.16 Reductions in Total Sum of Squares with Percent Urban and Sex Included for Example 7.2

Variable	Reduction
Age	138.31
Nonwhite	3.32
Poverty whites	34.16
Poverty nonwhites	50.00
Poverty ratio	14.62
Teens	30.11
Adults	18.57

TABLE 7.17 The ANOVA Table for Example 7.2 after Adding Age to the Model

Source of variation	df	ss	ms	
Percent urban	1	3559.12	3559.12	
Sex ratio	1	1348.06	1348.06	
Age	1	138.31	138.31	$F = 1.62$
Residual	46	3933.70	85.52	
Total	49	8979.19		

crime rate than does the simpler equation using only the percent living in urban areas and the sex ratio. Assuming this null hypothesis is true and that the apparent relationship is just chance, then the probability of an F-statistic with 1 and 46 degrees of freedom as large as or larger than 1.62 is approximatley 0.23. This probability is much too large to consider age useful in estimating crimes per 1000 population.

So of the original nine independent variables chosen for study, only two, percent urban and the sex ratio, are useful for estimating crimes per 1000 population. The fact that seven variables are discarded as useless implies that the researcher didn't know what to expect and perhaps engaged in a lot of unnecessary work. Many regression studies are like this, so a data analyst shouldn't be too surprised by such results.

The final regression equation relating the crime rate to population characteristics is given by

$$y = -103.62 + 0.4461x_1 + 0.1358x_2$$

where y = crimes per 1000 population
x_1 = the percent of the population living in urban areas
x_2 = the number of males per 1000 females in the population

As urbanization and the percentage of males in the population increases, the crime rate apparently increases. This relationship explains about 55% of the total sum of squares, i.e.,

$$\frac{\text{Reduction}}{\text{Total}} = \frac{4907.18}{8979.19} = 0.55$$

From the positive viewpoint, this isn't bad. Two variables apparently related to crime have been discovered and their discovery could be useful in deciding social and economic policy. The negative viewpoint focuses on the 45% of total variability that is unexplained. What might account for this large proportion? Certainly not the seven variables tried and discarded, at least not in a linear formulation. The researcher must look elsewhere for explanations.

SUMMARY

The ideas introduced in chapter 6 for examining the relation of a dependent variable to a single independent variable have been extended to more than one independent variable. The basic principles are the same but the magnitude of the work involved increased considerably. The step-forward procedure for choosing independent variables to add to the regression equations was used but three other methods were explained. Readers should use the residual sums of squares in table 7.10 to do example 7.1 by each of the methods described.

CHAPTER 8

More Complex Two-Variable Relationships

INTRODUCTION

Chapter 6 provided the basic structure for estimating and testing the relationship between two variables expressed in a linear equation. Chapter 7 extended the basic structure to more-than-two-variable relationships while setting the stage for more complex relationships.

Relationships between two variables are considered complex in this book if the algebraic equation expressing this relationship contains transcendental terms and/or higher powers of the independent variable. Some examples will make the definition clear.

- Transcendental

$$y = a + b \sin x$$
$$y = a + b \sin x + c \cos x$$
$$y = a + be^x$$

- Higher powers

$$y = b_0 + b_1 x + b_2 x^2 + b_3 x^3$$

- Combinations

$$y = b_0 + b_1 x + b_2 \sin x$$

In spite of their complexity, there are still only two variables, x and y, in each of these relationships and the relationships are linear in the unknown b coefficients.

The basic principle in this chapter is this. The independent variable is transformed to create new variables for each of the transcendental and power terms and then the procedures in chapter 7 are applied. For example, suppose the proposed regression equation is

$$y = b_0 + b_1 w + b_2 w^2 + b_3 \sin w$$

Then let

$$x_1 = w \qquad x_2 = w^2 \qquad x_3 = \sin w$$

and the equation becomes

$$y = b_0 + b_1 x_1 + b_2 x_2 + b_3 x_3$$

This equation is nothing more than the equations studied in chapter 7. Every complex two-variable relationship that is linear in the unknown coefficients, i.e., the b's, can be reduced to an equation like that above. Consequently, only one example is presented in this chapter.

RATIO AND INTERVAL MEASUREMENTS

Example 8.1

Vital capacity is an important physiological measurement in the study or evaluation of respiratory function. It is the maximum volume of air that can be forcibly inspired after a maximum expiration or that can be forcibly expired after a maximum inspiration. The inspiratory vital capacity is more reproducible even in diseased persons, but the expiratory vital capacity is typically measured in the evaluation of patients with respiratory deficiencies.

Vital capacity can be determined by simply collecting the air that can be forcibly expired into any device that does not present a backward or forward pressure to the patient and then measuring the volume. Accurate measurement of vital capacity requires a cumbersome positive-displacement device, such as the Stead-Wells spirometer or a bellows-wedge spirometer. Neither of these are very portable, so patients are routinely moved to the equipment rather than the equipment being moved to the patients. Consequently, less accurate portable devices depending on an indirect measure of volume have been developed. One such device makes use of an impeller placed in the expiratory air flow. By counting impeller revolutions during given time periods, the total air flow from a patient's lungs, i.e., the vital capacity, can be estimated. Such portable devices are usually calibrated by mechanically blowing measured quantities of air through the impeller and into a positive displacement device.

Table 8.1 presents 18 measurements using such an arrangement. The research problem is to determine a correction for the impeller device so that measurements made with it will more nearly approximate the positive-displacement mechanism. The mechanical device used to produce the "blow" for this calibration was a single-stroke piston that was moved at specified speeds to approximate human physiology. For each response, the total volume of air was an integral number of liters. The intended volume based on the setup of the mechanical device is probably the best measure of volume for each observation. The second-best measure is probably the positive-displacement measurement, while the flow measured by the impeller may be the worst. Conventional statistical methodology would dictate that the regression function should express impeller measurements as a function of the positive displacements but this doesn't address the research problem.

TABLE 8.1   Vital Capacities for
Example 8.1

PD	I	PD	I
1.00	0.95	4.20	3.97
0.97	0.93	4.01	3.87
1.01	1.06	4.15	3.80
2.05	2.13	5.07	4.72
2.09	2.10	5.09	4.89
2.02	2.05	5.14	4.83
3.16	3.21	6.33	5.64
3.08	3.11	6.20	5.38
3.22	3.18	6.13	5.34

NOTE: PD = positive displacement measurements; I = impeller measurements.

Restated, the research problem is to take an impeller measurement, and from this estimate a positive displacement. This is what is done here. Figure 8.1 is a plot of the data, and the straight line drawn beside the points makes it easy to see that a linear regression function will not be appropriate. Let y be the positive displacement measurements and x be the impeller measurements. A second degree equation, i.e., a quadratic,

$$y = b_0 + b_1 w + b_2 w^2$$

will probably be sufficient but just to be assured a higher power is not required, a cubic equation will be tried, i.e.,

$$y = b_0 + b_1 w + b_2 w^2 + b_3 w^3$$

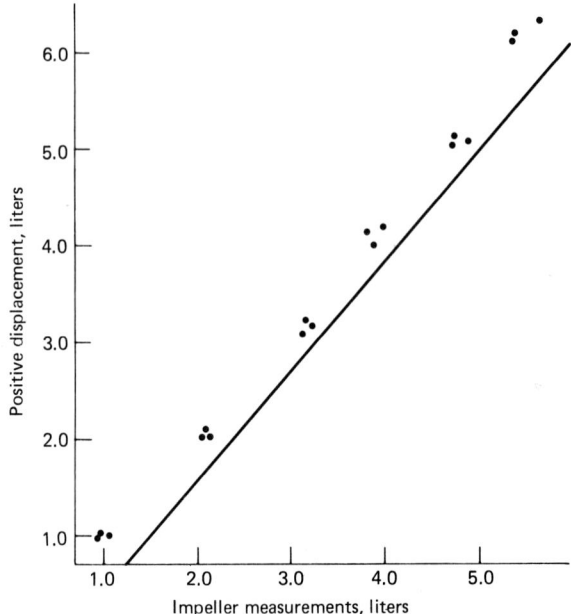

FIGURE 8.1   Plot of data for example 8.1.

The problem is placed in the context of chapter 7 by making the following substitutions. Let

$$x_1 = w \qquad x_2 = w^2 \qquad x_3 = w^3$$

The equation becomes

$$y = b_0 + b_1 x_1 + b_2 x_2 + b_3 x_3$$

and now looks like equations considered in chapter 7.

Sometimes researchers insist that polynomials such as the above include all terms up to the highest power that reduces the residual sum of squares by a significant amount. From the data analysis viewpoint, there is no reason for this and for parsimonious results, it may not be desirable.

Table 8.2 gives the positive-displacement measurements as a function of each of the three independent terms separately along with the residual sums of squares. x_1 reduces the total sum of squares the most. This reduction is summarized in the analysis of variance presented in table 8.3. The null hypothesis tested is that the positive displacements are not related to the impeller measurements. Assuming this to be true and that the apparent relationship is due to chance, then the probability of an F-statistic as large or larger than 1725 is so small it would be difficult to estimate. Hence, the first power of the impeller measurement should be in

TABLE 8.2 Positive Displacement
as a Function of Powers of Impeller
Measurements for Example 8.1

Equation	SS
$y = -0.298 + 1.149x_1$	0.5181
$y = 1.202 + 0.1731x_2$	1.1314
$y = 1.778 + 0.0293x_3$	4.3852

TABLE 8.3 The ANOVA Table for Example 8.1
for Positive Displacement

Source of variation	df	SS	ms	
Impeller	1	55.85	55.85	$F = 1725$
Residual	16	0.5181	0.0324	
Total	17	56.37		

the regression equations. The terms x_2 and x_3 are considered next in equations with x_1.

Table 8.4 gives the positive-displacement measurements as a function of x_1 and each of the remaining two independent variables. x_3 has the smallest residual, and the effect of adding this term is summarized in the analysis of variance in table 8.5. The null hypothesis is that after accounting for the relationship with the first power of the impeller measurement, the positive displacements are not related to the third power of the impeller measurements. Assuming this null hypothesis to be true and the

TABLE 8.4 Positive Displacement as a Function of
Powers of Impeller Measurements for Example 8.1

Equation	SS
$y = 0.2221 + 0.7329x_1 + 0.0644x_2$	0.2071
$y = 0.0878 + 0.9141x_1 + 0.0066x_3$	0.2042

TABLE 8.5 The ANOVA Table for Example 8.1
for Positive Displacement

Source of variation	df	SS	ms	
Impeller	1	55.58	55.85	
Impeller^3	1	0.3139	0.3139	$F = 23.1$
Residual	15	0.2042	0.0136	
Total	17	56.37		

apparent relationship is just chance, then the probability of an F-statistic as large or larger than 23.1 is less than 0.0004. This is a very small probability, so the third power of the impeller measurement should be considered for the regression equation.

Finally, all three terms are used to obtain this regression equation.

$$y = 0.1109 + 0.8834x_1 + 0.01072x_2 - 0.00553x_3$$

The analysis of variance summarizing the further reduction in the residual sum of squares is given in table 8.6. The F-statistic is very small, so the probability is large, about 0.90. Hence, the second power of the impeller measurement is not useful in an equation to estimate the positive-displacement measurement. Note that the reduction that is the result of adding the second power term is counterproductive in that the residual mean square actually increases.

TABLE 8.6 The ANOVA Table for Example 8.1 for Positive Displacement

Source of variation	df	ss	ms	
Impeller	1	55.85	55.85	
Impeller^3	1	0.3139	0.3139	
Impeller^2	1	0.00012	0.00012	$F = 0.008$
Residual	14	0.2040	0.0146	
Total	17	56.37		

It is a good idea to keep the results of data analyses as simple as possible, and this example provides an opportunity to make this point. The purpose of the analysis is to find a combination of the independent variables that will make the standard deviation about the regression a minimum. The standard deviation of y given x_1 and x_3 is 0.1167, while the standard deviation of y given x_1 and x_2 is 0.1175. These standard deviations are so nearly the same that it really does not make any difference which equation is used. In the interest of simplicity, this analyst would use the equation that expresses y in terms of x_1 and x_2, thus avoiding use of the third power of the impeller measurement. The results would be more acceptable and defendable to client investigators and/or management.

Substituting for the x's, the final regression equation is

$$y = 0.2221 + 0.7329w + 0.0644w^2$$

where y = the positive displacement
w = the impeller measurement

This equation is plotted in figure 8.2 along with the 18 data points.

At each step in the analysis, homogeneity of the deviations about the

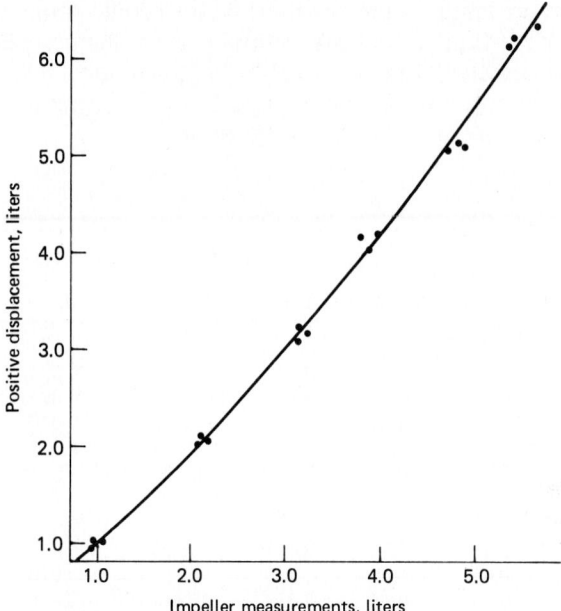

FIGURE 8.2 Plot of data and the final quadratic equation for example 8.1.

regression equation and normality should be checked. This particular example suggests comparing the variances for each of the little groups of three data (see table 8.1 and figure 8.1) using Bartlett's procedure. The reader can compute the 18 residuals for each equation and then find that Bartlett's M for the three models, i.e., for

$$y = -0.298 + 1.149w$$

$$y = 0.2221 + 0.7329w + 0.0644w^2$$

$$y = 0.1109 + 0.8834w + 0.01072w^2 + 0.00553w^3$$

are 2.42, 4.78, and 3.34, respectively, as terms are added to the regression equation. The sample size, 3, for estimating the variances is very small, but the procedure does indicate equivalent variances for the six populations is an acceptable assumption. The probability of a chi-square as large or larger than the largest M is almost 0.50, assuming the apparent differences are just chance.

Using these same residuals, the reader can verify that the Lilliefors statistics for the three regression equations are 0.167, 0.157, and 0.162, respectively. The null hypothesis tested is that the distribution of the deviations from the regression equation is not different from the normal distribution. Assuming this is true in each case and that the observable differences are just due to chance, then the probability of a Lilliefors statistic as large or larger is greater than 0.20 (refer to table M in part V). For guidance in duplicating this example, the Lilliefors calculations for the quadratic regression equation are presented in table 8.7. The standard deviation for converting the deviations from the regression equation to z's has only $n - 3 = 15$ degrees of freedom.

TABLE 8.7   The Lilliefors Test for Normality for Example 8.1

			Probabilities			
d	z	z sorted	Normal	Sample	da	db
0.02	0.20	−2.18	0.015	0.056	−0.041	
0.01	0.09	−1.06	0.144	0.111	0.033	0.089
−0.06	−0.52	−0.67	0.252	0.167	0.085	0.141
−0.03	−0.22	−0.64	0.262	0.222	0.040	0.096
0.04	0.38	−0.52	0.301	0.278	0.023	0.078
0.02	0.21	−0.40	0.346	0.333	0.013	0.068
−0.08	−0.67	−0.38	0.352	0.389	−0.037	0.019
−0.04	−0.38	−0.22	0.414	0.444	−0.030	0.025
0.02	0.13	−0.11	0.455	0.500	−0.045	0.011
0.05	0.45	0.09	0.536	0.556	−0.020	0.036
−0.01	−0.11	0.13	0.553	0.611	−0.058	−0.002
0.21	1.81	0.20	0.579	0.667	−0.087	−0.032
−0.05	−0.40	0.21	0.583	0.722	−0.139	−0.083
−0.26	−2.18	0.38	0.648	0.778	−0.129	−0.074
−0.12	−1.06	0.45	0.674	0.833	−0.159	−0.104
−0.07	−0.64	1.34	0.910	0.889	0.021	0.076
0.17	1.45	1.45	0.927	0.944	−0.018	0.038
0.16	1.34	1.81	0.965	1.000	−0.035	0.021

NOTE: The Lilliefors statistic is 0.159. Refer to Chapter 15 for details.

SUMMARY

A great many details have been omitted from this analysis since the procedure parallels exactly procedures described in Chapter 7. Example 8.1 is a very simple problem, but it illustrates the basic ideas and provides yet another opportunity to show that the purpose of the regression equation determines the functional relationship. The purpose in this study was to estimate positive-displacement measurements from impeller measurements. Therefore, the regression equation expresses positive displacement (the better of the two measurements) as a function of impeller measurements (the lesser of the two measurements).

Cause and effect is not an issue in this analysis. In fact, any thoughtful investigator would find it difficult to defend the thesis that the positive displacement was the cause and the impeller measurement was the effect in this problem. This would ignore the velocity of the air movement, the possible influence of air temperature and humidity, and probably a host of other variables not considered important enough by the investigator to measure. Calibration problems are always like this.

The impeller measurements could have been used to estimate a correction by making the dependent variable the difference between the positive displacement and the impeller measurement. The equation would be different but the result of estimating unknown vital capacities would be the same.

CHAPTER 9

More Complex More-Than-Two Variable Relationships

INTRODUCTION

Chapter 9 continues and concludes estimating and testing linear relationships. Relationships between more than two variables are considered complex in this book if the algebraic equation expressing this relationship contains transcendental terms and/or higher powers and/or products of the independent variables. Some examples will make the definition clear.

- Transcendental

$$y = b_0 + b_1 v_1 + b_2 \sin v_2$$

$$y = b_0 + b_1 v_1 + b_2 v_2 + b_3 e^{v_2}$$

- Higher powers

$$y = b_0 + b_1 v_1 + b_2 v_1^2 + b_3 v_2 + b_4 v_2^2$$

- Product terms

$$y = b_0 + b_1 v_1 + b_2 v_1 \sin v_2$$

$$y = b_0 + b_1 v_1 + b_2 v_2 e^{v_1^2} + b_3 v_2^2$$

The last two equations are given to demonstrate the wide variety of functional relationships that can be accommodated rather than to demonstrate typical data analysis problems.

9.1

As in chapter 8, the basic principal is to transform the independent variables to create new variables for each of the trancendental, power, and product terms and then to apply the procedures in chapter 7. Consider the last equation above and make the following transformations.

$$x_1 = v_1 \qquad x_2 = v_2 e^{v_1^2} \qquad x^3 = v_2^2$$

The equation becomes

$$y = b_0 + b_1 x_1 + b_2 x_2 + b_3 x_3$$

Any complex more-than-two variable relationship that is linear in the unknown coefficients can be simplified as this equation has been. Since this is true, only one example is presented in this chapter.

RATIO AND INTERVAL MEASUREMENTS

Example 9.1

Badger (1946) studied the factors which influence the removal of tar fog by high-speed turboexhausters from a gas stream. Inadequate removal of tar fog from carburetted water gas makes it very difficult to remove hydrogen sulfide, so Badger assessed the conditions under which turboexhausters alone might provide adequate tar fog removal.

The tests were made with a high-speed three-stage turboexhauster at rotor speeds from 2400 to 3900 rpm and at gas inlet temperatures ranging from 43 to 69°F. The data for 31 tests are presented in table 9.1 and are plotted in figure 9.1. Figure 9.1 is an example of one way that three variables can be plotted on a single two-dimensional figure. A perusal of figure 9.1 indicates that

1. Increased rotor speed reduces the tar content.
2. The relationship between rotor speed and tar content is probably not linear.
3. The tar content is less at lower inlet temperatures.

A reasonable regression function, the one proposed by Badger, would then be

$$y = b_0 + b_1 \cdot \text{speed} + b_2 \cdot \text{speed}^2 + b_3 \cdot \text{temp}$$

Let

$$x_1 = \text{speed} \qquad x_2 = \text{speed}^2 \qquad x_3 = \text{temp}$$

TABLE 9.1   Data for Example 9.1

Speed, rpm/100	Temp., °F	Tar, grains/ 100 ft³	Speed, rpm/100	Temp., °F	Tar, grains/ 100 ft³
24.00	54.50	60.00	30.75	57.00	33.50
24.50	56.00	61.00	31.00	57.50	34.00
24.50	58.50	65.00	31.50	64.00	44.00
25.00	43.00	30.50	32.00	57.00	33.00
25.00	58.00	63.50	32.00	64.00	39.00
25.00	59.00	65.00	32.00	69.00	53.00
27.00	62.50	44.00	32.25	68.00	38.50
27.00	55.50	52.00	32.50	62.00	39.50
27.00	58.00	54.50	32.50	64.50	36.00
27.50	45.00	30.00	32.50	48.00	8.50
27.75	45.50	25.00	35.00	60.00	30.00
28.00	48.00	23.00	35.00	59.00	29.00
28.00	63.00	54.00	35.00	58.00	26.50
29.00	58.50	36.00	36.00	58.00	24.50
29.00	64.50	53.50	39.00	61.00	26.50
30.00	66.00	57.00			

SOURCE: See Badger (1946). Reprinted with permission of the author.

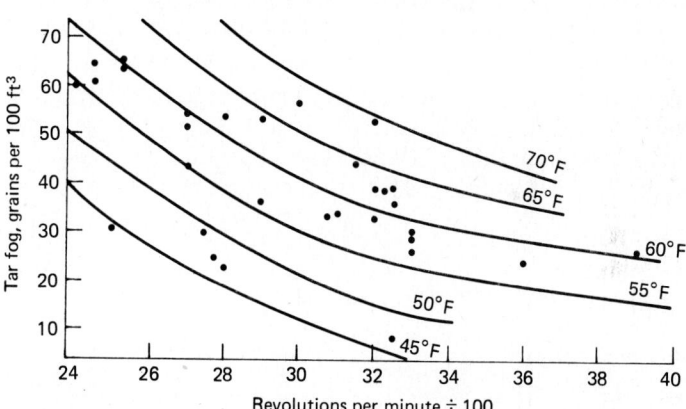

FIGURE 9.1   Tar fog in carburetted water gas at outlet turboexhausters.

Then the regression equation becomes

$$y = b_0 + b_1 x_1 + b_2 x_2 + b_3 x_3$$

and this is precisely the kind of regression function studied in chapter 7.

Table 9.2 gives the tar content measurements as a function of each of the independent variables separately. The first independent variable, speed, reduces the total sum of squares the most, i.e., has the smallest

TABLE 9.2 Tar Content in Example 9.1 as a
Function of the Three Independent Variables
Separately

Equation	SS
$y = $ 112.34 − 2.39 speed	3952.44
$y = $ 75.87 − 0.0384 speed^2	4043.51
$y = $ −13.43 + 0.935 temp.	5391.93

residual variance. This result is given in the analysis of variance in table
9.3. The null hypothesis is that there is no relationship between tar content and the speed of the turboexhauster. Assuming this is true and that
the apparent relationship in the data is just due to chance, then the probability of an F-statistic with 1 and 29 degrees of freedom as large or larger
than 18.86 is approximately 0.0012. This is a very small probability, so
turboexhauster speed should be in the regression equation for tar content.

The remaining two independent variables are now considered in equations with speed. The equations and the residual sum of squares for each
are given in table 9.4. Temperature of the gas reduces the residual sum
of squares the most, so it is considered next. The analysis of variance is
given in table 9.5. The null hypothesis now is that the tar content is not
related to temperature after accounting for the tar content relationship
with speed. The probability of an F-statistic as large or larger than 83.3,
assuming the null hypothesis is true, is so small that tables do not cover
it, so gas temperature should also be in the regression equation.

TABLE 9.3 The ANOVA Table for Tar Fog for
Example 9.1

Source of variation	df	SS	ms	
Speed	1	2570.2	2570.2	$F = 18.86$
Residual	29	3952.4	136.3	
Total	30	6522.7		

TABLE 9.4 The Content in Example 9.1 as a
Function of Pairs of the Three Independent Variables

Equation	SS
$y = 222.82 − 9.763 \cdot$ speed $+ 0.1212 \cdot$ speed^2	3829.87
$y = 46.72 − 3.325 \cdot$ speed $+ 1.611 \cdot$ temp.	993.90

TABLE 9.5   The ANOVA Table for Tar Fog for Example 9.1

Source of variation	df	ss	ms	
Speed	1	2570.2	2570.2	
Temp.	1	2958.5	2958.5	$F = 83.3$
Residual	28	993.9	35.50	
Total	30	6522.7		

There remains only to test x_2, the speed squared. The equation is

$$y = 248.41 - 17.13 \cdot \text{speed} + 0.2257 \cdot \text{speed}^2 + 1.712 \cdot \text{temp}$$

and the effect of adding this third term to the regression equation is summarized in the analysis of variance in table 9.6. The null hypothesis here is that the tar content is not related to the speed squared after accounting for the relationship with the other two variables. Assuming this null hypothesis is true and that the apparent relationship in the data is just chance, then the probability of an F-statistic with 1 and 27 degrees of freedom as large or larger than 19.2 is approximately 0.0012. Therefore the regression function should include all three of the terms originally proposed.

TABLE 9.6   The ANOVA Table for Tar Fog for Example 9.1

Source of variation	df	ss	ms	
Speed	1	2570.2	2570.2	
Temp.	1	2958.5	2958.5	
Speed^2	1	413.6	413.6	$F = 19.2$
Residual	27	580.3	21.49	
Total	30	6522.7		

Discussion

Failure to exclude any of the independent variables initially proposed for the regression function implies substantial insight, blind luck, or that there might be other terms not considered. Many data analysts, including this writer, feel more secure if one or two logically important independent variables are rejected. For example, if the third power of speed was considered and rejected, then the analyst would have more confidence that tar content depended on a quadratic function of speed, not on a higher polynomial. This isn't always true, of course, but experience indicates that many polynomials include all terms up to the highest significant power.

Data analysts also know that nature doesn't always add effects, it sometimes multiplies or divides them. For example, the effects of speed and temperature on tar content are simply summed. More complex terms such as temperature multiplied by powers of speed have not been considered.

It is an interesting exercise to use the data for example 9.1 to examine this regression equation:

$$y = b_0 + b_1x_1 + b_2x_2 + b_3x_3 + b_4x_4 + b_5x_5 + b_6x_6$$

where x_1 = speed
x_2 = speed^2
x_3 = speed^3
x_4 = temperature
x_5 = temperature multiplied by speed
x_6 = temperature multiplied by speed^2

Table 9.7 presents the residual sums of squares that an analyst would examine in a strict step-forward procedure to find the best regression equation. Beginning with these variables, step-forward will select a regression function with only two independent variables, speed and the product of temperature and speed squared. The residual for this regression equation, however, is larger than the residual for the equation developed in

TABLE 9.7 Residuals Using Step-Forward Procedure for the Expanded Regression Equation in Example 9.1

Independent	Residual	
Speed	3952.4	$F = 18.9$; df $= 1,29$
Speed^2	4043.5	
Speed^3	4154.5	
Temp.	5391.9	
Temp. · speed	6307.3	
Temp. · speed^2	5614.0	
Speed, speed^2	3829.9	
Speed, speed^3	3837.2	
Speed, temp.	993.9	
Speed, temp. · speed	997.3	
Speed, temp. · speed^2	785.6	$F = 112.9$; df $= 1,28$
Speed, temp. · speed^2, speed^2	728.6	$F = 2.11$; df $= 1,27$
Speed, temp. · speed^2, speed^3	730.4	
Speed, temp. · speed^2, temp.	782.6	
Speed, temp. · speed^2, temp. speed	781.3	

example 9.1. This is a good example where the all-combinations procedure (see chapter 7) would be superior to a step-forward procedure using
these six independent variables.

CONCLUSIONS

Earlier in part II, it was noted that regression analysis involved a bit of
art. Example 9.1 illustrates this assertion. The data analyst must sometimes try many approaches before being comfortable with the results, i.e.,
before being willing to stake a professional reputation on the results. In
the vernacular of derision, this is often called "massaging the data." Data
analysts know that nature is not simple and that is why researchers and
management often have difficulty understanding their own data. Perhaps
it is only natural in ignorance to poke fun at analyses not always understood. Analysts do sometimes overanalyze data, but then discovering a
relationship that no one knew before is exciting, especially if subsequent
investigations substantiate the results.

PART III

Comparison of Relationships

Part I discusses comparisons and part II presents techniques for relationships. Part III combines these two problems and presents techniques for comparing relationships. This problem is presented under the general title of "analysis of covariance" in statistical texts, but like many other topics in statistics, the treatment can vary considerably between books.

In part I, the data were samples from two or more populations. All of the data measured the same phenomenon and might be samples from independent populations as in chapters 2 and 4 or samples from dependent populations as in chapters 3 and 5. Then in part II, the data were related but the responses measured different phenomena. The relationship was the important thing in part II since the data analysis problem was to understand one of the responses, called the dependent variable, through knowledge of the other response or responses, called the independent variables.

In comparing relationships, part III considers data like that displayed in table III.1 for samples from independent populations and like table III.2 for samples from dependent populations. Consider the following as an example. The responses to competitive drugs, i.e., similar drugs for the same indication, are often compared. The responses y could depend on some characteristic x of the subjects used in the study, e.g., on their weight or on an estimate of blood volume. Such data would be displayed as table III.1. The blocks in table III.2 could arise from some secondary way of classifying the subjects. Human subjects might be classified by eth-

TABLE III.1   Typical Display for Data from Independent Populations

Sample 1	Sample 2	Sample 3
$y_{11}\, x_{11}$	$y_{21}\, x_{21}$	$y_{31}\, x_{31}$
$y_{12}\, x_{12}$	$y_{22}\, x_{22}$	$y_{32}\, x_{32}$
$y_{13}\, x_{13}$	$y_{23}\, x_{23}$	$y_{33}\, x_{33}$
$y_{14}\, x_{14}$		$y_{34}\, x_{34}$
		$y_{35}\, x_{35}$

TABLE III.2   Typical Display for Data from Dependent Populations

	Sample 1	Sample 2	Sample 3
	$y_{111}\, x_{111}$	$y_{211}\, x_{211}$	$y_{311}\, x_{311}$
Block 1	$y_{112}\, x_{112}$	$y_{212}\, x_{212}$	$y_{312}\, x_{312}$
	$y_{113}\, x_{113}$	$y_{213}\, x_{213}$	$y_{313}\, x_{313}$
	$y_{121}\, x_{121}$	$y_{221}\, x_{221}$	$y_{321}\, x_{321}$
Block 2	$y_{122}\, x_{122}$	$y_{222}\, x_{222}$	$y_{322}\, x_{322}$
	$y_{123}\, x_{123}$	$y_{223}\, x_{223}$	$y_{323}\, x_{323}$
	$y_{131}\, x_{131}$	$y_{231}\, x_{231}$	$y_{331}\, x_{331}$
Block 3	$y_{132}\, x_{132}$	$y_{232}\, x_{232}$	$y_{332}\, x_{332}$
	$y_{133}\, x_{133}$	$y_{233}\, x_{233}$	$y_{333}\, x_{333}$

nic origin or geographic location of residence, if either was suspected of having an influence on the responses or on the relationship between y and x. Laboratory animals might be classified on strain, source, or even species, if the species were expected to respond similarly. Completely different species might better be the subject of separate analyses.

Research questions arising from data this complex can be numerous, so no effort will be made in this introduction to suggest specific research hypotheses. That will be saved for the examples in chapters 10 and 11. Instead, consider this. The general problem which is the subject of part III can be described as a simple comparison of responses that depend on another variable x. From this viewpoint, part III is an extension of part I. The general problem can also be described as a comparison of regression equations such as discussed in part II. Both viewpoints are used in part III depending on the research hypothesis or on the data conditions being considered. Some statistics textbooks approach the subject from one or the other viewpoints and hence omit much that the data analyst needs.

While part III presents the comparison of simple relationships only, it will lay a foundation for the comparison of more complex relationships. The reader is advised, however, to seek consultation with a senior statistician much experienced in data analysis before attempting anything beyond that described in part III. The problems can quickly exceed the scope planned for this book.

The problem of comparing relationships has not received much attention from statisticians who specialize in nonparametric procedures, so techniques generally suitable for data analysis of complex situations in the ordinal measurement scale have not been developed. Consequently, chapters 10 and 11 are almost exclusively for ratio and interval data. Some suggestions for avoiding normality assumptions are nevertheless made.

Comparison of Two Relationships

INTRODUCTION

The comparison of two simple relationships of the form

$$y = a + bx$$

is considered in chapter 10. Slightly more complex forms like

$$y = a + b \sin x$$

or

$$y = a + b \log x$$

are included but relationships like

$$y = a + bx + cx^2$$

and

$$y = a + bx_1 + cx_2$$

are not considered in part III.

Research questions usually involve differences between slopes, i.e., the regression coefficient b, and differences between the intercepts, i.e., the coefficient a, in the above equations. As will be demonstrated, tests in differences between intercepts depend on whether or not the regression lines can be considered parallel.

AN INTRODUCTION FOR EXAMPLES 10.1 AND 10.2

One way drugs and sometimes food additives are studied for long-term and/or high-dose effects is the following. A group of some small animal, often the rat, will be fed a standard diet with the test product mixed in it. A particular daily intake of the test product is desired for each animal, but since these studies often use many, many units, individual dosing would be very expensive to manage. Instead, the test product is mixed in a standard diet before the diet is pelleted so that the estimated consumption of each animal will on the average contain the desired quantity of the test product. When explained like this, the problem of supplying the proper dose to each animal seems easy.

A quantity of the diet and test product more than sufficient for 1 week is prepared and weighed. The animals are fed from this batch for 1 week. At the end of the week, the amount of food not consumed is weighed. Laboratory animals are not known for tidiness, so some of their food is often contaminated with urine or knocked out of the feeding mechanism to be mixed with feces. The urine-contaminated pellets may be dried and weighed while that mixed with the feces can only be estimated. Laboratory personnel have developed many ingenious devices to limit the amount of uneaten contaminated food. For example, a common device dispenses one pellet on demand which the animal usually eats immediately.

The food preparation procedure is repeated weekly for a long period of time, often near the life expectancy of the animals. Sometimes the food consumed is recorded for the animals in a study as a group, but it is not unusual for the colony to be subgrouped by location with separate food-consumed weights recorded for even very small subgroups . The number of animals in a subgroup is always the number that are cared for as a unit, so this subgroup size is a logistic decision rather than a scientific decision. In other words, the cages for a subgroup are cleaned together; the animals are fed and watered at the same time; they are weighed and examined at the same time; etc.

The problem is this. The addition of large amounts of the test product changes the taste of the diet, and this causes the animals to eat less than they normally would. By the end of the second week, management will be aware that the food intake was not as high during the first week as expected, so instructions will go to the animal handlers to increase the concentration of the test product for the next week in order to maintain the desired average dose. This is a self-defeating strategy because it alters the taste of the diet even more so that the animals eat even less. The food consumption will decrease as the concentration of the test product increases until the animals accept the funny or objectionable taste in order to survive, or until they die.

Laboratories often compare subgroups in a colony for assurance that the experience is uniform over the colony. Sometimes proximity to outside walls, doorways, or air vents will influence food intake in spite of random rearrangement of subgroup storage during the study. Furthermore, it is not unusual for laboratories to try different diets in an effort to mask objectionable tastes of test products and hence minimize the fall off in food intake. These considerations create many diverse research problems, only two of which are considered in this chapter. The first, example 10.1, examines the differences in food consumed by two seemingly identical subgroups during the initial 8 weeks of a study. This is only part of a larger group of data, more of which will be used in an example for chapter 11. The second, example 10.2, examines the differences in food consumed by two seemingly identical subgroups fed different diets.

After a short time, the need for the animals to survive seems to overcome their dislike for the strange or repugnant taste, so that the food intake somewhat stabilizes without ever returning to normal. These examples use only the initial data when the drop off is still relatively uniform, i.e., can be represented by a simple linear function in time.

Example 10.1

Table 10.1 presents the food consumed by two equal subgroups of rats during each of the first 8 weeks of a study. The two subgroups both received the same diet and the same concentration of the test product in the diet during each of the 8 weeks. The research question here is this: Did the groups respond in a similar manner, i.e., did the food consumption fall off about the same for the two subgroups and did they consume roughly the same quantity of food?

The first step in the analysis is to plot the data for the two samples. This is done as presented in figure 10.1. Both of the plots show considerable variability, but they also drift downward with time. With only

TABLE 10.1   Data for Example 10.1

Subgroup 1		Subgroup 2	
x	y	x	y
1	7.77	1	7.91
2	8.30	2	7.31
3	7.61	3	7.96
4	7.60	4	7.00
5	7.62	5	7.00
6	7.05	6	6.76
7	7.05	7	7.26
8	6.99	8	6.84

NOTE: x = study week, y = total food consumed (in kilograms).

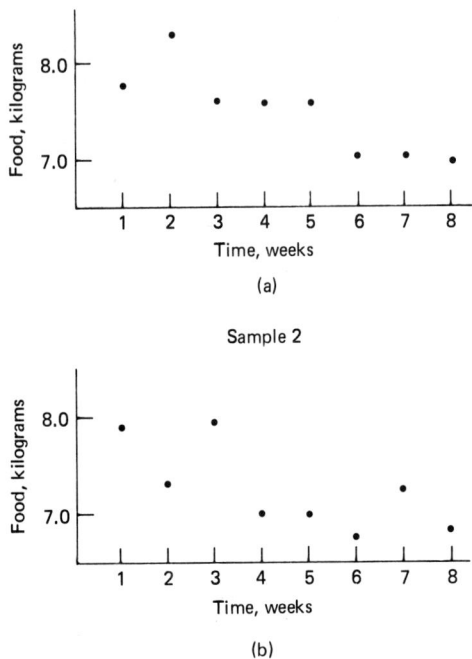

FIGURE 10.1 Example 10.1. Weekly food consumption for (*a*) subgroup 1; (*b*) subgroup 2.

eight observations for each, the general patterns do not suggest a regression function more complicated than a simple linear equation, i.e.,

$$y = a + bx$$

where y is the food consumed and x is time measured in weeks. Time is the independent variable and food consumed is the dependent variable for this regression analysis but cause and effect is not suggested. In fact, the cause for the decline probably depends on, among other things, the quantity of the test product mixed in the food and this is not considered in this analysis.

The next step in the analysis is to estimate the coefficients for the regression equations for the two samples. Some useful sums for these calculations are given below.

Sample 1	Sample 2
$\Sigma y = 59.99$	$\Sigma y = 58.04$
$\Sigma y^2 = 451.2645$	$\Sigma y^2 = 422.5566$
$\Sigma x = 36.00$	$\Sigma x = 36.00$
$\Sigma x^2 = 204.0000$	$\Sigma x^2 = 204.0000$
$\Sigma xy = 263.27$	$\Sigma xy = 255.51$
$n = 8$	$n = 8$

The slope and intercept are calculated for each sample so as to minimize the sum of the squares of the deviations of the points from the regression equation. From chapter 6, these are obtained using the following equations.

$$b = \frac{\Sigma xy - \Sigma x \Sigma y / n}{\Sigma x^2 - (\Sigma x)^2 / n}$$

$$a = \frac{\Sigma y - b \Sigma x}{n}$$

The slope and intercept for the two regression equations are

Sample 1	Sample 2
$b_1 = -0.1592$	$b_2 = -0.1350$
$a_1 = 8.215$	$a_2 = 7.863$

The negative signs of the slopes, i.e., the b's, indicate that the food consumed decreased with time in both samples as expected from the plots of the data in figure 10.1.

In chapter 6, a t-statistic for testing the difference between a regression slope and any arbitrary constant was given. It was

$$t = \frac{b - c}{se_b}$$

where

$$se_b = \frac{sd_{y|x}}{\sqrt{\Sigma(x - \overline{x})^2}}$$

and

$$sd_{y|x} = \sqrt{\frac{\Sigma(y - a - bx)^2}{n - 2}}$$

Two useful calculation formulas for the above, first given in chapters 6 and 2, respectively, are:

$$\Sigma(y - a - bx)^2 = \Sigma y^2 - a\Sigma y - b\Sigma xy$$

and

$$\Sigma(x - \overline{x})^2 = \Sigma x^2 - \frac{(\Sigma x)^2}{n}$$

The slopes of both of the regression equations should be compared to zero. The standard deviation of y given x, i.e., $sd_{y|x}$; se_b; and the t-statistic for both equations are

	Sample 1	Sample 2
$sd_{y|x}$	0.2417	0.3442
se_b	0.0373	0.0531
t	−4.268	−2.542

The food consumption was expected to decline so these are one-tailed t-tests. The null hypothesis is that the consumption did not decline through time. Assuming that this null hypothesis is true and that the observable decline is just chance, then the probability of observing t-statistics this large or larger in absolute value is determined from table B of part V to be 0.004 for sample 1 and 0.023 for sample 2. The slopes can be considered significantly smaller than zero even though the probability for the second sample is not very small.

The first research question: Did the food consumption fall off about the same for the two subgroups? can be restated in terms of these sample statistics. Are the two slopes about the same or do they differ by a significant amount? If the slopes are equivalent, then the question: Did the two subgroups consume roughly the same quantity of food? can be translated into a simple comparison of the intercepts. Any comparison of food consumed when the slopes are not equivalent is dependent on time. Assuming the subgroups ate about the same amount initially but that the slopes are different, an appropriate question might be: When did the two subgroups first consume significantly different amounts of food? For contrast, the data sets have been plotted together in figure 10.2 along with the regression equations.

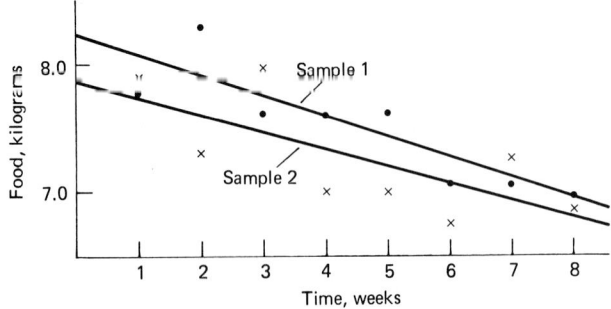

FIGURE 10.2 Example 10.1. Weekly food consumption for subgroups 1 and 2 showing plots of the two regression equations.

The t-statistic for comparing two intercepts is

$$t = \frac{b_1 - b_2}{\text{sd} \sqrt{1/\Sigma(x_1 - \bar{x}_1)^2 + 1/\Sigma(x_2 - \bar{x}_2)^2}}$$

where

$$\text{sd}^2 = \frac{\Sigma(y_1 - a_1 - b_1 x_1)^2 + \Sigma(y_2 - a_2 - b_2 x_2)^2}{n_1 + n_2 - 4}$$

In chapter 2, a t-statistic was given for testing the difference between the averages of two independent samples. It was

$$t = \frac{\bar{x}_1 - \bar{x}_2}{\text{sd} \sqrt{1/n_1 + 1/n_2}}$$

where

$$\text{sd}^2 = \frac{\Sigma(x_1 - \bar{x}_1)^2 + \Sigma(x_2 - \bar{x}_2)^2}{n_1 + n_2 - 2}$$

In chapter 6, a t-statistic for comparing a regression slope to a constant was explained. It was

$$t = \frac{b - c}{\text{se}_b}$$

where b = the regression coefficient
$\quad\quad\ c$ = the arbitrary constant

$$\text{se}_b = \frac{\text{sd}_{y|x}}{\sqrt{(\Sigma(x - \bar{x})^2}}$$

The reader should contrast the t-statistics for comparing a regression slope to a constant, for comparing two averages, and for comparing two regression slopes. The relationship among these t-statistics is not unique to regression slopes but occurs for intercepts as well. Understanding the similarities and differences here will make understanding the comparable relationship for intercepts easier.

The relationship of the above t-statistics suggests the conditions that must be true for the t-test of slopes to be valid. They are

1. The samples must be independent.
2. The samples must be random.
3. The variances of the residuals about the regression equations should be equivalent, i.e., the difference be not significant.
4. The residuals about the regression equations should be normally distributed.

As elsewhere in this book, independence and randomness are left to the data analyst's judgment, but the equivalence of the residual variances and the normality of the residuals are not easy to judge and should be examined analytically with the procedures used in chapter 2 for two independent samples. Each residual is

$$d = y - a - bx$$

For example, the residual for week 5 in subgroup 1 is

$$d = 7.62 - 8.215 - (-0.1592)5 = 0.201$$

and the residual for week 2 in subgroup 2 is

$$d = 7.31 - 7.863 - (-0.1350)2 = -0.283$$

Table 10.2 presents the details of a Lilliefors test of normality with the residuals combined into one set as was done for small samples in chapter 2. The z-column in the table is obtained by dividing each d by $sd_{y|x}$ for its sample. The other columns are standard for the procedure. Lilliefors' statistic for this data set is 0.140. The null hypothesis tested here is, of course, that the samples are from normal distributions. Assuming the null hypothesis is true and that observable deviations from normality are just chance, then the probability of a larger test statistic is something greater than 0.20 (see table M, part V). The residuals can be accepted as normally distributed.

The variances of the residuals can now be compared with a two-tailed

TABLE 10.2   The Lilliefors Test for Normality for Example 10.1

Sample	d	z	Sample	z sorted	Probabilities Normal	Probabilities Sample	da	db
1	−0.286	−1.183	1	−1.183	0.118	0.062	0.056	
1	0.403	1.667	2	−0.935	0.175	0.125	0.050	0.113
1	−0.128	−0.530	1	−0.869	0.192	0.188	0.004	0.067
1	0.022	0.091	2	−0.848	0.198	0.250	−0.052	0.010
1	0.201	0.832	2	−0.819	0.206	0.312	−0.106	−0.044
1	−0.210	−0.869	2	−0.546	0.293	0.375	−0.082	−0.019
1	−0.051	−0.211	1	0.530	0.298	0.438	−0.140	−0.077
1	0.048	0.199	1	−0.211	0.416	0.500	−0.084	−0.022
2	0.182	0.529	1	0.091	0.536	0.562	−0.026	0.036
2	−0.282	−0.819	2	0.168	0.567	0.625	−0.058	0.005
2	0.502	1.458	1	0.199	0.579	0.688	−0.109	−0.049
2	−0.322	−0.935	2	0.529	0.702	0.750	−0.048	0.014
2	−0.188	−0.546	1	0.832	0.797	0.812	−0.015	0.047
2	−0.292	−0.848	2	0.994	0.840	0.875	−0.035	0.028
2	0.342	0.994	2	1.458	0.928	0.938	−0.010	0.053
2	0.058	0.168	1	1.667	0.952	1.000	−0.048	0.014

NOTE: The Lilliefors statistic is 0.140. Refer to Chapter 15 for details.

F-statistic exactly as was done for two independent samples in chapter 2, remembering that the degrees of freedom for the estimates are $n - 2$, not $n - 1$,

$$F = \frac{0.3442^2}{0.2417^2} = 2.028$$

The null hypothesis is that these samples come from populations that have equal variances. Using linear interpolation with table D in part V, the probability of a two-tailed F-statistic this large or larger is about 0.44 (twice the probability for a one-tailed test with the same magnitude F-statistic), assuming the null hypothesis is true and the observable difference is just chance. This probability is large so the analysis can proceed as if the variances were equal.

The analysis is now ready for the test of slope differences. The null hypothesis for this test is that the samples come from populations that have the same slopes. Using the equations given above,

$$sd^2 = \frac{0.3505 + 0.7109}{8 + 8 - 4} = 0.08845$$

and

$$\Sigma(x_1 - \bar{x}_1)^2 = \Sigma(x_2 - \bar{x}_2)^2 = 42$$

so that

$$t = \frac{-0.1592 + 0.1350}{\sqrt{0.08845} \sqrt{2/42}} = \frac{-0.0242}{0.0649} = -0.373$$

This t-statistic has $n_1 + n_2 - 4 = 12$ degrees of freedom and should be compared with table B in part V. The test is two-tailed since there was no apparent prestudy reason to believe one subgroup would have a larger slope than the other. This means that the probability obtained from the table for the absolute value of the computed t must be doubled. Assuming the null hypothesis is true and that the observable difference between the slopes is just chance, then the probability, approximately 0.72, is very large. Consequently, there is no reason to believe the falloff in food consumption was any more for one of the subgroups than for the other.

Since the slopes are equivalent, an estimate of the common slope is required. There are now two regression relationships with a common slope β

$$y_1 = \alpha_1 + \beta x_1$$

and

$$y_2 = \alpha_2 + \beta x_2$$

The least-squares problem is to estimate α_1, α_2, and β such that the sum of squared deviations about the two regression equations is a minimum. The sum to be minimized is

$$D = \Sigma(y_1 - \alpha_1 - \beta x_1)^2 + \Sigma(y_2 - \alpha_2 - \beta x_2)^2$$

To find the minimum, the derivatives of D with respect to α_1, α_2, and β are set equal to zero and at the same time roman letters, indicating estimates, are substituted for α_1, α_2, and β. The result is the following set of three equations.

$$n_1 a_1 \qquad\qquad + b\Sigma x_1 \qquad\quad = \Sigma y_1$$

$$n_2 a_2 \qquad\qquad + b\Sigma x_2 \qquad\quad = \Sigma y_2$$

$$a_1\Sigma x_1 + a_2\Sigma x_2 + b(\Sigma x_1^2 + \Sigma x_2^2) = \Sigma x_1 y_1 + \Sigma x_2 y_2$$

Then these equations are solved for a_1, a_2, and b to get

$$b = \frac{\Sigma x_1 y_1 - \Sigma x_1 \Sigma y_1/n_1 + \Sigma x_2 y_2 - \Sigma x_2 \Sigma y_2/n_2}{\Sigma x_1^2 - (\Sigma x_1)^2 / n_1 + \Sigma x_2^2 - (\Sigma x_2)^2 /n_2}$$

$$a_1 = \frac{\Sigma y_1 - b\Sigma x_1}{n_1}$$

$$a_2 = \frac{\Sigma y_2 - b\Sigma x_2}{n_2}$$

which are not unexpected results. Readers familiar with elementary calculus should fill in the details between the equation for D and these equations for b, a_1, and a_2.

The sum of the squares of the deviations of the data about the two regression equations is given by

$$\Sigma(y_1 - a_1 - bx_1)^2 + \Sigma(y_2 - a_2 - bx_2)^2$$

so that the standard deviation of y given x is

$$\text{sd}_{y|x} = \sqrt{\frac{\Sigma(y_1 - a_1 - bx_1)^2 + \Sigma(y_2 - a_2 - bx_2)^2}{(n - 3)}}$$

(Note that the divisor is now $n - 3$ because only three statistics, a_1, a_2, and b, are estimated from the data.)

For example 10.1,

$$b = \frac{-12.3550}{84.0000} = -0.1471$$

$$a_1 = 8.161$$

$$a_2 = 7.917$$

$$sd_{y|x} = \sqrt{1.07367/13} = \sqrt{0.08259} = 0.28738$$

The individual slopes were both significantly different from zero, but it is still instructive to compare the joint slope b with zero. The null hypothesis for this test is that the slope computed from the combined samples is not less than zero. The t-statistic for this one-tailed test is computed as follows.

$$se_b = \frac{sd_{y|x}}{\sqrt{\Sigma(x_1 - \bar{x}_1)^2 + \Sigma(x_2 - \bar{x}_2)^2}}$$

$$= \frac{0.28738}{9.165} = 0.03136$$

$$t = \frac{-0.1471}{0.03136} = -4.691$$

The absolute value of the t-statistic should be compared to table B in part V. Assuming the null hypothesis is true and that the observable slope is just due to chance, then the probability of a t-statistic with 13 degrees of freedom this small or smaller is approximately 0.0003. This is a smaller probability than for either of the lines individually and is due to the increase in the degrees of freedom.

Attention can now turn to the intercepts. The t-statistic for comparing two intercepts is

$$t = \frac{a_1 - a_2}{sd_{y|x} \sqrt{1/n_1 + 1/n_2 + (\bar{x}_1^2 - \bar{x}_2^2)/[\Sigma(x_1 - \bar{x}_1)^2 + \Sigma(x_2 - \bar{x}_2)^2]}}$$

The null hypothesis tested is that the two samples come from populations that have equal intercepts, assuming that the populations have equal slopes.

A t-statistic for testing the difference between a regression intercept and any arbitrary constant was presented in chapter 6. It was

$$t = \frac{a - c}{se_a}$$

where

$$se_a = sd_{y|x} \sqrt{\frac{1}{n} + \frac{\bar{x}^2}{\Sigma(x - \bar{x})^2}}$$

The *t*-statistic for comparing two intercepts (when the slopes are equivalent) has characteristics of this *t*-statistic and the *t*-statistic for comparing the averages of two independent populations already repeated in this chapter from chapter 2. A similar relationship existed for the *t*-statistic for testing differences in slopes.

It should be obvious that the same four conditions for a valid test of slopes must also be true for a test of intercepts. In addition, the slopes must be equivalent. Using the equations above, the *t*-statistic for comparing the two intercepts in example 10.1 is

$$t = \frac{8.161 - 7.917}{0.28738\sqrt{\frac{1}{8} + \frac{1}{8}}} = 1.696$$

since $\bar{x}_1 = \bar{x}_2$. This *t*-statistic has $n_1 + n_2 - 3 = 13$ degrees of freedom and its absolute value should be compared with table B in part V. The probability obtained from the table must be doubled, since there was no prestudy reason to believe one subgroup would have a larger intercept than the other. Assuming the null hypothesis is true and that the observable intercept difference is just chance, then this probability is approximately 0.118 and indicates there is no reason to believe the intercepts of the two populations are different.

This *t*-test of the difference between the two intercepts depends on equivalence of the regression slopes. If they are not equivalent, the difference between the two regression lines depends on *x*, and so for a particular significance level may be considered significant at one value of *x* but not at another. Comparison of points on nonparallel regression lines is not a common problem but occasionally it does come up and statistics texts ignore it. However, something the data analyst can do when the *x* data are identical, as they are in the two samples for example 10.1, is discussed in the next example.

The analysis for example 10.1 can be summarized this way. The food consumption for the two subgroups declined significantly at an equivalent rate and on the average they consumed equivalent quantities over the 8-week period.

Example 10.2

Example 10.2 is very similar to example 10.1. The differences between the two examples, however, permit a new set of issues to be addressed. Table 10.3 presents the food consumed by two equal subgroups during the first 10 weeks of a study. The subgroups received different diets in an effort to find one that would mask the taste of the test product better than another. The diets were otherwise very similar in nutrition, bulk, appearance, etc. The two subgroups of course received the same initial concentration of the test product, but as in example 10.1 this concentration was changed as the food consumed fell off. The research questions for this

TABLE 10.3 Data for Example 10.2

Subgroup 1		Subgroup 2	
x	y	x	y
1	12.40	1	12.42
2	12.31	2	12.31
3	11.85	3	12.42
4	11.35	4	12.18
5	11.97	5	12.16
6	11.39	6	12.33
7	11.42	7	12.08
8	10.99	8	11.88
9	11.10	9	11.69
10	10.88	10	11.98

NOTE: x = study week, y = total food consumed (in kilograms).

example are the same as they were for example 10.1: Did the food consumption fall off about the same for the two subgroups and did they consume roughly the same quantity of food?

The first step in this analysis, as in all regression problems, is to plot the data. This is done as presented in figure 10.3. The plots show consid-

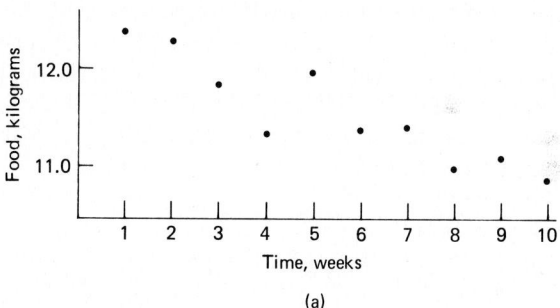

(a)

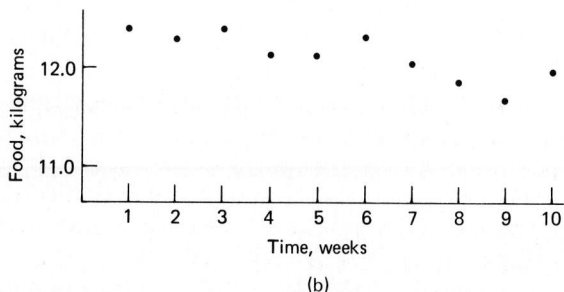

(b)

FIGURE 10.3 Example 10.2. Weekly food consumption for (*a*) subgroup 1; (*b*) subgroup 2.

erable variability with a downward drift, but the first plot does seem to have a steeper decline than the second. The data plots do not suggest a downward drift more complex than a simple linear function, i.e.,

$$y = a + bx$$

where y is the food consumed and x is time so that function is used in the analysis. As in example 10.1, no cause and effect is assumed here.

The coefficients for the two regression equations must be estimated first. The usual useful summations for these calculations are given below.

	Sample 1	Sample 2
Σy	115.66	121.45
Σy^2	1340.3350	1475.5331
Σx	55.00	55.00
Σx^2	385.0000	385.0000
Σxy	622.72	662.41
n	10	10

The slope and intercept are calculated for each sample so as to minimize the sum of the squares of the deviations of the data from the regression equation. Formulas for computing the slope b and the intercept a have already been given in this chapter. The results of the calculations are

Sample 1	Sample 2
$b_1 = -0.1625$	$b_2 = -0.0675$
$a_1 = 12.460$	$a_2 = 12.516$

As expected by an inspection of the data plots in figure 10.3, the food consumed declined in both groups, i.e., the slopes have negative signs. The questions now are these:

1. Are the residuals about the regression equations normally distributed?
2. Are these declines significantly different from zero?
3. Can the variances of the residuals about the regression equations be considered equivalent?

To answer the first question, each residual is calculated,

$$d = y - a - bx$$

for a Lilliefors test of normality. The null hypothesis, of course, is that the samples are from normal distributions. The details of the procedure are presented in table 10.4 where the test statistic is found to be 0.132. Assuming the null hypothesis is true and that the observable deviations from normality are just due to chance, then the probability of a Lilliefors statistic this large or larger (table M, part V) is somewhat greater than 0.20 so normality can be accepted.

The second question is answered by comparing the slopes of the individual data sets to zero. The null hypothesis for each test is that the slope is not less than zero.

	Sample 1	Sample 2
$sd_{y|x}$	0.2323	0.1358
se_b	0.02558	0.01495
t	−6.356	−4.513

The food consumed was expected to decline so these are one-tailed t-tests. The absolute values of these t-statistics are compared to table B in part V where the probability for sample 1 is found to be approximately 0.0002 and the probability for sample 2 is found to be approximately 0.01, assuming for each test that the null hypothesis is true and the

TABLE 10.4 The Lilliefors Test for Normality for Example 10.2

Sample	d	z	Sample	z sorted	Normal	Sample	da	db
1	0.103	0.443	1	−1.980	0.024	0.050	−0.026	
1	0.175	0.753	2	−1.613	0.053	0.100	−0.047	0.003
1	−0.122	−0.525	1	−0.732	0.232	0.150	0.082	0.132
1	−0.460	−1.980	2	−0.707	0.240	0.200	0.040	0.090
1	0.323	1.390	1	−0.525	0.300	0.250	0.050	0.100
1	−0.095	−0.409	2	−0.523	0.301	0.300	0.001	0.051
1	0.098	0.422	2	−0.486	0.314	0.350	−0.036	0.014
1	−0.170	−0.732	1	−0.409	0.341	0.400	−0.059	−0.009
1	0.103	0.443	2	−0.214	0.415	0.450	−0.035	0.015
1	0.046	0.198	2	−0.140	0.444	0.500	−0.056	0.006
2	−0.029	−0.214	1	0.198	0.578	0.550	0.028	0.078
2	−0.071	−0.523	2	0.265	0.604	0.600	0.004	0.054
2	0.106	0.781	1	0.422	0.664	0.650	0.014	0.064
2	−0.066	−0.486	1	0.443	0.671	0.700	−0.029	0.021
2	−0.019	−0.140	1	0.443	0.671	0.750	−0.079	−0.029
2	0.219	1.613	1	0.753	0.774	0.800	−0.026	0.024
2	0.036	0.265	2	0.781	0.783	0.850	−0.067	−0.017
2	−0.096	−0.707	2	1.016	0.845	0.900	−0.055	−0.005
2	−0.219	−1.613	1	1.390	0.918	0.950	−0.032	0.018
2	0.138	1.016	2	1.613	0.947	1.000	−0.053	−0.003

NOTE: The Lilliefors statistic is 0.132. Refer to Chapter 15 for details.

observable slopes are due to chance. These are small probabilities so the analysis indicates a falloff in food consumed for both diets.

The variances about the regression equations are compared with a two-tailed F-test to answer the third question. The null hypothesis for this test is that the samples come from populations with equal residual variances. There is no reason to believe that the variance for one population is any larger than the other so this is a two-tailed test. The F-statistic for this test is

$$F = \frac{0.2323^2}{0.1358^2} = 2.9275$$

Assuming the null hypothesis is true and that the observable difference is just chance, then the probability of an F-statistic this large or larger is approximately 0.16. Remember that this is a two-tailed test and so the apparent probability from table D in part V must be doubled. This is a respectably large probability, so one may proceed in the analysis as if the variances are equivalent.

All of the conditions for comparing the two slopes have been satisfied so the next step in the analysis is the comparison. Using equations given above,

$$sd^2 = \frac{0.4317 + 0.1475}{10 + 10 - 4} = 0.03620$$

and

$$\Sigma(x_1 - \bar{x}_1)^2 = \Sigma(x_2 - \bar{x}_2)^2 = 82.5$$

so that

$$t = \frac{-0.1625 + 0.0674}{\sqrt{0.03620}\sqrt{2/82.5}} = \frac{-0.0951}{0.02962} = -3.210$$

This t-statistic has $n_1 + n_2 - 4 = 16$ degrees of freedom and is two-tailed since there was no prestudy reason for believing one particular diet would have a slower falloff than the other. Assuming the null hypothesis is true and that the difference is just due to chance, then the probability of a t-statistic as large as or larger than the absolute value of this t statistic is approximately 0.008. This is small, so the two slopes cannot be considered equivalent.

The analysis to this point can be summarized as follows. The two subgroups both have significantly declining food consumptions during the first 10 weeks. The variances of the residuals about the regression equations seem to be equivalent and normally distributed. However, the falloff in food consumption was not the same for the two subgroups. Subgroup 2 falls off at a significantly lower rate than subgroup 1, implying

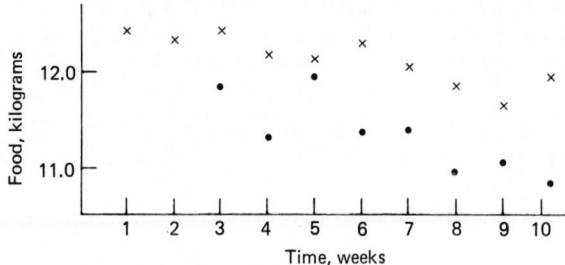

FIGURE 10.4 Example 10.2. Weekly food consumption for subgroups 1 and 2.

that diet 2 was preferred over diet 1. If the only difference between the subgroups was the diet to which the test product was added, then diet 2 seems to have masked the taste of the test product better than diet 1.

Since the falloff was not equivalent for the two groups, subgroup 2 obviously consumed more food than subgroup 1. The researcher might at this point ask the data analyst, "When did the food consumption become significantly different for the two diets?" A resourceful data analyst might develop several ways to answer this question, but only one, a procedure that depends on the x data being the same for both subgroups, is demonstrated.

Table 10.5 presents the raw difference between the food consumed by the two groups for each of the 10 weeks. Since there was a significant difference between slopes, these weekly differences, which can be called w, increase with time as shown in figure 10.5. The plot indicates that a function of the form

$$w = a + bx$$

could satisfactorily represent the relationship, so a direct application of techniques presented in chapter 6 is suggested for this analysis. The plan is this:

1. Use the data in table 10.5 to estimate a linear regression equation relating the difference in food consumed by the two subgroups with time.

TABLE 10.5 Data for Example 10.2

x	w	x	w
1	0.02	6	0.94
2	0.00	7	0.66
3	0.57	8	0.89
4	0.83	9	0.59
5	0.19	10	1.10

NOTE: x = study week, w = difference in food consumed (in kilograms).

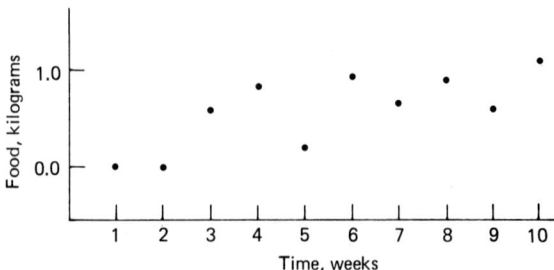

FIGURE 10.5 Example 10.2. Weekly differences in the total food consumed by subgroups 1 and 2.

2. Since the probability associated with the comparison of the individual subgroup slopes was low, 0.008, the slope of this new equation should be significantly different from zero. It should be tested anyway using the null hypothesis that the slope is not different from zero.

3. Compute a lower confidence limit for this regression equation using whatever confidence is desired. Plot this on the graph.

4. The point, or time, when this crosses the line $w = 0$, i.e., the x-axis, is the time when the food consumed differed significantly at a probability level equal to that used for the confidence limit.

Letting $y_2 - y_1 = w$, then some useful summations are

$\Sigma w = \quad 5.79$ $\qquad\qquad$ $\Sigma x = \quad 55.00$ $\qquad\qquad$ $\Sigma xw = \quad 39.69$

$\Sigma w^2 = \quad 4.7197$ $\qquad\qquad$ $\Sigma x^2 = 385.0000$ $\qquad\qquad$ $n = \quad 10$

The regression equation is

$$w = 0.056 + 0.0951x$$

and the standard deviation of the residuals about this regression line is

$$sd_{w|x} = \sqrt{\Sigma(w - a - bx)^2/(n - 2)} = 0.2787$$

The standard error of the slope is

$$se_b = \frac{sd_{w|x}}{\sqrt{\Sigma(x - \bar{x})^2}} = \frac{0.2787}{\sqrt{82.5}} = 0.0307$$

The residuals about this regression equation are examined for normality in Table 10.6. Assuming that the residuals are from a normal distribution and that the observable deviations from normality are just due to chance, then the probability of a Lilliefors statistic as large or larger than 0.116 (table M, part V) is greater than 0.20, so normality can be accepted for the analysis.

TABLE 10.6 The Lilliefors Test for Normality for Example 10.2: $d = w - 0.056 - 0.0951x$

Sample	d	z	Sample	z sorted	Probabilities Normal	Probabilities Sample	da	db
1	-0.131	-0.470	5	-1.227	0.110	0.100	0.010	
2	-0.246	-0.882	9	-1.155	0.124	0.200	-0.076	0.024
3	0.229	0.821	2	-0.882	0.184	0.300	-0.116	-0.016
4	0.394	1.413	1	-0.470	0.312	0.400	-0.088	0.012
5	-0.342	-1.227	7	-0.222	0.412	0.500	-0.088	0.012
6	0.313	1.123	8	0.262	0.603	0.600	0.003	0.103
7	-0.062	-0.222	10	0.334	0.631	0.700	-0.069	0.031
8	0.073	0.262	3	0.821	0.794	0.800	-0.006	0.094
9	-0.322	-1.155	6	1.123	0.869	0.900	-0.031	0.069
10	0.093	0.334	4	1.413	0.921	1.000	-0.079	0.021

NOTE: The Lilliefors statistic is 0.116. Refer to Chapter 15 for details.

10.23

The t-statistic for comparing the slope to zero is

$$t = \frac{0.0951}{0.0307} = 3.099$$

Assuming the null hypothesis is true and the observed slope is just due to chance, then the probability of a t-statistic with 8 degrees of freedom this large or larger is 0.016. Like the comparison of subgroup slopes, this is a two-tailed t-test because there was no prestudy expectation favoring one diet over the other.

The next step is to compute a lower confidence limit for this regression equation. Details for this were discussed in chapter 6. The formula for the lower confidence limit is

$$w = a + bX - t \cdot \text{sd}_{w|x} \sqrt{\frac{1}{n} + \frac{(X - \bar{x})^2}{\Sigma(x - \bar{x})^2}}$$

Points along this confidence limit are calculated in table 10.7 for a confidence level of 0.95, i.e., the probability that the confidence limit is below the true regression equation at any particular time x is 0.05. These points, column 6, are plotted in figure 10.6 and a curve drawn between them. The curve crosses $w = 0$ at about $x = 2.09$. From time $= 2.09$ forward, the probability is 0.05 or less that w does not differ from zero. Before time $= 2.09$, the probability is 0.05 or more that w does not differ from zero.

Confidence limits and statistical tests are kind of like the two sides of

TABLE 10.7   Confidence Limits for the Regression Equation $w = \alpha + \beta x$ in Example 10.2

X	(2) $(X - \bar{x})^2$	(3) $\dfrac{col\ (2)}{\Sigma(x - \bar{x})^2}$	(4) $\sqrt{\dfrac{1}{n} + col\ (3)}$	(5) $t \cdot sd_{w\mid x} \cdot col\ (4)$	(6) $a + bX - col\ (5)$
1	20.25	0.245	0.587	0.304	−0.154
2	12.25	0.148	0.498	0.258	−0.012
2.09	11.628	0.1409	0.4909	0.2545	0.0003
3	6.25	0.076	0.420	0.218	0.124
4	2.25	0.027	0.356	0.185	0.251
5	0.25	0.003	0.321	0.166	0.365
6	0.25	0.003	0.321	0.166	0.460
7	2.25	0.027	0.356	0.185	0.537
8	6.25	0.076	0.420	0.218	0.599
9	12.25	0.148	0.498	0.258	0.654
10	20.25	0.245	0.587	0.304	0.702

$\bar{x} = 5.5$ $\Sigma(x - \bar{x})^2 = 82.5$ $\text{sd}_{w|x} = 0.2787$ $a = 0.0560$ $b = 0.0951$
confidence level $= 0.95$ $t = 1.860$ $n = 10$

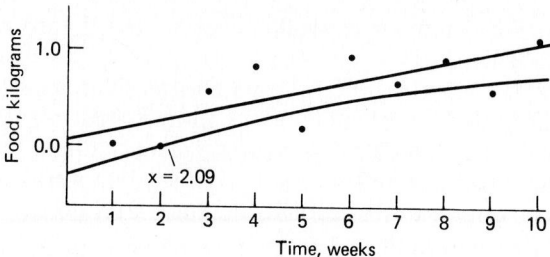

FIGURE 10.6 Example 10.2. Weekly differences in the total food consumed by subgroups 1 and 2 showing plots of the regression equation and the lower 95% confidence limit for the regression equation.

a single coin and the example above illustrates this fact. A one-sided confidence limit has been used for this reason—after determining which subgroup had the greatest fall off in food consumption, the determination of when the consumption differed significantly was a one-tailed problem.

In summary, the two subgroups experienced a decline in food consumption but subgroup 1 declined at a significantly greater rate than did subgroup 2. The analysis then focused on a new research question: When did subgroup 1 consume significantly less food than subgroup 2? The answer was found to be about $x = 2.09$. If the client asks how this value of x was obtained, the easiest explanation might be through use of figure 10.6. The analyst does not know the true regression function

$$w = \alpha + \beta x$$

but does know that the lower confidence limit lies below the true regression equation with the probability chosen for the limit. Since the 0.95 confidence limit crosses $w = 0$ at $x = 2.09$, the analyst knows that above $x = 2.09$ the probability is less than 0.05 that the difference w is not greater than zero.

Example 10.3

Infants normally experience a weight loss during the first 4 or 5 days after birth. This loss is greatest at about 3 days and within a week, the weight has usually been regained. Physicians have observed that this weight loss seems to be greater in breast-fed than in formula-fed infants. Example 10.3 compares the weight loss of two groups of newborns, one group breast-fed, the other group formula-fed. The causes and effects of this weight loss, or the advantages of one feeding method over another, are not the subject of example 10.3. Only the simple research hypothesis that formula-fed newborns have a smaller weight loss than breast-fed newborns is investigated.

Table 10.8 presents the birth weights and weight losses 72 hours post partum for 22 breast-fed infants and 22 formula-fed infants. Characteristics of the *x* data distinguish this example from examples 10.1 and 10.2. In the first two examples in this chapter, the *x* data were equally spaced points in time, while in this problem they are a random sample from a population of birth weights. Incidentally, these were all male infants.

TABLE 10.8   Data for Example 10.3

Sample 1: Breast-fed		Sample 2: Formula-fed	
x	*y*	*x*	*y*
2740	100	3360	150
3760	250	3240	130
3550	280	3850	230
3570	150	3060	0
3360	200	3160	0
2720	110	3300	140
3090	300	3640	140
3360	200	3280	−60
3790	170	2780	110
3750	310	3210	30
3590	250	2240	70
2960	90	2970	−70
3000	20	3110	100
3330	180	3390	220
3960	140	2490	30
3150	120	3260	160
3690	230	2670	0
3670	100	3200	50
3390	10	3700	220
3470	300	3040	20
3200	100	2920	50
3190	160	3450	110

NOTE: *x* = birth weight, *y* = weight loss.

Now the weight losses for the two groups could be compared directly using procedures from chapter 2, and in fact, the sample average weight losses are different by most commonly accepted statistical criteria. The situation is not as simple as this implies, however, for two reasons. First, there is also a small difference ($p = 0.046$) in sample birth weight averages. Second, weight loss is related to birth weight and so the apparent weight loss difference could be enhanced or diminished by this difference in birth weights. A comparison of the weight losses using the birth weights as a covariate will be a comparison independent of the birth weight differences. The analysis will compare regressions where *y*, the dependent variable, will be the weight loss and x, the independent variable, will be the birth weight.

The first step in the analysis is to plot the two samples and this is done in figure 10.7. There is a lot of scatter in these data and the scales chosen for the plots tend to emphasize it. There does seem to be a trend in both sample plots but relationships more complex than a simple linear function, i.e.,

$$y = a + bx$$

if they exist, are obliterated by the scatter.

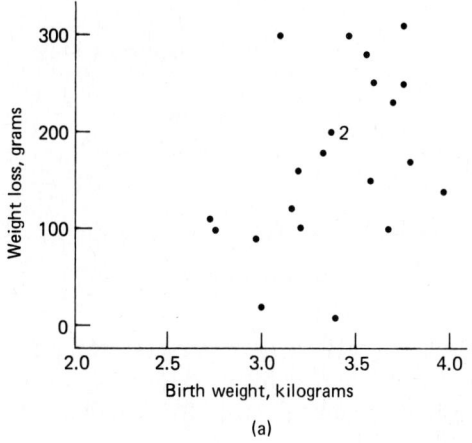

(a)

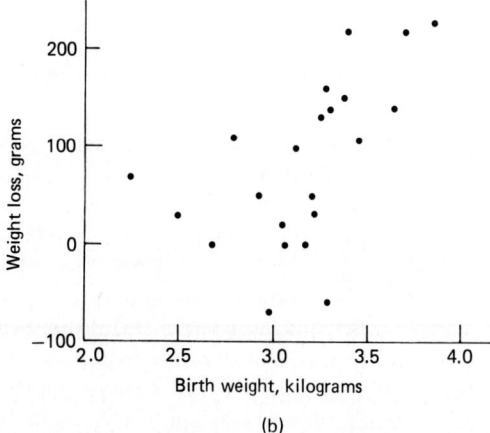

(b)

FIGURE 10.7 Example 10.3. Weight loss, in grams, as a function of birth weight, in grams, for (a) sample 1; (b) sample 2.

The usual useful summations for estimating the coefficients in the regression equations are given below.

	Sample 1	Sample 2
Σy	3770	1830
Σy^2	806,500	308,700
Σx	74,290	69,320
Σx^2	253,321,700	221,449,200
Σxy	12,997,200	6,149,900
n	22	22

The average y and average x for the samples will also be useful and they are

Sample 1	Sample 2
$\bar{y}_1 = 171$	$\bar{y}_2 = 83$
$\bar{x}_1 = 3377$	$\bar{x}_2 = 3151$

The simple difference between the y averages is 88 grams.

The slope and intercept are calculated for each sample using the usual least-squares formulas already given in this chapter. The results are

Sample 1	Sample 2
$b_1 = 0.1085$	$b_2 = 0.1267$
$a_1 = -195$	$a_2 = -316$

As suggested by the plots, the slopes of the regression equations are positive indicating that the weight lost by newborns is, in general, larger for large infants than it is for small infants.

The next step in the analysis is to verify that there is a relationship between weight lost and birth weight, i.e., to test the null hypothesis that no relationship exists between weight loss and birth weight. To do this, the residuals about the regression equations should be examined for normality and then the individual slopes compared to zero with the usual t-tests. Now in making these t-tests, the null hypothesis that a slope is not less than zero can be rejected at a moderately large probability level, even 0.10 or 0.20, for the following reason. The purpose for using the birth weight in the analysis is to adjust the t-statistic for the differences in birth weight averages. If an error is made in the analysis, it is better to have erred on the side of making an unnecessary adjustment than in not making an adjustment that was important for the overall decision-making process.

The samples are sufficiently large to test individually for normality and this is done in tables 10.9 and 10.10. Assuming the samples did come from normal populations and that the sample deviations from this null hypothesis are just chance variation, then the probability of Lilliefors' statistics as large as or larger than 0.092 and 0.121 is more than 0.20, the limit of table M in part V.

The individual slopes can now be compared with zero. The standard deviation of y given x, i.e., $sd_{y|x}$; se_b; and the t-statistic for both equations are given below.

	Sample 1	Sample 2
$sd_{y|x}$	81.10	73.43
se_b	0.05173	0.04220
t	2.097	3.003
p	0.025	0.004

TABLE 10.9 The Lilliefors Test for Normality for Example 10.3 for Breast-fed Infants

d	z	z sorted	Probabilities Normal	Probabilities Sample	da	db
−2.29	−0.03	−2.06	0.020	0.045	−0.026	
37.07	0.47	−1.40	0.081	0.091	−0.010	0.036
89.85	1.14	−1.30	0.096	0.136	−0.040	0.005
−42.32	−0.53	−1.20	0.116	0.182	−0.066	−0.020
30.46	0.38	−0.66	0.255	0.227	0.028	0.073
9.88	0.12	−0.58	0.280	0.273	0.007	0.053
159.75	2.02	−0.53	0.296	0.318	−0.022	0.024
30.46	0.38	−0.46	0.324	0.364	−0.040	0.006
−46.18	−0.58	−0.34	0.368	0.409	−0.041	0.004
98.16	1.24	−0.03	0.488	0.455	0.034	0.079
55.51	0.70	0.11	0.545	0.500	0.045	0.090
−36.15	−0.46	0.12	0.550	0.545	0.004	0.050
−110.49	−1.40	0.17	0.569	0.591	−0.022	0.023
13.71	0.17	0.31	0.622	0.636	−0.014	0.031
−94.62	−1.20	0.38	0.650	0.682	−0.032	0.013
−26.76	−0.34	0.38	0.650	0.727	−0.077	−0.032
24.67	0.31	0.47	0.680	0.773	−0.092	−0.047
−103.16	−1.30	0.70	0.758	0.818	−0.060	−0.014
−162.79	−2.06	1.14	0.872	0.864	0.008	0.054
118.53	1.50	1.24	0.893	0.909	−0.017	0.029
−52.18	−0.66	1.50	0.933	0.955	−0.022	0.024
8.90	0.11	2.02	0.978	1.000	−0.022	0.024

NOTE: The Lilliefors test statistic is 0.092. Refer to Chapter 15 for details.

TABLE 10.10   The Lilliefors Test for Normality for Example 10.3 for Formula-fed Infants

d	z	z sorted	Probabilities Normal	Probabilities Sample	da	db
40.32	0.56	−2.23	0.013	0.045	−0.032	
35.53	0.50	−1.82	0.035	0.091	−0.056	−0.011
58.23	0.81	−1.18	0.120	0.136	−0.017	0.029
−71.66	−1.00	−1.00	0.159	0.182	−0.023	0.022
−84.33	−1.18	−0.85	0.199	0.227	−0.029	0.017
37.93	0.53	−0.69	0.247	0.273	−0.026	0.019
−5.16	−0.07	−0.55	0.291	0.318	−0.027	0.018
−159.54	−2.23	−0.31	0.378	0.364	0.015	0.060
73.82	1.03	−0.15	0.439	0.409	0.029	0.075
−60.67	−0.85	−0.07	0.471	0.455	0.017	0.062
102.25	1.43	−0.05	0.478	0.500	−0.022	0.024
−130.26	−1.82	0.31	0.621	0.545	0.075	0.121
22.00	0.31	0.43	0.665	0.591	0.074	0.120
106.52	1.49	0.50	0.690	0.636	0.054	0.099
30.57	0.43	0.53	0.702	0.682	0.020	0.065
62.99	0.88	0.56	0.713	0.727	−0.014	0.031
−22.24	−0.31	0.81	0.792	0.773	0.019	0.064
−39.40	−0.55	0.88	0.810	0.818	−0.008	0.038
67.24	0.94	0.94	0.826	0.864	−0.038	0.008
−49.13	−0.69	1.03	0.849	0.909	−0.061	−0.015
−3.92	−0.05	1.43	0.923	0.955	−0.031	0.014
−11.08	−0.15	1.49	0.931	1.000	−0.069	−0.023

NOTE: The Lilliefors test statistic is 0.121. Refer to Chapter 15 for details.

The weight loss was expected to be larger for heavier infants so these are one-tailed t-tests. Assuming the null hypothesis is true and that the observable slopes are due to chance, then the probabilities of t-statistics as large as or larger than these are given below the t-statistics themselves.

Finally, the variance about the regression equations must be equivalent for a comparison of the slopes. A two-tailed F-statistic is used to test the null hypothesis that these samples are from populations with equal residual variances. The statistic is

$$F = \frac{81.10^2}{73.43^2} = 1.220$$

Referring to table D in part V, the probability of a two-tailed F-statistic this large or larger is about 0.69, assuming the null hypothesis is true and that the sample difference is just chance variation. This is a large probability, so the evidence contrary to the null hypothesis does not appear to be important.

Having satisfied the conditions for comparing the slopes, the next step is to compute the *t*-statistic for that comparison. The null hypothesis tested will be that there is no difference between the slopes for the two populations represented by the samples. Using equations already given in this chapter,

$$\text{sd}^2 = \frac{131,543 + 107,850}{22 + 22 - 4} = 5984.8$$

and

$$\Sigma(x_1 - \bar{x}_1)^2 = 2,457,877$$
$$\Sigma(x_2 - \bar{x}_2)^2 = 3,028,182$$

so that

$$t = \frac{0.1085 - 0.1267}{\sqrt{5984.8}\sqrt{1/2,457,877 + 1/3,028,182}} = -0.274$$

This is a very small *t*-statistic. It has $n_1 + n_2 - 4 = 40$ degrees of freedom and should be compared as a two-tailed statistic with table B in part V. Assuming the null hypothesis is true and that the difference between the sample slopes is just chance, then the probability of a *t*-statistic as large as or larger than the absolute value of this *t*-statistic is approximately 0.79. This is very large, so the slopes of the two populations can be considered equivalent and the data pooled to estimate a common slope.

The procedure for estimating a common slope was explained earlier in this chapter so for this example

$$b = \frac{266,595 + 383,736}{2,457,877 + 3,028,181} = 0.1185$$

$$a_1 = \frac{3770 - 0.1185 \cdot 74,290}{22} = -229$$

$$a_2 = \frac{1830 - 0.1185 \cdot 69,320}{22} = -290$$

These statistics define two new parallel regression equations. The standard deviation of weight loss given birth weight is computed from the sum of squares of the deviations of the data from the separate regression equations.

$$\text{sd}_{y|x} = \sqrt{\frac{128,856 + 110,988}{41}} = \sqrt{5849.86} = 76.48$$

Note that the degrees of freedom are $n_1 + n_2 - 3$.

As in example 10.1, the common slope is compared to zero. The t-statistic for this test is computed as follows.

$$se_b = \frac{76.48}{\sqrt{2,457,877 + 3,028,182}} = 0.03265$$

$$t = \frac{0.1185}{0.03265} = 3.630$$

Assuming that the slope is not greater than zero and the observable slope is just due to chance, then the probability of a one-tailed t-statistic with degrees of freedom equal to 41 this large or larger is less than 0.0005. Like the common slope in example 10.1, this smaller probability is due to the increase in degrees of freedom.

With a common slope, the intercepts can now be compared.

$$t = \frac{-229 + 290}{76.48 \sqrt{\frac{1}{22} + \frac{1}{22} + (3377 - 3151)/(2,457,877 + 3,028,182)}} = 2.66$$

This t-statistic has $n_1 + n_2 - 3 = 41$ degrees of freedom and only one tail since the breast-fed infants were expected to lose more weight. The probability of a t-statistic this large or larger is about 0.006, so that the intercepts are different, assuming the null hypothesis is true and the observable difference between the intercepts is just due to chance.

This analysis can be summarized as follows. There is a relationship between weight loss and birth weight and since the birth weights are not quite the same for the two samples, this relationship must be considered in a comparison of the weight losses. The formula-fed infants lost significantly less during the first 72 hours post partum than did the breast-fed infants, but this difference, 61 grams, was not as large as a simple comparison of the weight losses, 88 grams, without adjusting for birth-weight differences.

SUMMARY

Samples can have identical x data as in examples 10.1 and 10.2 or the x data can be different random samples as in example 10.3. Furthermore, samples can be classified by the relationship of their slopes.

1. The slopes can be equivalent and different from zero.
2. They can be equivalent and not different from zero.
3. They can be equivalent and only one different from zero.
4. They can be not equivalent.

These classifications create eight situations for data analysts to consider. Suppose the slopes are equivalent and different from zero with the x

data identical. The y data can be paired for common x and procedures presented in chapter 3 used instead of an intercept comparison. If the x data are not identical, the y data cannot be paired, so the procedures in this chapter must be used.

Now suppose the slopes are both equivalent to zero with the x data identical. The y data can again be paired for common x and procedures presented in chapter 3 used in lieu of intercept comparisons. If the x data are not identical, the y data can be compared without regard to x using the procedures of chapter 2. Incidentally, when both slopes are equivalent to zero, there are no regression relationships so the y data should be compared as if the x data never existed.

Next, suppose the slopes are equivalent but one slope can be considered different from zero while the other cannot. There are no clear-cut, scientifically correct procedures for this not-too-unusual situation. If the x data are identical so that the y data are pairable, procedures explained in chapter 3 can certainly be used. If the data cannot be paired, procedures of this chapter can always be used to compare the intercepts even though one of the intercepts is for a regression line whose slope may be zero.

Finally, suppose the slopes are not equivalent. Example 10.2 shows how procedures explained in chapter 6 can be used to compare the two samples when the x data are identical. When the x data are not identical, the y data cannot be paired, so the regression equation must be compared at specific values of x. These recommendations are probably valid even when one sample slope is equivalent to zero, but the analyst's understanding of each particular problem should be the final procedural guide.

In all of these situations, the analyst must report with care so that researchers and managers are fully aware of the limitations of the analysis. Consider, for example, that two slopes are equivalent but only one can be considered significantly different from zero. The analyst might report that both slopes suggest the y responses either increase or decrease (as the case may be), that one sample suggests the apparent x, y relationship may not be due to chance while the other doesn't so indicate, and that when the samples are pooled for a common slope estimate, this esti-

TABLE 10.11 Recommendation Summary

	x Data pairable	x Data not pairable
Slopes equal and both different from zero	Pair y data, use chapter 3	Use chapter 10
Slopes both equivalent to zero	Pair y data, use chapter 3	Use chapter 2
Slopes equivalent while one is equivalent to zero and the other is not	Pair y data, use chapter 3	Use chapter 10
Slopes not equal	Pair y data, use chapter 6	Use chapter 10

mate is either equivalent to zero or it isn't, depending on the criteria established by the data analyst. The researcher or manager should probably be encouraged to accept the result of the test of the common estimate. If all of this sounds complex, it is; but it is typical of data analysis issues.

Recommendations for each of the eight situations described above are summarized in table 10.11, but like all data analysis, the particular situation dictates the course to follow. Practicing data analysts shouldn't have any difficulty in finding situations where these recommendations are absurd.

CHAPTER 11

Comparison of More Than Two Relationships

INTRODUCTION

Chapter 11 is an extension and generalization of Chapter 10 to more than two relationships. As in chapter 10, comparison is limited to relationships of the general form

$$y = a + bx$$

Simple transformation of the x or y metameter (see chapter 14 for a discussion of metameters) does not complicate this basic relationship so that slightly more complex forms like

$$y = a + b \cos x$$

and

$$\log y = a + bx$$

are included. Relationships like

$$y = a + bx + cx^2$$
$$y = a + bx + c \cos x$$

and

$$y = a + bx_1 + cx_2$$

11.1

are not considered, though the reader may be able to develop some insight into methods for comparing such relationships by working through this chapter.

Basic research questions usually involve differences between slopes, i.e., the regression coefficient b, and differences between the intercepts, i.e., the coefficient a, but these spawn a number of other questions for the data analyst seldom considered by the client. These adjunct issues include normality of the residual data, homogeneity of the residual variances, and equivalency of regression slopes, and all are discussed in chapter 11 for specific examples.

The three examples considered in this chapter are the same three used in chapter 10. For chapter 10, two samples were chosen from the problems as they were originally presented for analysis. The same problems with additional samples are used in this chapter so the reader can better understand the similarities and differences and the new issues that more than two samples raise.

AN INTRODUCTION FOR EXAMPLES 11.1 AND 11.2

There seems to exist a belief that if a drug product or food additive is hazardous, then a large intake over a short period of time will have the same effect as a small intake over a long period of time. The principal responses investigated concern tumors, reproductive system changes, birth defects, etc. This assumption is probably true for chemicals that tend to deposit in the body but seems questionable for chemicals that wash out quickly. The long-term and/or high-dose effects of promising new drugs and food additives are often assessed by feeding the product to a group of small animals, sometimes for two or even three generations. To create worst-possible conditions and hence minimize the time for testing, very high dose levels of the product are mixed in the food prepared for the animals.

Many chemical products have a strange or an unpleasant taste in the high concentrations used for testing, while the very low concentrations planned for human consumption either have a pleasant taste or no perceptible taste at all. Anyone who has compared a concentrated bit of artificial sweetener on their tongue with the normally used dilution has a case in point. Strange or unpleasant tastes in the food for laboratory animals usually influence their food intake adversely. As food consumed decreases, the concentration of the test product is increased to maintain the required daily dose. This strategy is self-defeating, because as the concentration is increased, the taste of the diet is further altered and food consumption decreases more.

Sometimes the animals will adjust to the strange or unpleasant taste even though they will generally consume less food. More often their intake decreases to the bare survival level with accompanying debilitation and some decrease in life expectancy. Occasionally, but not very often, the life expectancy is reduced dramatically. The assumed need for high doses and the practicality of administering these doses creates never-ending problems for researchers in this field.

These dosing problems create stresses for the animals that seem to be exacerbated by even trivial things. Homogeneity in the responses is, of course, desirable, so observable differences are much studied. Most studies require more animals than can be attended or studied as a single group, so it is not unusual to define subgroups for logistic reasons. For example, all of the animal cages on a single cart, or even a particular shelf on a cart, may define a subgroup. Food intake and medical examination records are maintained separately for each subgroup and the subgroups compared in an effort to establish homogeneity. Failure to reject null hypotheses of no subgroup differences is usually accepted as evidence, perhaps erroneously, of homogeneity.

Examples 11.1 and 11.2 are two problems coming out of the situations described above.

Example 11.1

Animal cages in these studies are commonly placed on movable carts or racks to facilitate care and record keeping. Animals on different carts will consume different quantities of food and sometimes exhibit considerable differences in other secondary responses such as vitality, aggressiveness, animation, morbidity, and even mortality. While the causes often remain unknown, laboratory attendants make conscious efforts to minimize subgroup differences. Cages are systematically relocated on carts and the carts rearranged within storage rooms. Such procedures probably help make the animal experiences uniform, but the subgroup responses are nevertheless compared to detect unusual situations that might affect the principal experimental responses.

Example 11.1 is concerned with the food consumed by four subgroups in one experiment all fed the same diet, but the test product concentration adjusted individually for the subgroups. Experience with similar test products indicates that the food consumption will fall off for several months, so there is a prestudy reason to believe the same will happen for this investigation. The research questions for example 11.1 are these: Did the food consumption fall off about the same for the four subgroups and did they consume roughly the same quantity of food during the first two months? The food consumed during the relatively stable period following falloff would be the subject of a different analysis.

Four samples were chosen for this example from a somewhat larger routine investigation of food consumption and are presented in table 11.1. The first two samples were used in example 10.1 so that the reader may contrast procedures in the two situations. Incidentally, the procedures presented in this chapter for more than two samples are all applicable to the two-sample situation and give results identical to those found in chapter 10.

TABLE 11.1   Data for Example 11.1

Sample 1		Sample 2		Sample 3		Sample 4	
x	*y*	*x*	*y*	*x*	*y*	*x*	*y*
1	7.77	1	7.91	1	8.03	1	8.13
2	8.30	2	7.31	2	7.77	2	7.75
3	7.61	3	7.96	3	7.92	3	7.43
4	7.60	4	7.00	4	7.57	4	7.52
5	7.62	5	7.00	5	7.63	5	6.65
6	7.05	6	6.76	6	7.79	6	6.40
7	7.05	7	7.26	7	6.66	7	6.73
8	6.99	8	6.84	8	6.82	8	5.99

NOTE: x = time in weeks; y = food consumed (in kilograms).

As in all regression studies, the data should be plotted first and this is done in figure 11.1. The plots all show a downward trend, the last one perhaps a little steeper than the first three. With only eight observations in each sample, it is difficult to see enough form in the plots to justify any regression function more complex than a simple linear equation, i.e.,

$$y = a + bx$$

The usual useful summations are presented in table 11.2. Then the least-squares slope and intercept are calculated for each sample using equations from chapter 6.

$$b = \frac{\Sigma xy - \Sigma x \Sigma y / n}{\Sigma x^2 - (\Sigma x)^2 / n}$$

$$a = \frac{\Sigma y - b \Sigma x}{n}$$

The results are given below.

Sample 1	Sample 2	Sample 3	Sample 4
$b_1 = -0.1592$	$b_2 = -0.1350$	$b_3 = -0.1708$	$b_4 = -0.2862$
$a_1 = 8.215$	$a_2 = 7.863$	$a_3 = 8.293$	$a_4 = 8.363$

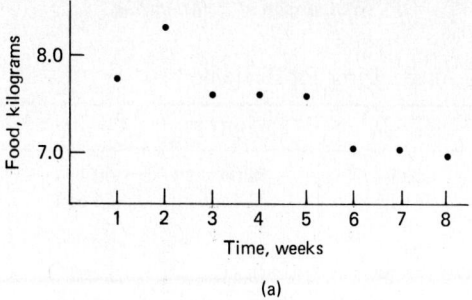

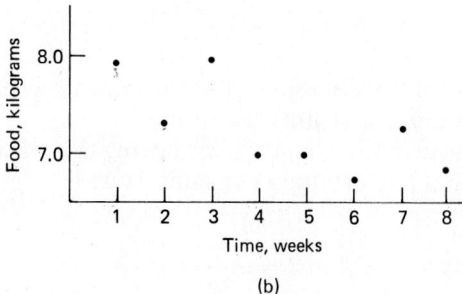

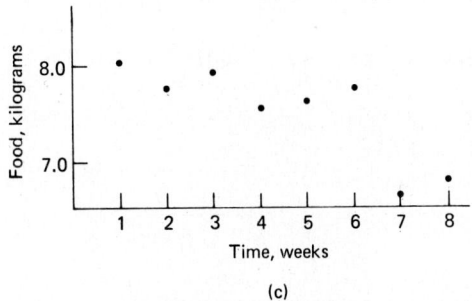

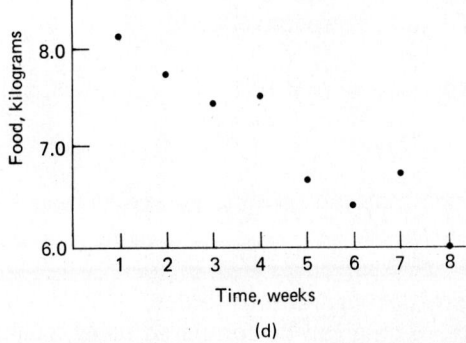

FIGURE 11.1 Example 11.1. Weekly food consumption for (*a*) subgroup 1; (*b*) subgroup 2; (*c*) subgroup 3; (*d*) subgroup 4.

11.5

TABLE 11.2 Data Sums for Exmaple 11.1

	Sample 1	Sample 2	Sample 3	Sample 4
Σy	59.99	58.04	60.19	56.60
Σy^2	451.2645	422.5566	454.6541	404.2702
Σx	36	36	36	36
Σx^2	204	204	204	204
Σxy	236.27	255.51	263.38	242.68
n	8	8	8	8

The negative signs of the slopes, i.e., the b's, indicate as expected that the food consumed decreased slightly with time.

A t-statistic was given in chapter 6 for testing the difference between a regression slope and any arbitrary constant. It was

$$t = \frac{b - c}{se_b}$$

where

$$se_b = \frac{sd_{y|x}}{\sqrt{\Sigma(x - \bar{x})^2}}$$

and

$$sd_{y|x} = \sqrt{\frac{\Sigma(y - a - bx)^2}{n - 2}}$$

The summations in the above equations may explain what is needed, but they are not appropriate for computing, so calculation formulas first given in chapters 6 and 2, respectively, are

$$\Sigma(y - a - bx)^2 = \Sigma y^2 - a\Sigma y - b\Sigma xy$$

and

$$\Sigma(x - \bar{x})^2 = \Sigma x^2 - \frac{(\Sigma x)^2}{n}$$

If the slopes are not significantly different from zero, there are no regressions and the y data can be compared using techniques presented in chapters 4 and 5. So the next step is to formally compare the slopes with zero. The null hypothesis for each test is that the sample came from a population where the slope is zero. The standard deviation of y given x, i.e., $\text{sd}_{y|x}$; se_b; and the t-statistic for the four equations are

	Sample 1	Sample 2	Sample 3	Sample 4
$\text{sd}_{y|x}$	0.2417	0.3442	0.3092	0.2534
se_b	0.0373	0.0531	0.0477	0.0391
t	−4.268	−2.542	−3.580	−7.320

These are all one-tailed tests since the food consumption was expected to decrease. Assuming the null hypothesis is true and that the observable slopes are just due to chance, then the probability of observing t-statistics this large or larger is determined by comparing them with table B in part V. The probabilities are

	Sample 1	Sample 2	Sample 3	Sample 4
p	0.004	0.023	0.006	0.0003

All of these probabilities are small so it is not unreasonable to proceed with the analysis on the basis that food consumption will generally decline with the subject test product at the dose levels fed. This raises some questions, however.

1. What if food consumption generally declined but the p-value for one or a few samples was not small, e.g., as high as 0.25?
2. What if food consumption generally declined but one or two samples actually increased?

The first question is probably an issue for the data analyst only; the second may be a management problem, not a data analysis problem at all. The data analyst is often the first to discover the scenarios described by these questions and so must deal with them or bring them to management's attention.

Consider the first question. Nothing guarantees that just by chance a sample or two will not have relatively high p-values. If the other p-values for this investigation are small and if experience with this and similar test products in the dose levels fed indicates food consumption should decline, then it is not unreasonable in data analysis to ignore the larger probability and use all samples in the next phase of the analysis.

Now consider the second question in contrast with the first. If most samples show a decline in consumption, samples that increase could represent excess food contamination or poor record management. The condition may be first detected by the data analyst but the problem may be management. The analyst should consider contrasting the samples that conform as expected with the samples that give unexpected results. In this way, the analyst accepts responsibility for measuring and reporting the difference, but the responsibility for the difference remains with managers of the investigation.

Given the results above, the analyst can now investigate the similarity of the regression functions. Can they be considered equivalent, that is, do they differ more than one would expect when compared to the way the data scatter about the regression equations themselves? In chapter 10, this meant a comparison of two slopes, or rather a comparison of the difference between two slopes with zero, using a t-test, most likely two-tailed. In this chapter, an F-test is used to examine the group of slopes. If the slopes are essentially equivalent, then the intercepts are examined for equivalency with another F-test. If the slopes are not essentially equivalent, a subset of samples with equivalent slopes can be used in an F-test for intercepts. Multiple comparisons of the response for a specific x can be made whenever the slopes are not equivalent, provided the conditions necessary for valid testing are met. This outlines the analysis used in this chapter.

The focus of the analysis is on the y responses and in particular on the overall sum of squares for the y data, i.e., on

$$\Sigma(y - \bar{\bar{y}})^2$$

where $\bar{\bar{y}}$ is the average for all the y data in the analysis. The symbol $\bar{y}$ will be used as necessary to denote a sample average. The notion for dividing a sum of squares like this into components for an analysis of variance was first introduced in chapters 4 and 5 for comparing more than two samples. It was extended to regression analyses in part II where interest focused on dividing a total sum of squares into a part due to differences in the x data and a remainder that was inherent in the y responses.

In procedures that combine most of the aspects of earlier chapters, the sum of squares of all the y data is divided into components that can be associated with various classifications of the responses or associated with the x variable. The somewhat complex procedure is presented in steps such that the patient reader can duplicate the analysis with nothing more than a hand-held calculator.

The algebraic procedure for dividing a sum into component parts is demonstrated in chapter 20 for a very simple case. A similar algebraic procedure for this much more complex case is not presented in this how-to book. After all, only the results of the algebraic breakup of the total sum of squares is needed, and that is described. Readers interested in the algebraic problem may have difficulty finding an easily understood presentation. One of the more complete is in chapter 11 of Brownlee (1965).

The usual conditions for a valid F-test in this situation are

1. The samples must be independent.
2. The samples must be random.
3. The variances of the residuals about the sample regression equations should be equivalent, i.e., the differences be not significant.
4. The residuals about the sample regression equations should be normally distributed.

Independence and randomness are left to the analyst's judgment, based on what is known about the conditions of the experimental situation. Equivalence of the residual variances and normality of the residuals can be examined analytically and this should be done. The sample sizes are small so it is debatable whether the samples should be examined individually or pooled. They were examined both ways during the preparation of this example.

Table 11.3 presents the Lilliefors procedure for these four small samples combined. The residuals are normalized by sample since the $sd^2_{y|x}$ are not yet known to be equivalent, i.e., to get the unsorted z column, each residual d is divided by the $sd_{y|x}$ for its sample. The null hypothesis tested here is that the data come from normally distributed populations. Any deviations from normality are just chance variation. The probability of a test statistic as large or larger than 0.117, assuming this null hypothesis is true, is greater than 0.20, so there is probably no reason not to accept normality. Incidentally, the Lilliefors statistics for the individual samples, 0.164, 0.222, 0.179, and 0.193, are all less than the tabulated value, 0.233, for a sample size of 8 and a probability of 0.20.

A problem with individual sample tests is, of course, that the probability of rejecting the null hypothesis of no difference from normality increases as the number of tests increases and this invites errors. In technical statistical terminology, multiple tests increase the probability α of a type I error, i.e., rejecting the null hypothesis when no difference exists.

A two-tailed F-statistic is used to compare the variances of two independent samples before making a t-test. Now in this example there are more than two variances to compare so the F-statistic is not appropriate. Bartlett's (1937) procedure is appropriate for comparing more than two variances and gives results equivalent to the two-tailed F-statistic when there are only two samples. The null hypothesis tested is that the samples come from normal populations with equal variances.

Bartlett's procedure is particularly sensitive to nonnormality of the samples. In fact, nonnormality will cause a significant test statistic about as often as will differences between variances. This is not important to the data analyst in this situation since either condition invalidates the F-test in an analysis of variance. Also, the Lilliefors procedure indicates the samples probably did come from normal populations.

TABLE 11.3   The Lilliefors Test for Normality for Example 11.1

d	z	z sorted	Probabilities Normal	Probabilities Sample	da	db
−0.286	−1.277	−1.525	0.064	0.031	0.032	
0.403	1.803	−1.277	0.101	0.062	0.038	0.069
−0.128	−0.570	−1.202	0.115	0.094	0.021	0.052
0.022	0.097	−1.047	0.147	0.125	0.022	0.054
0.201	0.898	−1.012	0.156	0.156	0.000	0.031
−0.210	−0.939	−0.939	0.174	0.187	−0.014	0.018
−0.051	−0.227	−0.918	0.179	0.219	−0.039	−0.008
0.048	0.216	−0.886	0.188	0.250	−0.062	−0.031
0.182	0.573	−0.632	0.264	0.281	−0.017	0.014
−0.282	−0.886	−0.588	0.278	0.312	−0.034	−0.003
0.502	1.577	−0.570	0.284	0.344	−0.059	−0.028
−0.323	−1.012	−0.370	0.356	0.375	−0.019	0.012
−0.188	−0.588	−0.355	0.361	0.406	−0.045	−0.014
−0.292	−0.918	−0.320	0.374	0.437	−0.063	−0.032
0.342	1.075	−0.317	0.376	0.469	−0.093	−0.062
0.057	0.180	−0.227	0.410	0.500	−0.090	−0.059
−0.092	−0.320	−0.173	0.432	0.531	−0.100	−0.068
−0.181	−0.632	−0.137	0.446	0.562	−0.117	−0.086
0.140	0.489	0.097	0.539	0.594	−0.055	−0.024
−0.039	−0.137	0.180	0.572	0.625	−0.053	−0.022
0.192	0.669	0.216	0.586	0.656	−0.071	−0.039
0.522	1.825	0.227	0.590	0.687	−0.098	−0.066
−0.437	−1.525	0.489	0.688	0.719	−0.031	0.000
−0.106	−0.370	0.573	0.717	0.750	−0.033	−0.002
0.053	0.227	0.669	0.748	0.781	−0.033	−0.002
−0.040	−0.173	0.898	0.815	0.812	0.003	0.034
−0.074	−0.317	1.075	0.859	0.844	0.015	0.046
0.302	1.287	1.287	0.901	0.875	0.026	0.057
−0.282	−1.202	1.577	0.943	0.906	0.036	0.068
−0.246	−1.047	1.579	0.943	0.937	0.005	0.037
0.370	1.579	1.803	0.964	0.969	−0.004	0.027
−0.083	−0.355	1.825	0.966	1.000	−0.034	−0.003

NOTE: The Lilliefors test statistic is 0.117. Refer to Chapter 15 for details.

    The Bartlett test statistic for k variances is computed as follows. Let df_i be the degrees of freedom and sd_i be the standard deviation for sample i. Also for convenience, let df be the sum of the degrees of freedom for the samples, i.e.,

$$df = df_1 + df_2 + \cdots + df_k$$

and

$$sd^2 = \Sigma \frac{df_i sd_i^2}{df}$$

Then the test statistic M is

$$M = \frac{1}{c} (\text{df} \ln \text{sd}^2 - \Sigma \text{df}_i \ln \text{sd}_i^2)$$

where ln is the natural logarithm function and

$$c = 1 + \frac{\Sigma(1/\text{df}_i) - 1/\text{df}}{3(k - 1)}$$

M is approximately distributed as chi-square with $k - 1$ degrees of freedom. Bartlett's test statistic is somewhat complex but it can be easily computed with most scientific pocket calculators.

Bartlett's M using the $\text{sd}_{y|x}^2$ for the four samples in this example is 1.073. Assuming the null hypothesis is true and that the observable differences between the sample variances are just chance variation, then the probability of a chi-square statistic with 3 degrees of freedom this large or larger is approximately 0.78, so there is no reason, based on these samples, not to assume the variances about the individual regression equations, $\text{sd}_{y|x}^2$, are equivalent for the analysis of variance.

The analysis thus far can be summarized to keep the procedure in perspective. The four samples all show a decrease in food consumed with time. Least-squares analysis was used to determine the coefficients for a simple linear regression equation for each sample. The slopes of these regression equations are all significantly different from zero. The deviations of the responses from the regression equations appear to be normally distributed and to have equivalent sample variances. This satisfies the requirements for a valid F-test in an analysis of variance.

The calculations for the analysis of variance are made systematic in table 11.4 where the columns have been numbered and labeled to facilitate explanations. A number in parentheses stands for an item in that column and will be used to explain how items in subsequent columns are obtained. For example, each number in column 8 is the corresponding number in column 3 minus the number in column 7. The letter S followed by a number, e.g., S3, will stand for the sum of that column or a result of calculations using these sums. Note that the first six columns repeat sample sums from table 11.2. Six additional calculations, defined at the bottom of table 11.4 in terms of the column sums, complete the preliminary calculations necessary for the ANOVA table.

Since the x data are identical for all four samples in this investigation, several shortcuts in the calculations could have been made. For example, TYX = S12, TXX = S10, and INTERCEPTS = S20. These equalities do not exist for investigations with nonidentical x data, so the procedure demonstrated in table 11.4 does not take advantage of these shortcuts.

TABLE 11.4 Calculations for the Analysis in Example 11.1

	(1) n	(2) Σy	(3) Σy^2	(4) Σx	(5) Σx^2
1	8	59.99	451.2645	36	204
2	8	58.04	422.5566	36	204
3	8	60.19	454.6541	36	204
4	8	56.60	404.2702	36	204
	32	234.82	1732.7454	144	816

	(6) Σxy	(7) $\dfrac{(\Sigma y)^2}{n} \dfrac{(2)(2)}{(1)}$	(8) $\Sigma(y-\bar{y})^2$ $(3)-(7)$	(9) $\dfrac{(\Sigma x)^2}{n} \dfrac{(4)(4)}{(1)}$	(10) $\Sigma(x-\bar{x})^2$ $(5)-(9)$
1	263.27	449.8500	1.41448	162	42
2	255.51	421.0802	1.47640	162	42
3	263.68	452.8545	1.79959	162	42
4	242.68	400.4450	3.82520	162	42
	1025.14	1724.2297	8.51566	648	168

	(11) $\dfrac{\Sigma y\Sigma x}{n} \dfrac{(2)(4)}{(1)}$	(12) $\Sigma(y-\bar{y})(x-\bar{x})$ $(6)-(11)$	(13) $\dfrac{[\Sigma(y-\bar{y})(x-\bar{x})]^2}{\Sigma(x-\bar{x})^2} \dfrac{(12)(12)}{(10)}$	(14) $ss_{y\mid x}$ $(8)-(13)$
1	269.9550	−6.6850	1.06403	0.35045
2	261.1800	−5.6700	0.76545	0.71095
3	270.8550	−7.1750	1.22573	0.57386
4	254.7000	−12.0200	3.44001	0.38519
	1056.6900	−31.5500	6.49522	2.02045

S15 = TYY   = S3 − S2 · S2/S1   = 9.60688

S16 = TYX   = S6 − S2 · S4/S1   = −31.54999

S17 = TXX   = S5 − S4 · S4/S1   = 168.0

S18 = ABOUT   = S15 − S16 · S16/S17 = 3.68187

S19 = SLOPES   = S13 − S12 · S12/S10 = 0.57020

S20   = S7 − S2 · S2/S1   = 1.09121

S21 = INTERCEPTS = S20 − S16 · S16/S17 − S12 · S12/S10 = 1.09121

The procedure presented here is applicable to both situations, where the x data are identical and where they are not.

Preparation of the ANOVA table is divided into two steps to better help the reader understand what is accomplished. First, the data are considered without regard to the sample classification. As one data set, the

32 x, y data pairs can be used to determine the least-squares coefficients for a single regression equation. All of the necessary sums are in table 11.4.

$$b = \frac{S6 - S2 \cdot S4/S1}{S5 - S4 \cdot S4/S1} = \frac{-31.55}{168} = -0.18780$$

$$a = \frac{S2 - b\,S4}{S1} = 8.183$$

The reader may note that b could also be determined as S12/S10, but this coincidence occurs only when the x data are identical for all the samples.

The 32 x, y data pairs have been plotted in figure 11.2 along with this regression equation. An analysis of variance like those in chapter 6 divides the total sum of squares of the y responses into two parts, one that is due to the regression relationship and the other that measures the differences of the data from the regression equation. Table 11.5 presents this ANOVA in terms of the summations in table 11.4, while table 11.6 presents the ANOVA table for this particular example. The reader should remember from chapter 6 that the mean square about regression is just

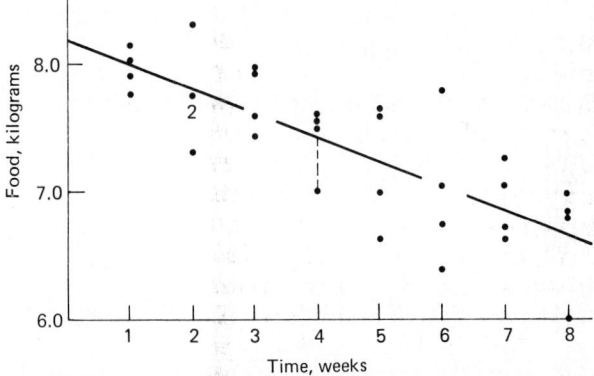

FIGURE 11.2 Example 11.1. Weekly food consumption for the four subgroups showing the plot of the regression equation computed without regard for the subgroup classifications.

TABLE 11.5 The ANOVA Table

Source of variation	df	ss
Due to regression	1	S15 − S18
About regression	S1 − 2	S18
Total	S1 − 1	S15

TABLE 11.6   The ANOVA Table for Example 11.1

Source of variation	df	SS	ms
Due to regression	1	5.9250	5.925
About regression	30	3.6819	0.1227
Total	31	9.6069	

the variance of the deviations of the data about the regression equation. One of the deviations for week 4,

$$d = y - a - bx$$

is identified in figure 11.2. The 2 by a point on the graph for week 2 indicates two coincident points.

The principal research question as stated earlier for this example concerns the four individual samples, not this overall set, so why was an analysis of variance table prepared for the overall data set? A baseline must be established for comparing the individual samples and this analysis of variance provides that baseline.

The analysis now focuses on the separate samples and figure 11.3 associates the plotted points with these samples. In a procedure that combines elements of chapters 4 and 6, the sum of squares about regression is divided into three components.

1. A component associated with the sample locations on the y measurement scale, i.e., the intercepts

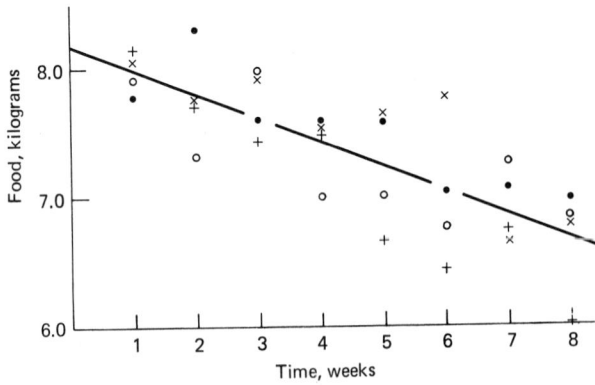

FIGURE 11.3   Example 11.1. Weekly food consumption for subgroups 1 (•), 2 (o), 3 (X), and 4 (+) showing the plot of the regression equation computed without regard for the subgroup classifications.

2. A component associated with the sample x, y relationships, i.e., the orientation or slopes of the sample regression equations

3. A residual, or leftover, portion

All of the preliminary calculations to do this are contained in table 11.4; all that remains is to assemble the ANOVA table.

The construction of the complete analysis of variance is explained in table 11.7 and the table for this example is given in 11.8. Note that the sum of squares about the overall regression line has been divided into three components corresponding to intercepts, slopes, and residual.

TABLE 11.7 The ANOVA Table

Source of variation	df	ss
Between intercepts	$k-1$	S21
Between slopes	$k-1$	S19
About the individual regression lines	$S1-2k$	S14
About the overall regression line	$S1-2$	S18
Due to the overall regression line	1	$S15-S18$
Total	$S1-1$	S15

TABLE 11.8 The ANOVA Table for Example 11.1

Source of variation	df	ss	ms	F
Between intercepts	3	1.0912	0.3637	
Between slopes	3	0.5701	0.1900	2.257
About individual lines	24	2.0206	0.0842	
About the overall line	30	3.6819		
Due to the overall line	1	5.9250		
Total	31	9.6069		

The F-statistic in table 11.8 measuring slope differences tests the null hypothesis that the samples came from populations of x, y data pairs all having the same slope. Assuming this is true, the probability of an F-statistic with 3 and 24 degrees of freedom as large as or larger than 2.257 as a result of chance variation is approximately 0.11. This probability is certainly not small enough to reject the null hypothesis that the slopes are equivalent.

The analysis of example 11.1 has now progressed to a critical point, i.e., the test of the intercepts. Within statistics as a science, there is debate about the validity of a test of intercepts whenever the slopes are not equivalent. The problem was discussed in chapter 10 and the conclusions stated there are restated here. Unless the slopes are equivalent, a test of intercept equivalency is probably an unwise test to make since the intercepts then depend on the location of zero in the x measurement scale. The simple addition of a constant to the x data will produce different results.

Now a second point to consider is this. The procedures in this chapter should be valid for two samples as a special case and should reduce to procedures given in chapter 10 when only two samples are analyzed. Some statistics books advise comparing the intercept mean square with the mean square about the individual regression lines. For example 11.1, the calculation would be

$$F = \frac{S21/(k-1)}{S14/(S1-2k)} = \frac{0.3637}{0.0842} = 4.319$$

This procedure is not equivalent for two samples to procedures given in chapter 10. For the procedures to be equivalent, the mean square for intercepts must be compared to the mean square obtained by pooling the sum of squares for slopes and the sum of squares about the individual regression lines after accepting the null hypothesis of equivalent slopes. This pooling produces a sum of squares for all samples about individual regression equations having a common slope but different intercepts. The reader can review chapter 10 to verify that this was the procedure used there. The F-statistic for a test of intercept equivalency is then properly computed as follows.

$$F = \frac{S21/(k-1)}{(S14 + S19)/(S1 - 2k + k - 1)}$$

$$= \frac{0.3637}{(2.0206 + 0.5701)/(24 + 3)} = 3.790$$

This is summarized in the analysis of variance in table 11.9. The null hypothesis tested here is that the samples came from populations that not only have equal slopes but also have equal intercepts. Assuming the observable sample differences are just due to chance, the probability of

TABLE 11.9 The ANOVA Table for Example 11.1

Source of variation	df	ss	ms	F
Between intercepts	3	1.0912	0.3637	3.790
About the individual lines	27	2.5907	0.0960	
About the overall line	30	3.6819		
Due to the overall line	1	5.9250		
Total	31	9.6069		

an *F*-statistic with 3 and 27 degrees of freedom as large or larger than 3.790 is approximately 0.023.

To assist the reader in following these procedures for samples 1 and 2 only, i.e., example 10.1, three tables are presented. Table 11.10 gives the preliminary calculations corresponding to table 11.4. Tables 11.11 and 11.12 give the analysis of variance corresponding to tables 11.8 and 11.9 for example 11.1. In comparing the *F*-statistics given in these tables with the *t*-statistics in chapter 10, remember that the square of a *t*-statistic with *n* degrees of freedom is identical to an *F*-statistic with 1 and *n* degrees of freedom.

TABLE 11.10 Calculations for the Analysis for Example 11.1

	(1) n	(2) Σy	(3) Σy^2	(4) Σx	(5) Σx^2
1	8	59.99	451.2645	36	204
2	8	58.04	422.5566	36	204
	16	118.03	873.8211	72	408

	(6) Σxy	(7) $\dfrac{(\Sigma y)^2}{n}$ $\dfrac{(2)(2)}{(1)}$	(8) $\Sigma(y - \bar{y})^2$ $(3) - (7)$	(9) $\dfrac{(\Sigma x)^2}{n}$ $\dfrac{(4)(4)}{(1)}$	(10) $\Sigma(x - \bar{x})^2$ $(5) - (9)$
1	263.27	449.8500	1.41448	162	42
2	255.51	421.0802	1.47640	162	42
	518.78	870.9302	2.89088	324	84

	(11) $\dfrac{\Sigma y \Sigma x}{n}$ $\dfrac{(2)(4)}{(1)}$	(12) $\Sigma(y - \bar{y})(x - \bar{x})$ $(6) - (11)$	(13) $\dfrac{[\Sigma(y - \bar{y})(x - \bar{x})]^2}{\Sigma(x - \bar{x})^2}$ $\dfrac{(12)(12)}{(10)}$	(14) $ss_{y|x}$ $(8) - (13)$
1	269.9550	−6.6850	1.06403	0.35045
2	261.1800	−5.6700	0.76545	0.71095
	531.1350	−12.3550	1.82948	1.06140

S15 = TYY   = 3.12854

S16 = TYX   = −12.35499

S17 = TXX   = 84.00000

S18 = ABOUT   = 1.31132

S19 = SLOPES   = 0.01226

S20   = 0.23766

S21 = INTERCEPTS = 0.23766

TABLE 11.11   The ANOVA Table for Example 11.1

Source of variation	df	ss	ms	F
Between intercepts	1	0.2377	0.2377	
Between slopes	1	0.0123	0.0123	0.139
About individual lines	12	1.0614	0.0885	
About the overall line	14	1.3113		
Due to the overall line	1	1.8172		
Total	15	3.1285		

TABLE 11.12   The ANOVA Table for Example 11.1

Source of variation	df	ss	ms	F
Between intercepts	1	0.2377	0.2377	2.878
About the individual lines	13	1.0737	0.0826	
About the overall line	14	1.3113		
Due to the overall line	1	1.8172		
Total	15	3.1285		

The analysis can be summarized to this point as follows. Coefficients for a linear regression equation were estimated for the four samples. The individual slopes were compared to zero and the null hypothesis of no difference was rejected for all four equations. An analysis of variance (table 11.8) was used to compare these slopes. The F-test did not reject the null hypothesis of no differences between the slopes. Assuming the x, y relationships were equivalent, an analysis of variance (table 11.9) was used to examine the equivalency of the intercepts. This F-test rejected the null hypothesis of no intercept differences.

Multiple Comparison of Intercepts, Duncan

It is usually not sufficient to determine that sample intercepts are not equivalent. Managers and researchers need to know something about the nature of the differences, i.e., which samples have equivalent intercepts and which have intercepts that may be considered different. Intercepts take the place of sample averages in the comparison of more than two x, y relationships. If the intercepts cannot be considered equivalent, then there are differences in the location of the samples on the y measurement scale. The nature of these differences is the focus of the remainder of this analysis.

Steps in the procedure, after determining that the sample slopes are equivalent, are as follows.

1. Determine a common slope.
2. Determine the sample intercepts using the common slope.
3. Determine the pooled $sd^2_{y|x}$ of the responses about their respective regressions.
4. Using the above statistics, do multiple paired comparisons of the intercepts.

The common slope, in terms of the notation used in table 11.4, is S12/S10, i.e., $b = -0.1878$. Then using this slope, the intercepts for each sample are

$$\frac{(2) - b(4)}{(1)}$$

i.e.,

$$\frac{59.99 + 0.1878 \cdot 36}{8} = 8.344 \qquad \text{for sample 1}$$

$$\frac{58.04 + 0.1878 \cdot 36}{8} = 8.100 \qquad \text{for sample 2}$$

$$\frac{60.19 + 0.1878 \cdot 36}{8} = 8.369 \qquad \text{for sample 3}$$

$$\frac{56.60 + 0.1878 \cdot 36}{8} = 7.920 \qquad \text{for sample 4}$$

The four parallel regression equations are

$$y_1 = 8.344 - 0.1878x_1$$

$$y_2 = 8.100 - 0.1878x_2$$

$$y_3 = 8.369 - 0.1878x_3$$

$$y_4 = 7.920 - 0.1878x_4$$

Finally, the pooled $sd^2_{y|x}$ about these four parallel regression equations is simply the mean square for "about the individual regression lines" in table 11.9. The quantity at the same location in table 11.8 is the pooled $sd^2_{y|x}$ about the four regression equations using the individual slopes, not the common slope.

Any of the multiple comparison procedures from chapters 4 and 5 can be used, depending on the particular characterstics the analyst wants. Duncan's is chosen for this example. To begin Duncan's procedure, the intercepts are written down in either ascending or descending order. A critical range (CR) is computed for pairs of intercepts as follows:

$$CR = q \sqrt{\frac{ms}{n}}$$

where ms = the residual ms from the ANOVA table, i.e., $sd^2_{y|x}$
       n = the sample size
       q = Duncan's special studentized range statistic for r samples, as explained below; df, the degrees of freedom for ms; and the significance level chosen for the tests.

This particular form of the Duncan procedure is applicable for intercepts, only when the x data are identical for the samples. A different formulation is given later in this chapter for use when the x are not identical.

The q-statistics are found in table Q of part V. The number of samples spanned by a comparison including the two samples compared is called r. For ordered samples that are adjacent, $r = 2$; $r = 4$ when there are two samples between the two being compared.

The sample intercepts are compared in the following order: the largest to the smallest, the largest to the second smallest, and so on until an observed range is less than the critical range. When this happens, the comparisons with the largest cease and all intercepts between the two in this last test are declared not significantly different from each other. The process is repeated in turn for the second largest, the third largest, and so on until an intercept not significantly different from the smallest is found. At this point, all comparisons cease.

The four intercepts are presented in ascending order in table 11.13. Using a significance level of 0.05, the critical range for samples 4 and 3 is computed as follows.

$$CR = 3.15 \sqrt{\frac{0.095952}{8}} = 0.345$$

TABLE 11.13   Multiple Comparison of Intercepts for Example 11.1 (Duncan's procedure: significance at 0.05 noted)

	Sample Intercept	
4	7.920	
2	8.100	
1	8.344	
3	8.369	

The intercepts for samples 4 and 3 differ by more than their critical range so the difference is significant at the 0.05 level. The critical range for samples 2 and 3 is computed next.

$$CR = 3.07 \sqrt{\frac{0.095952}{8}} = 0.336$$

The intercepts for samples 2 and 3 differ by less than their critical range so the difference is not significant at the 0.05 level. Duncan's rule is to stop comparison with the largest, i.e., assume the intercepts for samples 1, 2, and 3 are equivalent, and then to compare sample 4 with sample 1.

The difference between intercepts for samples 4 and 1 is 0.424. This exceeds the critical range for three intercepts computed above, 0.336, so these intercepts are significantly different at the 0.05 level.

The intercept for sample 2 is not compared with the intercept for sample 1 since Duncan's procedure is that no difference between two intercepts can be declared significant if the two intercepts concerned are both contained in a subset of intercepts which has a nonsignificant range. This is the case for the difference between samples 2 and 1 since it is contained in the nonsignificant range for samples 2 and 3.

Finally, the critical range for samples 4 and 2 is computed as follows.

$$CR = 2.92 \sqrt{\frac{0.095952}{8}} = 0.320$$

The difference between the intercepts, 0.180, is less than this critical range, so the intercepts for samples 4 and 2 do not differ significantly at the 0.05 level.

The result of all this is summarized by the vertical lines on the right of table 11.13. These lines join intercepts found not significantly different at the 0.05 level. First, it was determined that sample 4 but not sample 2 differed from sample 3. The first vertical line denotes this result. Then it was determined that sample 4 differed significantly at the 0.05 level from sample 1 but not from sample 2. This last equivalency is denoted by the second vertical line.

It may seem a bit strange that sample 2 does not differ significantly from any sample above or below it but that both samples above differ significantly from the one below sample 2. This situation can be rationalized this way. The populations probably have different intercepts, but the sample data are not sufficiently different when compared to the residual variance to justify this conclusion. Presenting the intercepts ordered as in table 11.13 with the vertical lines to designate groups that do not differ significantly at the chosen probability level makes a complex situation easier to comprehend and perhaps easier to explain.

A final summary of the analysis will help to put the procedure in perspective. The four samples all show a decrease in food consumed with time. Least-squares analysis was used to determine the coefficients for a simple linear regression equation for each sample. The slopes of these regression equations are all significantly different from zero. The deviations of the responses from the regression equations appear to be normally distributed and to have equivalent sample variances. This satisfies the requirements for a valid F-test in an analysis of variance. An F-statistic was used to measure the differences among the slopes and these were found to be equivalent. The ANOVA table was modified to reflect this and an F-statistic was used to measure the differences among the intercepts. Unlike the test of slopes, this test indicated significant differences among the intercepts. The nature or structure of these differences was determined by making multiple paired comparisons of the intercepts, assuming a common slope, using Duncan's procedure.

This completes the analysis for answering most questions associated with the comparison of more than two regressions, i.e., how do the functions differ and how are they equivalent with respect to slope and location. Other issues could be raised such as confidence limits for population parameters and for the differences between populations but these questions occur infrequently. More important problems arise with respect to unsatisfied conditions for the t- and F-tests.

1. What can the analyst do if the variances about the regression equations are not equivalent?
2. What can be done if the residuals are not normally distributed?
3. What should be done when there is a secondary classification for the samples as in chapter 5?
4. What should be done when the slopes are not equivalent?

The last of these questions is addressed in example 11.2.

Example 11.2

When subgroups like those considered in example 11.1 are generally homogeneous for many studies, managers can evaluate ways for improving their studies. One desirable improvement would be a smaller decrease in food consumption. Example 11.2 utilizes four subgroups fed different diets in an effort to mask the taste of the test product. Actually subgroup 1 was one of three subgroups fed the standard diet, but to keep this example manageable, only one of the three is used. The other three subgroups for the example were fed nonstandard diets during the study to determine, as a secondary purpose, if a different diet might mask the taste of the test product better than the standard one. The diets were very similar in nutrition, bulk, appearance, etc.

The subgroups received the same concentration of the test product initially, but as food consumption decreased, the concentration for each diet was increased to keep the dose more or less constant. The data for this example are given in table 11.14. The research questions for this secondary investigation are the same as they were for example 11.1, i.e., did the food consumption fall off about the same for the subgroups?

TABLE 11.14   Data for Example 11.2

Sample 1		*Sample 2*		*Sample 3*		*Sample 4*	
x	*y*	*x*	*y*	*x*	*y*	*x*	*y*
1	12.40	1	12.42	1	12.64	1	12.39
2	12.31	2	12.31	2	12.43	2	12.17
3	11.85	3	12.42	3	11.70	3	11.89
4	11.35	4	12.18	4	11.71	4	11.91
5	11.97	5	12.16	5	11.65	5	11.35
6	11.39	6	12.33	6	11.68	6	11.48
7	11.42	7	12.08	7	11.17	7	11.20
8	10.99	8	11.88	8	11.07	8	10.85
9	11.10	9	11.69	9	10.62	9	10.96
10	10.88	10	11.98	10	10.62	10	10.90

NOTE: x = time in weeks; y = food consumed (in kilograms).

The data for example 11.2 are plotted in figure 11.4. These plots show considerable variability, but all have a downward drift. With nothing more, the analyst can see the direction the analysis will take. Sample 2 seems to have a slightly less steep decline than do the other samples. The question is, could this decline have occurred because of chance or random causes, or is the difference from the other three sufficient to indicate that diet 2 just might mask the test product better? Eventually a test of the null hypothesis that the slopes are equivalent will be made. The difference that can be easily seen in these plots may be sufficient to cause the null hypothesis to be rejected.

It is hard to see enough form in these plots to justify any regression function more complex than a simple linear equation, i.e.,

$$y = a + bx$$

The usual useful sample-summarizing summations are presented in table 11.15. Then the least-squares slope and intercept are calculated for each sample using the equations given in example 11.1. These results are

Sample 1	*Sample 2*	*Sample 3*	*Sample 4*
$b_1 = -0.1625$ $a_1 = 12.460$	$b_2 = -0.0675$ $a_2 = 12.516$	$b_3 = -0.2157$ $a_3 = 12.715$	$b_4 = -0.1762$ $a_4 = 12.479$

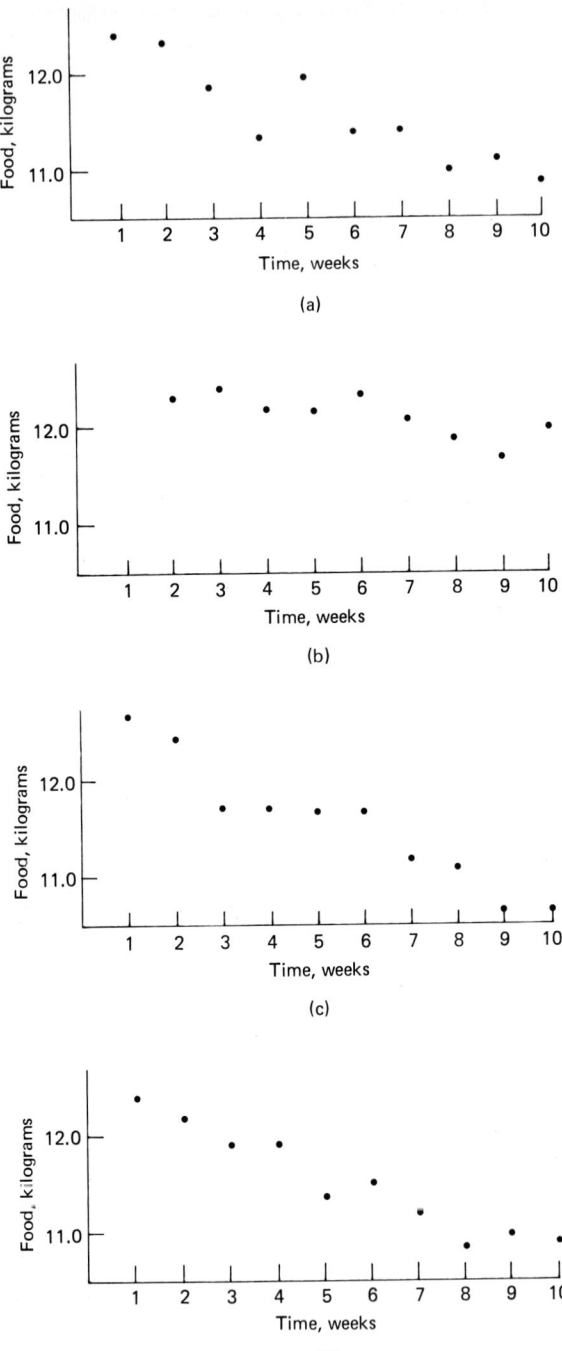

FIGURE 11.4 Example 11.2. Weekly food consumption for (*a*) subgroup 1; (*b*) subgroup 2; (*c*) subgroup 3; (*d*) subgroup 4.

11.24

TABLE 11.15   The Usual Useful Sample Summarizing Summations for Example 11.2

	Sample 1	Sample 2	Sample 3	Sample 4
Σy	115.66	121.45	115.29	115.10
Σy^2	1340.3350	1475.5331	1333.3161	1327.5482
Σx	55	55	55	55
Σx^2	385	385	385	385
Σxy	622.72	662.41	616.30	618.51
n	10	10	10	10

The food consumed declined in all the groups, but it declined in subgroup 2 less than half as fast as it declined in the other subgroups, and declined somewhat faster in subgroup 3 than in subgroups 1 and 4. At this point, one is tempted simply to do multiple comparisons of the slopes, but for this example, a more complete procedure employing an analysis of variance will be used.

The next step is to measure the deviation of the data points about the regression equations defined by the coefficients given above. For example, the deviation of the fifth week for subgroup 1 is

$$d = y - a_1 - b_1 x$$
$$= 11.97 - 12.46 - (-0.1625)5$$
$$= 0.32$$

The formula for calculating the standard deviation of these deviations, i.e., for calculating $sd_{y|x}$, has been presented earlier in this chapter. The results along with t-statistics for comparing each subgroup slope to zero are

	Sample 1	Sample 2	Sample 3	Sample 4
$sd_{y|x}$	0.2323	0.1357	0.1934	0.1519
se_b	0.02588	0.01495	0.02130	0.01673
t	−6.353	−4.513	−10.127	−10.532

Much prestudy experience indicates that food consumption will fall off with time no matter what diet is fed, so there is good reason to believe these will do the same. The null hypothesis tested for each of the samples is that the sample is from a population with slope not less than zero. Consequently, these are one-tailed t-statistics that, when compared to table B in part V, give the following probabilities, assuming the null hypothesis is true and that the apparent relationships are just due to chance.

	Sample 1	Sample 2	Sample 3	Sample 4
p	0.0007	0.001	0.000005	0.000004

All of these probabilities are very, very small, so it is reasonable to assume, on the basis of these samples, that food consumption declines for all diets.

Given the above results, the next logical question concerns the similarity of the slopes. This is the question anticipated by the plots of the data in figure 11.4. Can they be considered equivalent, and if not, how do they differ? In example 10.2, this meant a comparison of the difference of two slopes with zero using a two-tailed test. There are more than two slopes in this example, so an *F*-test must be used for examining the equivalency of the slopes as a group.

The analysis now focuses on the overall sum of squares for the *y* data, i.e., on

$$\Sigma(y - \bar{\bar{y}})^2$$

where $\bar{\bar{y}}$ is the average of all the *y* data in all the samples. As with example 11.1, $\bar{y}$ will denote a sample average. This total sum of squares is first partitioned into two components that can be identified with an overall regression equation. These components are a sum of squares resulting from the relationship of the responses to time and a residual sum of squares about this overall regression equation. The sum of squares about the overall regression is then further divided into components identified with intercept differences, slope differences, and a leftover residual. This division into components is an analysis of variance.

The only data condition required for an analysis of variance is that the data be measured in a scale where addition and multiplication are defined. Even ranks satisfy this condition, provided the results are not then transformed to the ratio or interval scales. The *F*-tests that are always associated with an analysis of variance, however, are another matter. *F*-tests of the components require that these conditions be true.

1. The samples must be independent.
2. The samples must be random.
3. The variances of the residuals about the sample regression equations should be equivalent, i.e., the differences be not significant.
4. The residuals about the sample regression equations should be normally distributed.
5. Then the *F*-test for the intercepts requires that the slopes be equivalent.

Independence and randomness are left to the analyst's judgment based on what is known about the conditions of the experimental situation that produced the data, but the equivalence of the residual variances and the normality of the residuals are difficult even for much experienced analysts to judge so they should be examined analytically.

The Shapiro-Wilk (1965, 1968) test for normality introduced in chapter 3 will be used for this example. The only requirement for using the Shapiro-Wilk procedure is that the data, in this case the differences of the y responses from the regression equations, be a random sample. The null hypothesis tested is that the sample is from a normal distribution with an unspecified average and variance.

The Shapiro-Wilk procedure requires the calculation of three summations after first sorting the data in ascending order. These summations are

$$D = \sum_{i=1}^{n} (d_i - \bar{d})^2$$

$$c_1 = \sum_{i=1}^{k} a_i d_{n-i+1}$$

and

$$c_2 = \sum_{i=1}^{k} a_i d_i$$

where d = the differences between the y responses and the regression equations sorted
k = $n/2$ rounded down
a = Shapiro-Wilk factors from table N1 of part V.

The test statistic is then

$$\pi = \frac{(c_1 - c_2)^2}{D}$$

Table N2 in part V is used to determine the probability of a π smaller than an observed π, assuming the null hypothesis is true and the observable differences from normality are just chance variation.

Most test statistics are defined so that significance increases as the test statistic increases, i.e., large test statistics are evidence against the null hypothesis. The Shapiro-Wilk statistic is one that behaves in the opposite way. This test statistic was defined so that significance increases as the test statistic decreases, i.e., small test statistics are evidence against the null hypothesis. Consequently, the analyst must be particularly careful in using table N2 of part V.

Table 11.16 presents the Shapiro-Wilk calculations. The test statistics are all large, implying little or no evidence against the null hypothesis. The probabilities of smaller π for each sample, assuming only chance variation, are given in the last line of table 11.16.

Bartlett's M for comparing the variances about the regression equations, i.e., comparing the $sd^2_{y|x}$, is 2.644. M is distributed approximately as chi-squared with degrees of freedom equal to one less than the number of samples. Assuming the population variances are equal and the observable sample differences are just due to chance, then the probability of a chi-squared statistic with 3 degrees of freedom larger than 2.644 is 0.46. This is a large probability so there is no reason to reject the hypothesis that the variances are equivalent.

Thus far, the analysis has shown that all the subgroups had a highly significant decrease in the food consumed. Furthermore, the deviations about the individual regression equations are normally distributed and the variances are equivalent. This satisfies the conditions for the F-tests, analysis of variance, so the ANOVA table can be constructed.

The calculations necessary for preparing the table are detailed in table 11.17. In the column headings, a number in parentheses stands for an item in the column with that number. Thus each item in column 8 is the corresponding item in column 3 minus the item in column 7, i.e., (8) = (3) − (7). A number preceded by S stands for the sum of the column with that number. The total sum of squares for all the y data is

$$S3 - S2 \cdot \frac{S2}{S1}$$

The column sums are used at the bottom of table 11.17 to complete the preliminary calculations for the analysis of variance. Using the instructions contained in table 11.7, the ANOVA table is constructed and presented as table 11.18.

The F-statistic for slopes in the ANOVA table is very large. The probability of a larger value is less than 0.005. Assuming that the observable slope differences are just chance variation, this rejects the null hypothesis that the slopes are equivalent.

Multiple Comparison of Slopes, Newman-Keuls

After plotting the data at the example's beginning, the slopes did not appear equivalent and the analysis of variance verifies that they are not. An analyst might at this point assume the slopes, except for b_2, are equivalent, but this hypothesis is examined analytically here. To do this, multiple comparisons are made using the Newman-Keuls procedure.

TABLE 11.16 Shapiro-Wilk Test for Normality for Example 11.2

Shapiro-Wilk statistic	Sample 1			Sample 2			Sample 3			Sample 4		
	x	d	Product	x	d	Product	x	d	Product	x	d	Product
0.5739	4	−0.460	−0.264	9	−0.219	−0.126	3	−0.368	−0.211	5	−0.248	−0.142
0.3291	8	−0.170	−0.056	8	−0.096	−0.032	9	−0.154	−0.051	8	−0.219	−0.072
0.2141	3	−0.122	−0.026	2	−0.071	−0.015	4	−0.143	−0.031	3	−0.061	−0.013
0.1224	6	−0.095	−0.012	4	−0.066	−0.008	7	−0.035	−0.004	7	−0.046	−0.006
0.0399	10	0.045	0.002	1	−0.029	−0.001	5	0.013	0.001	2	0.043	0.002
		$c_1 = -0.356$			$c_1 = -0.182$			$c_1 = -0.296$			$c_1 = -0.231$	

Shapiro-Wilk statistic	Sample 1			Sample 2			Sample 3			Sample 4		
	x	d^-	Product	x	d	Product	x	d	Product	x	d	Product
0.0399	7	0.098	0.004	5	−0.019	−0.001	10	0.062	0.002	6	0.058	0.002
0.1224	1	0.103	0.013	7	0.036	0.004	8	0.080	0.010	9	0.067	0.008
0.2141	9	0.103	0.022	3	0.106	0.023	1	0.140	0.030	1	0.087	0.019
0.3291	2	0.175	0.058	10	0.139	0.046	2	0.146	0.048	4	0.136	0.045
0.5739	5	0.323	0.185	6	0.219	0.126	6	0.259	0.149	10	0.183	0.105
		$c_2 = 0.282$			$c_2 = 0.198$			$c_2 = 0.239$			$c_2 = 0.179$	
		SSD = 0.432			SSD = 0.148			SSD = 0.299			SSD = 0.185	
		$\pi = 0.940$			$\pi = 0.976$			$\pi = 0.957$			$\pi = 0.912$	
		$p = 0.52$			$p = 0.93$			$p = 0.72$			$p = 0.35$	

TABLE 11.17   Calculations for the Analysis in Example 11.2

	(1) n	(2) Σy	(3) Σy^2	(4) Σx	(5) Σx^2
1	10	115.66	1340.3350	55	385
2	10	121.45	1475.5331	55	385
3	10	115.29	1333.3161	55	385
4	10	115.10	1327.5482	55	385
	40	467.50	5476.7325	220	1540

	(6) Σxy	(7) $\dfrac{(\Sigma y)^2}{n}$ $\dfrac{(2)(2)}{(1)}$	(8) $\Sigma(y-\bar{y})^2$ (3) − (7)	(9) $\dfrac{(\Sigma x)^2}{n}$ $\dfrac{(4)(4)}{(1)}$	(10) $\Sigma(x-\bar{x})^2$ (5) − (9)
1	622.72	1337.7236	2.61144	302.5	82.5
2	662.41	1475.0102	0.52287	302.5	82.5
3	616.30	1329.1784	4.13770	302.5	82.5
4	618.51	1324.8010	2.74720	302.5	82.5
	2519.94	5466.7133	10.01921	1210	330

	(11) $\dfrac{\Sigma y \Sigma x}{n}$ $\dfrac{(2)(4)}{(1)}$	(12) $\Sigma(y-\bar{y})(x-\bar{x})$ (6) − (11)	(13) $\dfrac{[\Sigma(y-\bar{y})(x-\bar{x})]^2}{\Sigma(x-\bar{x})^2}$ $\dfrac{(12)(12)}{(10)}$	(14) $ss_{y \mid x}$ (8) − (13)
1	636.1300	−13.4100	2.17973	0.43171
2	667.9750	−5.5650	0.37538	0.14748
3	634.0950	−17.7950	3.83833	0.29937
4	633.0500	−14.5400	2.56256	0.18464
	2571.2500	−51.3100	8.95601	1.06320

S15 = TYY   = S3 − S2 · S2/S1   = 12.82618

S16 = TYX   = S6 − S2 · S4/S1   = −51.31

S17 = TXX   = S5 − S4 · S4/S1   = 330.0

S18 = ABOUT   = S15 − S16 · S16/S17 = 4.84826

S19 = SLOPES   = S13 − S12 · S12/S10 = 0.97808

S20   = S7 − S2 · S2/S1   = 2.80697

S21 = INTERCEPTS = S20 − S16 · S16/S17 − S12 · S12/S10 = 2.80697

A Newman-Keuls critical range is calculated for each pair of slopes using the formula

$$CR = q \sqrt{\frac{ms}{2}\left(\frac{1}{\Sigma(x_i-\bar{x}_i)^2} + \frac{1}{\Sigma(x_j-\bar{x}_j)^2}\right)}$$

TABLE 11.18 The ANOVA Table for Example 11.2

Source of Variation	df	ss	ms	F
Between intercepts	3	2.8070	0.9357	
Between slopes	3	0.9781	0.3260	9.813
About individual lines	32	1.0632	0.03323	
About the overall line	38	4.8483		
Due to the overall line	1	7.9779		
Total	39	12.8262		

where ms = the residual mean square from the ANOVA table
i and *j* = the samples being compared
q = the studentized range statistic for *r* samples, as explained below; df, the degrees of freedom for ms; and the significance level chosen for the tests.

This is the general form of the Newman-Keuls critical range for slopes. When the *x* data are identical for all of the samples as they are in this example, the formula can be simplified to the following.

$$\text{CR} = q \sqrt{\frac{\text{ms}}{\Sigma(x - \bar{x})^2}}$$

The *q*-statistics are found in table R of part V. The number of samples spanned by a comparison including the two samples compared is called *r*. For ordered samples that are adjacent, $r = 2$; $r = 4$ when there are two samples with slopes between the two being compared.

The critical ranges are computed using the formula above, $\text{sd}^2_{y|x}$ from the analysis of variance, and factors from table R in part V. The residual sum of squares, $\text{sd}^2_{y|x}$, from table 11.18 has 32 degrees of freedom, so linear interpolation must be used with table R to estimate factors for this example. For a confidence level of 0.05, these are

2.884   for $r = 2$
3.480   for $r = 3$
3.838   for $r = 4$

To illustrate the procedure, the critical range for $r = 3$ is

$$\text{CR} = 3.48 \sqrt{\frac{0.0332}{82.5}} = 0.070$$

The three critical ranges needed for the slopes are
0.058   for $r = 2$
0.070   for $r = 3$
0.077   for $r = 4$

The actual slope differences in table 11.19 are all smaller than their respective critical ranges except for all comparisons with subgroup 2. All ranges with subgroup 2 exceed their respective critical ranges.

TABLE 11.19 Multiple Comparison of Slopes for Example 11.2 (Newman-Keuls procedure: significance at 0.05 noted)

Sample	Slope
2	−0.0675
1	−0.1625
4	−0.1762
3	−0.2157

The analysis has now verified what was suspected when the plots were prepared, i.e., subgroup 2 had a smaller decline in food consumed than did the other three samples. One more thing is known. After computing the individual slopes, subgroup 3 looked as if it might be declining faster than subgroups 1 and 4. This apparently is not the case, since the critical range for subgroup 4 and 3 is 0.058 while the actual range, 0.0395, is smaller.

Since the slopes are not all equivalent, it would be inappropriate to compare intercepts for the four subgroups, but sample 2 can be dropped and intercepts for the remaining three compared. The sums in table 11.17 are modified by subtracting out sample 2 and then making new calculations for S15 through S21.

$$S15 = TYY \qquad = S3 - S2 \cdot \frac{S2}{S1} \qquad = 9.51257$$

$$S16 = TYX \qquad = S6 - S2 \cdot \frac{S4}{S1} \qquad = -45.74499$$

$$S17 = TXX \qquad = S5 - S4 \cdot \frac{S4}{S1} \qquad = 247.5$$

$$S18 = ABOUT \qquad = S15 - S16 \cdot \frac{S16}{S17} = 1.05760$$

$$S19 = SLOPES \qquad = S13 - S12 \cdot \frac{S12}{S10} = 0.12566$$

$$S20 \qquad\qquad = S7 - S2 \cdot \frac{S2}{S1} \qquad = 0.01622$$

$$S21 = INTERCEPTS = S20 - S16 \cdot \frac{S16}{S17} - S12 \cdot \frac{S12}{S10} = 0.01622$$

The analysis of variance prepared from the above is given in table 11.20. This time the slopes are equivalent, as evidenced by the small F-statistic for slopes ($p = 0.23$). The sum of squares for slopes is pooled with the sum of squares about the individual lines for the analysis of variance in table 11.21. This F-statistic measuring differences between intercepts is very, very small so subgroups 1, 3, and 4 may be represented by a single regression function.

TABLE 11.20   The ANOVA Table for Example 11.2

Source of variation	df	ss	ms	F
Between intercepts	2	0.01622	0.00811	
Between slopes	2	0.12566	0.06283	1.647
About individual lines	24	0.91572	0.03816	
About the overall line	28	1.0576		
Due to the overall line	1	8.4550		
Total	29	9.5126		

TABLE 11.21   The ANOVA Table for Example 11.2

Source of variation	df	ss	ms	F
Between intercepts	2	0.01622	0.00811	0.202
About individual lines	26	1.04138	0.04005	
About the overall line	28	1.0576		
Due to the overall line	1	8.4550		
Total	29	9.5126		

The common slope computed from table 11.17 sums after subgroup 2 has been removed is

$$b = \frac{S12}{S10} = \frac{-45.7450}{247.5} = -0.1848$$

and the intercept is

$$a = \frac{S2 - b \cdot S4}{S1} = \frac{346.05 + 0.1848 \cdot 165}{30} = 12.55$$

The analysis for example 11.2 has been summarized as it progressed, but a final summary will help the reader in reporting results. Four subgroups were fed different diets during an investigation of adverse reactions to a test product. The purpose for using different diets was to determine which one, if any, masked the taste of the test product best. The response for this secondary investigation was the food consumed per week by each of the subgroups. The food consumption for all subgroups was expected to decline and this the analysis shows. One subgroup, 2, declined at a lower rate than the other three and the difference was statistically significant. Subgroup 2 was dropped, so the intercepts could be compared for the other three. This final test indicated the intercepts are equivalent, so a common slope and intercept were calculated for the three equivalent subgroups. If the only differences between subgroups were the diets, then it appears the diet fed subgroup 2 did mask the test product better than the other three. One can now see why a history of homogeneous responses is essential. Without such a history, who would be willing to bet on the diet fed subgroup 2?

Since a single regression equation can be used to represent the decrease in food consumption for three of the subgroups, the researcher might at this point ask the data analyst, "At what time did the food consumed by subgroup 2 differ significantly from the other three subgroups?" The food consumed by these subgroups can be averaged for each week and then the differences with subgroup 2 computed. Finally, these differences w can be analyzed as the differences in example 10.2 were analyzed.

Example 11.3

Infants normally experience a weight loss during the first 4 or 5 days after birth. This loss is greatest at about 3 days and is usually regained within a week. Physicians have observed that this weight loss is generally larger in breast-fed than in formula-fed infants. Example 11.3 compares the weight loss of four groups of newborns. The first group was breast-fed; the other three were formula-fed with slightly different formulas being used. The causes and effects of this weight loss, or the advantages of one feeding method over another, are not the subject of this investigation. The principal research issue concerns the weight loss for the formula-fed newborns as compared to the weight loss for breast-fed newborns.

Some points on randomization should be discussed before describing the study more completely. The assignment of newborns to the breast-fed or to the formula-fed group was not random. The mothers made this decision, perhaps with the advice of their physicians, but it was not random assignment in the usual sense. The result of this assignment procedure was that the breast-fed newborns may have differed from the group of formula-fed newborns in more ways than simple feeding technique. For example, mothers without the potential to feed adequately were probably advised to use formula. Thus there may have been marginally fed newborns in the breast-fed group that could tend to pull the breast-fed group as a whole down. The probability that an infant would be marginally formula-fed is much smaller.

Once the decision to formula-feed was made, assignment to one of the three formulas being investigated could have been random, but it was unclear in this study that assignment was entirely at random. It was suspected that at least some of the infants were assigned to a particular formula for infant-related reasons. This could account for group 2 birth weights averaging over 150 grams less than any other group. While an analysis of variance to compare the birth weights for the four groups indicates equivalency ($p = 0.24$), a simple t-test comparing sample 2 to the other 65 newborns without regard for groups almost contradicts the analysis of variance ($p = 0.054$). Any interpretation of the results should be qualified with this finding.

Table 11.22 presents the birth weights and the weight losses 72 hours post partum for the four groups. Two characteristics of the data distinguish this example from the other two in this chapter. First, the x data are not identical for the samples, and second, the number of data are not the same in each group. The study protocol called for 25 newborns in each group, but every study involving humans experiences drop outs, and this occurred for three of the four groups. The reasons are unknown, and

TABLE 11.22   Data for Example 11.3

Sample 1 Breast-fed		Sample 2 Formula 1		Sample 3 Formula 2		Sample 4 Formula 3	
x	y	x	y	x	y	x	y
274	10	336	15	394	23	367	20
376	25	324	13	318	3	261	3
355	28	385	23	344	-3	323	-4
357	15	306	0	345	20	311	10
336	20	316	0	327	7	237	-2
272	11	330	14	362	5	318	14
309	30	364	14	306	17	328	19
336	20	328	-6	306	6	370	8
379	17	278	11	351	13	318	13
375	31	321	3	361	11	387	12
359	25	224	7	287	6	347	9
296	9	297	-7	383	25	353	10
300	2	311	10	335	5	367	5
333	18	339	22	239	0	345	5
396	14	249	3	343	17	305	2
315	12	326	16	371	17	374	11
369	23	267	0	281	2	266	1
367	10	320	5	328	3	327	21
339	1	370	22			342	11
347	30	304	2			317	12
320	10	292	5			396	13
319	16	345	11			335	12
						281	-4
						317	2
						370	13

NOTE: x = birth weight in decigrams; y = weight loss in decigrams.

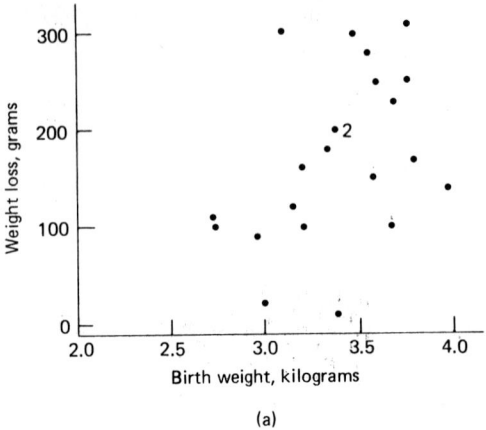

(a)

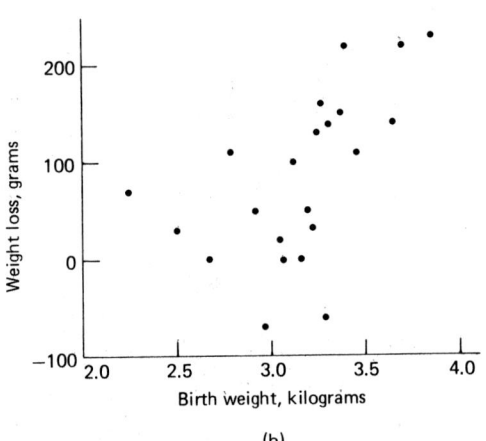

(b)

FIGURE 11.5 Example 10.3. Weight loss, in grams, as a function of birth weight, in kilograms, for (a) sample 1; (b) sample 2; (c) sample 3; (d) sample 4.

while important for an overall study interpretation, they do not influence a simple group comparison.

The weight losses for the four groups could be compared without regard to birth weights using procedures from chapter 4, but weight loss is related to birth weight and the birth weights do show a difference that shouldn't be ignored. An analysis without regard to birth weight could be criticized. An analysis using birth weight as a covariate will be an analysis independent of birth-weight differences.

The first step in the analysis is to plot the four samples, and this is done

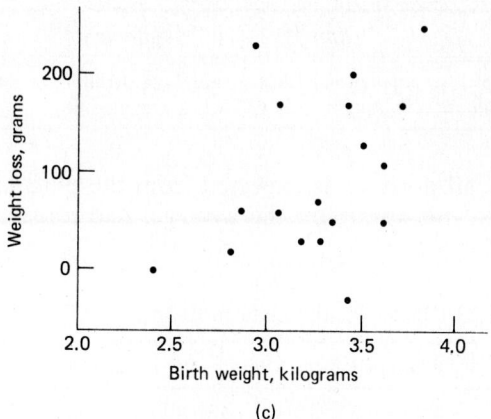

(c)

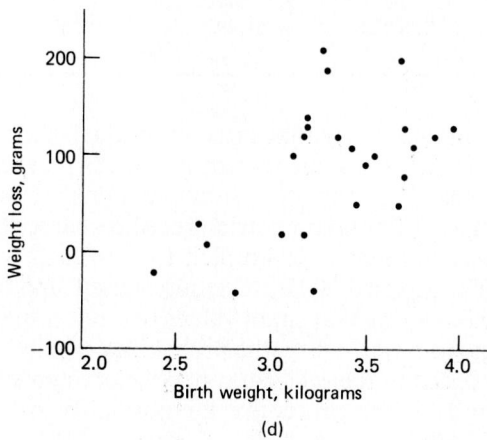

(d)

FIGURE 11.5 (*Continued*)

in figure 11.5. The scatter is so much that relationships more complex than simple linear functions, i.e.,

$$y = a + bx$$

are not suggested by the data. Table 11.23 presents summations of the raw data that will be useful in all of the calculations for the analysis. The reader may want to use these summations to do the analysis of variance of the birth weights discussed above.

The next step in the analysis is to confirm that the weight loss in each group is related to birth weight. Using formulas given earlier in this chapter, the slopes and intercepts for the four samples are

	Sample 1	Sample 2	Sample 3	Sample 4
	$b_1 = 0.1085$	$b_2 = 0.1267$	$b_3 = 0.1339$	$b_4 = 0.09484$
	$a_1 = -19.49$	$a_2 = -31.61$	$a_3 = -34.66$	$a_4 = -22.70$

The slopes are all positive as expected from the plots and indicate that larger babies in this study, on the average, had more weight loss than smaller babies.

TABLE 11.23   Sums for the Data in Example 11.3

	Sample 1	Sample 2	Sample 3	Sample 4
Σy	377	183	177	216
Σy^2	8065	3087	2913	3028
Σx	7429	6932	5981	8262
Σx^2	2,533,217	2,214,492	2,012,727	2,768,692
Σxy	129,972	61,499	62,211	75,013
n	22	22	18	25

The analysis should verify that these individual slopes are not due to chance by comparing each slope to zero. Student's t-test can be used provided the residuals about the regression equations are normally distributed, or one of the two nonparametric procedures described in chapter 6 can be used. Since Student's t is simpler, the residuals will be examined for normality. The Shapiro-Wilk procedure is used, and table 11.24 gives the details. Remembering that small values of π reject the null hypothesis of no differences from normality, the probability of π less than the values computed is included in table 11.24. It may be of interest to note that the p-values for the Lilliefors procedure for normality are all greater than 0.20, the limit of table M in part V.

The sample slopes can now be compared with zero to test the null hypotheses that each sample came from a population with a slope equal to zero. The standard deviation of y given x, i.e., $\mathrm{sd}_{y|x}$; se_b; and the t-statistic for all four samples are given below.

	Sample 1	Sample 2	Sample 3	Sample 4
$\mathrm{sd}_{y|x}$	8.110	7.343	6.696	5.962
se_b	0.05173	0.04220	0.04204	0.03048
t	2.097	3.003	3.185	3.112

These are one-tailed statistics since the prestudy expectation was for a positive relationship between weight loss and birth weight. Assuming the

TABLE 11.24 Shapiro-Wilk Test for Normality for Example 11.3

Shapiro-Wilk statistic	Sample 1			Shapiro-Wilk statistic	Sample 2		
	x	y	Product		x	y	Product
0.4590	19	−16.2	−7.4358	0.4590	8	−15.9	−7.2981
0.3156	13	−10.9	−3.4400	0.3156	12	−12.9	−4.0712
0.2571	18	−10.2	−2.6224	0.2571	5	−8.3	−2.1339
0.2131	15	−9.4	−2.0031	0.2131	4	−7.1	−1.5130
0.1764	21	−5.1	−0.8996	0.1764	10	−6.0	−1.0584
0.1443	9	−4.5	−0.6493	0.1443	20	−4.8	−0.6926
0.1150	4	−4.1	−0.4715	0.1150	18	−3.8	−0.4370
0.0878	12	−3.5	−0.3073	0.0878	17	−2.1	−0.1844
0.0618	16	−2.6	−0.1607	0.0618	22	−1.0	−0.0618
0.0368	1	−0.1	−0.0037	0.0368	7	−0.4	−0.0147
0.0122	22	0.9	0.0110	0.0122	21	−0.3	−0.0037

$$c_1 = -17.9826 \qquad c_1 = -17.4689$$

Shapiro-Wilk statistic	Sample 1			Shapiro-Wilk statistic	Sample 2		
	x	y	Product		x	y	Product
0.0122	6	1.0	0.0122	0.0122	13	2.2	0.0268
0.0368	14	1.4	0.0515	0.0368	15	3.1	0.1141
0.0618	17	2.5	0.1545	0.0618	2	3.6	0.2225
0.0878	8	3.0	0.2634	0.0878	6	3.8	0.3336
0.1150	5	3.0	0.3450	0.1150	1	4.0	0.4600
0.1443	2	3.7	0.5339	0.1443	3	5.8	0.8369
0.1764	11	5.6	0.9878	0.1764	16	6.3	1.1113
0.2131	3	9.0	1.9179	0.2131	19	6.7	1.4278
0.2571	10	9.8	2.5196	0.2571	9	7.4	1.9025
0.3156	20	11.9	3.7556	0.3156	11	10.2	3.2191
0.4590	7	16.0	7.3440	0.4590	14	10.7	4.9113

$$c_2 = 17.8855 \qquad c_2 = 14.5660$$

SSD = 1303.39 $\pi = 0.987$ $p = 0.98$ SSD = 1066.95 $\pi = 0.962$ $p = 0.52$

Shapiro-Wilk statistic	Sample 3			Shapiro-Wilk statistic	Sample 4		
	x	y	Product		x	y	Product
				0.4450	3	−11.8	−5.2510
				0.3069	23	−7.8	−2.3938
				0.2543	13	−7.0	−1.7801
0.4886	3	−14.3	−6.9870	0.2148	24	−5.3	−1.1384
0.3253	6	−8.7	−2.8301	0.1822	14	−4.9	−0.8928
0.2553	18	−6.2	−1.5829	0.1539	8	−4.3	−0.6618
0.2027	13	−5.1	−1.0338	0.1283	15	−4.1	−0.5260
0.1587	2	−4.8	−0.7618	0.1046	10	−1.9	−0.1987
0.1197	10	−2.6	−0.3112	0.0823	21	−1.8	−0.1481
0.0837	5	−2.0	−0.1674	0.0610	16	−1.7	−0.1037
0.0496	17	−0.9	−0.0446	0.0403	5	−1.7	−0.0685
0.0163	8	−0.2	−0.0033	0.0200	17	−1.4	−0.0280

$$c_1 = -13.7220 \qquad c_1 = -13.1910$$

TABLE 11.24 (*Continued*)

Shapiro-Wilk statistic	Sample 3			Shapiro-Wilk statistic	Sample 4		
	x	y	*Product*		x	y	*Product*
0.0163	9	0.7	0.0114	0.0000	11	−1.1	0.0000
0.0496	16	2.0	0.0992	0.0200	12	−0.7	−0.0140
0.0837	11	2.2	0.1841	0.0403	25	0.6	0.0242
0.1197	14	2.7	0.3232	0.0610	2	0.9	0.0549
0.1587	1	4.9	0.7776	0.0823	19	1.3	0.1070
0.2027	15	5.7	1.1554	0.1046	22	2.9	0.3033
0.2553	12	8.4	2.1445	0.1283	4	3.2	0.4106
0.3253	4	8.5	2.7651	0.1539	20	4.6	0.7079
0.4886	7	10.7	5.2280	0.1822	9	5.5	1.0021
				0.2148	6	6.5	1.3962
				0.2543	1	7.9	2.0090
				0.3069	7	10.6	3.2531
				0.4450	18	12.7	5.6515

$$c_2 = 12.6885 \qquad\qquad\qquad\qquad c_2 = 14.9058$$

SSD = 709.64 $\pi = 0.983$ $p = 0.96$ SSD = 805.21 $\pi = 0.980$ $p = 0.88$

null hypothesis is true and that the observable relationships for breast feeding and for the formulas are just chance, then the probabilities for the four samples are

	Sample 1	Sample 2	Sample 3	Sample 4
p	0.025	0.004	0.004	0.004

These probabilities are all small, so it is reasonable to assume, on the basis of these samples, that weight loss is related to birth weight.

Given these results, the analysis moves to a comparison of the slopes. The question is: Are the slopes essentially equivalent or do they differ more than the deviation about the regression equations indicate they might? The total sum of squares, i.e.,

$$\Sigma(y - \bar{\bar{y}})^2$$

where $\bar{\bar{y}}$ is the average of all 87 weight losses, is partitioned into components that can be identified first with an overall regression equation and then with the four separate equations. This dividing of a total sum of squares is an analysis of variance. It is nonparametric in the sense no assumptions are made about the distribution of the responses or the residuals about the regression equations. F-tests associated with the analysis of variance require for validity, however,

1. That the samples be independent
2. That the samples be random
3. That the variances of the residuals about the individual regression equations be equivalent to each other
4. That the residuals about the individual regression equations be normally distributed

The analyst must be satisfied that the samples are independent and random or qualifications must be attached to all interpretations of the results. At least one qualification concerning randomness has already been mentioned for this investigation, and this will be repeated when the anlaysis is summarized.

The Shapiro-Wilk procedure has been used to examine normality, but the variances have not been examined for equivalency. Bartlett's M-statistic can be used to test the null hypothesis that the samples come from populations all having the same variance. Using the procedure presented in connection with example 11.1, M for this example is 2.215. M is distributed approximately as chi-squared and has degrees of freedom equal to one less than the number of groups. Assuming the null hypothesis is true and that the sample differences are just chance variation, then the probability of a chi-square with 3 degrees of freedom larger than 2.215 is approximately 0.53. This, of course, is a large probability, so the variances about the regression equations may be accepted as equivalent.

The preliminary calculations for the analysis of variance are explained and presented in table 11.25. The shortcuts that were possible in examples 11.1 and 11.2 because of identical x data are not possible here. TYX is not equal to S12, TXX is not equal to S10, and INTERCEPTS (S21) is not equal to S20.

The ANOVA table is prepared in two steps. The total sum of squares, 6653.81, is first divided into two components, one that is identified with deviations of the responses about an overall regression equation, 4759.70, and another that is due to the overall regression equation, 1894.11. Then the sum about the regression equation is divided into components that are identified with differences in intercepts and slopes, and a remainder identified with deviations about the individual regression equations. This is all summarized in table 11.26.

The F-statistic in table 11.26 is very small, so the slopes are essentially equivalent. A common slope can be determined from the summations in table 11.25. It is

$$b = \frac{S12}{S10} = \frac{13,530.47}{118,500.44} = 0.1142$$

TABLE 11.25   Calculations for the Analysis in Example 11.3

	(1) n	(2) Σy	(3) Σy^2	(4) Σx	(5) Σx^2
1	22	377	8065	7429	2,533,217
2	22	183	3087	6932	2,214,492
3	18	177	2913	5981	2,012,727
4	25	216	3028	8262	2,768,692
	87	953	17,093	28,604	9,529,128

	(6) Σxy	(7) $\dfrac{(\Sigma y)^2}{n}$ $\dfrac{(2)(2)}{(1)}$	(8) $\Sigma(y-\bar{y})^2$ $(3)-(7)$	(9) $\dfrac{(\Sigma x)^2}{n}$ $\dfrac{(4)(4)}{(1)}$	(10) $\Sigma(x-\bar{x})^2$ $(5)-(9)$
1	129,972	6460.41	1604.59	2,508,638.23	24,578.77
2	61,499	1522.23	1564.77	2,184,210.18	30,281.82
3	62,211	1740.50	1172.50	1,987,353.39	25,373.61
4	75,013	1866.24	1161.76	2,730,425.76	38,266.24
8	328,695	11,589.38	5503.62	9,410,627.56	118,500.44

	(11) $\dfrac{\Sigma y \Sigma x}{n}$ $\dfrac{(2)(4)}{(1)}$	(12) $\Sigma(y-\bar{y})(x-\bar{x})$ $(6)-(11)$	(13) $\dfrac{[\Sigma(y-\bar{y})(x-\bar{x})]^2}{\Sigma(x-\bar{x})^2}$ $\dfrac{(12)(12)}{(10)}$	(14) $ss_{y\mid x}$ $(8)-(13)$
1	127,306.05	2665.95	289.16	1315.43
2	57,661.64	3837.36	486.28	1078.50
3	58,813.17	3397.83	455.01	717.49
4	71,383.68	3629.32	344.22	817.54
	315,164.53	13,530.47	1574.67	3928.95

S15 = TYY	= S3 − S2 ·	S2/S1	= 6653.82
S16 = TYX	= S6 − S2 ·	S4/S1	= 15,366.13
S17 = TXX	= S5 − S4 ·	S4/S1	= 124,658.85
S18 = ABOUT	= S15 − S16 · S16/S17		= 4759.70
S19 = SLOPES	= S13 − S12 · S12/S10		= 29.75
S20	= S7 − S2 ·	S2/S1	− 1150.19
S21 = INTERCEPTS = S20 − S16 · S16/S17 − S12 · S12/S10 = 801.00			

The intercepts for each of the four groups can be computed from the items in columns 1, 2, and 4 of table 11.25 and the common slope as follows:

$$a = \frac{(2) - b(4)}{(1)}$$

TABLE 11.26 The ANOVA Table in Example 11.3

Source of variation	df	ss	ms	F
Between intercepts	3	801.00	267.00	
Between slopes	3	29.75	9.917	0.199
About individual lines	79	3928.95	49.73	
About the overall line	85	4759.70		
Due to the overall line	1	1894.11		
Total	86	6653.81		

The four regression equations with the common slope are

$$y_1 = -21.43 + 0.1142x_1$$

$$y_2 = -27.66 + 0.1142x_2$$

$$y_3 = -28.11 + 0.1142x_3$$

$$y_4 = -29.10 + 0.1142x_4$$

The next step in the analysis is to determine if these intercepts are equivalent, that is, to test the null hypothesis that the samples all came from populations that not only have equal slopes but also have equal intercepts.

A test of intercept equivalency requires that the sum of squares about the overall regression equation be partitioned into a component due to these intercepts and a component about the four parallel regression equations. The component about the regression equations can be obtained from sums in table 11.25 as

$$S8 - S12 \cdot \frac{S12}{S10} = 3958.70$$

but an easier way is simply to sum, or pool, the sum of squares for slopes and the sum of squares about the individual lines in table 11.26. The between intercepts component can be obtained as the difference between this quantity and the sum of squares about the overall regression line, or it can simply be copied from table 11.26. The result is table 11.27.

TABLE 11.27 The ANOVA Table for Example 11.3

Source of variation	df	ss	ms	F
Between intercepts	3	801.00	267.00	5.531
About the individual lines	82	3958.70		
About the overall line	85	4759.70		
Due to the overall line	1	1894.11		
Total	86	6653.81		

Multiple Comparison of Intercepts, Dunnett

The F-statistic for intercepts in table 11.27 is large. Assuming the null hypothesis given above is true and that the observable sample differences are just chance variation, then the probability of a larger F-statistic is less than 0.005, the limit of table D in part V, so the intercepts are not all equivalent. To understand how the intercepts differ, Dunnett's multiple comparison procedure will be used to compare the three formula-fed groups with the breast-fed group. A critical difference (CD) is computed for the difference between the breast-fed group and each of the formula-fed groups. These are calculated as follows:

$$ CD = q \sqrt{ ms \left[\frac{1}{n_i} + \frac{1}{n_j} + \frac{(\bar{x}_i - \bar{x}_j)^2}{\Sigma(x_i - \bar{x}_i)^2 + \Sigma(x_j - \bar{x}_j)^2} \right] } $$

where ms = the mean square about the individual regression equations from table 11.27

n_i and n_j = the sample sizes

q = a factor from Dunnett's table for r, the number of samples in the study; df, the degrees of freedom for ms; and the significance level chosen for the tests

The q-statistics are found in table S of part V. The reader will recognize the quantity under the square root sign as the variance used in chapter 10 in a t-statistic to measure the difference between two intercepts.

Dunnett's procedure, like many others in statistics, was developed for equal sample sizes. Most data analyses do not involve equal-size samples, so to make the procedures applicable in real data analysis problems, they are modified in ways that analysts generally accept and consider reasonable. The formula for the critical difference given above has been modified in accordance with procedures for the t-test of two slopes given in chapter 10. Dunnett notes that if the numbers of observations in the samples are unequal, his tables may still be used, but the associated probabilities will be only approximate.

For this example, $r = 4$ and df = 82, so that $q = 2.09$ for a significance level of 0.05. The sum of squares for x and the sample sizes are different for the four samples so a different critical difference must be computed for each of the three comparisons. For comparing the breast-fed group with group 2,

$$ CD = 2.09 \sqrt{ 48.28 \left[\frac{1}{22} + \frac{1}{22} + \frac{(337.7 - 315.1)^2}{24{,}578.77 + 30{,}281.82} \right] } = 4.60 $$

The difference between the intercepts, 6.24, exceeds 4.60, so it is significant at the 0.05 level.

For comparing the breast-fed group with group 3,

$$CD = 2.09 \sqrt{48.28 \left[\frac{1}{22} + \frac{1}{18} + \frac{(337.7 - 332.3)^2}{24,578.77 + 25,373.61} \right]} = 4.63$$

The difference between the intercepts, 6.68, is larger than 4.63, so this difference is also significant at the 0.05 level. Finally, for comparing the breast-fed group with group 4,

$$CD = 2.09 \sqrt{48.28 \left[\frac{1}{22} + \frac{1}{25} + \frac{(337.7 - 330.5)^2}{24,578.77 + 38,266.24} \right]} = 4.27$$

The difference between the intercepts is 7.67, so it is significant at the 0.05 level.

This completes the analysis for example 11.3. The analysis has shown that the four groups all have weight losses related to birth weight and that the weight-loss rate with respect to birth weight was equivalent for the four groups. The difference between the weight lost by the breast-fed group and each of the formula-fed groups was significant. The assignment of the newborns to the three formula-fed groups produced a rather low average birth weight for group 2. It is impossible, from the data alone, to know if some kind of bias entered into the assignment process or not. Before completing this brief summary of the analysis, it must be noted that the assignment of infants to the four groups was not random in the usual sense.

SUMMARY

A class of problems that combine characteristics of problems considered in both parts I and II has been considered in part III, but the analysis procedures presented here are limited to ratio and interval measurements. In a sense the book has reached a level of complexity for which there are no generally available easy-to-use data analysis procedures. This certainly introduces a lot of things that statistical theorists can do for data analysts.

Several times, relationships have been noted between formulas used in part III and those presented in parts I and II. It is hoped that by noting such relationships, resourceful data analysts will be encouraged to synthesize procedures for use when the samples are not normally distributed and/or the variances are not equivalent. For example, part III has not considered any ordinal scale techniques, but analysts should be able to combine procedures presented in chapter 6 with any of the multiple comparison procedures to develop techniques applicable to ordinal data. The possibilities are numerous; the results can be important.

PART IV

Supporting Topics

In some respects, statistics describes the world as how it might be rather than how it is. Investigators and managers don't always understand the idealizations of statistics, so explanations to them must be made relative to their immediate problems. Consequently, data analysts need a prospective of statistical concepts and principles that are understandable and useful to their clients. This is one of the objectives of part IV.

Another objective of part IV is to review and summarize some elementary statistical concepts and procedures from the data analysis viewpoint. The reader may find these useful and helpful in parts I, II, and III. It is not my intention to teach or to reteach statistics. There are too many good statistics texts available. Recommended personal favorites are

Allen and Cady (1982)

Brownlee (1965)

Conover (1980)

Crow et al. (1960)

Hollander and Wolfe (1973)

Kendall (1955)

Koopmans (1981)

Siegel (1956)

CHAPTER 12

Random Samples

DEFINITIONS

Every science begins with certain self-evident and generally accepted principles called axioms. These axioms are the foundation on which everything else is built. The idea of a random sample is almost an axiom in data analysis and statistics. For most statistical procedures, the data are regarded as a random sample of a parent population or universe. The analyst is then interested in learning, through a study of the sample, characteristics of the population. Consequently, data analysts need a good understanding of what statisticians mean by random sample.

In *A Dictionary of Statistical Terms,* Kendall and Buckland (1957) give these definitions,

> Random sample: A sample which has been selected by a method of random selection.

> Random selection: A method of selecting sample units such that each possible sample has a fixed and determinate probability of selection. Ordinary haphazard or seemingly purposeless choice is generally insufficient to guarantee randomness when carried out by human beings. Devices such as random sampling numbers, or analogue machines, are used to remove subjective biases inherent in personal choice.

> Random: This word may be taken as representing an undefined idea, or, if defined, must be expressed in terms of the concept of probability. A process of selection applied to a set of objects is said to be random if it gives to each one an equal chance of being chosen. Generally, the use of the word "random" implies that the process under consideration is in some sense probabilistic.

This definition of random sample is more abstract than it is practical or useful to the data analyst. Each member of a population must be identifiable so that it can be given its equal and independent chance of being included in the sample. Most populations of interest to data analysts are not so well defined. In many analyses, the idea of a population is so vague that the sample may seem like the entire population. The samples of patients in a clinical trial are examples.

This definition gives the data analyst little help in how to obtain a random sample or in how to test a given sample for randomness. Both of these data analysis problems are discussed below.

An Illustration

Sometimes basic concepts are easier to illustrate than to define. The following paraphrases Ernest Nagel (1961, p. 326) to illustrate randomness.

> Suppose a man leaves his home in order to purchase the evening newspaper and on the way is felled by a brick displaced from the roof of a building. The man's misfortune is then said to be a random occurrence not because it is "uncaused" (indeed, the description of the event indicates the cause), but because it occurs at the "juncture" of two independent causal sequences. One sequence terminates in the man's being beside the building at a given time and the other sequence terminates in the brick's motion at that same time. These causal series are said to be "independent" in the sense that the events in one do not determine the events in the other. Had the brick not fallen, the man would have proceeded on his journey to purchase the evening newspaper and had the man not been at the particular spot, the brick would have struck the ground.

The interesting thing from the data analysis viewpoint is that these two events occurred at the right time to cause the brick to hit the man. No one would be surprised to hear of another brick hitting another person, especially if the pedestrian traffic was rather dense near the building and nothing was done to stabilize the structure. One or more additional hits would certainly appear to be random events. If, however, the same man was hit on several different occasions, the event would no longer be considered random. The interesting question for the data analyst would now be: Why was this poor, unfortunate person hit several times?

Nagel's description of a random event provides a mechanism for obtaining random samples. Before considering this, consider some nonrandom sampling techniques.

NONRANDOM SAMPLING TECHNIQUES

A *purposive* sample is one selected according to one or more criteria established by the data collector to make the sample more "typical" of the parent population than a random sample would be. The reliability and usefulness of such samples depend on the data collector's skill and judgment. Another data collector, with different skills and judgment, will choose a sample with different properties. Purposive samples are not comparable because there is no general way of assessing the skill and judgment of different data collectors. While summary statistics for different random samples from a population will be different, the reliability and usefulness of random samples are comparable and assessable. This fundamental characteristic of a random sample is implied in the definition but not specifically stated, of course.

Individuals in finite populations may be arranged in some logical order, e.g., alphabetically, chronologically, or geographically, and then numbered. A *systematic* sample is produced by choosing every kth individual starting with the ith, i being some number from 1 to k chosen at random. Panels from which petit jurors are selected are typically chosen in this manner. If the logical order is independent of the attributes desired for the sample, the systematic sample may be practically equivalent to a random sample.

A natural periodicity of period k in the parent population will bias a systematic sample. Households ordered geographically provides the typical, if somewhat contrived, example of how bias can occur in systematic samples. Suppose every eighth house is chosen along a street, and by chance in the sampling sequence an eighth house is on a corner lot in a large subdivision with eight houses per block. This systematic sample may have a preponderance of houses on the usually more expensive corner lots. Consequently, the sample may not be representative of the population of households with respect to any economic status characteristic.

There is a variation of the systematic sample that will nullify any natural periodicity. First create a sequence of random numbers with expected value equal to k and call them

$$n_1, n_2, n_3, \ldots$$

Then choose for the sample the individuals with the following positions in the ordered population.

$$n_1$$

$$n_1 + n_2$$

$$n_1 + n_2 + n_3$$

$$\cdot$$

$$\cdot$$

$$\cdot$$

Thus instead of adding k each time to determine the index of the next sample member, a random number whose expectation is k is added to the previous index.

Until someone has created a sequence of random numbers with a specific expected value, it may not be clear how to do this. Tables of random numbers are available in many handbooks and many computer programming languages have intrinsic random-number-generating subroutines. The tables and the subroutines can be classified according to the frequency distribution of the random numbers tabulated or generated. The two most common distributions are rectangular, where every number is equally likely, and normal, where the frequency distribution of the random numbers is given by a normal distribution with a specified mean and standard deviation.

Table T of part V contains rectangularly distributed random numbers. These numbers are printed in groups of four digits but may be divided or combined to produce random numbers with any number of digits.

Implicit in the systematic sample described above is the notion that every member of the population should be equally likely to be chosen for the sample. Hence, rectangularly distributed random numbers, rather than random numbers distributed some other way, must be used to generate the sequence. To do this, move vertically or horizontally through table T in part V and select numbers that are between 0 and $2k$. Do not use either of these limits.

Many numbers will not be in this range, so to avoid skipping nearly all of the table, devise a simple transformation to create numbers with the range. For example, if $2k$ is 16, use the left two digits and then the right two digits in each group of 4. Next subtract 0, 20, 40, 60, or 80 from the two digits to create numbers in the range 0 through 19. Now discard all zeros and the numbers 16 through 19. The remaining numbers will be the numbers needed.

Since all of the numbers selected in this way from table T in part V are equally likely, the expected value will be simply the average of the numbers in the range, i.e., k. Don't be surprised if the average for a particular sequence of random numbers is not exactly k, but be surprised if the averages for many sequences with expected value k do not form a normally distributed sample with average very close to k.

RANDOM SAMPLING

Nagel's description of a random event provides a mechanism for obtaining a random sample from naturally ordered populations. Periodicity problems are avoided by choosing individuals at the juncture points of the naturally occurring sequence and an independent sequence of random numbers. Numbers that for all practical purposes satisfy the definition for random can be computer-generated, "drawn from a hat," or obtained from tables such as table T in part V.

If the population can be delineated, random numbers greater than the actual size of the population are ignored and replaced by others within the range. The sample then consists of individuals from the population corresponding to the random numbers, i.e., at the juncture points of the two independent sequences. When a population cannot be delineated, an arbitrary starting point in the population sequence is selected and individuals corresponding to the random numbers from that starting point are placed in the sample. Population members that occur outside the area of juncture points are simply ignored, with the unstated and sometimes unrecognized assumption that no chronological or geographic trends exist to cause biases.

Sometimes populations can be divided into natural subclasses that are more homogeneous than the populations taken as a whole. Often then, it is appropriate to classify the members of the population and to select a random sample from each subclass. Such subclasses are called *strata* and the sample is called a *stratified random sample*. Usually the same proportion is taken from each strata, but unequal proportions may be taken to accomplish some specific objective for the investigator. If equal proportions are taken, then the strata will be represented in the sample in the same proportion as they exist in the population.

Stratification may be employed to assure that each stratum is properly represented. For example, a river or railroad or freeway might divide a city into two or more natural geographic regions, which in this context would be called strata. A stratified random sample of housing units for the city would consist of a random sample from each of the geographic regions. If equal proportions were taken from the strata, no one region would be underrepresented in the total sample. Deliberately disproportionate sampling might be used to gain a more precise picture of housing in low-income strata than in middle or high-income strata.

The stratification variable, i.e., the basis for the subclassifications, can be used directly in an analysis to remove strata differences that interfere with the interpretation of the data. Removing strata differences enables an analyst to see smaller differences, so if strata are used, the report is not complete without a full discussion of the stratification process.

EXPERIMENTS

In experiments, hypothetical populations are created by assigning homogeneous units randomly to groups that will be treated differently. The sample analyzed is all of the hypothetical population that exists in practice, but in theory, more of the population could be created, subject only to the number of homogeneous units available for assignment. This is a special kind of randomization and care must be taken to interpret results properly. Strictly speaking, the results of the experiment apply only to the hypothetical populations created for the experiment. Investigators and management always want wider applicability, so the analyst's role in the process is to provide guidance, less unwarranted extensions are made. A clinical trial in pharmaceutical research provides a good illustration.

In a clinical trial, the investigator wants to compare one or more active medications with a placebo. A "homogeneous" group of patients is chosen and assigned at random to subgroups that will receive an active medication or the placebo. This creates samples from hypothetical populations for comparison. Usually the homogeneous group of patients is supplied by a single physician. Another physician could also supply a group of homogeneous patients but they would probably differ from the first. The data analyst must caution against a natural tendency to believe that results from the first homogeneous group apply to all others. They may apply, and then they may not.

TESTS OF RANDOMNESS

The members of a sample are not all selected or measured at the same time. They are selected and/or measured one at a time, and this creates a natural chronological ordering of the members. Several tests of randomness utilizing this property have been proposed.

The mean square successive difference test [von Neumann et al. (1941)] is particularly sensitive in detecting long-term trends, both periodic and aperiodic, or excessively rapid oscillation in observations from a normal population. The test is based on estimating the population variance in two different ways. The first is the usual unbiased estimate

$$\mathrm{sd}^2 = \frac{\Sigma y^2 - (\Sigma y)^2/n}{n - 1}$$

The second is based on successive differences,

$$d^2 = \frac{\Sigma(y_{i+1} - y_i)^2}{n - 1}$$

von Neumann showed that

$$z = \frac{(d^2/2)/\mathrm{sd}^2 - 1}{\sqrt{(n - 2)/(n^2 - 1)}}$$

is approximately distributed as a unit normal deviate. Long-term trends or slow oscillations produce negative values of z, whereas rapid oscillations produce positive values of z. The exact distribution, assuming the observations are from a normal population, has been tabulated by Hart (1942) for n over the range 4 to 60. The normal approximation is probably satisfactory for n as small as 20. See table O in part V.

The mean square successive difference test is limited to ratio and inter-val data. The theory of runs provides a valid test in the nominal scale and for any data in the other measurement scales transformed to the nominal. The procedure is particularly effective in detecting trends, oscil-lations, and discontinuities, common manifestations of nonrandomness.

Ratio, interval, and ordinal data are customarily transformed to the nominal scale in one of two ways for these tests. One transformation changes consecutive pairs of observations into pluses and minuses while ignoring contiguous observations that are equal. There should be $n - 1$ pluses, minuses, and zero differences. The other common transformation changes each datum into a plus or a minus depending upon its relation to some arbitrary constant, often the median. If the median is used and n is odd, the median itself is dropped from the sequence.

After transforming the observations, the number of runs u is counted. A run is defined as a contiguous group of pluses or a contiguous group of minuses. It may be easier to count the number of transitions, i.e., where a plus is followed by a minus or a minus is followed by a plus, since the number of runs is one more than the number of transitions.

The expected number of runs in a random sample and the variance of the number of runs is a function of the number of pluses p and minuses m only. They are

$$\text{Expected } u = 1 + \frac{2mp}{m + p}$$

$$\text{Variance } u = \frac{2mp(2mp - m - p)}{(m + p)^2(m + p - 1)}$$

Furthermore, the statistic

$$z = \frac{u - 1 - 2mp/(m + p)}{\sqrt{\text{var}(u)}}$$

is approximately normal for large m and p. As with the mean square suc-cessive difference test, long-term trends produce negative values of z while rapid oscillations produce positive values of z.

This statistic was studied by Stevens (1939) and Wald and Wolfowitz (1940). Swed and Eisenhart (1943) computed exact values of the distri-bution of the test statistic for small m and n. See table P in part V.

DATA ANALYSIS IMPLICATIONS

Many data analysis problems do not involve sampling. They begin with all of the observations available on a particular problem. Additional observations on the same problem can be obtained, making it convenient to think of the observations as a sample of some hypothetical population. The population is created as the observations are made, thus the obser-vations depend on the creation mechanism. The observations are not a random sample in the strict sense of the term. Hence the sample may be considerably different from the hypothetical population the data analyst wants to study.

Consider an example to clarify these ideas. This example concerns an experiment, i.e., a clinical trial in pharmaceutical research, but the principles are the same for many other data analyses. An investigator wants to compare use of a drug with a placebo. To administer the two treatments, the investigator needs a physician who seems likely to have a group of patients with approximations of the disease the drug is expected to treat. Now in typical double-blind fashion, the patients are given either the drug or the placebo. Two hypothetical populations are created and a sample from each is made available for analysis. First, the investigator would like to think that the group of patients represented the population of patients with the particular disease. Second, the investigator would like to think that the two samples represent hypothetical populations with the particular disease that differ only by the drug and placebo treatment.

The problem in this scenario is of course different physicians with different patient populations will produce different samples. These samples are so different sometimes that they cannot be considered from any one hypothetical population. The drug and placebo samples created by a single physician may be random samples from hypothetical populations but they may not, and indeed are probably not, the hypothetical populations the sponsor wants and expects.

The question then is this: Can statistical procedures that depend on random samples be used to analyze the data? Thoughtful data analysts might find it hard to defend these samples as random on the basis of the formal definitions in Kendall and Buckland but a paraphrase of Nagel will describe most such samples. Since the data enter the study in a chronological sequence, the tests of randomness can even be applied.

SUMMARY

This discussion of random samples has not been given specifically as an aid in selecting random samples but rather to focus attention on the real data analysis issue. Most data analyzed are not random samples from parent populations in the sense of Kendall and Buckland's definition, so the data analysis issues are these:

1. Were the data produced by a mechanism such that they can be considered a random sample from some population?
2. If they were, was this the correct population for the purpose of the research?

The problem is really one of assessing the limitations of data sets. While there are tests for randomness or the lack thereof, statistics doesn't give the data analyst much help in making assessments or judgments in real data analyses.

Data analysts who don't recognize these problems make errors.

CHAPTER 13

Distributions

INTRODUCTION

The notion of one or more characteristics of the members of a population distributed in a particular way is fundamental to all data analysis. Kendall and Buckland (1957) define *population* as follows:

> In statistical usage the term population is applied to any finite or infinite collection of individuals . . . and does not necessarily refer to a collection of living organisms.

A measurable characteristic of a population is often itself called a population. For example, the population might be the male residents of Kansas City and the measurable characteristic be their weights, so one might study the population of weights of male residents of Kansas City. Issues on a stock exchange are another population and their prices at a given time, e.g., the closing, would be the population of a measurable characteristic.

Frequency is a term meaning the number of occurrences of a given type of event, or the number of members of a population falling into a specified class or range of measurements. Frequencies can be raw counts or the counts can be expressed as proportions of the total population, in which case they are called *relative* or *proportional* frequencies.

The term *distribution* is used in two different ways with populations. A frequency distribution specifies the frequencies of a measurable characteristic of members of a population along the measurement scale. Tables and histograms are used to present the actual frequencies. For example, table 13.1 and figure 13.1 present a frequency distribution of

TABLE 13.1   Frequencies of Adult Males by Weight Categories

Range, pounds	Count	Range, pounds	Count
<134	65	174–178	1027
134–138	58	178–182	880
138–142	90	182–186	818
142–146	165	186–190	689
146–150	280	190–194	488
150–154	387	194–198	380
154–158	562	198–202	264
158–162	740	202–206	191
162–166	835	206–210	117
166–170	952	210–214	58
170–174	980	>214	59

weights such as might be determined for a sample of the adult male population of Kansas City.

The distribution function may not be as familiar a concept to data analysts as is the concept of a frequency distribution. It is, nevertheless, an important idea that must be understood. The distribution function for a population gives the total frequency of measurements, or the proportion of the total frequency, less than or equal to specific values along the measurement scale. Table 13.2 and figure 13.2 present the distribution func-

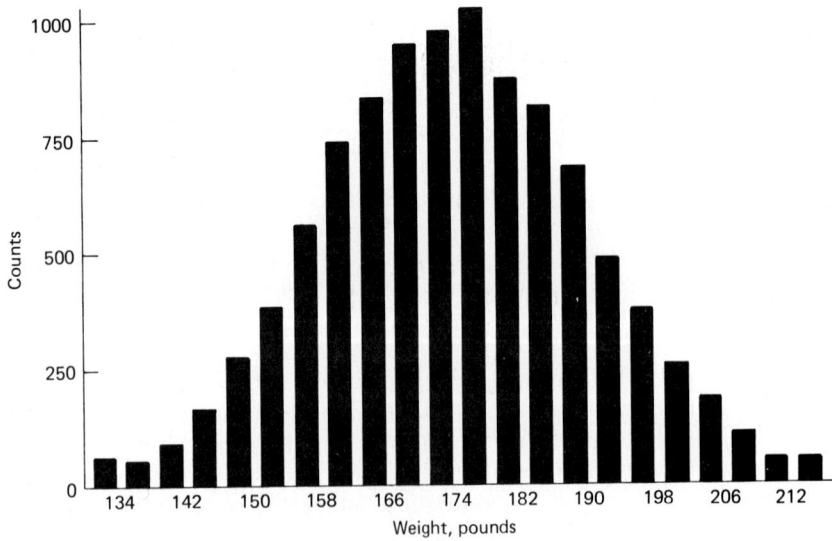

FIGURE 13.1   Frequency distribution of weights of adult males.

TABLE 13.2 Cumulative Count of
Adult Males by Weight Categories

Range, pounds	Count	Range, pounds	Count
<134	65	174–178	6141
134–138	123	178–182	7021
138–142	213	182–186	7839
142–146	378	186–190	8528
146–150	658	190–194	9016
150–154	1045	194–198	9396
154–158	1607	198–202	9660
158–162	2347	202–206	9851
162–166	3182	206–210	9968
166–170	4134	210–214	10,026
170–174	5114	>214	10,085

tions of the weights from table 13.1 and figure 13.1. Note the characteristic S shape of the tops of the bars in figure 13.2.

The frequency distribution and distribution function are ways of describing or presenting populations. Furthermore, they are used to describe the characteristics of samples as well as to present the raw data in samples. For example, an average of a sample is itself a datum from a population of averages that may be described by frequency distributions and distribution functions. With this statement it should now be evident

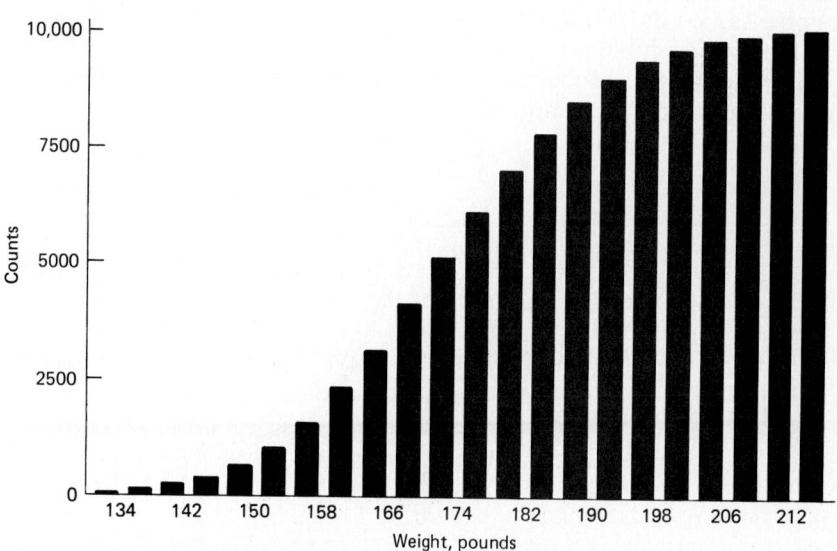

FIGURE 13.2 Distribution function of weights of adult males.

that every raw measurement and every summary of raw measurements are members of populations, albeit different populations.

The term population has been used above to denote any finite or infinite collection of individual things or the collection that is a particular measurement of the individuals. Certain mathematical functions are used as approximations for populations. This collection of mathematical functions includes such well-known functions as the normal or Gaussian, binomial, chi-square, Student's *t*, and *F*. These are commonly called distributions, but this invites confusion with the way the term distribution has already been used in this chapter. Furthermore, the group of mathematical functions really has no name, at least not one in common enough usage to be in Kendall and Buckland (1957). Thus the group of such functions will be called, descriptively enough, *theoretical distributions,* while each individual will be given its specific name, e.g., *t* distribution, chi-square distribution, etc.

The notion of frequency distributions and distribution functions exists for theoretical distributions, but the names are different and smooth curves may replace the step functions used for populations. Frequency distribution is replaced by probability distribution or probability density function but the idea is the same. Figure 13.3 presents the probability density functions for the *t*-statistic with 2 and with 4 degrees of freedom superimposed on the probability density function for the normal distribution. For easier comparison, the scale has been adjusted so that the curves cover the same area. Note the similarity of these curves with the tops of the bars in figure 13.1. Vertical lines can be drawn in figure 13.3 to create bars as in figure 13.1 to demonstrate that these two figures convey the same kind of information.

Distribution function is replaced by cumulative probability distribution or cumulative probability density for theoretical distributions. Figure 13.4 presents the cumulative probability density function for the *t*-

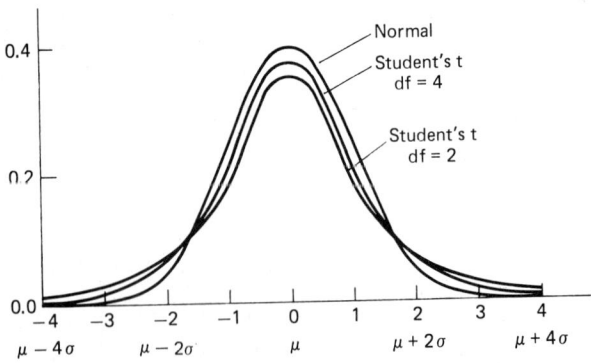

FIGURE 13.3 Probability distributions for Student's *t*-statistic with 2 and with 4 degrees of freedom and the normal.

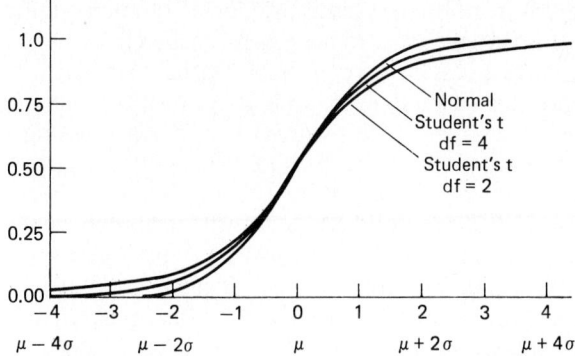

FIGURE 13.4 Cumulative probability density function for Student's *t*-statistic with 2 and with 4 degrees of freedom and the normal.

statistic with 2 and with 4 degrees of freedom, along with the density function for the normal distribution. Note the characteristic S shapes are similar to the tops of the bars in figure 13.2. Vertical lines can also be drawn for the cumulative probability density function to demonstrate that it conveys the same kind of information as the distribution function in figure 13.3.

The probability density functions for theoretical distributions can be expressed as mathematical equations and, indeed, this distinguishes them from the frequency distributions of real things such as the weights of the male residents of Kansas City. The frequency distribution and the distribution function of real things can be approximated by these mathematical functions and depicted as smooth curves instead of histograms, but the data analyst must not confuse the two in so doing. The mathematical functions are still only approximations for the real distributions.

Readers familiar with the calculus will recognize the cumulative probability density as the integral of the probability density from minus infinity to points on the measurement scale. Similarly, the distribution function is the finite mathematical analogue of the integral of the frequency distribution, i.e., the sum from the smallest measurement to points on the measurement scale.

RELATION OF POPULATIONS TO DATA ANALYSIS

The notion of populations should be conceptually real for analysts even though details of the populations in any particular data analysis will probably be unknown. Nevertheless, every population has specific properties, and one purpose of data analysis is to infer these properties through a study of samples from the populations. Correctly and adequately defined populations contribute much to successful research; incorrectly or inadequately defined populations are probably useless, since no one will know how to interpret or where to apply analytical results from such studies.

Sample frequency distributions provide one way of using sample data to infer what the population may be like. Whenever the sample frequency distribution can be approximated by a theoretical distribution, it is customary to think of the population as being distributed in that way. The data analyst then has an analytical basis for inferences about the unknown population, but this is only as good as the approximation. Invalid assumptions about the adequacy of the approximation must be avoided at all costs.

The remainder of this chapter discusses the relationship of several theoretical distributions to sample populations.

THE NORMAL DISTRIBUTION

The principal distribution in all data analysis is the normal distribution. The name does not imply that all other distributions are abnormal, but it is interesting that the distribution of many kinds of observations can be approximated with the normal distribution.

The normal distribution is important as a "limiting" distribution. For example, as the sample size increases, the t and the binomial distributions, as well as many others, become more and more like the normal. The mathematical study of sampling for a normal population is comparatively simple and has been highly developed.

These three facts somewhat explain why the normal holds center place in statistics and data analysis. The normal is sometimes called the Gaussian distribution. Widespread use of this name or any other name would eliminate the possibility of anyone assuming all nonnormal, i.e., non-Gaussian, distributions are abnormal.

Because the normal is comparatively simple to study and has been highly developed, there is a tendency to assume a normal distribution for data without sufficient investigation to justify the assumption. On the other hand, statisticians have shown that moderate departures from normality do not seriously affect the validity of many of the procedures based on the normal distribution. It is difficult for statisticians to define what they mean by moderate departures, so data analysts must rely on specific tests for normality such as the Lilliefors and the Shapiro-Wilk.

The normal probability density is given by

$$f(x) = \frac{1}{\sigma\sqrt{2\pi}}\, e^{-(x-\mu)^2/2\sigma^2}$$

The factor $1/\sigma\sqrt{2\pi}$ makes the integral of $f(x)$ from $-\infty$ to $+\infty$ equal to 1, i.e.,

$$\int_{-\infty}^{+\infty} f(x)\, dx = 1$$

Consequently, the area under the curve between any two points is the proportion of the total between those points, i.e., it is a probability.

There are infinitely many normal probability density functions, one for each combination of values for μ and σ. The most common is the *standard* or *unit* normal where $\mu = 0$ and $\sigma = 1$. Any other normal probability density function can be easily transformed to this function. Let $y = (x - \mu)/\sigma$, then $f(y)$ is the unit normal probability density function. A graph of this function is presented in figure 13.5.

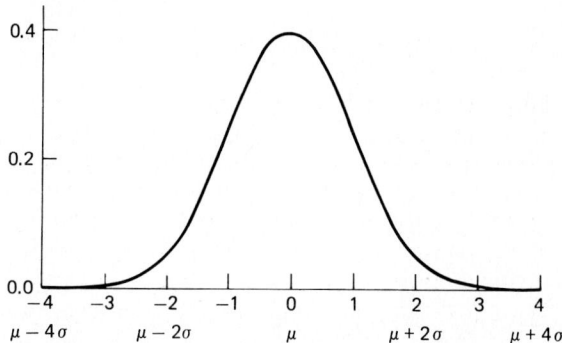

FIGURE 13.5   The normal or Gaussian probability density.

Note that the variable x may have any value from positive infinity to negative infinity, i.e., $f(x)$ is never zero except for these two values. Note, however, that more than 99.99% of the area under the curve lies between -4σ and $+4\sigma$, that approximately 99.73% lies between -3σ and $+3\sigma$, and that approximately 95.45% lies between -2σ and $+2\sigma$. Thus when the normal distribution is an adequate approximation for a finite sample, very few observations should be beyond the interval defined by the average plus and minus 3 or 4 times σ.

Not all symmetrical, bell-shaped distributions are normal. For example, the t distribution is symmetrical and bell-shaped. As the sample size increases, the binomial becomes symmetrical and bell-shaped for any value of the binomial parameter p.

A linear combination of variables is any sum of this general form

$$w = a_1 x_1 + a_2 x_2 + \cdots + a_n x_n$$

where the a's are constants and the x's are variables. Now if the x's are normally distributed variables, then w will itself be normally distributed. Furthermore, the unknown average of the population of w's will be the linear combination of the unknown averages of the populations of the x's. The unknown variance of the population of w's will be the linear combination of the unknown variances with each constant squared, i.e.,

$$\text{Average } w = \Sigma a_i \mu_i$$

$$\text{Variance } w = \Sigma a_i^2 \sigma_i^2$$

Finally, if $w = \Sigma x$ (i.e., all of the a's are 1), then the average $w = n\mu$ and the variance of $w = n\sigma^2$. It follows from this that if $w = \Sigma x/n$ (i.e., all of the a's are $1/n$), then the average $w = \mu$ and the variance of $w = \sigma^2/n$. Estimates of the unknown average and unknown variance of the population of w's are obtained from sample statistics using the above relationships.

The Central Limit Theorem

One possible explanation for why so many variables seem to follow the normal distribution may be the central limit theorem. This theorem has many forms, but the one most important to data analysts is probably the simplest.

Consider the sum

$$w = x_1 + x_2 + \cdots + x_n$$

where $x_1, x_2, \ldots, x_n$ are independent and all come from any single distribution that has a finite variance σ^2 and whose average is μ. The most common form of the central limit theorem states that the distribution of

$$\frac{w - n\mu}{\sigma \sqrt{n}}$$

approaches the unit normal distribution, i.e., the normal distribution which has a mean of zero and a standard deviation of one.

Other forms of the central limit theorem relax even these minimal conditions. For example, the theorem is also true when the variables are not identically distributed provided their variances are not too different. Another form of the theorem states that not all of the variables must be independent. Finally, there is a form of the central limit theorem which shows that the variables do not have to be independent nor do they have to come from a single distribution provided they are from a population of size $2n$ with a finite variance, where n is the sample size.

Essentially, this theorem states that in almost any data analysis situation, a sum of data is normally distributed. Except as noted above, the central limit theorem places no requirements on the distributions of the x variables; it only describes the distribution of the sum w.

A problem of concern to data analysts is the number of data n sufficient for a reasonable approximation to the normal distribution. A reasonable approximation might be one that passes a Lilliefors test of normality with a probability of 0.10 or more. The number n depends to a great extent on the shape of the distribution of the original x variables. If the x come from a bell-shaped but not necessarily symmetrical distribution, then the sum of as few as five or six independent observations can be considered normally distributed for most practical purposes.

A rectangular distribution is one in which every number is equally likely. Two examples are the outcome of flipping a coin and the outcome of rolling a single die. Another example is a table of random numbers. Only about 12 independent observations from any of these distributions are needed for the sum to be normally distributed.

The widespread applicability of the central limit theorem makes it a very useful tool. In fact, it is the justification for tests based on statistics that are *asymptotically normal,* i.e., approach the normal distribution as the sample size n increases. Examples are any of the rank procedures for large sample sizes. Since these tests are based on ranks, μ and σ for the distribution of the ranks are known. Such tests are certainly an important application of the central limit theorem.

Many measurements may be considered the result of many factors all acting together. For example, the yield of corn in a particular field is the result of soil conditions, weather conditions, type and amount of fertilizer used, cultivation practices, etc. If these influences on the yield are somewhat additive, then the central limit theorem offers a possible explanation why the distribution of such measurements can often be approximated by the normal distribution. A similar argument can be made for many other variables that data analysts study. Nonnormality, when it occurs, may really be abnormal after all.

THE CHI-SQUARE DISTRIBUTION

Let $x_1, x_2, \ldots, x_n$ be n independent observations from a normal distribution with mean μ and variance σ^2. The distribution of the sum of squares of the standard normal variables

$$z = \frac{x - \mu}{\sigma}$$

is called chi-square, i.e.,

$$\chi^2 = z_1^2 + z_2^2 + \cdots + z_n^2$$

has the chi-square probability distribution for n degrees of freedom,

$$f(x^2) = \frac{1}{2^{n/2}\Gamma(n/2)}\, e^{-x^2/2}(x^2)^{(n/2)-1}$$

where $\Gamma(n/2)$ is the gamma function for $n/2$.

The range of the chi-square distribution is zero to plus infinity since negative values of chi-square are impossible. The distribution is highly skewed for small values of n, but since chi-square is the sum of n independent identically distributed variables, it follows from the central limit theorem that chi-square approaches the normal distribution for large n. Graphs for this function when $n = 2$, 4, and 10 are given in figure 13.6. Note the approach to symmetry as n increases.

Except in a few special situations, data analysts do not know μ and σ, so let

$$t = \frac{x - \bar{x}}{\sigma}$$

Then

$$t_1^2 + t_2^2 + \cdots + t_n^2$$

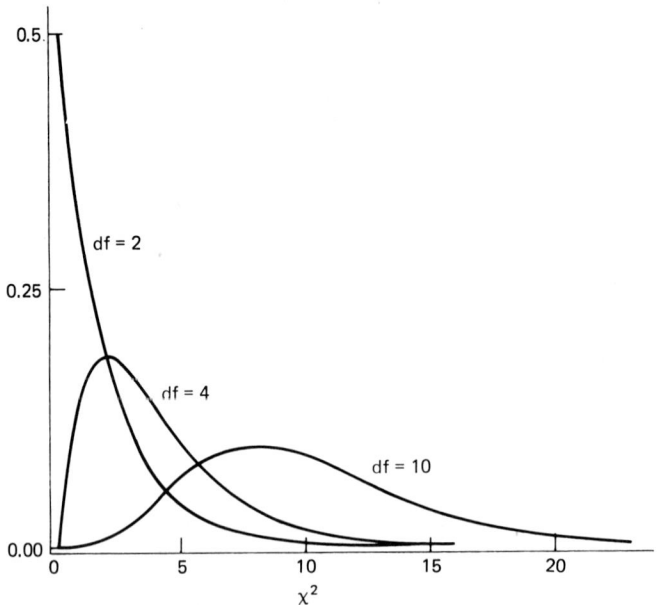

FIGURE 13.6   The chi-squared probability density.

also has a chi-square distribution but with $n - 1$ degrees of freedom. The above can be written as

$$\Sigma t^2 = \frac{(n - 1)sd^2}{\sigma^2}$$

then $(n - 1)sd^2$ is distributed as $\sigma^2 \chi^2$, a fact that will be used in connection with the F distribution below.

The principal use made of the chi-square distribution in this book is for testing the independence of two classifications in a contingency table or the equivalency of variances in Bartlett's procedure. The test statistics for these purposes, however, are distributed only approximately as the chi-square described above.

THE *t* DISTRIBUTION

Let u and v be two independent variables, u normally distributed with mean μ and variance σ^2, and v distributed as chi-square with $n - 1$ degrees of freedom. The variable

$$t = \frac{(u - \mu)/\sigma}{\sqrt{v/(n - 1)}}$$

has Student's probability distribution with $n - 1$ degrees of freedom,

$$f(t) = \frac{\Gamma(n/2)}{\sqrt{(n - 1)\pi}\Gamma(n/2 - 1/2)}\left(1 + \frac{t^2}{n - 1}\right)^{-n/2}$$

where Γ is the gamma function. Since v is distributed as chi-square,

$$\frac{(n - 1)sd^2}{\sigma^2}$$

can be substituted for v, and t written as

$$t = \frac{u - \mu}{sd}$$

Thus in data analysis, Student's t is the ratio of a difference from the population mean to the sample standard deviation in samples from a normal population. It is independent of the population standard deviation, i.e., σ does not appear in the definition of t or in the probability density function for t, $f(t)$, above.

Student's t-statistic has been adapted for use in many data analysis situations. All one must do is show that the numerator is normally distributed while the denominator is distributed as chi-square with $n - 1$ degrees of freedom. One of the most important uses of Student's t is to test the null hypothesis that a population mean is not different from a specified constant. To do this, u is a sample average, $\bar{x}$, μ is the specified constant, and sd is the standard deviation of the sample average, sometimes called the standard error, i.e., the sample $sd/\sqrt{n}$. The t-statistic for this test is, of course,

$$t = \frac{\bar{x} - \mu}{sd/\sqrt{n}}$$

Student's t-statistic can also be used to set confidence intervals for the population mean μ independently of the population standard deviation.

This distribution was first studied by W. S. Gosset in connection with problems he had as a chemist employed by Guinness Breweries. He wrote under the pseudonym of Student, hence the name usually given the distribution. Initially he conducted a large-scale sampling experiment from which he deduced the independence of the sample average and sample variance. After obtaining the distribution of sd^2 empirically, he deduced the exact distribution of

$$t = \frac{(\bar{x} - \mu)}{sd}$$

Student's papers have been collected for publication [see Pearson and Wishart (1958)]. They are interesting to read from a historical viewpoint to see how some of the pioneer data analysts solved their problems.

The range of Student's t distribution is from minus infinity to plus infinity just like the normal distribution. The distribution is symmetrical for all degrees of freedom and rather quickly approaches the normal distribution as a limit (see figure 13.3). Most tables of the t distribution do not go beyond 30 degrees of freedom because the normal is considered an adequate approximation for all degrees of freedom greater than 30. This is often used erroneously as the basis for assuming that a sample size of 30 is adequate for all purposes.

THE F DISTRIBUTION

It was noted above that $(n - 1)sd^2$ was distributed as $\sigma^2 \chi^2$ where $n - 1$ is the degrees of freedom. Let sd_1^2 and sd_2^2 be independent sample estimates of σ^2 based on k and n degrees of freedom, respectively. Then

$$F = \frac{sd_1^2}{sd_2^2}$$

is the ratio of two chi-square-distributed variables each divided by its degrees of freedom. The F-statistic has the probability density function

$$f(F) = \frac{k^{k/2}n^{n/2}}{\beta(k/2,\, n/2)}\, F^{(k/2)-1}(n + kF)^{-(k+n)/2}$$

where β is the beta function. Note that the function for F depends only on the degrees of freedom for the two variables; it does not depend on σ.

The range of the F distribution, like chi-square, is zero to plus infinity, since negative values of F are impossible. The distribution is highly skewed for small values of k and n but becomes more symmetrical as either increases. It can be shown that the square of t for n degrees of freedom is identical to F for 1 and n degrees of freedom. Figure 13.7 pictures the F distribution for several different combinations of degrees of freedom.

NONPARAMETRIC STATISTICS

The distribution of most nonparametric statistics is quite complex and considered beyond the scope of this book. Discussions of these distribu-

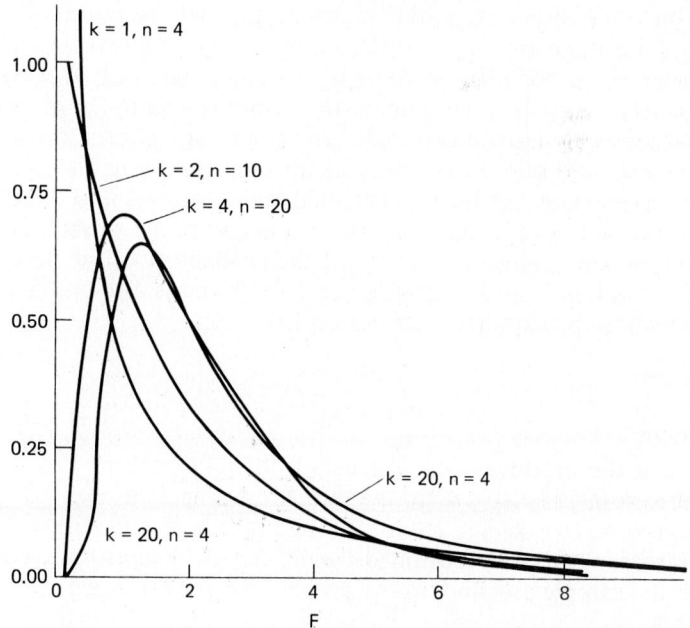

FIGURE 13.7 The F probability density.

tions are not easy to find in statistical literature, and when two or more references can be found, it often takes considerable effort for a data analyst even to see that the references are indeed on the same nonparametric statistic.

Statisticians create a mystique for nonparametric statistics by presenting results without presenting some of the very simple assumptions that would help data analysts understand the statistician's rationale. A good example of this is the very complex and almost incomprehensible methods usually described for ties in the Wilcoxon tests. Procedures presented in chapters 2 and 3 using ordinary sums of squares of ranks seem to this writer easier to comprehend and to use.

Data analysts interested in learning about the exact distributions of some of the more common nonparametric statistics may refer to Owen (1962) and Johnson and Kotz (1969) for references that may be helpful.

As the sample size increases, the distribution for most nonparametric statistics can be approximated by one of the distributions discussed in this chapter. These approximations have been described as the tests were presented in chapters 2 through 11 and so are not presented again here.

THE BINOMIAL DISTRIBUTION

Consider data measured with a nominal scale that has only two classes and let the probability be p that an observation will be in one of the two classes. Since there are only two classes and all data must be in one or the other, then the probability of being in the other class is $1 - p$. Such situations occur routinely in data analysis. Some examples of two-class nominal scales are female and male, Texan and not Texan, lose and win, A.M. and P.M., and pluses and minuses for the sign test in chapter 3.

For convenience, call the two nominal scale classes A and B. Now suppose that a sample of n data is observed and that the probability of an observation being in class A is p while the probability of the observation being in class B is $1 - p$. The number of data x in class A is itself a random variable whose probability distribution is

$$p(x) = \frac{n!}{x!(n - x)!} \, p^x(1 - p)^{n-x}$$

where $n!$ is the product of the integers from 1 through n. The values of $p(x)$ for $x = 0, 1, 2, \ldots, n$ are the terms resulting from expanding $(p + q)^n$ where $q = 1 - p$, i.e., from multiplying $p + q$ times itself n times. For this reason, this distribution is usually called the binomial. It depends only on the sample size and on the probability p of the observation being in class A.

The average number of data assigned to class A in binomial samples is np, while the variance is $np(1 - p)$. From these, the proportion of the data in class A is p and the variance of the proportion is $p(1 - p)/n$. There is a different distribution for each combination of n and p, so tables of the binomial distribution can be very large. The National Bureau of Standards published an extensive table of the binomial in 1950 under the title "Tables of the Binominal Probability Distribution." This publication contains both individual terms and partial sums for all combinations of $n = 2(1)49$ and $p = 0.01(0.01)0.50$. The notation $a(i)b$ used here is standard for describing tables and should be read "all values from a to b inclusive with increment $= i$." Thus the values of p tabulated are

$$0.01, 0.02, 0.03, \ldots 0.47, 0.48, 0.49, 0.50$$

The binomial for a limited range of n and p is presented in part V as table H. This table is useful for finding confidence limits for proportions. For example, suppose that census tracts are to be represented in a survey by 15 households and that in a pilot survey, 6 households in 120 refuse to participate in the survey. From the subtable for $n = 15$ and $p = 6/120 = 0.05$, there is a probability of 0.8290 that the number of households per census tract refusing to participate in the total survey will be less than 2. Hence an upper 82.9% confidence limit for p is $2/15 = 13.3\%$.

Table G in part V is derived from the binomial for $p = 0.5$ but presented in a form more useful for the sign test discussed in chapter 3.

CHAPTER 14

Transformations

INTRODUCTION

Measurement is the process of assigning numbers to objects or observations. In chapter 1, measurement scales were defined on the basis of relationships appropriate for each. It is necessary to extend the ratio and interval definition somewhat for a succinct discussion of transformations in this chapter.

A ruler can be graduated in inches or in centimeters, but it could just as well be graduated, for example, in logarithms of inches or centimeters. This scale would not be linear, so measurement would be a bit difficult, but such a ruler could measure distances directly in logarithms. Figure 14.1 presents two such rulers for measuring inches and centimeters in natural logarithms. Anyone using a logarithmic ruler must remember that any distance measured starts at minus infinity, since from the calculus

$$\lim_{y \to 0} \ln y = -\infty$$

where ln is the natural logarithm function.

The purpose of this discussion is to demonstrate that measurement scales do not have to be linear. In fact, there are many nonlinear scales in daily use. The Richter magnitude scale for measuring earthquakes is logarithmic. The scale on logarithmic graph paper is really an exponential scale. It may be difficult to understand, but by plotting measurements made with a linear scale on this exponential scale, we end up with logarithmic displacements on the graph paper. Many chemical analyses, particularly in the parts-per-million (ppm) range, are based on comparing

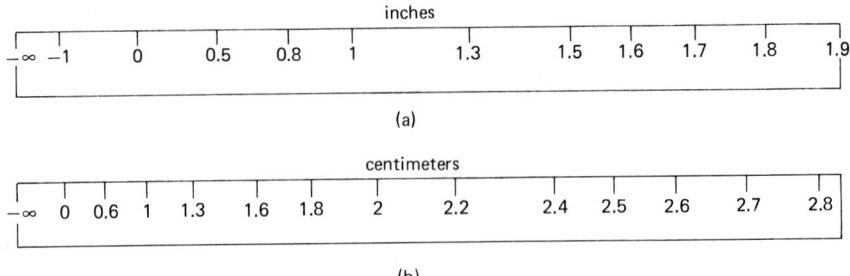

FIGURE 14.1 The logarithmic metameter. (*a*) Inches in the natural logarithmic meta-meter; (*b*) centimeters in the natural logarithmic metameter.

samples with standards that differ by a factor of 2, e.g., 1 ppm, 2 ppm, 4 ppm, 8 ppm.

The word *metameter* has been used in a very special area of statistics, i.e., bioassay, to denote measurement scale in this sense. For example, common rulers have a linear metameter, the Richter metameter is logarithmic, while the metameter for logarithmic graph paper is exponential. A transformation in statistics can be defined as any procedure, formula, or table for changing measurements made with one metameter to measurements made with another metameter. A couple of examples will make the concept clear.

Consider the data in table 14.1. Column 1 contains raw data as observed for a study. Column 2 contains the same data after transforming to natural logarithms, i.e.,

$$y = \ln x$$

If the metameter for column 1 is linear, then the metameter for column 2 is logarithmic. Now suppose that raw data less than 8.0 are considered

TABLE 14.1 Transformations Explained

Raw data, x	$\ln, y$	Class, z
5.7	1.74	1
8.3	2.12	2
11.9	2.48	2
12.8	2.55	3
7.0	1.95	1
9.7	2.27	2
14.0	2.64	3
10.2	2.32	2
8.8	2.17	2
9.4	2.24	2

class 1 responses, greater than 12.0 are class 3 responses, and between 8.0 and 12.0 inclusive are class 2 responses. Then the z's in table 14.1 are the classes of the x-data according to this scheme. The metameter for column 3 is an arbitrary one as defined above.

PURPOSE

There is no universally applicable statistical procedure for any purpose. Specific conditions are required for every statistical procedure, and prudent data analysts always check their data for satisfying these conditions. The central position that the normal distribution and all of the procedures derived for normally distributed data plays in data analysis means that the data analyst usually wants data to be normally distributed. Very frequently an apparently nonnormal data set can be made approximately normal by a transformation, i.e., by a change in metameter. The most frequently used are some combination of the logarithmic, exponential, square, and/or square root transformations. Also algebraic, trigonometric, and/or inverse trigonometric functions are used sometimes. The ultimate metametric change moves data from one measurement scale to another as in the class definitions for table 14.1. It is always difficult to obtain the best transformation for any particular situation, but a few rules should help data analysts make good choices.

1. The transformation should make sense, e.g., the measurement method may hold a clue.
2. The same transformation should be applicable to repeated experiments, i.e., different sets of data employing the same experimental units and measurement procedures.

Consider, for example, chemical analyses determined by comparing samples with standards that are 1, 2, 4, 8, ... ppm with guesstimates made between the standards. The distribution of such data is likely to have a long tail toward larger measurements simply because larger measurements are made less precise than smaller measurements. Some form of the logarithmic transformation might be useful for pulling the tail in and thus producing a more symmetrically, and perhaps normally, distributed sample. Table 14.2 contains data from such a chemical analysis and figure 14.2 shows the sample distribution as a histogram. Note that there is a longer tail toward larger results. Table 14.2 also contains the natural logarithmic transforms of the data and figure 14.3 shows the new sample distribution. This histogram is decidedly more symmetrical than the first.

The logarithmic transformation would probably be appropriate for all experiments yielding chemical measurements such as those described above, but it might not be correct for some other analytical method, e.g., titrations. This is not the point of rule 2 above. Rule 2 is intended to caution against the following situation.

TABLE 14.2   Chemical Analyses

y, ppm	ln y	y, ppm	ln y
22	3.09	19	2.94
33	3.50	22	3.09
21	3.04	18	2.89
19	2.94	22	3.09
19	2.94	28	3.33
14	2.64	16	2.77
19	2.94	17	2.83
21	3.04	22	3.09
23	3.14	15	2.71
17	2.83	14	2.64
13	2.56	23	3.14
24	3.18	27	3.30
25	3.22	17	2.83
20	3.00	19	2.94

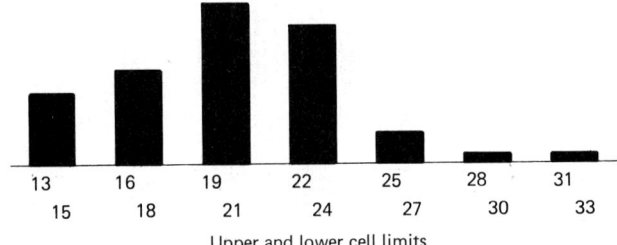

FIGURE 14.2   Raw data.

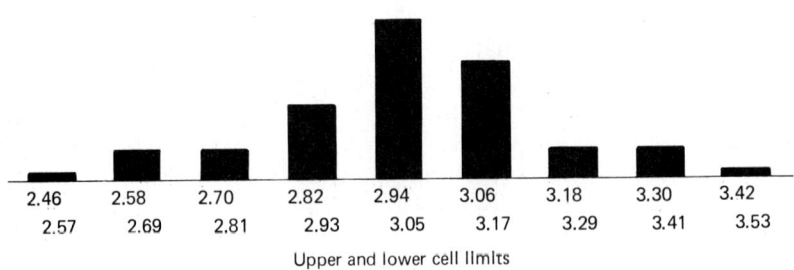

FIGURE 14.3   Transformed data.

    A very long term study yields a dozen or so observations every year. In eagerness to show some results and hence to justify the costs of the study, preliminary analyses are made about every 6 months. The data sample is obviously not from a normally distributed population, so the best transformation in the sense of normalizing the sample data is determined by an elaborate computer program trying all of the basic transformations; square, square root, exponential, logarithmic, trigonometric, inverse trigonometric, and even some combinations. No thought is given to the transformation making sense, i.e., having a logical basis, yet out of the scores tried, a best one is chosen and used for a preliminary analysis. Now another 6 months passes and it is time for an update report. The same process is repeated but, of course, a different transformation is now best. The update reports an analysis based on the new transformation which makes no more sense than did the first.

This scenario really did occur, but when I tried to explain the absurdity of the procedure, my client went off muttering something about the consultant not understanding the situation. The consultant may not have appreciated the politics of the situation but did understand the data and what constitutes a reasonable analysis. I hope this book will help analysts find better alternatives. It is perhaps a good rule to choose transformations on the basis of inherent data characteristics rather than sample characteristics.

EXAMPLES

Sometimes the logarithmic or square root transformations are appropriate with negative data for which the logarithm and square root are not defined. It is customary in such cases to add a large constant to the raw data to eliminate all negative values. A scaling factor may also be used for a more general transformation, e.g.,

$$y = \ln (a + bx)$$
$$y = \sqrt{a + bx}$$

Sometimes the exponential or square transformations are appropriate for data which already have rather large magnitudes. It is customary in these cases to divide the raw data by a constant so that the transformed data is in a more manageable range. The transformations would then be

$$y = e^{x/a}$$
$$y = \left(\frac{x}{a}\right)^2$$

Of course, other terms could be added for even more flexibility. For example,

$$y = c + e^{x/a+b}$$
$$y = c + \left(\frac{x}{a} + b\right)^2$$

The square and square root have been suggested as transformations but a more general power function might be used, such as

$$y = x^t$$

where t is any real number. For example, $t = 0.5$ would be the square root transformation, but $t = 0.8$ might be appropriate in some special situation.

Analysts should not hesitate to try unusual transformations or even combinations of transformations provided the two rules given above are not violated and the results of the analysis can be adequately explained to the client.

LIMITATIONS

A viewpoint adopted consistently within this book is that analyses and the results of analyses should be easily explained to and understood by managers and researchers. Transformations that do not satisfy the two rules given above are usually difficult to explain and often misunderstood, as was discussed in chapter 2 for an actual problem in analysis. Whenever a truly satisfactory transformation cannot be found within the ratio or interval measurement scale, analysts are well advised to transform to the next lower measurement scale, e.g., to ranks. A change in measurement scale is usually easy to explain and the advantages of the normal distribution are just not worth incomprehensible results.

Throughout this chapter, the implied purpose for transformations has been to normalize sample data, but this is not the only reason for using transformations. Perhaps a more important and certainly a more difficult use of transformations is to satisfy conditions necessary for some of the more complex analyses of variance. Two important papers on this subject appeared in volume 3 of *Biometrics*. Eisenhart (1947) discusses the assumptions underlying the analysis of variance and then Bartlett (1947) discusses satisfying these assumptions through transformations. This problem, however, is beyond the scope intended for this book.

CHAPTER 15

Statistical Tests

INTRODUCTION

Statistical tests can only be understood in terms of hypotheses and hypotheses arise from management and/or research problems. So beginning with problems, the sequence from research hypothesis through null hypothesis to statistical test is described. The purpose and characteristics of tests in general are discussed and then it is shown how several specific tests satisfy the general purpose and possess the characteristics described. Not all tests used in chapters 2 through 11 are discussed in this chapter; only representative tests sufficient to present the general notion of statistical tests are included.

PURPOSE

A research or management objective to document the existence of desirable or undesirable differences generates data analysis problems. Statistical tests provide a formal framework for evaluating differences which may be between two or more averages (chapters 2 to 5), differences of regression slopes from zero (chapters 6 to 9), or differences between regression slopes (chapters 10 and 11). Initially such problems are not described in terms of hypotheses so the first service a data analyst often must do for a client is to negotiate a research hypothesis formally stating the anticipated or existent difference.

Dictionaries define a *hypothesis* as "an assumption or concession made for the sake of argument," and that is the meaning adopted here. It is impossible, except in pure mathematics, to prove a hypothesis, so no attempt is ever made to prove a research hypothesis, but the research hypothesis does provide a clear statement of what is believed to be the problem. In contrast, it is always possible, in the real world of data analysis, to improve measurement methods and thereby to demonstrate that things are different. This inability to prove equality and the universal ability to demonstrate inequality provides the basis for all statistical tests.

The logic is this. Examine the data for consistency with a null hypothesis that no difference exists between averages, slopes, etc. This null hypothesis stands in contrast with the research hypothesis which states a suspected or anticipated difference. If the data are consistent with the null hypothesis, then it is assumed that the research hypothesis is probably not true. The null hypothesis has not been proved. The data analyst has just failed to find evidence contrary to it. If, on the other hand, the data are inconsistent with the null hypothesis, then the null hypothesis assumption of no difference is rejected. The data analyst has still not proved the research hypothesis but has found evidence contrary to the null hypothesis. The analyst may then want to estimate the difference that appears to exist. Chapter 19 discusses this subject.

This examination of the data is the statistical test. The test consists of comparing a measure of data dispersion that includes the difference anticipated by the research hypothesis to a measure of dispersion that specifically excludes the difference. All comparisons are ratios (rather than differences), so that most test statistics are dimensionless numbers. Since the difference described by the research hypothesis is always in addition to random or chance data differences, the magnitude of the test statistic's numerator should be larger than the denominator, and the larger the ratio, the larger the difference. The principal test defined the other way around is the Shapiro-Wilk test for normality. Another is Wilks' (1951) criteria for multivariate analysis of variance, a subject not considered in this book.

Fisher (1973) notes that the common tests of significance were all conceived and/or made mathematically precise by data analysts for immediate problems in their own research.

ALTERNATIVES

A researcher should know the kind of difference expected, e.g., the desired change or the way something has changed. Very seldom is one looking for change that might occur in either direction. The same is true for management problems. Consequently, research hypotheses are usually positive statements of specific differences. The only alternatives to the null hypothesis that are of interest are differences in the direction described by the research hypothesis. Interest in differences opposite to that may indicate the problem and/or the research objectives are not adequately understood.

Statistical tests, for the situation described above, are called one-tailed tests. Two-tailed tests are in contrast to one-tailed tests and imply an interest in differences in either direction from the null hypothesis. They usually occur at some intermediate step in an analysis when the analyst would really like to prove a null hypothesis. Of course, null hypotheses are never proven so the next best thing to do is show that a two-tailed test statistic has a high probability of occurrence because of random or chance differences in the data.

Since most data analysis tests are one-tailed tests, most of the tables in part V have been arranged for one-tailed tests. This creates a uniformity that many sets of statistical tables do not possess. The exceptions to this uniformity are the tables of multiple comparison factors. No a priori assumptions are made concerning which sample will have the larger mean, slope, or intercept in Duncan or Newman-Keuls comparisons, so these tests are two-tailed. Such a priori assumptions may or may not be made for Dunnett comparisons. Consequently, Dunnett provided tables of factors for both one- and two-tailed tests.

t TESTS

In chapter 13, the *t*-statistic was defined as the ratio of a sample mean measured from the population mean to the standard error in samples from a normal population. Many problems can be formulated this way once normality has been established. The most common *t*-statistic is the test of the mean of a sample being different from a constant, which may be zero. The problem of assessing the difference between two dependent samples can be put in this context by creating the population of data pair differences (chapter 3) and then comparing the average difference to zero, after determining that the distribution of the differences can be approximated by the normal. In both situations, the null hypothesis would be that the average is not different from the constant.

The standard error is the sample standard deviation divided by the square root of *n*, the sample size. The formula for the test statistic can be written as

$$t = \frac{\bar{x} - c}{\text{sd}/\sqrt{n}}$$

where
$\bar{x}$ = the average of the sample
c = the constant to which $\bar{x}$ is compared
sd = the sample standard deviation
n = the sample size.

This formulation satisfies the definition of a *t*-statistic. The numerator is the deviation of a sample mean, whose normality is assured by the central limit theorem, from a population mean assumed to be c. The denominator is the standard error computed from the standard deviation of the sample. The sample mean and standard deviation have the same metameter, so the test statistic is dimensionless.

The test is completed by comparing the calculated value of the test statistic to a table of the t-statistic that assumes the sample mean does not differ from c. This aspect of the procedure is discussed in chapter 17.

Data analysts also compare averages of samples from two independent populations (chapter 2). To do this, the analyst in effect creates a new population whose members are the differences between averages of samples of n_1 and n_2 data from the independent populations. A data analyst usually has only one datum from this new population, the one computed from the two samples. Nevertheless, the population is real and must be understood by the data analyst.

When the standard deviations of the two independent populations can be assumed equal, an estimate of the standard deviation of the new population can be obtained by "pooling" the two sample standard deviations. The pooled variance is

$$sd^2 = \frac{(n_1 - 1)\, sd_1^2 + (n_2 - 1)\, sd_2^2}{n_1 + n_2 - 2}$$

The pooled standard deviation of the new population, often called the standard error of the difference, is

$$sd \sqrt{\frac{1}{n_1} + \frac{1}{n_2}}$$

The formula for the test statistic can now be written as

$$t = \frac{\bar{x}_1 - \bar{x}_2 - c}{sd \sqrt{1/n_1 + 1/n_2}}$$

where $\bar{x}_1, \bar{x}_2$ = the sample averages
c = a constant to which the difference in averages is compared
sd = the pooled sample standard deviation
n_1, n_2 = the sample sizes

This formulation satisfies the definition of a t-statistic. The numerator is the deviation of a sample datum whose normality is assured by the central limit theorem from a population mean assumed to be c. The denominator is the standard error of the difference between the sample means computed from the sample standard deviations. Assuming the same metameter was used for the two samples, the test statistic is dimensionless. Null hypotheses such as

The means of the two populations are the same except for a constant displacement c

would not be possible without c in the formula for t. When the populations are assumed to have equal averages, then c is zero.

It is an entirely different problem when the standard deviations of the two populations cannot be assumed equal. This famous and historically important statistical problem has been given a special name, the Behrens-Fisher problem, denoting two pioneers who studied it. The Behrens-Fisher problem is not discussed in this book. Instead, a transformation of the sample data to the ordinal scale and then a rank procedure is recommended. Readers interested in a discussion of the problem are referred to Fisher (1973).

F TESTS

The F-statistic was defined in chapter 13 as the ratio of two independent estimates of a variance. Sometimes the two estimates are for separate populations where the null hypothesis is that these populations have the same variance (see chapter 2). In this situation, the analyst has no preconception about which of the two estimates is likely to be the largest. The research hypothesis may even make the analyst hope for equality or equivalence.

Such an F-test is said to be two-tailed, since if the null hypothesis of equality cannot be accepted, the variance of the first population may be either greater than or less than the variance of the second. F-tables are arranged for one-tailed tests with the larger variance in the numerator. Consequently, the larger of the two variances in a two-tailed test must be the numerator of the F-statistic and the analyst must remember to double the apparent p-values for all two-tailed tests.

The usual F-test compares two independent estimates of the same variance (see chapters 4 to 11). The research hypothesis is that a specific one of these estimates will be larger than the other, while the null hypothesis states that there is no difference between the two estimates. Such F-tests are one-tailed in the sense that if the two variances are not equal, the differences of interest to the analyst can occur in only one direction as anticipated by the research hypothesis.

Typically the numerator of the F-statistic measures the variance as if the difference anticipated by the research hypothesis existed, while the denominator measures the variance as if the null hypothesis was true. The F-statistic defined this way should always be greater than 1.0, but in actual practice, data analysts occasionally encounter one-tailed F-statistics that are less than 1.0. If the statistic is not much larger than 1.0 or if it is less than 1.0, then there is little evidence against the null hypothesis and the differences specified by the research hypothesis probably don't exist. An F-statistic much larger than 1.0 is evidence that the null hypothesis is not tenable and so it is rejected. What follows in the analysis depends on the problem and how the analyst pursues it.

The one-way analysis of variance is the simplest data analysis situation requiring an F-test (chapter 4). The situation is this. There are more than two samples from populations that have equal variances. The research hypothesis asserts that the means of the populations are not all equal while the null hypothesis, in contrast, asserts that they are. The numerator and denominator of the F-statistic must measure the common population variance under these respective assumptions.

Consider the denominator first. Before computing F, the analyst will have established that the variances of the populations can be considered equal. An estimate of this common variance based on all of the samples is obtained by pooling.

$$\text{sd}^2 = \frac{(n_1 - 1)\,\text{sd}_1^2 + (n_2 - 1)\,\text{sd}_2^2 + \cdots + (n_k - 1)\,\text{sd}_k^2}{n_1 + n_2 + \cdots + n_k - k}$$

This is just an extension of pooling variance estimates for the t-tests discussed in the previous section and is algebraically equivalent to the denominator used for the analysis of variance in chapter 4. To compute the pooled estimate of variance, the deviations of the data from their respective sample means are used to estimate each sample standard deviation separately. All differences between the sample means are removed by this process. Thus the denominator is an estimate of the common variance such that differences between the means of the populations, should they exist, will have no effect.

Now consider the numerator. It must be an estimate of the common population variance that includes differences between the populations, if any exist. The variance of the sample averages will measure between population differences.

$$\frac{\Sigma(\bar{x}_i - \bar{\bar{x}})^2}{k - 1}$$

where $\bar{x}_i$ = sample averages
$\bar{\bar{x}}$ = the overall average
k = the number of samples

Now the variance of sample means is the standard error squared, so it must be multiplied by the sample size for an estimate of the common population variance. If all of the samples are the same size, this is

$$\frac{n\Sigma(\overline{x}_i - \overline{\overline{x}})^2}{k - 1}$$

The sample sizes can be different with

$$\frac{\Sigma n_i(\overline{x}_i - \overline{\overline{x}})^2}{k - 1}$$

This last form is algebraically equivalent to the formula used for between samples in chapter 4. The F-statistic constructed from this and the pooled estimate of variance given earlier in this section is identical to the F-statistic used in chapter 4.

The F-statistics used in chapter 5 are the same as the above once the differences due to the secondary classification are removed. The reader is invited to verify this by comparing the above with procedures examined in chapter 5.

Linear regression provides an example of an F-statistic that may at first appear completely different from the above but, on closer inspection, is quite similar. In the simplest situation, the research hypothesis is that there is a relationship between two variables and that knowing one of these variables, which is generally called x, helps in estimating or understanding the other, called y, i.e., $y = \alpha + \beta x$. Now the null hypothesis disputes this assertion. It states simply that x does not help in estimating or understanding y. This is equivalent to saying that $\beta = 0$.

To explore the controversy created by these opposing hypotheses, α and β are estimated. Then the variance of the data is calculated in two different ways, first using the estimates of α and β and then using an estimate of α, assuming $\beta = 0$. The two estimates are

$$\text{Variance about regression} = \frac{\Sigma(y - a - bx)^2}{n - 2}$$

$$\text{Total variance} = \frac{\Sigma(y - \overline{y})^2}{n - 1}$$

where a and b are the estimates of α and β and n is the sample size. These estimates violate a first principle for F-tests—they are not independent. Statisticians can show, however, that

$$\text{Variance due to regression} = \frac{\Sigma(y - \overline{y})^2 - \Sigma(y - a - bx)^2}{1}$$

is independent of the first estimate above and, furthermore, that the ratio

$$\frac{\text{Variance due to regression}}{\text{Variance about regression}}$$

satisfies all the conditions to be a one-tailed F-statistic with 1 and $n - 2$ degrees of freedom.

All of the F-statistics used for tests in chapters 6 through 9 are essentially the same as the above ratio.

MULTIPLE COMPARISONS

After a significant F-test comparing more than two samples, analysts still have no information on the nature and size of the contributory differences. Multiple comparisons of averages, regression slopes, or regression intercepts for the sample pairs can often add much useful information for researchers and managers, so the procedures illustrated in chapters 4, 5, and 11 are discussed in more general terms and a little more detail in this section.

There is a serious problem with multiple comparisons, and that is, the more comparisons conducted, the more often a true null hypothesis will be rejected unless a specific effort is made to control this kind of error. For any given set of k samples, there can be only $k - 1$ independent, i.e., orthogonal, comparisons. So when all possible sample pairs are compared, many comparisons are simple functions of the others. For example, consider three samples compared as follows:

$$A - B \qquad A - C \qquad B - C$$

The third comparison is just the second, $A - C$, minus the first, $A - B$, i.e.,

$$(A - C) - (A - B) = B - C$$

The number of comparisons possible increases rapidly as the number of samples increases. There are 3 independent comparisons for 4 samples in a total of 6 two-sample comparisons. For 6 samples there are 5 independent comparisons in 15 possible comparisons. For 10 samples, 45 two-sample comparisons can be made, but only 9 of these are independent.

An "experimentwise error rate" can be estimated, assuming that all observable sample differences are just random variation. This error rate is

$$\alpha_{\text{ER}} = 1 - (1 - \alpha)^k$$

where α = the significance level for the comparisons
 k = the number of comparisons, not the number of samples

All possible comparisons are *not* independent and multiple comparisons are never made unless the F-test in an analysis of variance indicates the sample differences are *not* all due to chance. The actual experimentwise error rate in any particular situation depends on the true population differences, which the analyst never knows. Nevertheless, the formula for α_{ER} is often used, with k defined as the number of comparisons, to demonstrate how fast the experimentwise error rate can increase.

Assuming α = 0.05, then the experimentwise error rate for k = 3 (all possible pairs for three samples) is 0.14. For k = 6 (all possible pairs of four samples), it is 0.26 and for k = 10 (all possible pairs for five samples), it is 0.40. Assuming all the restrictions and conditions discussed above apply, then the experimentwise error rate goes up rather dramatically as the number of samples increases. Such high error rates would be disastrous in an analysis, so some control, i.e., reduction, is desirable.

All of the multiple comparison procedures are attempts to control, through special limits for sample differences, the probability of rejecting true null hypotheses. If a sample pair differs by more than the appropriate limit, then the pair is said to be *significantly* different. If they differ by less than the appropriate limit, then the pair is *not significantly* different. Multiple comparison procedures differ in the method and effectiveness of their control of the experimentwise error rate.

Fisher's least-significant difference (LSD) is defined as

$$LSD = t_{df,\alpha/2}\ \text{sd}\ \sqrt{\frac{1}{n} + \frac{1}{m}}$$

where sd = the residual standard deviation from the ANOVA table
 n, m = the two sample sizes
 df = the degrees of freedom for sd
 α = the probability of rejecting a true null hypothesis
 t = Student's t-statistic for α and df degrees of freedom

No effort is made with the LSD to control the experimentwise error rate. It would be at least as large as that described for independent samples. In contrast, the Tukey and Scheffè procedures control the experimentwise error rate as if all possible comparisons were being made. All other multiple comparison procedures, e.g., Duncan, Newman-Keuls, Dunnett, Dunn, etc., lie somewhere between these two extremes. Several of these intermediate procedures have been used in parts I and III.

Duncan (1955) proposed a procedure that adjusts the size of the comparison limit according to the number of samples spanned by a comparison. For example, suppose three samples are ordered so that $A < C < B$, then a comparison of A and B will have a slightly longer limit than the limit for a comparison of A and C or C and B. Duncan's critical range is

$$\text{CR}_{\text{Duncan}} = q_{r,\text{df}} \text{ sd } \sqrt{\frac{1}{n}}$$

where sd = the residual standard deviation from the ANOVA table
n = the sample size
r = the number of samples spanned by the comparison plus the two being compared
df = the degrees of freedom for sd
q = a special studentized range statistic

Table Q in part V provides values of q for combinations of r, df, and α, the significance level for the tests.

The term *studentized* requires explanation. Kendall and Buckland (1957) define *studentization* as

The process of removing complications due to the existence of an unknown parent scale-parameter by constructing a statistic whose sampling distribution is independent of it.

In multiple comparison problems, the differences or ranges between sample means have dimensions, e.g., meters, degrees, tons, etc., depending on the measurement metameter (see chapter 14). A studentized range is a range divided by the sample standard deviation. Since this standard deviation will have the same measurement dimensions as the range itself, studentization makes the range dimensionless while at the same time standardizes the measurement scale. W. S. Gosset, whose nom-de-plume was Student, introduced the idea in 1907 when he discussed the distribution of the mean divided by the standard deviation.

The Newman-Keuls procedure is similar to Duncan's procedure but a little more conservative, i.e., provides slightly larger limits for comparison with sample pair differences. It also varies as the number of samples spanned by the comparison and is given by

$$\text{CR}_{\text{Newman-Keuls}} = q_{r,\text{df}} \text{ sd } \sqrt{\frac{1}{n}}$$

where all terms are the same as in the equation for Duncan's procedure except that the q statistics are given in table R of part V. A comparison of tables Q and R for r greater than two will show just how much larger are the Newman-Keuls factors than the Duncan factors.

The comparisons for the Duncan and Newman-Keuls procedures are made in a specific order: the largest to the smallest, the largest to the second smallest, and so on until a pair has a range less than the critical range for the number of statistics still spanned. When this happens, the comparisons with the largest cease and all remaining statistics between the two in the last test are considered not significantly different, at the α level, from the largest statistic. The process is repeated in turn for the second largest, the third largest, and so on until a statistic not significantly different from the smallest is found. At this point, all comparisons cease.

Several texts describe a procedure proposed by Tukey (1953) in an unpublished manuscript. This procedure holds the experimentwise error rate at α for all possible sample pair comparisons. The Tukey critical range is given by

$$\text{CR}_{\text{Tukey}} = q_{r,\text{df}} \, \text{sd} \sqrt{\frac{1}{n}}$$

where q = the studentized range given in table R in part V
r = the total number of samples

All other terms are the same as defined for the other multiple comparison procedures. Thus the Tukey critical range is the maximum Newman-Keuls critical range for any particular analysis, i.e., any specific number of samples.

Scheffè (1953) proposed a procedure that holds the experimentwise error rate at α for all possible comparisons of the samples, not necessarily limited to pair comparisons. Thus comparisons of more than two samples such as

$$A + B - 2C \qquad A + B - C - D$$

etc. may be included in the comparisons made for any one analysis. The critical range for the Scheffè procedure is given by

$$\text{CR}_{\text{Scheffe}} = \sqrt{(r-1)F_{r-1,\text{df}}} \, \text{sd} \sqrt{\frac{2}{n}}$$

where r = the total number of samples in the analysis
F = the F-statistic from table D in part V for $r - 1$ and df degrees of freedom

All other terms are as previously defined for the other critical ranges. The Scheffè procedure is unduly conservative, especially when researchers and managers only need pairwise comparisons.

Two special kinds of critical ranges are used in parts I and III. These are the Dunn (1964) and Dunnett (1955) procedures. Dunnett's procedure compares several treatment samples with a single control sample and should be used for this situation only. The formula for calculating the critical difference for the comparison of any treatment with the control is

$$CD_{Dunnett} = q_{r,df} \text{ sd } \sqrt{\frac{2}{n}}$$

where q is a statistic from table S in part V and all other terms are as previously defined in this section. A comparison of the factors in tables Q, R, and S will show that Dunnett's critical difference falls between Duncan's and Newman-Keuls' critical range in size.

Dunn (1961) proposed dividing the experimentwise error rate among the comparisons in any arbitrary manner so that the sum of the error rates for the comparisons equaled the desired experimentwise rate. The division can be done so as to emphasize the importance of one comparison over another. If all comparisons are equally important, which often is the case in data analysis, then the experimentwise error rate would be divided by the total number of pair comparisons to be made. Dunn's critical range is determined from

$$CR_{Dunn} = d_{c,df} \text{ sd } \sqrt{\frac{2}{n}}$$

where $c = $ the number of comparisons
$d = $ a quantity from a special table prepared by Dunn

All other quantities are as previously defined. This procedure for ratio and/or interval data was not used in part I, but a related nonparametric procedure given by Dunn (1964) was used in chapter 4. In this nonparametric procedure,

$$CR_{Dunn} = q_{\alpha/c} \text{ sd } \sqrt{\frac{1}{n} + \frac{1}{m}}$$

where $c = $ the number of comparisons to be made
$\alpha = $ the experimentwise error rate (e.g., 0.05)
$q = $ the standard normal deviate (table A in part V) for α/c
$sd = $ the standard deviation of the ranks when all of the data in all of the samples are ranked together
$n, m = $ the sample sizes

Dunn's nonparametric procedure depends, as do many other nonparametric procedures, on the fact that, according to the central limit theorem, averages of ranks tend to be normally distributed. So the samples shouldn't be too small. Twelve or more are sufficient for each sample, but the procedure is probably used successfully with smaller samples. If there are no ties in the N ranks (R) where N is the total number of data in all the samples, then

$$\Sigma R = \frac{N(N + 1)}{2}$$

$$\Sigma R^2 = \frac{N(N + 1)(2N + 1)}{6},$$

so that

$$sd_R^2 = \frac{N(N + 1)}{12}$$

This formula cannot be used whenever there are tied ranks.

Sometimes the multiple comparison procedures create rather unusual structures among the statistics compared, making it difficult to explain to researchers and managers which summary statistics can be considered equivalent and which should be considered different. Data analysts typically sort the statistics compared from smallest to largest and arrange these in a column. Then to the right of the column, straight lines or brackets are drawn to indicate groups that may be considered equivalent at the chosen α level. Examples of this kind of presentation may be seen in tables 4.8, 4.14, 5.8, 5.15, 5.20, 5.24, 11.13, and 11.19.

Most of the multiple comparison procedures, like many other procedures in statistics, were developed for equal sample sizes. Most data analyses do not involve equal-size samples, so to make the procedures applicable for real data analysis problems, they are modified in ways that analysts generally accept and consider reasonable. In the Duncan, Newman-Keuls, and Tukey procedures, the statistic multiplied by the q-factor is the denominator in a t-statistic for comparing an average to a constant. The typical adjustment for these procedures is that required in the denominator of a t-statistic to compare the difference of two averages to a constant. In other words,

$$\frac{1}{n}$$

is replaced by

$$\frac{\dfrac{1}{n_i} + \dfrac{1}{n_j}}{2}$$

where n_i and n_j are the unequal sample sizes.

In the Scheffè, Dunnett, and Dunn procedures, the statistic multiplied by the q- or the F-factor is the denominator in a t-statistic for comparing the difference of two averages of equal size samples with a constant. The typical adjustment, consequently, is to replace

$$\frac{2}{n}$$

with

$$\frac{1}{n_i} + \frac{1}{n_j}$$

where n_i and n_j are the unequal sample sizes. If Fisher's LSD above had been defined for equal sample sizes, this would be the adjustment required for it too.

RANK TESTS

The test statistics for analysis of ordinal data throughout chapters 2 to 5 are explained in terms analogous but not equivalent to t- and F-statistics. In all cases, the numerator is a measure of the differences described by the research hypothesis, while the denominator is a measure specifically excluding these differences. The denominator depends on the sample size and can sometimes be calculated more easily with a special formula, but that doesn't help the analyst understand what is being done. The reader should compare the nonparametric tests in chapters 2 to 5 with the t- and F-tests, as presented in this chapter, for further understanding of how tests in general are constructed.

BARTLETT'S TEST

Bartlett (1937) proposed using the quantity

$$M = -\ln sd^2 \Sigma(n_i - 1) - \Sigma(n_i - 1) \ln sd_i^2$$

where i identifies the samples from 1 to k

sd_i^2 = the variance of sample i
n_i = the number of data in sample i
$\ln$ = logarithms to the base e
sd = the pooled standard deviation assuming the null hypothesis that the samples all come from populations having the same variance, i.e.,

$$sd^2 = \frac{\Sigma(n_i - 1) \, sd_i^2}{\Sigma(n_i - 1)}$$

as a test for differences among a group of sample variances from normal populations.

It can be shown that M is approximately distributed as chi-square with k degrees of freedom where k is the number of samples. A closer approximation can be obtained by using the statistic M/c where

$$c = 1 + \frac{1}{3(k-1)}\left(\Sigma \frac{1}{n_i - 1} - \frac{1}{\Sigma(n_i - 1)}\right)$$

M/c is the statistic used throughout earlier chapters in this book.

THE CHI-SQUARE TEST FOR CONTINGENCY TABLES

Suppose nominal data can be classified in two or more ways. These ways are used to define subclasses and the number of data in each subclass constitutes the sample data. The arrangement of these data in a systematic geometric pattern is a contingency table. An example will clarify the concept.

Example 15.1

Medals awarded in the 1976 Winter Olympics (Innsbruck) can be classified several ways. One way could be by kind of medal, i.e., gold, silver, or bronze, and another way could be by country of the participant receiving the medal, e.g., East Germany, West Germany, U.S.S.R., U.S.A., and others. The 15 subclasses would be

U.S.S.R. gold, U.S.S.R. silver, other gold, East Germany silver, East Germany gold, U.S.A. silver, U.S.A. bronze, other silver, West Germany bronze, East Germany bronze, U.S.A. gold, other bronze, U.S.S.R. bronze, West Germany silver, and West Germany gold.

The number of medals awarded in each of these 15 subclasses can be counted. Table 15.1 shows how these counts can be placed systematically in a table. Such an arrangement of the data is called a contingency table.

The expected frequency in each of these categories can be calculated under the null hypothesis that the two ways of classifying the medals are independent. Then the chi-square goodness-of-fit statistic is calculated as follows:

$$\chi^2 = \Sigma \frac{(O_i - E_i)^2}{E_i}$$

TABLE 15.1 1976 Winter Olympics
Medals Data

	Gold	Silver	Bronze
East Germany	7	5	7
West Germany	2	5	3
U.S.A.	3	3	4
U.S.S.R.	13	6	8
Other	12	18	15

where $O =$ the observed count
$\quad\quad E =$ the expected count

An expected count for a subclass in a contingency table can be viewed two ways. Let j indicate the columns from 1 to c and let k indicate the rows from 1 to r. The expected count for cell j,k is the total count for column j apportioned among the cells in this column according to the row totals, i.e.,

$$E = n_j \frac{n_k}{n}$$

where $n_j =$ the total count for column j
$\quad\quad n_k =$ the total count for row k
$\quad\quad n =$ the total count for the table

Equivalently, the expected count for cell j,k is the total count for row k apportioned among the cells in this row according to the column totals, i.e.,

$$E = n_k \frac{n_j}{n}$$

These formulas usually result in fractional expected counts and should not be rounded to integers.

The summation for example 15.1 is over the 15 subclasses. The test is completed by comparing this statistic with the chi-square distribution for $(r - 1)(c - 1)$ degrees of freedom, where r and c are the number of rows and columns, respectively, in the contingency table.

Gibbons (1971) presents an excellent discussion of the chi-square goodness-of-fit test for contingency tables from the statistician's viewpoint. Assuming the classification categories, the frequencies constitute a set of random variables from the k-variate multinomial distribution.

Chi-square given above approximates the likelihood ratio statistic for a test of the null hypothesis. This approximation is good for large samples and is improved by combining categories so that the expected frequency for all cells is at least 5. Incidentally, some categories would have to be combined in table 15.1 to achieve expected values of 5 or more.

Now the two ways for classifying the 1976 Winter Olympics medals define nominal measurement scales. Sometimes data analysts transform ratio, interval, and ordinal data to a nominal scale, and as illustrated in chapter 4, the categories defined for the nominal scale can have a profound influence on the chi-square test. One purpose for this section is to emphasize again the importance of defining categories wisely or of avoiding contingency tables in such situations. Statistics texts do not consider this data analysis issue. In fact, they seem to avoid it.

The Chi-Square Test for a Two-by-Two Contingency Table

The two-by-two contingency table is used so much in analysis that it deserves a special section in this chapter. Suppose a random sample of n units can be classified as possessing or not possessing two attributes. The units can then be classified into one of four groups:

1. Has both attributes
2. Has attribute 1 but not attribute 2
3. Has attribute 2 but not attribute 1
4. Has neither attribute

The groups can be counted and the counts arranged in a table such as that presented in table 15.2.

TABLE 15.2 A Two-by-Two Contingency Table in Conventional Notation

Attribute 2	Attribute 1		Total
	Has	*Doesn't have*	
Has	a	b	$a + b$
Doesn't have	c	d	$c + d$
Total	$a + c$	$b + d$	n

Table 15.2 is a contingency table, so the independence of the two classifications can be determined with a chi-square test. The procedure has

been described above. Since there are only two classifications of two categories each, the procedure can be simplified.

$$\chi^2 = \frac{n(ad - bc)^2}{(a + b)(c + d)(a + c)(b + d)}$$

This little formula is easily memorized. The numerator is the difference between the products of the diagonal elements squared and then multiplied by the total sample size. The denominator is the product of the marginal totals. $(a + b)$ and $(c + d)$ are the row totals while $(a + c)$ and $(b + d)$ are the column totals.

Since the chi-square statistic is a continuous variable and the test statistic is a discrete variable, this test statistic is only approximately distributed as chi-square. The approximation by the chi-square distribution is excellent if a correction for continuity proposed by Yates (1934) is made. The above formula modified for the continuity correction is

$$\chi^2 = \frac{n(|ad - bc| - n/2)^2}{(a + b)(c + d)(a + c)(b + d)}$$

where the $|ad - bc|$ indicates the absolute value of $ad - bc$. This correction is recommended. It always reduces the test statistic and the reduction can be rather dramatic for small sample sizes.

GOODNESS-OF-FIT TESTS

Data analysts frequently must compare samples with a probability density function to determine if the conditions necessary for testing differences and relationships are reasonably well satisfied. Such tests are *goodness-of-fit* tests. The most common is the comparison of a sample with the normal probability distribution. Two tests for this purpose, the Lilliefors (1967) and the Shapiro-Wilk (1965, 1968), have been described and used elsewhere in this book. Goodness-of-fit tests are discussed in more general terms here.

Perhaps the oldest of the goodness-of-fit tests is an adaptation of the chi-square test of a contingency table. In this procedure, the measurement scale is divided into nc intervals like those used in preparing histograms, and the number of data in each interval counted. The intervals should be chosen so that no interval, or cell, will contain less than five data, this being the usual requirement for contingency tables. The sample size is then divided among the intervals according to the theoretical probability density function to become the expected counts for calculating a chi-square statistic. The data may be used to estimate the parameters of the probability density function or parameters may be assumed. In the first situation, it is customary to compare the chi-square statistic with the chi-square distribution with $nc - np - 1$ degrees of freedom, where np is the number of parameters estimated, while in the second situation, the chi-square distribution with $nc - 1$ degrees of freedom is used.

Example 15.2

The normally distributed data in table 15.3 will be used to demonstrate the procedure. Since there are only 28 data in the sample, the measurement interval can be divided into no more than four or five intervals. Four intervals were chosen, and table 15.4 gives the count in each. Table 15.4 also gives the expected count and chi-square assuming the data came from a population with $\mu = 14$ and $\sigma = 2$. Using table A in part V, the proportion of the normal probability distribution that is less than the mean minus one standard deviation is 0.15866. The expected count of

TABLE 15.3 Data for Example 15.2

10.28	17.60	15.73	15.96
14.28	10.13	12.39	14.27
17.89	17.50	16.12	11.59
11.82	14.10	21.33	12.82
16.93	16.80	14.48	15.09
13.15	13.10	19.32	11.82
15.73	10.00	10.54	13.28

data less than $14 - 2 = 12$ is then $28 \cdot 0.15866 = 4.44$. Likewise, the proportion of the normal probability distribution that is between the mean and the mean minus one standard deviation is $0.50000 - 0.15866 = 0.34134$. The expected count of data in the second interval is then $28 \cdot 0.34134 = 9.56$. The expected counts for the other intervals are computed in the same way.

The chi-square in table 15.4 should be compared with a table for 3 degrees of freedom since no population parameters were estimated. Assuming that the data came from a normally distributed population with $\mu = 14$ and $\sigma = 2$ and that the observable differences are just chance variation, then the probability of a chi-square as large as or larger than 5.28 is approximately 0.160.

The average of these data is 14.43 and the standard deviation is 2.90. Table 15.5 presents the goodness-of-fit calculations using these sample statistics for estimates of the population parameters. Assuming the same null hypothesis, the probability of a chi-square with 1 degree of freedom as large as or larger than 1.98 is approximately 0.168.

TABLE 15.4 Goodness of Fit Assuming $\mu = 14$ and $\sigma = 2$			
Intervals	<12	12–13	14–15	>16
Counts	7	5	8	8
Expected	4.44	9.56	9.56	4.44

NOTE: $\chi^2 = 5.28$

TABLE 15.5 Goodness of Fit Using $\bar{x} = 14.43$ and sd $= 2.90$			
Intervals	<12	12–13	14–15	>16
Counts	7	5	8	8
Expected	5.66	6.69	9.89	5.76

NOTE: $\chi^2 = 1.98$

This procedure has several disadvantages in data analysis. There must be at least four intervals on the measurement scale (or the degrees of freedom for chi-square will not be meaningful), and each interval should have a minimum of five data. Thus the procedure is not applicable with samples of less than 20 data. Furthermore, there is usually no reason for dividing the measurement scale in any particular way, so the choice for interval boundaries, or cut-points, is up to the data analyst. The test statistic, however, is very dependent on the choices made. To illustrate, let the cut-points for example 15.2 be 12.38, 14.00, and 16.11. These new cut-points will not change the observed counts in table 15.5, but the new chi-square statistic will be 0.102, a rather dramatic shift from 1.98.

The Kolmogorov-Smirnov test is a nonparametric test of goodness of fit in the sense that the sampling distribution of the test statistic does not depend on either the explicit form of, or the value of parameters in, the distribution of the population. Massey (1951) presents the Kolmogorov-Smirnov test for goodness of fit and gives the references reporting the original Kolmogorov and Smirnov research. The test can be described as follows.

Suppose that a sample is thought to come from a population having some specified cumulative probability density function. This means that the distribution function for the sample, expressed as proportions of the sample size, is expected to be fairly close to the specified function. A large departure would be evidence that the sample did not come from, or agree with, the specified cumulative probability density function. The Kolmogorov-Smirnov test statistic is the maximum departure of the sample distribution from the probability density function. The distribution of this maximum is independent of the probability density function if the density function is continuous. The idea is illustrated in figure 15.1. In this illustration, the step function is the distribution function for the sample, while the smooth function is the probability density for the normal distribution with $\mu = 14$ and $\sigma = 2$. The Kolmogorov-Smirnov statistic is the maximum vertical difference between the two functions.

Lilliefors (1967) noted that

> The standard tables used for the Kolmogorov-Smirnov test are valid when testing whether a set of observations are from a completely specified continuous distribution. If one or more parameters must be estimated from the

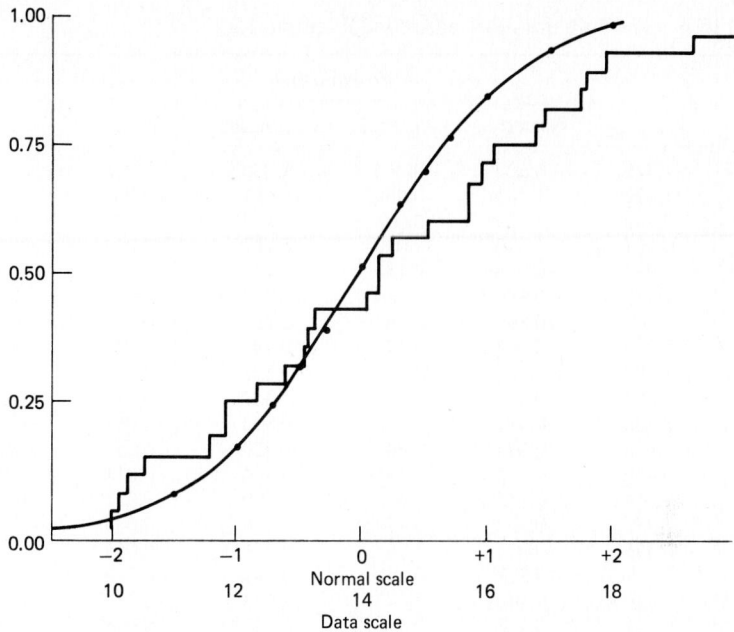

FIGURE 15.1   The Lilliefors statistic illustrated for example 15.2.

sample, then the tables are no longer valid. [He gives a table] for use with the Kolmogorov-Smirnov statistic for testing whether a set of observations is from a normal population when the mean and variance are not specified but must be estimated from the sample.

Lilliefors obtained the table by simulating for each sample size a thousand or more samples from a normal population and computing the test statistic for each. His table, which is table M in part V, gives values exceeded by specified proportions p of the Kolmogorov-Smirnov statistics generated during the simulation.

The details for calculating the Lilliefors statistic for example 15.2, i.e., the Kolmogorov-Smirnov statistic when the population mean and standard deviation are estimated from the sample, are given in table 15.6. The data are displayed in column 1. These data are normalized by subtracting the sample mean and then dividing by the sample standard deviation. These transformed data, often called z scores, are presented directly in column 2 and sorted in column 3. Column 4 contains the proportion of the unit normal distribution less than the corresponding z.

Table A of part V gives the proportion of the unit normal distribution greater than positive z or less than negative z. Consequently, the proportion less than the fourth sorted z, -1.340, is 0.09012, rounded to 0.090.

TABLE 15.6   The Lilliefors Procedure for Example 15.2

y	z	z sorted	Normal	Sample	da	db
			Probabilities			
10.280	−1.429	−1.526	0.064	0.036	0.028	
17.600	1.092	−1.481	0.069	0.071	−0.002	0.034
15.730	0.448	−1.429	0.076	0.107	−0.031	0.005
15.960	0.527	−1.340	0.090	0.143	−0.053	−0.017
14.280	−0.052	−0.978	0.164	0.179	−0.015	0.021
10.130	−1.481	−0.899	0.184	0.214	−0.030	0.006
12.390	−0.703	−0.899	0.184	0.250	−0.066	−0.030
14.270	−0.055	−0.703	0.241	0.286	−0.045	−0.009
17.890	1.191	−0.555	0.290	0.321	−0.032	0.004
17.500	1.057	−0.458	0.323	0.357	−0.034	0.002
16.120	0.582	−0.441	0.330	0.393	−0.063	−0.028
11.590	−0.978	−0.396	0.346	0.429	−0.083	−0.047
11.820	−0.899	−0.114	0.455	0.464	−0.010	0.026
14.100	−0.114	−0.055	0.478	0.500	−0.022	0.014
21.330	2.376	−0.052	0.479	0.536	−0.056	−0.021
12.820	−0.555	0.017	0.507	0.571	−0.065	−0.029
16.930	0.861	0.227	0.590	0.607	−0.017	0.018
16.800	0.816	0.448	0.673	0.643	0.030	0.066
14.480	0.017	0.448	0.673	0.679	−0.006	0.030
15.090	0.227	0.527	0.701	0.714	−0.013	0.022
13.150	−0.441	0.582	0.720	0.750	−0.030	0.005
13.100	−0.458	0.816	0.793	0.786	0.007	0.043
19.320	1.684	0.861	0.805	0.821	−0.016	0.020
11.820	−0.899	1.057	0.855	0.857	−0.002	0.033
15.730	0.448	1.092	0.862	0.893	−0.030	0.005
10.000	−1.526	1.191	0.883	0.929	−0.045	−0.010
10.540	−1.340	1.684	0.954	0.964	−0.010	0.025
13.280	−0.396	2.376	0.991	1.000	−0.009	0.027

NOTE: The Lilliefors test statistic is 0.083.

To find the proportion of the unit normal distribution less than a positive z, subtract the table value from 1.0. For example, the portion for the fifth largest z, 1.057, is found by interpolation to be

$$1.0 - 0.14526 = 0.85474$$

rounded to 0.855.

The sample probabilities for each z are just the counts divided by the sample size. For example, the sample probability for the seventh datum is

$$7/28 = 0.250$$

da is the difference between the normal and sample probabilities on the same line. *db* is the difference between the normal probability on the same line and the sample probability on the preceding line. *da* and *db* have the following interpretation. As one moves along the measurement scale in, for example, figure 15.1, *db* is the difference between the curve and the step function just before the step function increases. *da* is the difference immediately following the increase. The test statistic is just the largest *da* or *db* in absolute value.

(All of the tables illustrating the Lilliefors calculations were prepared with a DEC 11/23 computer. The columns were prepared as described above and rounded when printed. This procedure can cause *da* and *db* errors of 1 in the last place. For example, the first *db* in table 15.6 is not $0.069 - 0.036 = 0.033$, but 0.034 as given.)

Assuming the data are from a normal population and that the observable deviations from normality are just chance variation, then the test statistic 0.083 can be compared to table M in part V. The probability of a test statistic this large or larger is much more than 0.20, the limit of table M.

The normal density function plotted in figure 15.1 has $\mu = 14$ and $\sigma = 2$, while in table 15.6 the sample statistics 14.43 and 2.90 were used for μ and σ. However, the normal probabilities in table 15.6 correspond to this curve in figure 15.1. The sample probabilities in the table are plotted with the sorted y data in figure 15.1.

An advantage of this test over the chi-square test for data analysis is that smaller samples can be examined and each datum in the sample is examined rather than counts in intervals.

Shapiro and Wilk (1965) describe a goodness-of-fit test for normality that may be used instead of the Lilliefors test for samples of 50 or fewer data. The statistical theory of this test is too lengthy and complex for presentation here, so interested readers should refer to the Shapiro and Wilk papers listed in the bibliography.

The test statistic is obtained by dividing the square of an appropriate linear combination of the sample order statistics by the usual estimate of the variance. Table N1 in part V provides the coefficients for the linear combination, while table N2 gives percentiles of the test statistic, assuming that the sample came from a normally distributed population. The steps to compute the test statistic are these

1. Sort the sample data.
2. Using table N1 in part V, find the coefficients a_i for the sample size.
3. Associate the first coefficient with the largest and the smallest datum, the second coefficient with the second largest and the second smallest, etc., until every datum has been assigned a coefficient. See table 15.7.
4. Compute

$$c_1 = \sum_{i=1}^{n/2} a_i y_i \quad \text{and} \quad c_2 = \sum_{i=1}^{n/2} a_i y_{n-i+1}$$

These are the sums of the coefficient-datum products for the smallest and the largest halves of the sample.

5. The test statistic is

$$\pi = \frac{(c_1 - c_2)^2}{\Sigma(y - \bar{y})^2}$$

6. Assuming the null hypothesis of normality is true, then π can be compared with table N2 in part V to determine a probability resulting from chance deviations in the sample data. Table 15.7 presents these calculations for example 15.2.

TABLE 15.7 The Shapiro-Wilk
Normality Test for Example 15.2

0.4328	10.000	4.328
0.2992	10.130	3.031
0.2510	10.280	2.580
0.2151	10.540	2.267
0.1857	11.590	2.152
0.1601	11.820	1.892
0.1372	11.820	1.622
0.1162	12.390	1.440
0.0965	12.820	1.237
0.0778	13.100	1.019
0.0598	13.150	0.786
0.0424	13.280	0.563
0.0253	14.100	0.357
0.0084	14.270	0.120
		$c_1 = 23.395$
0.0084	14.280	0.120
0.0253	14.480	0.366
0.0424	15.090	0.640
0.0598	15.730	0.941
0.0778	15.730	1.224
0.0965	15.960	1.540
0.1162	16.120	1.873
0.1372	16.800	2.305
0.1601	16.930	2.710
0.1857	17.500	3.250
0.2151	17.600	3.786
0.2510	17.890	4.490
0.2992	19.320	5.781
0.4328	21.330	9.232
		$c_2 = 38.257$
		$\pi = 0.970$

The Shapiro-Wilk test statistic decreases as significance increases, so data analysts must use table N2 with extreme care. This table has been arranged like other tables in part V so that significance increases as one moves to the right in the table, but unlike other tables, this means that the test statistic decreases. In all other tables, values of the test statistic increase as significance increases.

The probability of π as small as or smaller than 0.970 is approximately 0.60. This is a very large probability, indicating no reason to reject the hypothesis that the sample came from a normal population.

Using an empirical sampling investigation, Shapiro and Wilk compared properties of this test statistic with other goodness-of-fit tests. This investigation indicated that the test statistic has as good and in many cases much better power to reject the null hypothesis of normality when the sample is from a population that is not normally distributed.

CHAPTER 16

p-Values

INTRODUCTION

Until one understands the purpose of data analysis, it can be very disconcerting for a manager or a researcher to say "so what" to a statistically significant difference, but every data analyst has found statistically significant differences that had no economic or strategic importance. These are so-what differences. This chapter attempts to put this kind of problem in perspective so the data analyst can deal with it effectively.

Data analysis provides information that can be useful when making a decision. The results of an analysis are seldom the only information a manager or a researcher will use, however. Statisticians have raised statistical testing to the level of decision making, but the process requires managers and researchers to have a more sophisticated understanding of statistics and the scientific method than most possess. Some very big decisions are perhaps worth the tedious education process required, but day-to-day problems cannot be made as formal as decision theory requires.

Data analysts attempt to provide a quantitative input to a decision-making process by reporting *p*-values. A *p*-value is nothing more than the probability of the test statistic assuming

1. The null hypothesis is true.
2. *All* necessary conditions for the statistical test are true.
3. Chance or random variation is the only reason for sample differences.

So the usefulness of *p*-values, like everything else in data analysis, is directly related to the care the data analyst exercised in verifying that the required conditions for the test were satisfied.

MEANING FOR DATA ANALYSTS

A *p*-value stands in contrast to a simple yes or no result of a statistical test by attempting to give a measure of "strength" to the results. The *p*-value is an attempt to say: If the null hypothesis is true, then here is a quantitative measure to feed into the decision making process. Such a quantitative measure has more meaning than does a simple yes or no in much the same way ratio and interval data have more meaning than nominal data.

Reporting *p*-values often gives the data analysis a greater impact on the decision-making process. For example, a small measurement difference that should be easily repeated in another study (i.e., has a very small *p*-value) may be worth as much to management as a larger difference that may not be easily repeated (i.e., has a significant but not very small *p*-value). This, of course, is a decision that the analyst can help management make in each situation by supplying information on interpreting the results of the analysis, especially *p*-values.

p-Values for specific situations must always be evaluated along with the entire analysis. For example, a *p*-value of 0.01 must be worth more, in the overall decision-making process, when all of the conditions for the statistical test are well satisfied than when some are a bit iffy. The analyst should, therefore, even qualify *p*-values by presenting them with discussions of how well the test conditions such as normality and equal variances were satisfied.

The analyst will find it necessary to explain *p*-values to managers and researchers in order to avoid misinterpretations and to avoid having the *p*-value simply compared to the de facto significance level of 0.05. This can often be done best by discussing or presenting the analysis in its entirety with *p*-values as just one important part of the total package.

AN EXAMPLE

Example 2.1 can be used to illustrate the limits of an analysis. The data are ratio measurements that are most likely from a normal population since the probability of the Lilliefors statistic is much larger than 0.20, the extent of table M in part V. The probability of the *t*-statistic, assuming the null hypothesis is true, is reasonably small (0.017). This, plus the normality of the sample data, implies that the observed difference can probably be easily repeated in another study, that management can rely on a difference in the neighborhood of 423 feet. The management decision now rests on such questions as: Is the difference medically important and is the difference a saleable product? The data analyst has shown that there is very probably a repeatable difference but this finding is not the only information needed to make a decision.

An Introduction to the Tables in Part V

Tables of the normal distribution are usually arranged to give the probability, i.e., the *p*-value, for specific unit normal deviates ($\mu = 0$ and $\sigma = 1$). This makes finding *p*-values very easy, since one enters the table with the test statistic and finds the *p*-value immediately for that value. Suppose, for example, a test statistic that is normally distributed has been calculated to be 0.93. Entering table A in part V with this number, the probability of a unit normal deviate this large or larger is very conveniently found to be 0.1762.

The ordinal scale analyses discussed in parts I, II, and III are all based on ranks. Since ranks are integers, test statistics based on ranks can have only a specific set of values when there are no tied ranks, and this set is rather small. Tables for these test statistics usually give the acceptance or the rejection probability for each value that the test statistic can have with these exceptions:

1. The distributions are symmetrical so only one tail needs to be presented.
2. The tail given is usually truncated at a point where the probability is considered inconsequentially small.
3. Intermediate values of the test statistic in a range of contiguous values all having the same probability are often omitted.

In part V, the tables for the Wilcoxon signed rank, the Wilcoxon rank sum, the Kruskal-Wallis, the Friedman, Kendall's tau, and Spearman's rho statistics all give upper-tail rejection probabilities very much like the table for the normal distribution. The binomial tables give upper-tail rejection probabilities for the entire distribution and are not truncated.

Most other statistical tables are organized for finding values of the test statistic for given probabilities. This arrangement makes it difficult to obtain *p*-values, since it requires interpolation between values of the test statistic for two specific probabilities. Suppose, for example, a test statistic that is distributed as Student's *t* with 10 degrees of freedom has been calculated to be 2.01. By rummaging through table B in part V, it can be determined that the probability of a *t*-statistic this large or larger is somewhere between 0.050 and 0.025. Linear interpolation is used to obtain a more precise approximation. Table 16.1 presents the necessary entries from table B in part V. The unknown *p*-value for $t = 2.010$ is assumed to divide the distance from 0.050 to 0.025 on the *p*-value scale the same

TABLE 16.1   Linear Interpolation for a *p*-Value

Tabled probabilities	0.050		0.025
Values of *t*	1.812	2.010	2.228

way that 2.010 divides the distance from 1.812 to 2.228 on the *t*-statistic scale. The following equation expresses this relationship.

$$\frac{p - 0.025}{0.050 - 0.025} = \frac{2.010 - 2.228}{1.812 - 2.228}$$

Solving the equation for *p* gives 0.038.

$$p = 0.025 + \frac{2.010 - 2.228}{1.812 - 2.228}(0.050 - 0.025) = 0.038$$

It is necessary sometimes to interpolate between degrees of freedom as well as between probabilities, as the following example illustrates. When interpolating between degrees of freedom, their reciprocals are customarily used to obtain untabled values. Suppose an estimate is needed for the probability of an *F*-statistic with 3 and 48 degrees of freedom that is equal to 2.79. Table D in part V does not contain *F*-statistics for 48 degrees of freedom, so the analyst must create a little subtable for the untabled degrees of freedom. Only the entries that bracket 2.79 need be created for this example, but sometimes it is difficult to tell just which ones those will be.

Table 16.2 presents the necessary entries from table D in part V for this example. Since it is not yet clear whether 2.79 will be slightly above or slightly below the 0.05 *p*-value, entries for three *p*-values have been copied from table D. Starting with $p = 0.05$, the equation for estimating *F* for 3 and 48 degrees of freedom is

$$\frac{F_{0.05,3,48} - 2.790}{2.839 - 2.790} = \frac{1/48 - 1/50}{1/40 - 1/50}$$

TABLE 16.2   Data Needed for Two-Way Linear Interpolation

P	df_2	$df_1 = 3$
0.10	40	2.226
0.05	40	2.839
0.02	40	3.667
0.10	48	
0.05	48	
0.02	48	
0.10	50	2.197
0.05	50	2.790
0.02	50	3.585

Solving this equation gives 2.7982. Since this is slightly larger than 2.79, the next table entry estimated should be smaller, i.e., have a larger *p*-value. The equation for $p = 0.1$ is

$$\frac{F_{0.10,3,48} - 2.197}{2.226 - 2.197} = \frac{1/48 - 1/50}{1/40 - 1/50}$$

Solving this equation gives 2.2018. Table 16.3 summarizes the problem to this point.

TABLE 16.3 Data Needed to
Complete Two-Way Linear
Interpolation

p	df_2	$df_1 = 3$
0.10	40	2.226
0.05	40	2.839
0.02	40	3.667
0.10	48	2.2018
0.05	48	2.7982
0.02	48	
0.10	50	2.197
0.05	50	2.790
0.02	50	3.585

These two values can now be used, as in the example for the *t*-statistic, to obtain an estimate of the *p*-value for 2.79. The equation for the *p*-value is

$$\frac{p - 0.05}{0.10 - 0.05} = \frac{2.79 - 2.7982}{2.2018 - 2.7982}$$

Solving this equation for *p* gives 0.0507. The procedure is easily extended for situations where neither degrees of freedom is given in table D. When interpolating between the last numerical entry and ∞, it is customary to define the reciprocal of ∞ as zero.

This arrangement for statistical tables sacrifices utility for succinctness since a table of Student's *t* arranged as the table of the normal is arranged would require a separate page comparable to the normal for each degree of freedom. A statistic such as *F*, with numerator and denominator degrees of freedom, would require a separate table for each combination of the two degrees of freedom. Tables would simply be too voluminous for practical use.

The columns in statistical tables may be headed with the probability of a test statistic being greater than values in the column, i.e., p, or it may be headed with the probability of a test statistic being less than values in the column, i.e., $1 - p$. The former are called *rejection regions,* while the latter are called *acceptance regions.* See figures 16.1 and 16.2.

The columns in symmetrical tables such as Student's t may be headed

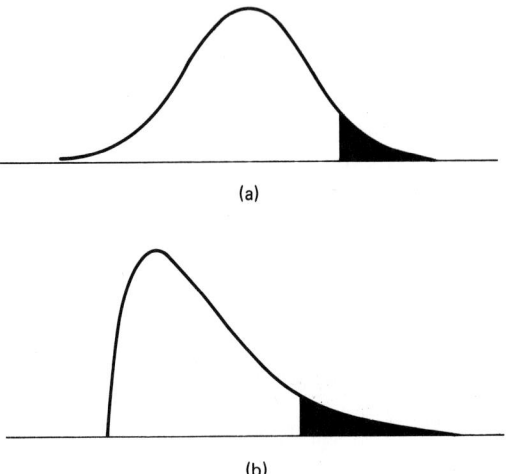

FIGURE 16.1 One-tailed rejection regions. (*a*) Symmetrical such as the normal and t; (*b*) Nonsymmetrical such as F and chi-squared.

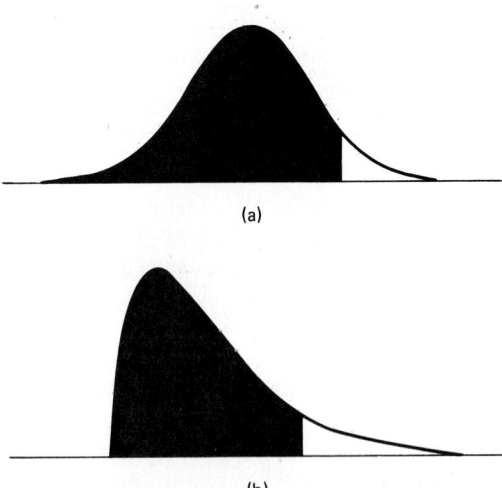

FIGURE 16.2 One-tailed acceptance regions. (*a*) Symmetrical such as the normal and t; (*b*) Nonsymmetrical such as F and chi-squared.

with the probability of a test statistic being greater than values in the column or less than the negative of values in the column, i.e., $2p$. These are called *two-tailed rejection regions* and, of course, columns can be headed with two-tailed acceptance regions, i.e., $1 - 2p$. See figure 16.3. Somewhat by necessity, tables of nonsymmetrical distributions such as F and chi-square usually have columns headed with one-tailed rejection regions or one-tailed acceptance regions.

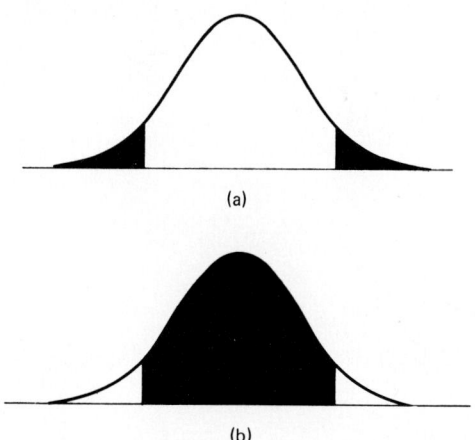

(a)

(b)

FIGURE 16.3 Two-tailed symmetrical regions such as the normal and t. (*a*) Rejection; (*b*) acceptance.

Sometimes the tables in a particular book are arranged in some unusual manner to suit the writer's explanations or fancy, and hence the tables are useless as general purpose tables. The analyst must be very familiar with the particular book in order for the tables to make sense. This condition seems to be particularly true for books on nonparametric statistics.

The explanation in this section is designed to tell the analyst that tables are arranged differently and that errors are sometimes made simply by using tables incorrectly. Data analysts do not need confusing or even difficult-to-use tables, so in an effort to be uniform, all tables in this book give one-tailed rejection regions. Table A for the normal distribution is entered with the unit normal deviate and the probability of that value or greater is given. All other tables are entered with the degrees of freedom or sample size and then values of the test statistic are given in columns headed by the probability of that value or larger. This uniformity should make these tables easier to use as a set then are most sets of tables.

Finally, as one moves to the right in the tables in this book, the table values increase while the probability decreases, i.e., becomes more significant. The one exception to this universal rule is the Shapiro-Wilk table for normality, table N2. Here, as one moves to the right for increased significance, the table values decrease. It is too bad for data analysts that this statistic was not defined as its reciprocal. Further cautions are included with this table in part V.

CHAPTER 17

Statistical Testing

INTRODUCTION

Statistical testing is best understood through hypotheses. Management and research problems can usually be stated in terms of a research hypothesis that describes an existing condition or a desirable goal perhaps not yet achieved. The research hypothesis is a statement of differences between two or more things or a relationship (e.g., regression) between two or more things and stands in contrast to the null hypothesis that there are no differences or that there is no relationship. Data are examined not for proving or disproving the research hypothesis but rather for consistency with the null hypothesis. If the data are not inconsistent with the null hypothesis, then any differences or relationships that might exist are probably of no practical importance since they cannot be distinguished from ordinary measurement uncertainties, i.e., random error. If the data are inconsistent with the null hypothesis, then differences or relationships are distinguishable from ordinary measurement uncertainties and the null hypothesis is rejected. The only data analysis purpose for statistical testing is the rejection of null hypotheses, not the proving or disproving of research hypotheses.

Statistical tests are said to be either one-tailed or two-tailed depending on the nature of the research hypothesis. A research hypothesis is ordinarily a positive statement of differences or of a relationship. For example,

1. A is 12 greater than B.
2. As x increases, y decreases twice as fast.

are positive statements that might be simple research hypotheses. The corresponding null hypotheses would assert that A was *not* greater than $B + 12$ and that as x increases, y does *not* decrease twice as fast but decreases less rapidly or perhaps even increases. The statistical test for either of these two situations would be a one-tailed test.

Now consider a contrasting situation. Suppose the research hypothesis was

1. A is different from B, but it is not known whether A is larger than B or smaller than B.
2. As x increases, y changes, but it is not known how this change compares with changes in x, not even whether it increases or decreases.

Such research hypotheses would represent an unusual management or research ignorance, but should they exist, the corresponding null hypotheses would assert that A was not different from B or that as x increases, y does not change. The statistical test for either of these two situations would be a two-tailed test.

The research hypothesis for a one-tailed test specifies algebraically the direction of the anticipated difference or the anticipated relationship, while the research hypothesis for a two-tailed test does not. The names one-tailed and two-tailed come from the number of null hypothesis rejection regions that are defined on the measurement scale. Consider, for example, the simple comparison of the location of two independent populations. For a one-tailed test, the null hypothesis may be rejected if sample A is much larger than sample B, or it may be rejected if sample A is much smaller than sample B, but before the data are examined, only one of these two possibilities is of interest. The other is of absolutely no consequence and is considered equivalent to no difference between the two. On the other hand, the null hypothesis for a two-tailed test may be rejected for either reason since the research hypothesis makes no distinction between the two kinds of deviation.

Statisticians have defined two kinds of errors associated with statistical testing and then have shown the data analyst that the probability of these errors can be controlled. When sample data are inconsistent with the null hypothesis but the populations are not, the data analyst, knowing only sample data, will reject the null hypothesis and assume the populations are different. The analyst controls this error of the first kind (called *type I*) through choice of the significance level α, for the statistical test. α is the probability of rejecting a null hypothesis on the basis of sample data when the populations are not inconsistent with the hypothesis of no difference or no relationship.

When sample data are not inconsistent with the null hypothesis but the research hypothesis or something similar is really true, the data analyst, again knowing only sample data, will not reject the null hypothesis. In effect, the research hypothesis is rejected. This is an error of the second kind (called *type II*) that the analyst can control through choice of a probability called β. β is the probability of the sample data not being inconsistent with the null hypothesis when a difference or relationship at least as strong as that specified in the research hypothesis is true.

Types I and II errors and the α and β probabilities have been defined in terms of rejecting or not rejecting the null hypothesis since this is all one can do in data analysis. The acceptance of any hypothesis is an arbitrary act that can never be supported in fact.

Power, defined as $1 - \beta$, is generally regarded as the probability of rejecting the null hypothesis when at least the difference or the relationship expressed by the research hypothesis is true. That is, $1 - \beta$ is the power of the test to detect differences or relationships at least as strong as those specified by the research hypothesis.

α, β, the research hypothesis, sample size, and population characteristics are functionally related. Given all but one of these, the remaining one can be determined uniquely. This is the basis for estimating sample sizes using previous experience for estimates of the population characteristics and specific α and β probabilities. After data have been collected, the sample size is determined so that α and β, the remaining quantities in the functional relationship, cannot both be chosen arbitrarily. The choice of a specific probability for one determines the other.

Consequently, the α and β probabilities are ordinarily specified when planning a study and then the sample size necessary for these probabilities determined. It has recently become popular to compute power after a study has been completed as a measure of the adequacy of the sample size employed. From a data anlaysis viewpoint, there is probably nothing wrong with this procedure, but it isn't quite the original use of power. A poststudy power above 0.80 is usually considered adequate, above 0.90 good, above 0.95 excellent, and above 0.99 unusual or improbable or "you spent too much on the sample."

For the reader familiar with terminology used in quality control, α is the producer's risk, i.e, the probability of deciding the product does not meet specifications when in fact it does, and β is the consumer's risk, i.e., the probability of deciding the product is acceptable when in fact it is unacceptable.

Sample sizes and power are not discussed in this book even though they are subjects of great importance to data analysts. Readers are referred to Cohen's (1969) book, *Statistical Power Analysis,* for a good presentation covering many data analysis situations.

ERRORS OF THE THIRD AND FOURTH KINDS

Kimball (1957) wrote an interesting paper on errors of the third kind, which he defined as the error of finding the right answer to the wrong problem. He gives several examples, all of which are beyond the level of problems used in this book, but the paper is, nevertheless, recommended reading. The third kind of error occurs because the data analyst does not understand a problem well enough to translate it into the rather restricted formal situations that statistics permits. So the data analyst must not only be adept at explaining results to clients but must also know how to question clients thoroughly in order to understand the problem initially. Communication is a two-way street.

A variation of this that can be called an error of the fourth kind is the error made by trying to answer the right question with the wrong data. In this situation, the problem is well understood and the research hypothesis is a proper statement of that problem, but for some reason, relevant data have not been or cannot be obtained. Related data not specific for the problem are available for analysis instead. A reasonable procedure would be to redefine the problem with a corresponding restatement of the research hypothesis. More often data analysts, not wanting to reveal their weaknesses, proceed with their analyses as if the client understood the problems and was willing to accept the limitations on interpretations. The client probably doesn't know there are problems and thinks the data analyst will answer all of the research questions with this one analysis.

Many readers may object to suggestions that such errors exist, but they do and this writer has corrected many of them. The data analyst who fails to recognize that these errors can occur, sometimes through ignorance but most often through a failure to communicate, will be doomed to repeat them time and again.

Statistics provides a formal but limited situation for the data analyst, one in which null hypotheses can be rejected. Often real problems are so complex that their simplification into comparisons within this framework is difficult to achieve. Perhaps the easiest way to avoid errors of the third and fourth kinds is to ask (of oneself) just before making a test such questions as

1. Are the populations relevant to the research hypothesis?
2. Is the statistical test chosen appropriate for the research hypothesis?
3. Does the regression model make sense?
4. Did the transformation make it more difficult or less difficult to explain results of the analysis?
5. Are the data pertinent?

The following hypothetical example is offered to help in understanding types III and IV errors. A bank expends considerable time and effort to develop a new kind of checking account with many "free" auxiliary services such as safe deposit boxes, travelers checks, etc. Then after much advertising and a reasonable time for the new account to be available, they want to know if the account has been a success or not. They interview a random sample of their customers, perhaps those who visit the bank in person, and ask

1. Have you heard about our new account called PLENTY?

If the answer is yes, then such questions as these might be asked

2. What do you like about it?
3. What don't you like about it?
4. Does it meet your needs?

If the answer to this last question is no, then

5. What other features would you like to see added? might be asked.

These seem like reasonable questions to ask bankers, but to ask customers who have little notion of what a bank does, or could do, the questions may be irrelevant and meaningless for the problem at hand. The real research questions were more likely

1. Did the new account increase use of bank services?
2. Did the new account increase bank revenues?

These research questions might be better answered by examining such issues as

1. When offered side by side, did the new account outsell the old account?
2. Did the new account increase, by some criteria, the use of safe deposit boxes, travelers checks, etc.?
3. Did many regular customers opt for the new accounts over their old accounts?

The real problem was not a popularity question but rather a question of substantive changes.

It is fascinating and even a little entertaining to evaluate in this respect the surveys that arrive almost daily in the mail. Assuming the sponsors are really trying to do research, are they going to get data appropriate for what is the apparent research problem? Some surveys seem designed to produce biased results and it is educational to identify these studies too. For example, a recent survey asked the respondent to check a box beside the appropriate statement:

I support Senator _____ in his stand on
_____.

I do not understand Senator _____'s stand on
_____.

It is obvious here that the sponsor did not want unbiased results since the second statement is not an alternative to the first.

LIMITATIONS AND INTERPRETATIONS

The limitations of statistical testing in data analysis can now be summarized. Management and research problems are usually very complex. These complex problems must be simplified in a meaningful way to one or more subproblems each within the narrow formality permitted by statistics. The simplication must be evaluated for truly representing the original complex problem. The statistical tests must add, in some minor way at least, to the available information for the necessary management and/ or research decision.

Managers and researchers must be made to realize that their problems have been reformatted into statistical problems. The results of the data analysis probably do not exactly address the original management or research concerns but are nevertheless useful in the overall decision process. Finally, it is most likely a disservice to let management base decisions entirely on data analyses without understanding these issues. It is the responsibility of the data analyst to understand the limitations of data analyses and to make a best effort to communicate this to the client.

CHAPTER 18

Least Squares

INTRODUCTION

Relationships between variables are often approximated with algebraic equations. The most widely used procedure for estimating from data the numerical coefficients in an approximating equation is the method of least squares.

The method of least squares is largely a result of the work of Karl Friedrich Gauss (1777–1855), a German mathematician and astronomer. The method does not require normality of the data, but the deviations of the data from the approximating equation must be normally distributed for valid tests and for confidence limits and intervals based on the normal distribution. The method of least squares is not the only procedure ever proposed for estimating parameters, but it is certainly the most popular procedure. The least-squares estimates of parameters are unbiased and, among all unbiased linear estimates, have minimum variance. When the deviations of the data from the approximating equation are normally distributed, the least-squares estimates are also the estimates of maximum likelihood. These desirable properties account for the universal popularity and almost exclusive use of the method. Data analysts can find discussions of maximum likelihood in many books on mathematical or theoretical statistics.

Data analysts should understand the least-squares procedure of estimation, but once the basic principles are understood, they only need formulas or a computer program for producing least-squares estimates. In this chapter, the estimation formulas for a simple linear regression are developed completely and used with a numerical example. Then to further illustrate the basic principles, least-squares estimates for two additional situations are calculated.

Example 18.1

To illustrate the method of least squares, consider the science achievement and intelligence quotient (IQ) data in table 18.1. These data, plotted
in figure 18.1, form an elliptical cluster, suggesting the relationship
between the two scores might be approximated by a linear equation, i.e.,

$$\text{Science achievement} = \alpha + \beta \cdot \text{IQ}$$

The parameters α and β define a relationship between science achievement and IQ. For succinctness, let y stand for the science achievement
score and let x stand for the IQ. Then

$$y = \alpha + \beta x$$

TABLE 18.1   Data for Example 18.1

IQ	Science achievement	IQ	Science achievement
130	70	125	68
134	54	104	46
101	43	100	48
96	39	121	54
119	67	94	42
137	74	114	59
134	79	107	58
93	51	128	72
95	45	104	59
109	73	107	42

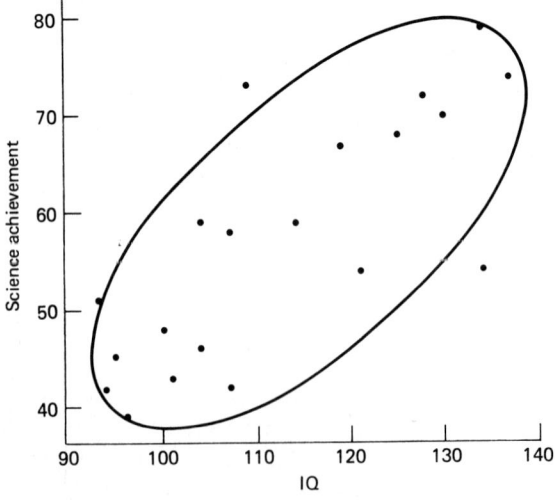

FIGURE 18.1   Plot of data for example 18.1.

This is the equation of a straight line such as drawn in figure 18.2. The slope or orientation of the line is given by β, while the position of the line is given by α. The line in figure 18.2 is an arbitrary line. It is not the result of analysis, yet it seems to fit fairly well. Such arbitrary lines often aid the data analyst in deciding what kind of algebraic equation to use to approximate the relationship between the variables.

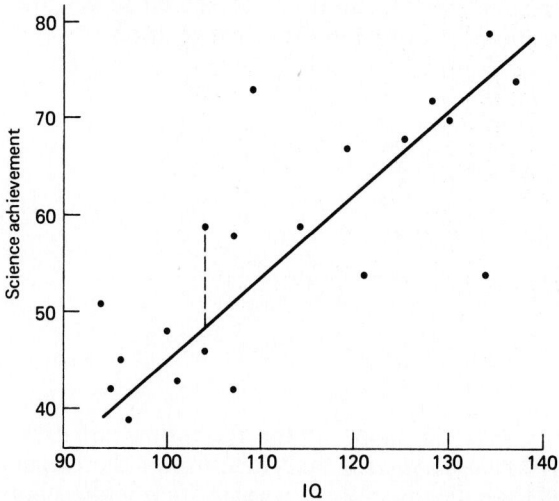

FIGURE 18.2 Plot of data for example 18.1; approximate regression line indicated.

The position and slope of the line are initially unknown. The least-squares problem is to determine estimates for α and β which minimize the sum of squares of the differences between observed values of y and the values of y determined by substituting the associated x into the least-squares equation. These calculated values of y are called Y and are points on the graph of the equation. The $y - Y$ differences are called *residual errors* and are denoted by d.

$$d = y - Y = y - \alpha - \beta x$$

One such difference from the arbitrary line in figure 18.2 is indicated.

The quantity that is minimized by the least-squares procedure is the sum of the d's squared, i.e.,

$$D = \Sigma d^2 = \Sigma(y - Y)^2 = \Sigma(y - \alpha - \beta x)^2$$

To find the estimates of α and β that minimize D, the partial derivatives (reference elementary calculus) of D with respect to α and to β are set equal to zero.

$$\frac{\partial D}{\partial \alpha} = -2\Sigma(y - a - bx) = 0$$

$$\frac{\partial D}{\partial \beta} = -2\Sigma(y - a - bx)x = 0$$

Once the partial derivatives are set equal to zero, estimates for α and β, indicated by Roman letters, are used in the equations. By rearranging these two equations slightly, they can be solved to get the well-known formulas for a and b in a one-term linear regression.

$$na - b\Sigma x = \Sigma y$$
$$a\Sigma x - b\Sigma x^2 = \Sigma xy$$

$$b = \frac{\Sigma xy - \Sigma x \Sigma y/n}{\Sigma x^2 - (\Sigma x)^2/n}$$

$$a = \frac{\Sigma y - b\Sigma x}{n}$$

The quantity b is the slope of the regression equation, sometimes called the *regression coefficient,* and determines the orientation of the graph of the regression equation. The quantity a is the y-intercept. This is the point where the graph of the equation crosses the y-axis, i.e., the value of y obtained from the regression equation when x is equal to zero.

The necessary sums for computing a and b for example 18.1 are

$$\Sigma y = 1143 \qquad \Sigma x = 2252$$
$$\Sigma y^2 = 68,345 \qquad \Sigma x^2 = 257,726$$
$$\Sigma xy = 131,439 \quad n \ = 20$$

Substituting these into the formulas gives

$$a = -17.1028$$
$$b = \quad 0.6594$$

The algebraic equation determined by least squares to approximate the relationship between science achievement (y) and IQ (x) is

$$y = -17.1 + 0.66x$$

The graph of this equation is plotted in figure 18.3.

Obviously this algebraic equation is only an approximation. Other-

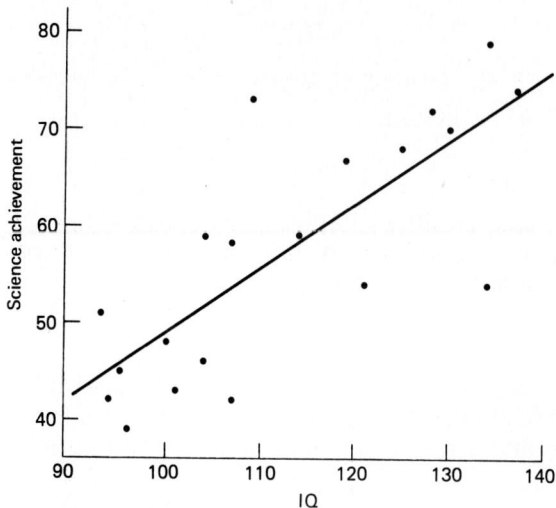

FIGURE 18.3 Plot of data for example 18.1; computed regression equation plotted.

wise, all of the points would fall on the graph of the equation. A simplistic explanation is that the deviations from the line are measurement errors, but this invites the data analysis question: Which variable has measurement error, or do they both have measurement error? The data analysis viewpoint is that this approximation is made in ignorance of all other factors that influence the relationship between science achievement and IQ. Therefore the deviations from the approximating relationship may simply be the effect of other, unknown factors. In fact, there may be no measurement errors at all.

In order to estimate science achievement from IQ, the approximating relationship was determined by minimizing the square of deviations measured parallel to the science achievement axis. There is no reason why the square of deviations measured parallel to the IQ axis cannot be minimized to determine a relationship for estimating IQ from science achievement. This problem is discussed in some detail in part II.

There is always a data analysis need to calculate D, the sum of the squares of the deviations of the data, from the approximate relationship. Each individual d can be calculated, squared, and summed. The following formula, however, is an easier way to calculate D for the simple linear relationship used here.

$$D = \Sigma y^2 - a\Sigma y - b\Sigma xy$$

For example 18.1, D, calculated by substituting into this equation, is 1217.53.

Example 18.2

In example 18.2, the least-squares procedure for estimating the parameters in a quadratic equation

$$y = \beta_0 + \beta_1 + \beta_2 x^2$$

is presented. Data for this example are given in table 18.2.

These data, plotted in figure 18.4, form a kind of inverted fishhook that implies an approximation linear in x, i.e.,

$$y = \beta_0 + \beta_1 x$$

would not fit very well. In the absence of information indicating a more complex formulation, e.g., logarithmic, the quadratic is a good approximation to try first. It is the only formulation used for this example.

TABLE 18.2 Data for Example 18.2

y	x	y	x	y	x
16.29	2.6	18.61	5.7	19.61	5.2
15.60	1.8	18.95	4.2	13.01	1.0
18.90	4.6	19.04	5.6	17.56	2.9
14.44	1.4	15.42	1.8	17.23	2.7
17.88	3.4	16.77	2.8	15.23	1.8
18.26	5.3	18.43	5.4	18.09	3.5
14.27	1.5	17.22	2.9	17.64	3.3

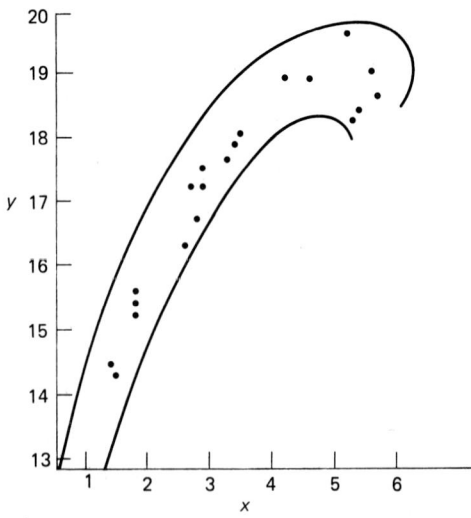

FIGURE 18.4 Plot of data for example 18.2.

An arbitrary line has been drawn in figure 18.5. The position, slope, and curvature of the approximation are initially unknown. The least-squares problem is to determine estimates for β_0, β_1, and β_2 which minimize the sum of squares of the differences between observed values of y and the values of y determined by substituting the associated x into the least-squares equation.

$$d = y - Y = y - \beta_0 - \beta_1 x - \beta_2 x^2$$

One such difference from the arbitrary line is indicated by a dashed line in figure 18.5. The quantity that is minimized by the least-squares procedure is the sum of the d's squared, i.e.,

$$D = \Sigma d^2 = \Sigma(y - Y)^2 = \Sigma(y - \beta_0 - \beta_1 x - \beta_2 x^2)^2$$

To find the values of β_0, β_1, and β_2 that minimize D, the partial derivatives (reference elementary calculus) of D with respect to β_0, β_1, and β_2 are set equal to zero.

$$\frac{\partial D}{\partial \beta_0} = -2\Sigma(y - b_0 - b_1 x - b_2 x^2) = 0$$

$$\frac{\partial D}{\partial \beta_1} = -2\Sigma(y - b_0 - b_1 x - b_2 x^2)x = 0$$

$$\frac{\partial D}{\partial \beta_2} = -2\Sigma(y - b_0 - b_1 x - b_2 x^2)x^2 = 0$$

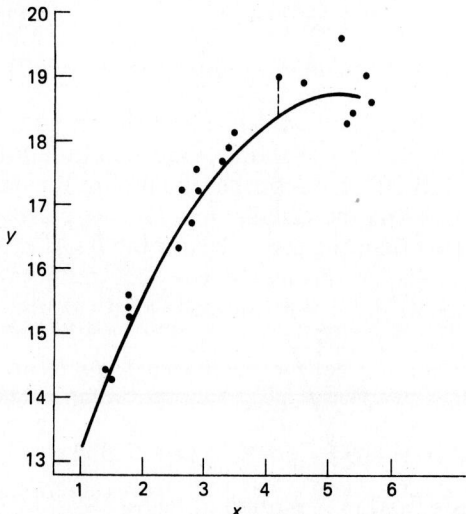

FIGURE 18.5 Plot of data for example 18.2; approximate regression line indicated.

In these equations, b's are used to imply estimates of the β's once the derivatives are set equal to zero. The three equations can be solved for b_0, b_1, and b_2 after rearranging them as follows.

$$b_0 n + b_1 \Sigma x + b_2 \Sigma x^2 = \Sigma y$$

$$b_0 \Sigma x + b_1 \Sigma x^2 + b_2 \Sigma x^3 = \Sigma xy$$

$$b_0 \Sigma x^2 + b_1 \Sigma x^3 + b_2 \Sigma x^4 = \Sigma x^2 y$$

Now it is possible to solve these three equations to obtain formulas for b_0, b_1, and b_2, but formulas are not practical for larger numbers of equations and unknowns. Instead, the numerical sums are substituted into the equations and then the equations are solved directly for the unknowns. This procedure is followed here to illustrate the general approach. When used this way, the set of equations are often called the estimation or estimating equations.

The necessary sums for example 18.2 are

$$\Sigma y = 358.45 \qquad \Sigma x = 69.4$$

$$\Sigma y^2 = 6183.7707 \qquad \Sigma x^2 = 275.28$$

$$\Sigma xy = 1234.396 \qquad \Sigma x^3 = 1229.902$$

$$\Sigma x^2 y = 5019.9004 \qquad \Sigma x^4 = 5913.9288$$

Substituting into the above gives the following set of equations.

$$21b_0 + 69.4b_1 + 275.28b_2 = 358.45$$

$$69.4b_0 + 275.28b_1 + 1229.902b_2 = 1234.396$$

$$275.28b_0 + 1229.902b_1 + 5913.9288b_2 = 5019.9004$$

The solution of these equations is a problem in numerical mathematics. Of the many procedures available, Gauss elimination is the simplest. The appendix to this chapter describes the procedure and Doolittle's systematization of it, while the calculations for this example are presented in table 18.3. After obtaining estimates for the b's, they were substituted back into the three equations as a check. The results were 358.450000, 1234.395999, and 5019.900398, indicating no calculation errors and a very good solution.

The algebraic equation determined by least squares to approximate the relationship is then

$$y = 9.971 + 3.552x - 0.354x^2$$

The graph of this equation is plotted in figure 18.6.

The relationship between x and y was determined by minimizing deviations parallel to the y-axis. Some other relationship, perhaps of a

TABLE 18.3   Doolittle Computations for Example 18.2

Computation	b_0	b_1	b_2	y	*Check*
(1)	21	69.4	275.28	358.45	724.1300
(2)		275.28	1229.902	1234.396	2808.9780
(3)			5913.9288	5019.9004	12439.0112
(4) $= \dfrac{(1)a_{12}}{a_{11}}$		229.350476	909.734857	1184.591905	2393.077238
(5) $=$ (2) $-$ (4)		45.929524	320.167143	49.804095	415.900762
Check:		$a_{22.1}$	$+ a_{23.1}$	$+ a_{20.1} =$	415.900762
(6) $= \dfrac{(1)a_{13}}{a_{11}}$			3608.527543	4698.767429	9492.309830
(7) $= \dfrac{(5)a_{23.1}}{a_{22.1}}$			2231.832393	347.176139	2899.175675
(8) $=$ (3) $-$ (6) $-$ (7)			73.568864	-26.043168	47.525690
Check:			$a_{33.12}$	$+ s_{30.12} =$	47.525696

$$(9)\ b_2 = \frac{-26.043168}{73.568864} = -0.353997$$

$$(10)\ b_1 = \frac{49.804095 - 320.167143 b_2}{45.929524} = 3.552015$$

$$(11)\ b_0 = \frac{358.45 - 275.28 b_2 - 69.4 b_1}{21} = 9.970881$$

NOTE: Refer to the appendix to this chapter for an explanation of the a's in the computation column.

quite different character, would be required for minimizing deviations measured parallel to the x-axis.

D, the sum of the squares of the deviations of the data from the approximate relationship, can be obtained from the individual d, but the following is easier.

$$D = \Sigma y^2 - b_0 \Sigma y - b_1 \Sigma xy - b_2 \Sigma x^2 y$$

Using all of the decimal places recorded in table 18.3, D for example 18.2 is 2.145.

Example 18.3

It is interesting and perhaps instructive to determine the least-squares estimate for a single number to approximate a sample. This is a very sim-

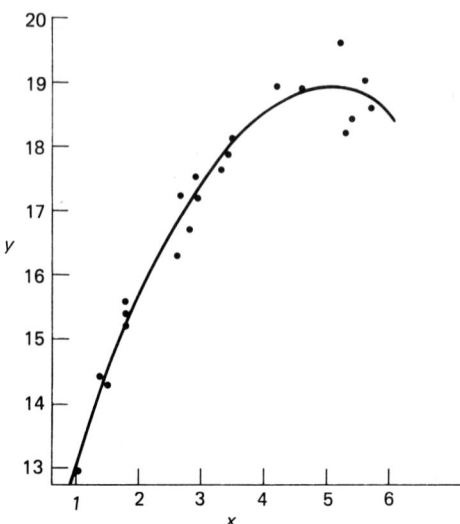

FIGURE 18.6 Plot of data for example 18.2; computed regression equation plotted.

ple problem, but until explained the solution may not be apparent. This problem is almost never discussed in connection with least squares, but it once was a favorite to ask beginning graduate students in statistics.

The y data in table 18.2 will be used for the illustration. Let μ be the single number to approximate the sample. Then

$$d = y - \mu$$

and

$$D = \Sigma d^2 = \Sigma(y - \mu)^2$$

Using elementary calculus, the derivative of D with respect to μ is determined and set equal to zero, i.e.,

$$\frac{dD}{d\mu} = -2\Sigma(y - m) = 0$$

where m is the estimate of μ.

This equation is now solved for m.

$$\Sigma m = \Sigma y$$

Since m is a constant for any particular sample, the summation of m is simply the number of m's in the sum multiplied by m, i.e.,

$$nm = \Sigma y$$

Then,

$$m = \frac{\Sigma y}{n}$$

Hence, the least-squares estimate of a single number to approximate a sample is simply the average of the data.

For the y data in table 18.2,

$$m = \frac{69.4}{21} = 3.3048$$

D, the sum of the squares of the deviations of the data from the single number representing the sample, can be obtained as follows.

$$D = \Sigma y^2 - m\Sigma y = 45.93$$

Note that this is the sum of squared deviations about the mean or the variance multiplied by $n - 1$.

SOME GENERALIZATIONS

For each of the examples in this chapter, the deviations of the data from the approximate relationship were defined, squared, and summed. Procedures from elementary calculus were used to obtain equations for estimating the unknown coefficients so that the sum of squared deviations D is a minimum. In a few very simple situations such as examples 18.1 and 18.3, the estimating equations can be solved to provide useful formulas. More generally, solution of the estimating equations is fairly complex and formulas are not practical. Example 18.2 is such a problem. There are many algebraic procedures available for obtaining solutions in complex situations and these have all been programmed for most computers.

Data analysts are always interested in calculating D, especially if the d are normally distributed. This is easily done with the estimated equation coefficients and the y sums in the estimation equations. The procedure used above without explanation is explained for the general situation in the appendix to this chapter.

Appendix 18.A

Gauss Elimination

Solution of least-squares estimation equations is a problem in numerical mathematics. There are many procedures for solving the estimation equations, but the simplest is one given by Karl Friedrich Gauss, who is also credited with the method of least squares. Gauss' procedure uses the first equation to eliminate the first unknown in the remaining equations. This results in one fewer equations in one fewer unknowns. The procedure is repeated until a numerical quantity is obtained for the last unknown. Then a process of systematic substitution into the equations provides numerical estimates for the other unknowns. This total process is often called *Gauss elimination.*

Consider a set of four equations in four unknowns b_0, b_1, b_2, b_3.

$$a_{11}b_0 + a_{12}b_1 + a_{13}b_2 + a_{14}b_3 = a_{10}$$

$$a_{21}b_0 + a_{22}b_1 + a_{23}b_2 + a_{24}b_3 = a_{20}$$

$$a_{31}b_0 + a_{32}b_1 + a_{33}b_2 + a_{34}b_3 = a_{30}$$

$$a_{41}b_0 + a_{42}b_1 + a_{43}b_2 + a_{44}b_3 = a_{40}$$

In data analysis, the a's are numerical quantities calculated from the data and $a_{ij} = a_{ji}$.

The first equation is multiplied by a_{21}/a_{11} and subtracted from the second. It is then multiplied by a_{31}/a_{11} and subtracted from the third. Finally, it is multiplied by a_{41}/a_{11} and subtracted from the fourth. This transforms the last three equations into

$$\left(a_{22} - \frac{a_{12}^2}{a_{11}}\right)b_1 + \left(a_{23} - \frac{a_{12}a_{13}}{a_{11}}\right)b_2 + \left(a_{24} - \frac{a_{12}a_{14}}{a_{11}}\right)b_3 = a_{20} - \frac{a_{12}a_{10}}{a_{11}}$$

18.13

$$\left(a_{32} - \frac{a_{12}a_{13}}{a_{11}}\right)b_1 + \left(a_{33} - \frac{a_{13}^2}{a_{11}}\right)b_2 + \left(a_{34} - \frac{a_{13}a_{14}}{a_{11}}\right)b_3 = a_{30} - \frac{a_{13}a_{10}}{a_{11}}$$

$$\left(a_{42} - \frac{a_{12}a_{14}}{a_{11}}\right)b_1 + \left(a_{43} - \frac{a_{13}a_{14}}{a_{11}}\right)b_2 + \left(a_{44} - \frac{a_{14}^2}{a_{11}}\right)b_3 = a_{40} - \frac{a_{14}a_{10}}{a_{11}}$$

The first coefficient in these reduced equations became zero, eliminating the first term. These zero terms have not been included in the reduced equations above, but to see that they really are zero, examine the coefficient for the first term in the first reduced equation calculated according to the directions given above. It is

$$a_{21} - a_{11}\frac{a_{21}}{a_{11}} = 0$$

Before repeating this process to eliminate b_1 in the last two equations, the notation for the coefficients in the three reduced equations must be simplified, otherwise it will become unmanageable. One way is to let $a_{jk.1}$ stand for the coefficient of b_1 in the first equation after b_0 has been eliminated. Thus let

$$a_{jk.1} = a_{jk} - \frac{a_{1j}a_{1k}}{a_{11}}$$

For example, let

$$a_{22.1} = a_{22} - \frac{a_{12}^2}{a_{11}}$$

and let

$$a_{23.1} = a_{23} - \frac{a_{12}a_{13}}{a_{11}}$$

The other coefficients in the reduced equations are similarly designated and the notation simplified to

$$a_{22.1}b_1 + a_{23.1}b_2 + a_{24.1}b_3 = a_{20.1}$$

$$a_{32.1}b_1 + a_{33.1}b_2 + a_{34.1}b_3 = a_{30.1}$$

$$a_{42.1}b_1 + a_{43.1}b_2 + a_{44.1}b_3 = a_{40.1}$$

Since $a_{ij} = a_{ji}$, then $a_{ij.1} = a_{ji.1}$.

Now consider these three equations. The first is used to eliminate b_1 from the second and third equations. To do this, the first equation is multiplied by $a_{23.1}/a_{22.1}$ and subtracted from the second equation. Then the first equation is multiplied by $a_{24.1}/a_{22.1}$ and subtracted from the third equation. This transforms the last two equations into:

$$\left(a_{33.1} - \frac{a_{23.1}^2}{a_{22.1}}\right) b_2 + \left(a_{34.1} - \frac{a_{23.1}a_{24.1}}{a_{22.1}}\right) b_3 = a_{30.1} - \frac{a_{23.1}a_{20.1}}{a_{22.1}}$$

$$\left(a_{43.1} - \frac{a_{23.1}a_{24.1}}{a_{22.1}}\right) b_2 + \left(a_{44.1} - \frac{a_{24.1}^2}{a_{22.1}}\right) b_3 = a_{40.1} - \frac{a_{24.1}a_{20.1}}{a_{22.1}}$$

The reader may want to verify that the first coefficient in these two new equations became zero as a result of this step. The notation in these two equations is simplified, as in the three above, by letting

$$a_{jk.12} = a_{jk.1} - \frac{a_{2j.1}a_{2k.1}}{a_{22.1}}$$

For example, let

$$a_{33.12} = a_{33.1} - \frac{a_{23.1}^2}{a_{22.1}}$$

and let

$$a_{34.12} = a_{34.1} - \frac{a_{23.1}a_{24.1}}{a_{22.1}}$$

and so on. Note that $a_{ij.12} = a_{ji.12}$ since $a_{ij.1} = a_{ji.1}$.

The last two equations can now be written as follows:

$$a_{33.12}b_2 + a_{34.12}b_3 = a_{30.12}$$

$$a_{43.12}b_2 + a_{44.12}b_3 = a_{40.12}$$

Repeating the process once more eliminates b_2 from the last equation, resulting in one equation in one unknown. Multiply the first of the two equations above by $a_{34.12}/a_{33.12}$ and subtract from the second equation to get

$$\left(a_{44.12} - \frac{a_{34.12}^2}{a_{33.12}}\right) b_3 = a_{40.12} - \frac{a_{34.12}a_{30.12}}{a_{33.12}}$$

Then after simplifying the notation in a way analogous to that done twice before for the reduced equation sets,

$$b_3 = \frac{a_{40.123}}{a_{44.123}}$$

By repeated elimination, a numerical value can be obtained for b_3. The other unknowns are easily obtained from the first equation in each reduced set and the first equation in the original set. Beginning with

$$a_{33.12}b_2 + a_{34.12}b_3 = a_{30.12}$$

substitute for b_3 and solve for b_2.

$$a_{33.12}b_2 + a_{34.12}\frac{a_{40.123}}{a_{44.123}} = a_{30.12}$$

$$b_2 = \frac{a_{30.12} - a_{34.12}(a_{40.123}/a_{44.123})}{a_{33.12}}$$

Similarly, b_1 is obtained from

$$a_{22.1}b_1 + a_{23.1}b_2 + a_{24.1}b_3 = a_{20.1}$$

and b_0 is obtained from

$$a_{11}b_0 + a_{12}b_1 + a_{13}b_2 + a_{14}b_3 = a_{10}$$

The above is the Gauss elimination procedure for four symmetrical equations in four unknowns. The procedure is easily generalized but quickly becomes very tedious for larger numbers of equations.

Example 18.A1

Data from chapter 9 will be used to illustrate the Gauss procedure. The first four equations in table 18.A1 are the least-squares equations for estimating the coefficients for the final model in chapter 9 and are the starting point for this example. The next three equations are the result of eliminating b_0 from the last three of the original equations. The first coefficient, 451.1935, for example, was computed according to instructions given above as follows:

$$28,186.44 - \frac{927.25^2}{31.00} = 451.19$$

The third coefficient in the second reduced equation, 15,488.20, was computed as follows:

$$1,654,393.16 - \frac{28,186.44 \cdot 1802.50}{31.00} = 15,488.20$$

TABLE 18.A1 Gauss Elimination Example 18.A1

$$31.00b_0 + 927.2500b_1 + 28{,}186.44b_2 + 1802.50b_3 = 1269.50$$
$$927.25b_0 + 28{,}186.44b_1 + 870{,}561.67b_2 + 54{,}177.88b_3 = 36{,}895.50$$
$$21{,}186.44b_0 + 870{,}561.67b_1 + 27{,}308{,}910.14b_2 + 1{,}654{,}393.16b_3 = 1{,}089{,}730.31$$
$$1802.50b_0 + 54{,}177.88b_1 + 1{,}654{,}393.16b_2 + 106{,}099.25b_3 = 75{,}024.25$$

$$451.19b_1 + 27{,}468.96b_2 + 262.77b_3 = -1076.88$$
$$27{,}468.96b_1 + 1{,}680{,}675.98b_2 + 15{,}488.20b_3 = -64{,}549.77$$
$$262.77b_1 + 15{,}488.20b_2 + 1292.60b_3 = 1208.98$$

$$8348.03b_2 - 509.66b_3 = 1011.58$$
$$-509.66b_2 + 1139.56b_3 = 1836.14$$

$$1108.44b_3 = 1897.90$$

The next two equations are the result of eliminating both b_0 and b_1 from the last two of the original equations. Just for illustration, the second coefficient in the first equation according to instructions given is

$$15{,}488.20 - \frac{27{,}468.96 \cdot 262.77}{451.19} = -509.66$$

The last equation is the result of eliminating b_0, b_1, and b_2 from the last of the original equations. It is easily solved for b_3.

$$b_3 = \frac{1897.90}{1108.44} = 1.712224$$

The remaining least-squares regression coefficients are obtained from the first equation in each reduced set.

$b_2 = (1836.14 - 1139.56 \cdot 1.712224)/-509.66 = 0.225710$

$b_1 = [1208.97 - (1292.60 \cdot 1.712224) - (15{,}488.20 \cdot 0.225710)]/262.77$

$\quad = -17.125293$

$b_0 = [75{,}024.25 - (106{,}099.25 \cdot 1.712224) - (1{,}654{,}393.16 \cdot 0.225710)$

$\quad - (54{,}177.88 \cdot -17.125293)]/1802.50 = 248.409404$

This completes example 18.A1. Readers are urged to duplicate sufficient of the solution to understand the process but not to expect results to agree exactly. The calculations reported above are the result of 15-place arithmetic with rounding at the end of a sequence.

THE DOOLITTLE PROCESS

Doolittle (1878) arranged the computations for Gauss elimination into a succinct system. Doolittle omitted the unknown b's and made use of the symmetry of the set of coefficients, i.e., the fact that $a_{ij} = a_{ji}$. His system is illustrated here for four equations in four unknowns. First, record the coefficients in a table as below.

$$
\begin{array}{ccccc}
a_{11} & a_{12} & a_{13} & a_{14} & a_{10} \\
 & a_{22} & a_{23} & a_{24} & a_{20} \\
 & & a_{33} & a_{34} & a_{30} \\
 & & & a_{44} & a_{40}
\end{array}
$$

Add a computation check column on the right which is the sum of all the coefficients in the original equation. For example, the sum for the third equation is

$$a_{13} + a_{23} + a_{33} + a_{34} + a_{30}$$

since $a_{13} = a_{31}$ and $a_{23} = a_{32}$. Call these sums a_{1s}, a_{2s}, a_{3s}, and a_{4s}.

The order of the computations in the Doolittle system is slightly different from the order presented for Gauss elimination. The first equation in each reduced set of equations is computed, while the others in each reduced set are not. The procedure is presented in table 18.A2 for four equations in four unknowns.

Any number in parentheses stands for the table entries in a row. To create a new row in the table, do the indicated computation using the items in the designated rows. For example, each item in row (7) is the negative product of the corresponding item in row (1) multiplied by a_{13} and divided by a_{11}. Each item in row (9) is row (3) minus rows (7) and (8).

After computing each reduced equation, the sum of the columns to the left of the check column should closely approximate the check column. This check on the computations should be made routinely.

It is usually necessary to use a large number of digits for the computations, for instance, two digits more than the number of digits in the largest of the a's. Furthermore, two decimal places more than the number desired in the b estimates should be used throughout. Results can always be truncated by rounding off excess decimal places at the end of the process.

Gauss elimination is not the only method for solving estimation equations, and the Doolittle system is not the only succinct system for accomplishing Gauss elimination. There are, if fact, situations where this method will not work well and some other procedure will be necessary. Discussion of additional procedures is a topic in numerical mathematics and so will not be pursued further here.

Usually, the sum of squares of the deviations of the data from the approximating equation must be calculated. The general procedure for doing this can be illustrated for four equations in four unknowns.

$$D = \Sigma y^2 - b_0 a_{10} - b_1 a_{20} - b_2 a_{30} - b_3 a_{40}$$

Note that D is the sum of the y's squared minus the b estimates multiplied, respectively, by the numbers in the y column in table 18.A2.

Example 18.A2

The system of equations used in example 18.A1 is used to illustrate Doolittle's systemization of Gauss elimination. The details are presented in

TABLE 18.A2 Doolittle Computations for Four Equations in Four Unknowns in Example 18.A2

Computation	b_0	b_1	b_2	b_3	y	Check
(1)	a_{11}	a_{12}	a_{13}	a_{14}	a_{10}	a_{1s}
(2)		a_{22}	a_{23}	a_{24}	a_{20}	a_{2s}
(3)			a_{33}	a_{34}	a_{30}	a_{3s}
(4)				a_{44}	a_{40}	a_{4s}

$(5) = \dfrac{(1)a_{12}}{a_{11}}$		$\dfrac{a_{12}^2}{a_{11}}$	$\dfrac{a_{12}a_{13}}{a_{11}}$	$\dfrac{a_{12}a_{14}}{a_{11}}$	$\dfrac{a_{12}a_{10}}{a_{11}}$	$\dfrac{a_{12}a_{1s}}{a_{11}}$
$(6) = (2) - (5)$		$a_{22.1}$	$a_{23.1}$	$a_{24.1}$	$a_{20.1}$	$a_{2s.1}$
Check		$a_{22.1}$	$+\,a_{23.1}$	$+\,a_{24.1}$	$+\,a_{20.1}$	

$(7) = \dfrac{(1)a_{13}}{a_{11}}$			$\dfrac{a_{13}^2}{a_{11}}$	$\dfrac{a_{13}a_{14}}{a_{11}}$	$\dfrac{a_{13}a_{10}}{a_{11}}$	$\dfrac{a_{13}a_{1s}}{a_{11}}$
$(8) = \dfrac{(6)a_{23.1}}{a_{22.1}}$			$\dfrac{a_{23.1}^2}{a_{22.1}}$	$\dfrac{a_{23.1}a_{24.1}}{a_{22.1}}$	$\dfrac{a_{23.1}a_{20.1}}{a_{22.1}}$	$\dfrac{a_{23.1}a_{2s.1}}{a_{22.1}}$
$(9) = (3) - (7) - (8)$			$a_{33.12}$	$a_{34.12}$	$a_{30.12}$	$a_{3s.12}$
Check			$a_{33.12}$	$+\,a_{34.12}$	$+\,a_{30.12}$	

$(10) = \dfrac{(1)a_{14}}{a_{11}}$				$\dfrac{a_{14}^2}{a_{11}}$	$\dfrac{a_{14}a_{10}}{a_{11}}$	$\dfrac{a_{14}a_{1s}}{a_{11}}$
$(11) = \dfrac{(6)a_{24.1}}{a_{22.1}}$				$\dfrac{a_{24.1}^2}{a_{22.1}}$	$\dfrac{a_{24.1}a_{20.1}}{a_{22.1}}$	$\dfrac{a_{24.1}a_{2s.1}}{a_{22.1}}$
$(12) = \dfrac{(9)a_{34.12}}{a_{33.12}}$				$\dfrac{a_{34.12}^2}{a_{33.12}}$	$\dfrac{a_{34.12}a_{30.12}}{a_{33.12}}$	$\dfrac{a_{34.12}a_{3s.12}}{a_{33.12}}$
$(13) = (4) - (10)$ $\quad\;\; - (11) - (12)$				$a_{44.123}$	$a_{40.123}$	$a_{4s.123}$
Check				$a_{44.123}$	$+\,a_{40.123}$	

$$(14)b_3 = \frac{a_{40.123}}{a_{44.123}}$$

$$(15)b_2 = \frac{a_{30.12} - a_{34.12}b_3}{a_{33.12}}$$

$$(16)b_1 = \frac{a_{20.1} - a_{24.1}b_3 - a_{23.1}b_2}{a_{22.1}}$$

$$(17)b_0 = \frac{a_{10} - a_{14}b_3 - a_{13}b_2 - a_{12}b_1}{a_{11}}$$

table 18.A3 and follow the instructions given above. The process begins with the triangular matrix in the first four lines of the table. The columns

TABLE 18.A3 Doolittle Calculations for Example 18.A2

b_0	b_1	b_2	b_3	y	Check
(1) 31.00	927.25	28,186.44	1802.50	1269.50	32,216.69
(2)	28,186.44	870,561.67	54,177.88	36,895.50	990,748.73
(3)		27,308,910.14	1,654,393.16	1,089,730.31	30,951,781.71
(4)			106,099.25	75,024.25	1,891,497.03
(5)	27,735.24	843,092.72	53,915.10	37,972.38	963,642.69
(6)	451.19	27,468.96	262.77	−1076.88	27,106.04
Check sum 1					27,106.04
(7)		25,628,234.16	1,638,904.95	1,154,280.08	29,292,698.34
(8)		1.672,327.95	15,997.86	−65,561.34	1,650,233.42
(9)		8348.03	−509.66	1011.58	8849.95
Check sum 2					8849.95
(10)			104,806.65	73,815.28	1,873,244.49
(11)			153.04	−627.17	15,786.50
(12)			31.12	−61.76	−540.30
(13)			1108.44	1897.90	3006.34
Check sum 3					3006.34

on the right are the sums of all the coefficients in the respective equations. The second number in line (5) is

$$\frac{927.25 \cdot 28,186.44}{31.0000} = 843,092.72$$

The third number in line (6) is

$$54,177.88 - 53,915.10 = 262.77$$

The first check sum is

$$451.19 + 27,468.96 + 262.77 - 1076.88 = 27,106.04$$

and this agrees with the number above it in the check sum column.

Note that the numbers in lines (1), (6), (9), and (13) are the coefficients of the equations used in example 18.A1 to obtain the estimates of b_0, b_1, b_2, and b_3. These lines are used in the Doolittle scheme to obtain the estimates.

This completes example 18.A2. Readers are urged to duplicate as much of the arithmetic as they can, but are cautioned not to expect their numbers to agree with this example exactly. Like example 18.A1, 15 places were carried in all calculations until rounded for the example.

CHAPTER 19

Estimation and Confidence Limits

INTRODUCTION

Data analysis can be generally divided into two different tasks or functions. One of these is the estimation of population parameters, while the other is the testing of hypotheses concerning the parameters and the distribution assumed for the population. Consequently, for completeness, this chapter begins with a definition of sample and population.

Every data analysis problem begins with data. It is usually possible, at least conceptually, to add a few more data to those already available. Consequently, the available data may be considered, and actually are, part of some larger group of data, i.e., they are a sample. This larger, unseen group of data is called a population. Populations may be either finite or infinite in size, but for most practical data analysis purposes, they are treated as if they are infinite.

Now no one has much interest in samples except as they mirror populations. Averages, standard deviations, histograms, and other summary statistics describe the particular sample, while similar measures for the population remain unknown. The sample statistics are only of interest insofar as they permit the data analyst to describe or to infer characteristics of the population. A description of the population provides managers and researchers with a picture of where the results of their research may be applicable.

Since the sample is the data analyst's principal source of information on the population, populations are described about the same way samples are summarized. The analyst needs to know where the population is located on the measurement scale and how wide it is. Its shape is important, especially if the data analyst wants to use procedures based on a particular distribution, e.g., the normal. The analyst will also want estimates of differences between populations and of relationships between two or more variables measured on individuals in a population.

The most common estimate of a population characteristic or of differences between populations is always the corresponding sample statistic, but this can provide a false sense of security. A sample that was not selected randomly probably does not represent the population. So the data analyst must first make a judgment on this point. This problem was discussed in chapter 12.

The mean of a small sample, e.g., two or three measurements, is obviously less desirable than the mean of a large sample, e.g., 10,000 measurements, but without some forceful way to emphasize the difference, managers and researchers will treat two such different means the same. Statistics provides the way to distinguish between two different estimates of a population parameter in the form of confidence intervals defined by confidence limits.

Until now, this introduction has been concerned with the characteristics of a single population, whereas most data analysis problems concern the comparison of the characteristics of two or more populations or the comparison of the characteristics of a single population computed in more than one way. Estimation and confidence limits for a single population will be discussed in this chapter and then analogous ideas for the comparison of populations will be presented.

ESTIMATOR CHARACTERISTICS

Many procedures have been proposed for estimating population parameters, often more than one for the same parameter, so statisticians have developed ways to compare and to evaluate competing estimation procedures. Three characteristics of these point estimators, *consistency, efficiency,* and *sufficiency,* are used in these evaluations. Data analysts should know what these terms mean and how they apply to statistics in common usage.

A statistic, e.g., an average, calculated from a very large sample is generally considered to be very precise in the sense that adding more data to the sample will not change the sample statistic very much. Assume that a number of very large samples can be made available and that the same statistic is calculated for each sample. It will usually, but not always, be true that the differences among the statistics for the samples will become smaller and smaller as the samples are made larger and larger. If the sample sizes could be increased without limit, the sample statistics will all tend to a fixed value that is characteristic of the population and, therefore, related to the parameters of the population. Such a statistic is said to be consistent in the sense that if the sample size is increased, the statistic will converge to a constant that is related to the population characteristic. In the simplest case, the sample average for ratio and interval measurements not only tends to the population mean, but the differences between sample averages and the population mean, for samples of given size, tend to be normally distributed (central limit theorem). The differences among the averages of the samples are measured by the variance, or its square root, the standard deviation, of the sample of sample averages.

It is usually possible to invent or define any number of statistics for estimating population parameters which are consistent in the sense defined above and which have a variance that decreases as the sample size increases. The arithmetic average, the median, and the mode are all estimates of the population mean, for example, but for large samples of a fixed size, the variance of these different estimates will generally be different. The estimator with the smallest variance for a given sample size holds a special place in data analysis because it is to this statistic and procedure that all others are compared.

The notion of efficiency can be best explained with an example. The average and the median are both consistent estimates of the mean of a normally distributed population. The variance of these estimates for samples of size n are, respectively,

$$\frac{\sigma^2}{n} \quad \text{for any sample size and}$$

$$\frac{\pi\sigma^2}{2n} \quad \text{for large sample sizes.}$$

The median is less efficient for large samples than the average because its variance, for a given sample size, will be $\pi/2$ greater than the variance of the sample average.

Efficiency is a way of measuring or comparing the precision of different estimates of a population parameter. Hence another way to view efficiency is illustrated by the following. Given samples of n, the variance of the medians is expected to be $\pi/2$ times as large as the variance of the sample averages. The medians would have to be calculated from samples of $n\pi/2$ data in order for the variance of the medians to be as small as the variance of averages of samples of n. The term *efficient statistic* is reserved for the consistent statistic that is generally the most efficient, i.e., has the smallest variance for all sample sizes, of all the estimators of a particular population parameter.

Statistics with efficiencies less than 1.0 are used routinely and legitimately for many purposes. For example, the efficiency of the Wilcoxon statistic for comparing samples from two normally distributed independent populations approaches $3/\pi = 0.955$ as the sample size increases and is near 0.95 for even moderate sample sizes when compared to the efficient statistic, Student's t. The sample size for a Wilcoxon test of normally distributed populations should then be increased about 1 in 20 over the sample size used for Student's t. This slight increase is all that is required to remove the normality restriction required for Student's t.

It is often desirable, as the above illustrates, to compare the efficiencies of comparable statistics used in two different measurement scales. Student's t is not appropriate for ordinal data, but samples from normally distributed populations can always be transformed to the ordinal scale. Consequently, the procedure is to compare Student's t with Wilcoxon's statistic for data that originated in the ratio and/or interval measurement scales and were then transformed to the ordinal. This procedure is typical for evaluating the performance of ordinal and even nominal statistics with their counterparts in the ratio and interval measurement scales.

It is enough for data analysis that statistics be consistent and efficient. Some statistics, however, are said to be sufficient in the sense that they include all of the relevant information contained in the sample data. Such statistics are of considerable theoretical interest, and when they exist, they are definitely superior to other efficient statistics for small samples. The average of samples from the normal and Poisson distributions are examples of sufficient statistics. The estimation method of maximum likelihood, discussed below, leads to sufficient statistics when they exist.

ESTIMATION METHODS

The three principal methods of estimation for data analysis are maximum likelihood, least squares, and nonparametric.

Maximum likelihood is parametric in the sense that a particular distribution is assumed for the population. The sample is then used to estimate the most likely value of the parameter or parameters that is necessary to specify the distribution. For example, the normal distribution has two parameters, usually designated μ and σ. The first, μ, is a measure of location along the measurement scale, while σ is a measure of the spread of the population. Now maximum likelihood estimates for μ and σ are those estimates, based on the sample, that are the most likely, assuming the population is normally distributed. Maximum likelihood is a particularly important method of estimation because if the distribution assumed for the population is correct, then parameters estimated this way are consistent, efficient, and, as noted above, sufficient. If the population is not actually distributed as assumed, the parameter estimates will be incorrect and projections from the sample to the population will be inappropriate.

Consider chi-square as another example. Chi-square is a family of distributions, one for each value of the degrees of freedom. Given the degrees of freedom, the distribution depends on one parameter only, chi-square. Maximum likelihood can be used to estimate this parameter from a sample distributed as chi-square. Least squares can also be used to estimate the parameter from a sample distributed as chi-square but, except for very large degrees of freedom, will not in general be near the maximum likelihood estimate.

The actual calculation of a maximum likelihood estimate requires an understanding of mathematics not assumed for this book.

Least-squares estimation is nonparametric in the sense that the procedure as described in chapter 18 requires no assumptions about the distribution of the population. Consequently, least-squares estimates are often reported as such without any distribution being specified. But before any test of a least-squares estimate can be made, a distribution assumption must be made. Normality is the principal distribution assumption data analysts try to make, and when this is appropriate, least-squares estimates are identical to maximum likelihood estimates.

The arithmetic average was shown in chapter 18 to be a least-squares estimate. If the population from which the sample was taken is normally distributed, then the arithmetic average is also a maximum likelihood estimate.

The usual nonparametric estimates, and confidence limits for those estimates, do not depend on knowledge of the distribution of the populations in the original measurement scale but rather depend on properties of the samples themselves, e.g., the Hodges-Lehmann sample differences in chapters 2 and 3, the Moses confidence limits in chapter 2, and the Tukey confidence limits in chapter 3.

INTERVAL ESTIMATION

Point estimates are probably never correct in the sense that it is very unlikely the arithmetic average of a sample is the same as the population mean. The three characteristics of point estimates discussed above contain no measure of how far a point estimate may be from the corresponding population parameter. Intuitively, a point estimate should be close to the unknown value of the parameter if the spread of the population is small and/or the sample size large. Confidence intervals, based on the sample size and spread and the point estimate itself, take estimation one step further by providing a clear measure of just how "good" a point estimate really is.

Let θ be a population parameter. If it is possible, using only sample data, to define two statistics, u and v, such that the interval from u to v contains θ with a probability greater than zero, then the interval is called a confidence interval and u and v are called confidence limits. Confidence intervals are either closed, i.e., have an upper and a lower limit as defined above, or they are open, i.e, have only one limit and extend indefinitely either above or below that limit.

Confidence intervals are characterized by the probability $1 - \alpha$ that they will include the population parameter. Like the probability chosen for a statistical test, this probability may be anything the data analyst wants it to be. Confidence intervals have a kind of self-correcting mechanism. If the data analyst wants a high degree of confidence that the interval will include the population parameter, then the interval will be long, but by being long, it will be a less precise definition of the population parameter. A more precise definition of a population parameter using a particular sample, i.e., a shorter confidence interval, is possible only at the expense of a reduced confidence that the interval will contain the population parameter. Since this is true, the principal way to increase the probability and/or to shorten the interval is to increase the sample size. The data analyst can use this approach for rather powerful examples of the effect of changes in sample size.

Data analysts commonly compute and report confidence limits for population means, but confidence limits for measures of population spread, e.g., the standard deviation, are not often reported. This may be because the idea of uncertainty in location seems to be a more easily grasped idea than is the idea of uncertainty in spread. It is a sobering experience in data analysis to calculate a few confidence limits for standard deviations, so every data analyst should. The range of possible populations from which a particular sample could have come is astonishing.

The whole idea of confidence intervals depends on the data analyst knowing the distribution of estimates of the parameter. The central limit theorem, for example, is assurance that the distribution of sample averages of data measured in a ratio or interval scale will be normally distributed. The confidence limits in chapters 2 and 3 for ordinal measurement scales may be considered confidence limits for the average rank, i.e., the median, where, of course, the distribution of the ranks is always known.

Confidence limits for a population mean or any other measure of population location must never be interpreted as including or in any way defining a range of individuals in the population. Intervals that will include a proportion of the population are called tolerance intervals and are much longer than confidence intervals. The interested reader is referred to A. H. Bowker, Tolerance Limits for Normal Distributions, in Eisenhart, Hastay, and Wallis (1947).

The value of a population parameter is always an unknown constant. It is therefore incorrect to describe a confidence level as the probability that the population parameter falls in the confidence interval. The interval is the variable, and so the correct interpretation of the confidence level is the probability that the confidence interval includes the population parameter. This semantic difference is important in interpreting data analyses.

Example 19.1

Example 19.1 illustrates the ideas presented above for a sample from a normally distributed population. Table 19.1 presents a sample of seven data whose average is 20.5. The sample standard deviation is slightly less than 4.47. From chapter 18, the mean is known to be a least-squares estimate. It is also the maximum likelihood estimate since the population is normally distributed.

TABLE 19.1   Data for Example 19.1

16.6	26.2
20.4	14.6
25.0	17.7
22.6	

Since the population is normally distributed,

$$\frac{(\bar{y} - \mu)}{\text{sd}} \sqrt{n}$$

is distributed as Student's t with 6 degrees of freedom and

$$\frac{\text{df} \cdot \text{sd}^2}{\sigma^2}$$

where df is the degrees of freedom, is distributed as chi-square with 6 degrees of freedom. These can be used to estimate confidence intervals for μ and σ respectively.

It follows that

$$\text{Prob}\left(-t_{n-1,\alpha/2} \leq \frac{\bar{y} - \mu}{\text{sd}/\sqrt{n}} \leq + t_{n-1,\alpha/2}\right) = 1 - \alpha$$

and that

$$\text{Prob}\left(\bar{y} - \frac{t_{n-1,\alpha/2}\,\text{sd}}{\sqrt{n}} \leq \mu \leq \bar{y} + \frac{t_{n-1,\alpha/2}\,\text{sd}}{\sqrt{n}}\right) = 1 - \alpha$$

The most commonly reported confidence interval for a population mean is one that has a 0.95 probability of including that mean. From table B in part V, $t = 2.447$, so the confidence limits are

$$20.5 - 2.447 \cdot \frac{4.47}{\sqrt{7}} = 16.37$$

$$20.5 + 2.447 \cdot \frac{4.47}{\sqrt{7}} = 24.63$$

This interval is wide and so it does not provide a very precise definition of the population mean.

A more precise definition of the population mean, i.e., a shorter interval, can be obtained by reducing the probability of including the mean in the interval. The *t*-statistic for a probability of 0.80 is 1.440, so the 0.80 confidence limits are 17.9 and 23.1. This interval is considerably shorter, i.e., a more precise definition, but the probability that the interval contains the population mean is considerably smaller too. Conversely, the probability can be increased, e.g., to 0.99, but this will lengthen the interval, so little is gained by juggling probabilities and limits. It is better to standardize probabilities for the confidence limits in a report, or a series of reports, so managers and researchers can compare intervals without having to translate their meanings.

Chi-square is tabled in part V such that

$$\text{Prob}\left(\chi^2_{n-1,1-\alpha} \le \frac{df \cdot sd^2}{\sigma^2}\right) = 1 - \alpha$$

Upon solving the inequality within the parentheses for σ^2, the following is obtained

$$\text{Prob}\left(\sigma^2 \le \frac{df \cdot sd^2}{\chi^2_{n-1,1-\alpha}}\right) = 1 - \alpha$$

From table C in part V, χ^2 for 6 degrees of freedom and $1 - \alpha = 0.95$ is 1.635, so that the upper 0.95 limit on σ^2 is 73.24. The upper limit on σ is 8.56. This is almost twice as large as the point estimate for the standard deviation. Figure 19.1 pictures normal distributions with standard deviations of 4.47 and 8.56 to demonstrate the considerable range of normally distributed populations that could have given rise to this sample. Of course, the population standard deviation could be less than 4.47 for an even more narrow distribution.

Example 19.2

These confidence intervals are large, so to reduce them, additional data were obtained and are presented with the first 7 in table 19.2. The average, 20.9, and the standard deviation, 4.18, of all 28 data agree well with the summary statistics for the first 7. The 0.95 confidence limits for the population mean are calculated as follows.

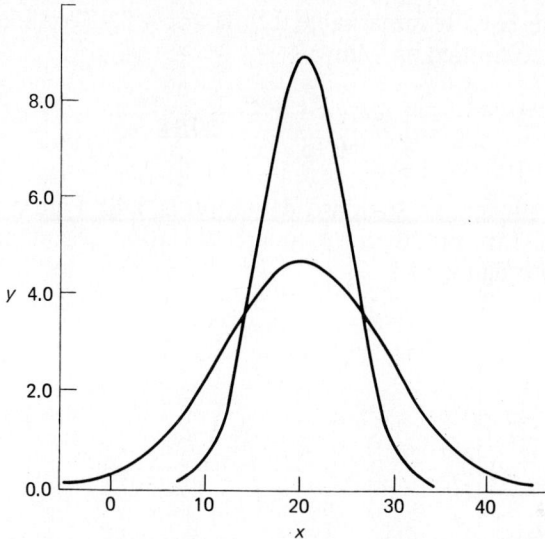

FIGURE 19.1 Normal probability densities illustrating the range of densities acceptable for the data in example 19.1.

$$20.9 - 2.052 \cdot \frac{4.18}{\sqrt{28}} = 19.28$$

$$20.9 + 2.052 \cdot \frac{4.18}{\sqrt{28}} = 22.52$$

This confidence interval is smaller than the first for three reasons. The standard deviation is not quite as large and the *t*-statistic is a little smaller, but more importantly, the number of data was increased by a factor of 4. This has the effect of dividing the length of the confidence interval by $\sqrt{4} = 2$.

TABLE 19.2 Data for Example 19.2

16.6	23.0	20.6	16.4
20.4	28.3	25.9	21.3
25.0	19.1	23.2	21.9
22.6	19.2	13.0	25.9
26.6	20.7	19.8	27.0
14.6	17.4	24.1	14.6
17.7	16.0	25.0	19.5

The effect is equally remarkable for the upper 0.95 confidence limit for the variance computed as follows.

$$\frac{27 \cdot 4.178^2}{16.151} = 29.18$$

The upper limit for the standard deviation is 5.40. Figure 19.2 pictures the normal distribution with a standard deviation of 5.40 and should be compared with figure 19.1.

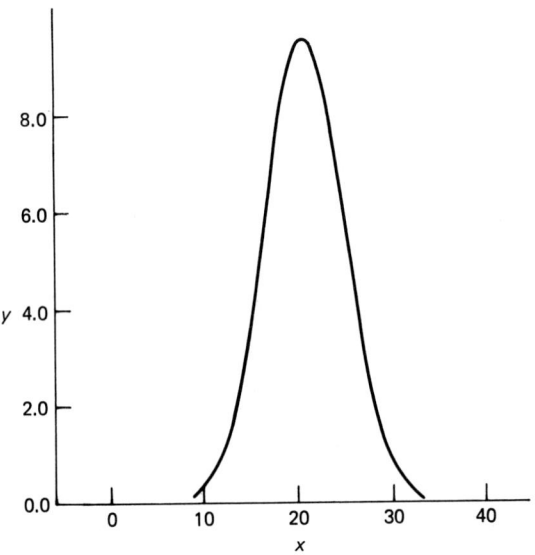

FIGURE 19.2   Normal probability density for example 19.2 using the upper 0.95 confidence limit for the variance. Compare spread with densities pictured in Figure 19.1.

COMPARISONS

The difference between the location of two independent populations in a ratio or interval measurement scale is usually estimated by the difference between the averages of samples from the two populations. In chapter 2, populations with equal variances were compared using

$$t = \frac{\bar{y}_1 - \bar{y}_2}{sd \, \sqrt{1/n + 1/m}}$$

where

$$sd^2 = \frac{\sum\limits_{i=1}^{n} (y_1 - \bar{y}_1)^2 + \sum\limits_{i=1}^{m} (y_2 - \bar{y}_2)^2}{n + m - 2}$$

Let the difference $\bar{y}_1 - \bar{y}_2$ be one observation from a new population consisting of the difference between averages of samples of size n and m, respectively. Then the procedure described above for estimating confidence limits for a single population can be used to estimate the limits for the difference between two populations. An estimate of the standard deviation of this new population is

$$sd \sqrt{\frac{1}{n} + \frac{1}{m}}$$

Let μ be the mean of this new population. Then by direct substitution,

$$\text{Prob}\left(-t_{n+m-2,\alpha/2} \leq \frac{\bar{y}_1 - \bar{y}_2 - \mu}{sd \sqrt{1/n + 1/m}} \leq + t_{n+m-2,\alpha/2} \right) = 1 - \alpha$$

and

$$\text{Prob}\left(\bar{y}_1 - \bar{y}_2 - t_{n+m-2,\alpha/2}\, sd \sqrt{\frac{1}{n} + \frac{1}{m}} \leq \mu \right.$$

$$\left. \leq \bar{y}_1 - \bar{y}_2 + t_{n+m-2,\alpha/2}\, sd \sqrt{\frac{1}{n} + \frac{1}{m}} \right) = 1 - \alpha$$

Example 19.3

Consider the ratio and interval data for samples from two populations as given in table 19.3. Using the above formula for sd^2, it is 9.1220. The estimate of the standard deviation of the new population is

$$sd \sqrt{\frac{1}{n} + \frac{1}{m}} = 3.02 \cdot 0.5563 = 1.68$$

The t-statistic for $7 + 6 - 2 = 11$ degrees of freedom and $\alpha = 0.05$ is 2.201. The 0.95 confidence limits for the population mean computed from the above are

$$15.69 - 18.77 - 2.201 \cdot 1.68 = -6.78$$
$$15.69 - 18.77 + 2.201 \cdot 1.68 = +0.62$$

TABLE 19.3 Data for Example 19.3

Sample 1	Sample 2
12.0	16.0
12.1	18.6
16.1	16.0
18.4	24.8
15.0	17.5
18.0	19.7
18.2	

The 0.95 confidence interval extends from -6.78 to $+0.62$ on the measurement scale. Since the interval includes zero, the null hypothesis that the populations have equal means cannot be rejected in a two-tailed test with $\alpha = 0.05$. One-sided confidence limits can be constructed with the above inequalities by replacing $\alpha/2$ with α on the appropriate side.

The situation for two dependent samples is exactly like example 19.1 once the data pair differences have been computed.

Example 19.4

The critical ranges in chapter 4 and the critical differences in chapter 5 may be viewed as half-lengths of confidence intervals computed in several different ways, each way possessing characteristics different from the others. Any of these may be used as the basis for confidence intervals having the respective special characteristics. Consider, for example, the data in table 19.4 for four related samples. Table 19.5 presents the analysis of variance needed to find the residual standard deviation that is required for all of the multiple comparison procedures. It is $\sqrt{6.3189} = 2.51$. Fisher's LSD

$$ t_{\mathrm{df},\alpha/2} \; \mathrm{sd} \; \sqrt{\frac{1}{n_1} + \frac{1}{n_2}} $$

TABLE 19.4 Data for Example 19.4

Sample 1	Sample 2	Sample 3	Sample 4
19.58	16.71	22.33	21.18
14.39	17.99	17.07	23.32
17.99	15.18	18.24	22.75
18.36	25.77	19.08	24.77
17.58	23.94	21.96	24.29
22.77	22.05	19.07	23.89

TABLE 19.5 The ANOVA Table for Example 19.5

Source of variation	df	ss	ms	
Between samples	3	79.4207	26.4736	F = 4.19
Between blocks	5	63.3639	12.6728	
Residual	15	94.7830	6.3189	
Total	23	237.5676		

The t-statistic for 15 degrees of freedom and $\alpha/2 = 0.025$ is 2.131 so that Fisher's LSD is

$$2.131 \cdot 2.51 \sqrt{\frac{1}{6} + \frac{1}{6}} = 3.09$$

The sample averages are given sorted in table 19.6. The difference for samples 1 and 3 is 1.18, so that confidence limits are $1.18 - 3.09 = -1.91$ and $1.18 + 3.09 = 4.27$. Note that this interval, -1.91 to 3.09, includes zero, so using Fisher's LSD for multiple comparisons would indicate no significance in this difference at a probability of 0.025 for a one-tailed test. In contrast, consider the difference between sample 4 and any other sample, e.g., sample 2. The difference here is 3.10, so that confidence limits are $3.10 - 3.09 = 0.01$ and $3.10 + 3.09 = 6.19$. Note that this interval, 0.01 to 6.19, does not include zero, so the difference is significant at a probability of 0.025 for a one-tailed test.

TABLE 19.6 Sample Averages for Example 19.4

Sample	$\bar{y}$
1	18.44
3	19.62
2	20.27
4	23.37

Multiple comparisons, as constructed or defined by statisticians, are two-tailed tests. Many data analysis problems, even in analyses of variance, are one-tailed, since the researcher usually expects or wants the sample averages to be in a particular order. Consequently, the data analyst must be particularly careful in interpretations lest one-tailed ranges or differences be calculated and interpreted as two-tailed or vice versa.

Confidence intervals based on Fisher's LSD will include the unknown population differences with probability $1 - \alpha$, but this does not mean that all population pair differences for a single problem, e.g., there are six in example 19.2, will be included with probability $1 - \alpha$. Furthermore, the analyst has no way of knowing which difference or differences may not be included. Excluded differences, if there are any, are not necessarily the largest or smallest or the middle one or any other difference in particular.

Confidence limits or intervals with characteristics different from Fisher's LSD can be estimated using any of the multiple comparison procedures. For example, confidence intervals based on Tukey's multiple comparison procedure will include all population pair differences with probability $1 - \alpha$. Of course to do this the intervals are considerably longer than Fisher's LSD. The data analyst should be particularly careful in interpreting intervals based on the intermediate compromises such as the Duncan, Dunn, Dunnett, or the Newman-Keuls procedures.

NONPARAMETRIC

Procedures for nonparametric estimation and confidence limits are specific for each particular data analysis situation. Consequently, these have been discussed as the need arose in parts I and II. In general, point estimates are based on the median, whereas confidence limits are based on the distribution of all possible data pair differences, one datum from each sample. Nonparametric intervals computed in this way are usually larger than parametric limits whenever both are appropriate but, of course, should be used when the necessary parametric assumptions, e.g., normality, are not tenable.

NOMINAL MEASUREMENTS

The principal summary statistic for nominal scale measurements is the proportion. Some writers, e.g., Fleiss (1973), have suggested that confidence limits for proportions are useless, but data analysts do need them often enough to justify a discussion in this chapter.

Even though there may be several nominal scale categories in an analysis, analysts can and often do view membership in each category as an either/or situation, i.e., the observation is either in a specific category or it isn't. From this viewpoint, the observations are binomially distributed. Let x be the number of observations in a sample of n that are in the specified category. Then the probability of a particular number c or less in the category, assuming x has a binomial distribution, is given by

$$\text{Prob} (x \le c) = \sum_{x=0}^{c} \binom{n}{x} p^x (1 - p)^{n-x}$$

where $\binom{n}{x}$ = the binomial coefficient $n!/x! \, (n - x)!$

p = the unknown population proportion for the category

A lower confidence limit may be found by solving

$$\sum_{x=c}^{n} \binom{n}{x} p_1^x (1 - p_1)^{n-x} = \frac{\alpha}{2}$$

for p_1. While solving this equation directly would be a rather difficult task, it is a simple matter to use table H in part V to find p_1 by interpolation. An upper confidence limit may be found by solving

$$\sum_{x=0}^{c} \binom{n}{x} p_2^x (1 - p_2)^{n-x} = \frac{\alpha}{2}$$

for p_2. It is a little more difficult to find p_2 from table H than it is to find p_1, but it can be done. Use the part of table H for n and $c - 1$. Then find p_2 for $1 - \alpha/2$.

The desired confidence limits for the population proportion are these values of p. Example 19.5 illustrates the procedure. If $c = 0$, the lower confidence limit is taken to be zero; if $c = n$, the upper confidence limit is taken to be one. One-sided confidence limits are computed by using the appropriate formula and α instead of $\alpha/2$.

Example 19.5

Suppose in a sample of eight observations, three are in the category and five are not. The observed proportion and the best estimate for the unknown population parameter is then $3/8 = 0.375$. The problem is to find confidence limits for this estimate such that the limits will include the parameter with a specified confidence, e.g., 0.90 where $\alpha = 0.10$. Referring to table H for $n = 8$ and $c = 3$, find p_1 so that

$$\text{Prob} (x \leq 3) = \frac{\alpha}{2} = 0.05$$

Interpolating between $p = 0.10$ and $p = 0.15$, p_1 is found to be 0.109. Now referring to table H for $n = 8$ and $c = 2$, find p_2 so that

$$\text{Prob} (x \leq 2) = 1 - \frac{\alpha}{2} = 0.95$$

Using interpolation, it is 0.474. The best estimate for the population parameter p is 0.375, and the 0.90 confidence interval is defined by the limits 0.109 and 0.474.

CHAPTER 20

The Analysis of Variance

INTRODUCTION

An analysis of variance is a generic name for a very important kind of analysis used in many, many situations. The idea underlying an analysis of variance is basically simple, but like many simple ideas, it is both powerful and elegant. The purpose of this chapter is to explain in simple terms the motivation for the analysis of variance and at the same time to describe what is accomplished by the procedure. Once understood, its use in specific situations will be more meaningful for the data analyst and results of an analysis can be more easily explained.

Since R. A. Fisher's pioneering work in the analysis of variance, this technique has been extended to many, many situations, so much so that the procedure is almost the universal tool for ratio and interval data analyses as well as the basis for many rank procedures. In this chapter, it is discussed not from the viewpoint of a specific application but from a general viewpoint, even though the ideas will be presented through use of an example. This chapter begins with a description of what the procedure is supposed to accomplish. Example 4.1 in chapter 4 represents the simplest application of analysis of variance, and it will be used here both to illustrate the procedure and to explain the underlying motivation. Each chapter where the procedure is used contains additional explanation specific for a particular application.

PURPOSE

The prudent researcher will insist on more than one observation. The result of the observations, i.e., the measurements, will not all be identical.

The differences between repeated measurements of the "same" phenomenon are called *variability,* and statisticians have provided data analysts with several ways to measure the magnitude of the variability. The most common measure is the sample variance

$$sd_y^2 = \frac{\Sigma(y - \bar{y})^2}{n - 1}$$

Given two samples with one sample scattered along the measurement scale more than the other, the sample with the bigger scatter will have the larger variance. This scatter is called the *total variability* and sd_y^2 is a measure of it.

Sometimes, because of ways the data can be classified or because of relationships between variables, the total variability can be divided into parts that are then associated with the classifications or the relationships. This process, along with the associated statistical tests, is called an analysis of variance.

Actually, the variance per se is not divided into components. The sum of squares

$$\Sigma(y - \bar{y})^2$$

and the degrees of freedom, $n - 1$, are the quantities divided into parts. In every analysis of variance, there will remain, after every logical and appropriate assignment of variability is made, a portion of the total variability that is unassignable. This is given various names including residual, experimental error, and within. All of these have special meanings for particular situations, but for the data analyst, this leftover variability

TABLE 20.1   Reductions in Diastolic
Blood Pressure for Example 20.1

	Study		
1	*2*	*3*	*4*
10	0	20	30
5	10	25	25
5	20	25	10
5	25	25	30
10	5	25	15
25	10	20	15
10	10	20	
25	10	20	
25	10	25	
15	14	20	
	16		
	8		

is simply the unassignable portion of the total that plays a very special role in the analysis.

As discussed in chapters 15 and 17, every statistical test is in some way a comparison of two summary statistics. One of these statistics will measure differences that are the subject or purpose of the analysis. The other measures data variability when the subject differences are not present. The statistical test is a determination of the probability of observing the two statistics when only chance variation is present. In the analysis of variance, the unassignable portion of the total variability is the second of these two statistics.

In summary, the purpose of the analysis of variance procedure, up to the time a statistical test is made, is to separate from the total variability every assignable portion so that the unassignable part will be as small as possible.

THE SIMPLEST SITUATION

The data for example 20.1 are presented in table 20.1. To begin, forget about the sample classification. The variance of these or any other group of data is defined by

$$sd_y^2 = \frac{\Sigma(y - \bar{y})^2}{n - 1}$$

$n - 1$ is called the degrees of freedom while

$$\Sigma(y - \bar{y})^2$$

is the sum of squares. The sum of squares is algebraically identical to

$$\Sigma y^2 - \frac{(\Sigma y)^2}{n}$$

This is easily demonstrated as follows:

$$\Sigma(y - \bar{y})^2 = \Sigma(y^2 - 2y\bar{y} + \bar{y}^2)$$
$$= \Sigma y^2 - 2\Sigma y\bar{y} + \Sigma\bar{y}^2$$

Now $\bar{y}$ is a constant insofar as this summation for the sample is concerned, i.e., it is the same for all of the y, so that

$$\Sigma\bar{y}^2 = \bar{y}^2\Sigma 1 = \bar{y}^2 n$$

Furthermore, on substituting for $\bar{y}$,

$$2\Sigma y\bar{y} = 2\Sigma y \frac{\Sigma y}{n} = \frac{2(\Sigma y)^2}{n}$$

and

$$n\bar{y}^2 = n\left(\frac{\Sigma y}{n}\right)\left(\frac{\Sigma y}{n}\right) = \frac{(\Sigma y)^2}{n}$$

Then the sum of squares can be written as

$$\Sigma(y - \bar{y})^2 = \Sigma y^2 - \frac{2(\Sigma y)^2}{n} + \frac{(\Sigma y)^2}{n}$$

$$= \Sigma y^2 - \frac{(\Sigma y)^2}{n}$$

This, of course, is the calculation form for the sum of squares.

The total sum of squares for the 38 data is calculated as follows.

$$\Sigma y = 623$$
$$\Sigma y^2 = 12{,}641$$
$$n = 38$$

Then

$$\Sigma(y - \bar{y})^2 = 12{,}641 - \frac{623 \cdot 623}{38}$$

$$= 2427.08$$

In an analysis of variance, the total sum of squares and the total degrees of freedom are divided according to carefully defined rules into components that can be identified with the *structure* or characteristics of the data. The 38 data in table 20.1 can be classified by sample and that classification defines the structure. If there were other ways to classify these data, then these classes would become part of a more complex structure. The total sum of squares and the total degrees of freedom for the data in table 20.1 can be divided into two components, one associated with the sample classification and the other associated with membership in the samples. The data within a sample cannot be classified into smaller groups, so this last component cannot be further subdivided. It is the residual.

Consider the degrees of freedom first. There are four samples. Sample 1 has 10 data and 9 degrees of freedom, sample 2 has 12 data and 11 degrees of freedom, and samples 3 and 4 have 10 and 6 data, respectively, with 9 and 5 degrees of freedom. The sum of these sample degrees of freedom is 34. A sum calculated in this way has several names within statistics books, among which are degrees of freedom for *within samples* and *residual* degrees of freedom. The explanation above justifies the first name. The second name will be justified later.

The remaining 3 degrees of freedom are associated with the four samples and are called the degrees of freedom *between samples* or simply *the between* degrees of freedom. This division of the total degrees of freedom into two parts can be represented in a table like table 20.2.

Next consider the total sum of squares. It can be divided into two parts corresponding to between and within samples just like the total degrees of freedom was divided. The procedure can be explained in several ways, and three of these will be used to reinforce the reader's understanding of what is being done. These different ways of explaining the procedure have no special names, so they will be given arbitrary names for this chapter.

TABLE 20.2 Degrees of Freedom for Example 20.1

Source	df
Between	3
Within	34
Total	37

The Within-First Procedure

The data in table 20.1 have been classified into samples. In this procedure, the sum of squares is computed for each sample separately and then summed for the four samples. Table 20.3 presents the calculations. This total, 1564.83, is called the *within samples sum of squares*. Any differences that exist between the samples will not influence this total. For example, suppose the six data in sample 4 were each 100 larger than they really are, i.e.,

$$130, 125, 110, 130, 115, \text{ and } 115.$$

If this were true, the data in sample 4 would in general be much larger than the data in the other three samples, but the sum of squares for sample 4 would still be 370.83.

TABLE 20.3 Calculation of the Within Sum of Squares for Example 20.1

Sample	Σy^2	Σy	n	$\Sigma y^2 - (\Sigma y)^2/n$
1	2475	135	10	652.5
2	2066	138	12	479.0
3	5125	225	10	62.5
4	2975	125	6	370.83
Total				1564.83

The difference between this total, 1564.83, and the sum of squares for all of the data, i.e., 2427.08, is the part of the total associated with the samples or structure of the data. This breakup of the total sum of squares into two components added to table 20.2 is shown in table 20.4. The column headings stand for degrees of freedom (df) and sum of squares (ss).

TABLE 20.4   The ANOVA Table for Example 20.1

Source	df	ss
Between	3	862.25
Within	34	1564.83
Total	37	2427.08

The Between-First Procedure

Since the sum of squares for within is unaffected by differences between samples, then differences between samples must be measured or accounted for in the sum of squares for between. It can be calculated from the sample averages as illustrated in table 20.5. This procedure is not recommended for routine use because it requires, as is evident in table 20.5, a great many decimal places to assure accuracy in the final sum. A calculation formula will be developed later in this chapter. Beginning with the sum of squares in table 20.5 for between and the sum of squares for total, table 20.4 can be constructed easily.

The influence of sample differences on the sum of squares for between can be illustrated dramatically with the same example used to illustrate lack of influence on the sum of squares for within. Suppose each datum in sample 4 was 100 larger. Then the sum of squares for between would be 66,188.5614 while, as was demonstrated, the sum of squares for within would remain the same.

TABLE 20.5   Calculations for the Between-First Procedure for Example 20.1

Sample	Sample mean	Diff	Diff^2	n_j	$n_j \cdot diff^2$
1	13.5000	−2.8947	8.3795	10	83.7950
2	11.5000	−4.8947	23.9584	12	287.5014
3	22.5000	6.1053	37.2742	10	372.7424
4	20.8333	4.4386	19.7011	6	118.2068
Total					862.2456

NOTE: Diff = difference.

The Algebraic Procedure

The total sum of squares for the data in table 20.1 is

$$\sum_{j=1}^{m} \sum_{i=1}^{n_j} (y_{ij} - \bar{\bar{y}})^2$$

where j counts the samples
$\quad m =$ the number of samples
$\quad i \quad$ counts the data within samples
$\quad n_j =$ the number of data in sample j
$\quad y_{ij} =$ datum i in sample j
$\quad \bar{\bar{y}} =$ the mean of all the data.

This sum is unchanged by adding and subtracting within the parentheses the sample mean $\bar{y}_j$ for each y_{ij}.

$$\sum_{j=1}^{m} \sum_{i=1}^{n_j} (y_{ij} - \bar{y}_j + \bar{y}_j - \bar{\bar{y}})^2$$

The quantity within the brackets can be squared to get

$$\sum_{j=1}^{m} \sum_{i=1}^{n_j} [(y_{ij} - \bar{y}_j)^2 + 2(y_{ij} - \bar{y}_j)(\bar{y}_j - \bar{\bar{y}}) + (\bar{y}_j - \bar{\bar{y}})^2]$$

Next consider the sum of each term inside the brackets separately. The first

$$\sum_{j=1}^{m} \sum_{i=1}^{n_j} (y_{ij} - \bar{y}_j)^2$$

is the sum of squares within the samples. The reader should verify this by comparing the above expression with the within-first procedure. The last

$$\sum_{j=1}^{m} \sum_{i=1}^{n_j} (\bar{y}_j - \bar{\bar{y}})^2$$

is the sum of squares between the samples. Since this may not be immediately evident, note that for every datum in a sample,

$$\bar{y}_j - \bar{\bar{y}}$$

is a constant. Therefore

$$\sum_{i=1}^{n_j} (\bar{y}_j - \bar{\bar{y}})^2 = n_j(\bar{y}_j - \bar{\bar{y}})^2$$

Then the last term can be written

$$\sum_{j=1}^{m} n_j \, (\bar{y}_j - \bar{\bar{y}})^2$$

which should look a lot like the between-first procedure.

The middle term was saved for last because it is simply zero. To understand this, write the sum as follows

$$2 \sum_{j=1}^{m} \left[\sum_{i=1}^{nj} (y_{ij} - \bar{y}_j) \right] (\bar{y}_j - \bar{\bar{y}})$$

and consider the part within the brackets. The sum of the data in a sample minus the sample mean,

$$\sum_{i=1}^{nj} (y_{ij} - \bar{y}_j)$$

is always zero, but just for the sake of completeness, sum the two terms separately. Then

$$\sum_{i=1}^{nj} y_{ij} - \sum_{i=1}^{nj} \bar{y}_j = \sum_{i=1}^{nj} y_{ij} - n_j \bar{y}_j$$

Finally note that

$$n_j \bar{y}_j = n_j \frac{\displaystyle\sum_{i=1}^{nj} y_{ij}}{n_j} = \sum_{i=1}^{nj} y_{ij}$$

Now the summation within the brackets is

$$\sum_{i=1}^{nj} y_{ij} - \sum_{i=1}^{nj} y_{ij} = 0$$

The algebraic procedure can be summarized.

$$\sum_{j=1}^{m} \sum_{i=1}^{nj} (y_{ij} - \bar{\bar{y}})^2 = \sum_{j=1}^{m} \sum_{i=1}^{nj} (y_{ij} - y_j)^2 + \sum_{j=1}^{m} n_j (\bar{y}_j - \bar{\bar{y}})^2$$

Calculation Formulas

The preceding equation is good for understanding the ideas behind analysis of variance, but it is not good for calculations. The generally accepted formulas for calculating the between and the within sum of squares are given below. The reader should verify these by completing their derivation using principles explained earlier in this chapter under The Simplest Solution.

For between,

$$\sum_{j=1}^{m} n_j(\bar{y}_j - \bar{\bar{y}})^2 = \sum_{j=1}^{m} \frac{\left(\sum_{i=1}^{n_j} y_{ij}\right)^2}{n_j} - \frac{\left(\sum_{j=1}^{m}\sum_{i=1}^{n_j} y_{ij}\right)^2}{N}$$

and for within,

$$\sum_{j=1}^{m}\sum_{i=1}^{n_j}(y_{ij} - \bar{y}_j)^2 = \sum_{j=1}^{m}\sum_{i=1}^{n_j} y_{ij}^2 - \sum_{j=1}^{m} \frac{\left(\sum_{i=1}^{n_j} y_{ij}\right)^2}{n_j}$$

where N = the sum of the n_j.

Summary

The data in table 20.1 can be classified in one way, the four samples, and this defines a structure for the data. The total sum of squares for the data has been divided into two parts. One has been associated with differences between samples while the other has been associated with data differences within samples. A simple illustration was used to show that differences between samples did not influence the within sum of squares. Intuitively, any differences between the samples must be in the between-samples sum of squares. This was also illustrated.

The ANOVA Table

An ANOVA table is prepared by computing the variance, usually called a mean square, for each of the components and adding these to table 20.4. The result is given in table 20.6. The degrees of freedom and the sum of squares have been defined; ms is, of course, the mean square, which is

TABLE 20.6   The ANOVA Table for Example 20.1

Source of variation	df	ss	ms	
Between samples	3	862.25	287.42	$F = 6.24$
Within samples	34	1564.83	46.02	
Total	37	2427.08		

obtained by dividing each sum of squares by the corresponding degrees of freedom.

The ratio of the two mean squares, when the data are all from the same normal distribution, has a distribution discovered by R. A. Fisher and was named F in his honor by George W. Snedecor (1934). This means that the probability of an F-statistic when only chance differences exist between samples can be determined. If this probability is small, then the data analyst may recommend proceeding with the research as if the populations have different mean values. If on the contrary this probability is large, the analyst may want to recommend proceeding with the research as if the populations have equivalent means. The F-statistic is often added to the ANOVA table in a column on the right as has been done in table 20.6.

DATA ANALYSIS IMPLICATIONS

A very simple problem has been used to illustrate analysis of variance calculations. Essentially the same end result, i.e., the division of the total sum of squares and the total degrees of freedom into component parts, can be accomplished in much more complex situations but not necessarily by using the simple procedures given here. A far more sophisticated procedure called *linear models* must be used. The study of linear models is beyond the scope of this book.

PART V

Tables

INTRODUCTION

Part V contains 20 tables useful to data analysts. Recognizing that difficult-to-interpret tables can interfere with the analysis process, these have been arranged as uniformly as possible. Most give upper one-tailed rejection probabilities. The exceptions are noted in the list below.

Table	*Page*
A. The normal or Gaussian distribution	T.4
B. Student's *t* distribution	T.6
C. The chi-square distribution	T.8
D. Fisher's *F* distribution	T.10
E. Wilcoxon's rank sum	T.24
This table gives the probability that the rank sum minus the expected value will be equal to or larger than *W*.	
F. Wilcoxon's signed rank	T.91
This table gives the probability that the signed rank minus the expected value will be equal to or larger than *W*.	
G. The sign test	T.95
The table for the sign test is in two parts. The first is used when the least-frequent attribute is counted and gives upper limits on counts for particular probabilities. The second is used when the most-frequent attribute is counted and gives lower limits on counts for particular probabilities.	
H. The binomial distribution	T.100
For given *p* and *n*, this table gives the probability of *c* or more units with the particular attribute.	
I. The Kruskal-Wallis test statistic	T.110
This table gives the probability that the Kruskal-Wallis test statistic will be equal to or greater than KW.	
J. Friedman's chi-square statistic	T.115
This table gives the probability that the test statistic will be equal to or larger than χ^2.	
K. Kendall's tau	T.118
This table gives the probability that the numerator *s* of Kendall's coefficient of correlation tau will be equal to or larger than *t*.	
L. Spearman's rho	T.119
This table gives the probability that Spearman's coefficient of correlation will be equal to or larger than rho given in the table. The table also gives *s* for each rho, a quantity much used in the social sciences. If $R(x)$ is the rank of *x* and $R(y)$ is the rank of *y*, then	

$$s = \Sigma R(x)^2 + \Sigma R(y)^2 - 2\Sigma R(x)R(y)$$

M. Lilliefors' test statistic	T.121
Lilliefors' table is used for a test of the null hypothesis that a random sample is from a normal distribution with unspecified mean and variants.	
N. The Shapiro-Wilk normality test	T.122
Table N1 gives coefficients for computing the test statistic. Table N2 gives lower limits on the test statistic for given probabilities. For	

Table	Page

TABLE A   The Normal or Gaussian Distribution

z	0,00	0,01	0,02	0,03	0,04	0,05	0,06	0,07	0,08	0,09
0,0	,50000	,49601	,49202	,48803	,48405	,48006	,47608	,47210	,46812	,46414
0,1	,46017	,45620	,45224	,44828	,44433	,44038	,43644	,43251	,42858	,42465
0,2	,42074	,41683	,41294	,40905	,40517	,40129	,39743	,39358	,38974	,38591
0,3	,38209	,37828	,37448	,37070	,36693	,36317	,35942	,35569	,35197	,34827
0,4	,34458	,34090	,33724	,33360	,32997	,32636	,32276	,31918	,31561	,31207
0,5	,30854	,30503	,30153	,29806	,29460	,29116	,28774	,28434	,28096	,27760
0,6	,27425	,27093	,26763	,26435	,26109	,25785	,25463	,25143	,24825	,24510
0,7	,24196	,23885	,23576	,23270	,22965	,22663	,22363	,22065	,21770	,21476
0,8	,21186	,20897	,20611	,20327	,20045	,19766	,19489	,19215	,18943	,18673
0,9	,18406	,18141	,17879	,17619	,17361	,17106	,16853	,16602	,16354	,16109
1,0	,15866	,15625	,15386	,15150	,14917	,14686	,14457	,14231	,14007	,13786
1,1	,13567	,13350	,13136	,12924	,12714	,12507	,12302	,12100	,11900	,11702
1,2	,11507	,11314	,11123	,10935	,10749	,10565	,10383	,10204	,10027	,09853
1,3	,09680	,09510	,09342	,09176	,09012	,08851	,08691	,08534	,08379	,08226
1,4	,08076	,07927	,07780	,07636	,07493	,07353	,07215	,07078	,06944	,06811
1,5	,06681	,06552	,06426	,06301	,06178	,06057	,05938	,05821	,05705	,05592

x	.00	.01	.02	.03	.04	.05	.06	.07	.08	.09
1.6	.05480	.05370	.05262	.05155	.05050	.04947	.04846	.04746	.04648	.04551
1.7	.04457	.04363	.04272	.04182	.04093	.04006	.03920	.03836	.03754	.03673
1.8	.03593	.03515	.03438	.03362	.03288	.03216	.03144	.03074	.03005	.02938
1.9	.02872	.02807	.02743	.02680	.02619	.02559	.02500	.02442	.02385	.02330
2.0	.02275	.02222	.02169	.02118	.02068	.02018	.01970	.01923	.01876	.01831
2.1	.01786	.01743	.01700	.01659	.01618	.01578	.01539	.01500	.01463	.01426
2.2	.01390	.01355	.01321	.01287	.01255	.01222	.01191	.01160	.01130	.01101
2.3	.01072	.01044	.01017	.00990	.00964	.00939	.00914	.00889	.00866	.00842
2.4	.00820	.00798	.00776	.00755	.00734	.00714	.00695	.00676	.00657	.00639
2.5	.00621	.00604	.00587	.00570	.00554	.00539	.00523	.00508	.00494	.00480
2.6	.00466	.00453	.00440	.00427	.00415	.00402	.00391	.00379	.00368	.00357
2.7	.00347	.00336	.00326	.00317	.00307	.00298	.00289	.00280	.00272	.00264
2.8	.00256	.00248	.00240	.00233	.00226	.00219	.00212	.00205	.00199	.00193
2.9	.00187	.00181	.00175	.00169	.00164	.00159	.00154	.00149	.00144	.00139
3.0	.00135	.00131	.00126	.00122	.00118	.00114	.00111	.00107	.00104	.00100
3.1	.00097	.00094	.00090	.00087	.00084	.00082	.00079	.00076	.00074	.00071
3.2	.00069	.00066	.00064	.00062	.00060	.00058	.00056	.00054	.00052	.00050
3.3	.00048	.00047	.00045	.00043	.00042	.00040	.00039	.00038	.00036	.00035
3.4	.00034	.00032	.00031	.00030	.00029	.00028	.00027	.00026	.00025	.00024
3.5	.00023	.00022	.00022	.00021	.00020	.00019	.00019	.00018	.00017	.00017
3.6	.00016	.00015	.00015	.00014	.00014	.00013	.00013	.00012	.00012	.00011
3.7	.00011	.00010	.00010	.00010	.00009	.00009	.00008	.00008	.00008	.00008
3.8	.00007	.00007	.00007	.00006	.00006	.00006	.00006	.00005	.00005	.00005
3.9	.00005	.00005	.00004	.00004	.00004	.00004	.00004	.00004	.00003	.00003
4.0	.00003	.00003	.00003	.00003	.00003	.00003	.00002	.00002	.00002	.00002

TABLE B Student's *t* Distribution

df	0,40	0,30	0,20	0,10	0,05	0,025	0,010
1	0,325	0,727	1,376	3,078	6,314	12,71	31,82
2	0,289	0,617	1,061	1,886	2,920	4,303	6,965
3	0,277	0,584	0,978	1,638	2,353	3,182	4,541
4	0,271	0,569	0,941	1,533	2,132	2,776	3,747
5	0,267	0,559	0,920	1,476	2,015	2,571	3,365
6	0,265	0,553	0,906	1,440	1,943	2,447	3,143
7	0,263	0,549	0,896	1,415	1,895	2,365	2,998
8	0,262	0,546	0,889	1,397	1,860	2,306	2,896
9	0,261	0,543	0,883	1,383	1,833	2,262	2,821
10	0,260	0,542	0,879	1,372	1,812	2,228	2,764
11	0,260	0,540	0,876	1,363	1,796	2,201	2,718
12	0,259	0,539	0,873	1,356	1,782	2,179	2,681
13	0,259	0,538	0,870	1,350	1,771	2,160	2,650
14	0,258	0,537	0,868	1,345	1,761	2,145	2,624
15	0,258	0,536	0,866	1,341	1,753	2,131	2,602
16	0,258	0,535	0,865	1,337	1,746	2,120	2,583
17	0,257	0,534	0,863	1,333	1,740	2,110	2,567
18	0,257	0,534	0,862	1,330	1,734	2,101	2,552
19	0,257	0,533	0,861	1,328	1,729	2,093	2,539
20	0,257	0,533	0,860	1,325	1,725	2,086	2,528
21	0,257	0,532	0,859	1,323	1,721	2,080	2,518
22	0,256	0,532	0,858	1,321	1,717	2,074	2,508
23	0,256	0,532	0,858	1,319	1,714	2,069	2,500
24	0,256	0,531	0,857	1,318	1,711	2,064	2,492
25	0,256	0,531	0,856	1,316	1,708	2,060	2,485
26	0,256	0,531	0,856	1,315	1,706	2,056	2,479
27	0,256	0,531	0,855	1,314	1,703	2,052	2,473
28	0,256	0,530	0,855	1,313	1,701	2,048	2,467
29	0,256	0,530	0,854	1,311	1,699	2,045	2,462
30	0,256	0,530	0,854	1,310	1,697	2,042	2,457
40	0,255	0,529	0,851	1,303	1,684	2,021	2,423
60	0,254	0,527	0,848	1,296	1,671	2,000	2,390
120	0,254	0,526	0,845	1,289	1,658	1,980	2,358
∞	0,253	0,524	0,842	1,282	1,645	1,960	2,326

TABLE B   Student's *t* Distribution   (*Continued*)

df	0,005	0,0025	0,001	0,0001	0,00001	0,000001
1	63,66	127,3	318,3	3183,	31831,	318310,
2	9,925	14,09	22,33	70,70	223,60	707,11
3	5,841	7,453	10,22	22,20	47,93	103,30
4	4,604	5,598	7,173	13,03	23,33	41,58
5	4,032	4,773	5,893	9,678	15,55	24,77
6	3,707	4,317	5,208	8,025	12,03	17,83
7	3,499	4,029	4,785	7,063	10,10	14,24
8	3,355	3,833	4,501	6,442	8,907	12,11
9	3,250	3,690	4,297	6,010	8,102	10,72
10	3,169	3,581	4,144	5,694	7,527	9,752
11	3,106	3,497	4,025	5,453	7,098	9,042
12	3,055	3,428	3,930	5,263	6,765	8,504
13	3,012	3,372	3,852	5,111	6,501	8,083
14	2,977	3,326	3,787	4,985	6,287	7,743
15	2,947	3,286	3,733	4,880	6,109	7,466
16	2,921	3,252	3,686	4,791	5,959	7,233
17	2,898	3,222	3,646	4,714	5,832	7,036
18	2,878	3,197	3,610	4,648	5,722	6,869
19	2,861	3,174	3,579	4,590	5,627	6,722
20	2,845	3,153	3,552	4,539	5,543	6,597
21	2,831	3,135	3,527	4,493	5,468	6,485
22	2,819	3,119	3,505	4,452	5,402	6,386
23	2,807	3,104	3,485	4,415	5,343	6,299
24	2,797	3,091	3,467	4,382	5,290	6,218
25	2,787	3,078	3,450	4,352	5,241	6,146
26	2,779	3,067	3,435	4,324	5,197	6,081
27	2,771	3,057	3,421	4,299	5,157	6,020
28	2,763	3,047	3,408	4,275	5,120	5,967
29	2,756	3,038	3,396	4,254	5,085	5,917
30	2,750	3,030	3,385	4,234	5,054	5,871
40	2,704	2,971	3,307	4,094	4,835	5,554
60	2,660	2,915	3,232	3,962	4,631	5,264
120	2,617	2,860	3,160	3,837	4,442	4,997
∞	2,576	2,807	3,090	3,719	4,265	4,753

TABLE C The Chi-Square Distribution

df	0,99	0,98	0,95	0,90	0,80	0,70	0,50	0,30
1	0,000	0,001	0,004	0,016	0,064	0,148	0,455	1,074
2	0,020	0,040	0,103	0,211	0,446	0,713	1,386	2,408
3	0,115	0,185	0,352	0,584	1,005	1,424	2,366	3,665
4	0,297	0,429	0,711	1,064	1,649	2,195	3,357	4,878
5	0,554	0,752	1,145	1,610	2,343	3,000	4,351	6,064
6	0,872	1,134	1,635	2,204	3,070	3,828	5,348	7,231
7	1,239	1,564	2,167	2,833	3,822	4,671	6,346	8,383
8	1,646	2,032	2,733	3,490	4,594	5,527	7,344	9,524
9	2,088	2,532	3,325	4,168	5,380	6,393	8,343	10,656
10	2,558	3,059	3,940	4,865	6,179	7,267	9,342	11,781
11	3,053	3,609	4,575	5,578	6,989	8,148	10,341	12,899
12	3,571	4,178	5,226	6,304	7,807	9,034	11,340	14,011
13	4,107	4,765	5,892	7,042	8,634	9,926	12,340	15,119
14	4,660	5,368	6,571	7,790	9,467	10,821	13,339	16,222
15	5,229	5,985	7,261	8,547	10,307	11,721	14,339	17,322
16	5,812	6,614	7,962	9,312	11,152	12,624	15,338	18,418
17	6,408	7,255	8,672	10,085	12,002	13,531	16,338	19,511
18	7,015	7,906	9,390	10,865	12,857	14,440	17,338	20,601
19	7,633	8,567	10,117	11,651	13,716	15,352	18,338	21,689
20	8,260	9,237	10,851	12,443	14,578	16,266	19,337	22,775
21	8,897	9,915	11,591	13,240	15,445	17,182	20,337	23,858
22	9,542	10,600	12,338	14,041	16,314	18,101	21,337	24,939
23	10,196	11,293	13,091	14,848	17,187	19,021	22,337	26,018
24	10,856	11,992	13,848	15,659	18,062	19,943	23,337	27,096
25	11,524	12,697	14,611	16,473	18,940	20,867	24,337	28,172
26	12,198	13,409	15,379	17,292	19,820	21,792	25,336	29,246
27	12,879	14,125	16,151	18,114	20,703	22,719	26,336	30,319
28	13,565	14,847	16,928	18,939	21,588	23,647	27,336	31,391
29	14,256	15,574	17,708	19,768	22,475	24,577	28,336	32,461
30	14,953	16,306	18,493	20,599	23,364	25,508	29,336	33,530
40	22,164	23,838	26,509	29,051	32,345	34,872	39,335	44,165
50	29,707	31,664	34,764	37,689	41,449	44,313	49,335	54,723
60	37,485	39,699	43,188	46,459	50,641	53,809	59,335	65,227
70	45,442	47,893	51,739	55,329	59,898	63,346	69,334	75,689
80	53,540	56,213	60,391	64,278	69,207	72,915	79,334	86,120
90	61,754	64,635	69,126	73,291	78,558	82,511	89,334	96,524
100	70,065	73,142	77,929	82,358	87,945	92,129	99,334	106,906

TABLE C The Chi-Square Distribution (*Continued*)

df	0,20	0,10	0,05	0,02	0,01	0,001	0,0001	0,00001	0,000001
1	1,642	2,706	3,841	5,412	6,635	10,828	15,137	19,512	23,936
2	3,219	4,605	5,991	7,824	9,210	13,816	18,421	23,026	27,631
3	4,642	6,251	7,815	9,837	11,345	16,266	21,108	25,901	30,669
4	5,989	7,779	9,488	11,668	13,277	18,467	23,513	28,473	33,377
5	7,289	9,236	11,070	13,388	15,086	20,515	25,745	30,857	35,885
6	8,558	10,645	12,592	15,033	16,812	22,458	27,856	33,107	38,258
7	9,803	12,017	14,067	16,622	18,475	24,322	29,877	35,259	40,516
8	11,030	13,362	15,507	18,168	20,090	26,124	31,828	37,332	42,701
9	12,242	14,684	16,919	19,679	21,666	27,877	33,720	39,340	44,810
10	13,442	15,987	18,307	21,161	23,209	29,588	35,564	41,296	46,863
11	14,631	17,275	19,675	22,618	24,725	31,264	37,367	43,206	48,870
12	15,812	18,549	21,026	24,054	26,217	32,909	39,134	45,076	50,825
13	16,985	19,812	22,362	25,472	27,688	34,528	40,871	46,911	52,632
14	18,151	21,064	23,685	26,873	29,141	36,123	42,579	48,716	54,635
15	19,311	22,307	24,996	28,259	30,578	37,697	44,263	50,491	59,011
16	20,465	23,542	26,296	29,633	32,000	39,252	45,925	52,245	58,324
17	21,615	24,769	27,587	30,995	33,409	40,790	47,566	54,000	76,030
18	22,760	25,989	28,869	32,346	34,805	42,312	49,189	55,683	61,914
19	23,900	27,204	30,144	33,687	36,191	43,820	50,795	57,321	76,030
20	25,038	28,412	31,410	35,020	37,566	45,315	52,386	59,045	65,421
21	26,171	29,615	32,671	36,343	38,932	46,797	53,964	76,030	76,030
22	27,301	30,813	33,924	37,659	40,289	48,268	55,525	62,341	68,856
23	28,429	32,007	35,172	38,968	41,638	49,728	57,062	76,030	76,030
24	29,553	33,196	36,415	40,270	42,980	51,179	58,613	65,581	72,229
25	30,675	34,382	37,652	41,566	44,314	52,620	70,000	76,030	76,030
26	31,795	35,563	38,885	42,856	45,642	54,052	61,657	68,771	75,547
27	32,912	36,741	40,113	44,140	46,963	55,475	70,000	76,030	76,030
28	34,027	37,916	41,337	45,419	48,278	56,892	64,662	71,917	78,817
29	35,139	39,087	42,557	46,693	49,588	58,300	70,000	76,030	76,030
30	36,250	40,256	43,773	47,962	50,892	59,703	67,633	75,023	82,044
40	47,269	51,805	55,758	60,436	63,691	73,402	82,062	90,079	97,653
50	58,164	63,167	67,505	72,613	76,154	86,661	95,969	104,542	112,608
60	68,972	74,397	79,082	84,580	88,379	99,607	109,503	118,581	127,096
70	79,715	85,527	90,531	96,388	100,425	112,317	122,755	132,300	141,229
80	90,405	96,578	101,879	108,069	112,329	124,839	135,783	145,764	155,081
90	101,054	107,565	113,145	119,648	124,116	137,208	148,627	159,019	168,701
00	111,667	118,498	124,342	131,142	135,807	149,449	161,319	172,099	182,127

TABLE D Fisher's *F* Distribution

df_2	pr	df_1 1	2	3	4	5	6	7	8	9	10
2	0,500	,6667	1,000	1,135	1,207	1,252	1,282	1,305	1,321	1,334	1,345
2	0,250	2,571	3,000	3,153	3,232	3,280	3,312	3,335	3,353	3,366	3,377
2	0,100	8,526	9,000	9,162	9,243	9,293	9,326	9,349	9,367	9,381	9,392
2	0,050	18,51	19,00	19,16	19,25	19,30	19,33	19,35	19,37	19,38	19,40
2	0,020	48,51	49,00	49,17	49,25	49,30	49,33	49,36	49,37	49,39	49,40
2	0,010	98,50	99,00	99,17	99,25	99,30	99,33	99,36	99,37	99,39	99,40
2	0,005	198,5	199,0	199,2	199,3	199,3	199,3	199,4	199,4	199,4	199,4
2	0,002	498,5	499,0	499,2	499,3	499,3	499,3	499,4	499,4	499,4	499,4
2	0,001	998,5	999,0	999,2	999,2	999,3	999,3	999,4	999,4	999,4	999,4
3	0,500	,5851	,8811	1,000	1,063	1,102	1,129	1,148	1,163	1,174	1,183
3	0,250	2,024	2,280	2,356	2,390	2,410	2,422	2,430	2,436	2,441	2,445
3	0,100	5,538	5,462	5,391	5,343	5,309	5,285	5,266	5,252	5,240	5,230
3	0,050	10,13	9,552	9,277	9,117	9,013	8,941	8,887	8,845	8,812	8,786
3	0,020	20,62	18,86	18,11	17,69	17,43	17,25	17,11	17,01	16,93	16,86
3	0,010	34,12	30,82	29,46	28,71	28,24	27,91	27,67	27,49	27,35	27,23
3	0,005	55,55	49,80	47,47	46,19	45,39	44,84	44,43	44,13	43,88	43,69
3	0,002	104,3	92,99	88,45	85,98	84,42	83,35	82,57	81,98	81,51	81,13
3	0,001	167,0	148,5	141,1	137,1	134,6	132,8	131,6	130,6	129,9	129,2
4	0,500	,5486	,8284	,9405	1,000	1,037	1,062	1,080	1,093	1,104	1,113
4	0,250	1,807	2,000	2,047	2,064	2,072	2,077	2,079	2,080	2,081	2,082
4	0,100	4,545	4,325	4,191	4,107	4,051	4,010	3,979	3,955	3,936	3,920
4	0,050	7,709	6,944	6,591	6,388	6,256	6,163	6,094	6,041	5,999	5,964
4	0,020	14,04	12,14	11,34	10,90	10,62	10,42	10,27	10,16	10,07	10,00
4	0,010	21,20	18,00	16,69	15,98	15,52	15,21	14,98	14,80	14,66	14,55
4	0,005	31,33	26,28	24,26	23,15	22,46	21,97	21,62	21,35	21,14	20,97
4	0,002	51,45	42,72	39,27	37,39	36,21	35,40	34,80	34,35	33,99	33,70
4	0,001	74,14	61,25	56,18	53,44	51,71	50,53	49,66	49,00	48,47	48,05
5	0,500	,5281	,7988	,9071	,9646	1,000	1,024	1,041	1,055	1,065	1,073
5	0,250	1,692	1,853	1,884	1,893	1,895	1,894	1,894	1,892	1,891	1,890
5	0,100	4,060	3,780	3,619	3,520	3,453	3,405	3,368	3,339	3,316	3,297
5	0,050	6,608	5,786	5,409	5,192	5,050	4,950	4,876	4,818	4,772	4,735
5	0,020	11,32	9,454	8,670	8,233	7,953	7,758	7,614	7,503	7,415	7,344
5	0,010	16,26	13,27	12,06	11,39	10,97	10,67	10,46	10,29	10,16	10,05
5	0,005	22,78	18,31	16,53	15,56	14,94	14,51	14,20	13,96	13,77	13,62
5	0,002	34,73	27,53	24,70	23,16	22,20	21,53	21,04	20,67	20,37	20,13
5	0,001	47,18	37,12	33,20	31,09	29,75	28,83	28,16	27,65	27,24	26,92
6	0,500	,5149	,7798	,8858	,9419	,9765	1,000	1,017	1,030	1,040	1,048
6	0,250	1,621	1,762	1,784	1,787	1,785	1,782	1,779	1,776	1,773	1,771
6	0,100	3,776	3,463	3,289	3,181	3,108	3,055	3,014	2,983	2,958	2,937
6	0,050	5,987	5,143	4,757	4,534	4,387	4,284	4,207	4,147	4,099	4,060
6	0,020	9,876	8,052	7,287	6,859	6,585	6,393	6,251	6,141	6,055	5,984
6	0,010	13,75	10,92	9,780	9,148	8,746	8,466	8,260	8,102	7,976	7,874
6	0,005	18,64	14,54	12,92	12,03	11,46	11,07	10,79	10,57	10,39	10,25
6	0,002	27,12	20,81	18,34	17,01	16,16	15,58	15,15	14,83	14,57	14,36
6	0,001	35,51	27,00	23,70	21,92	20,80	20,03	19,46	19,03	18,69	18,41

TABLE D   Fisher's *F* Distribution   (*Continued*)

df_1

10	12	15	20	24	30	40	60	120	∞	pr	df_2
1,345	1,361	1,377	1,393	1,401	1,410	1,418	1,426	1,434	1,443	0,500	2
3,377	3,393	3,410	3,426	3,435	3,443	3,451	3,459	3,468	3,476	0,250	2
9,392	9,408	9,425	9,441	9,450	9,458	9,466	9,475	9,483	9,491	0,100	2
19,40	19,41	19,43	19,45	19,45	19,46	19,47	19,48	19,49	19,50	0,050	2
49,40	49,42	49,43	49,45	49,46	49,47	49,47	49,48	49,49	49,50	0,020	2
99,40	99,42	99,43	99,45	99,46	99,47	99,47	99,48	99,49	99,50	0,010	2
199,4	199,4	199,4	199,5	199,5	199,5	199,5	199,5	199,5	199,5	0,005	2
499,4	499,4	499,4	499,5	499,5	499,5	499,5	499,5	499,5	499,5	0,002	2
999,4	999,4	999,4	999,4	999,5	999,5	999,5	999,5	999,5	999,5	0,001	2
1,183	1,197	1,211	1,225	1,232	1,239	1,246	1,254	1,261	1,268	0,500	3
2,445	2,450	2,455	2,460	2,463	2,465	2,467	2,470	2,472	2,474	0,250	3
5,230	5,216	5,200	5,184	5,176	5,168	5,160	5,151	5,143	5,134	0,100	3
8,786	8,745	8,703	8,660	8,639	8,617	8,594	8,572	8,549	8,526	0,050	3
16,86	16,76	16,66	16,55	16,50	16,45	16,39	16,34	16,29	16,23	0,020	3
27,23	27,05	26,87	26,69	26,60	26,50	26,41	26,32	26,22	26,13	0,010	3
43,69	43,39	43,08	42,78	42,62	42,47	42,31	42,15	41,99	41,83	0,005	3
81,13	80,56	79,97	79,38	79,08	78,78	78,48	78,17	77,87	77,56	0,002	3
129,2	128,3	127,4	126,4	125,9	125,4	125,0	124,5	124,0	123,5	0,001	3
1,113	1,126	1,139	1,152	1,158	1,165	1,172	1,178	1,185	1,192	0,500	4
2,082	2,083	2,083	2,083	2,083	2,082	2,082	2,082	2,081	2,081	0,250	4
3,920	3,896	3,870	3,844	3,831	3,817	3,804	3,790	3,775	3,761	0,100	4
5,964	5,912	5,858	5,803	5,774	5,746	5,717	5,688	5,658	5,628	0,050	4
10,00	9,894	9,783	9,670	9,612	9,554	9,495	9,436	9,376	9,315	0,020	4
14,55	14,37	14,20	14,02	13,93	13,84	13,75	13,65	13,56	13,46	0,010	4
20,97	20,70	20,44	20,17	20,03	19,89	19,75	19,61	19,47	19,32	0,005	4
33,70	33,26	32,82	32,36	32,13	31,90	31,66	31,43	31,19	30,95	0,002	4
48,05	47,41	46,76	46,10	45,77	45,43	45,09	44,75	44,40	44,05	0,001	4
1,073	1,085	1,098	1,111	1,117	1,123	1,130	1,136	1,143	1,149	0,500	5
1,890	1,888	1,885	1,882	1,880	1,878	1,876	1,874	1,872	1,869	0,250	5
3,297	3,268	3,238	3,207	3,191	3,174	3,157	3,140	3,123	3,105	0,100	5
4,735	4,678	4,619	4,558	4,527	4,496	4,464	4,431	4,398	4,365	0,050	5
7,344	7,235	7,123	7,009	6,951	6,893	6,833	6,773	6,712	6,650	0,020	5
10,05	9,888	9,722	9,553	9,466	9,379	9,291	9,202	9,112	9,020	0,010	5
13,62	13,38	13,15	12,90	12,78	12,66	12,53	12,40	12,27	12,14	0,005	5
20,13	19,77	19,40	19,02	18,83	18,64	18,45	18,25	18,05	17,85	0,002	5
26,92	26,42	25,91	25,39	25,13	24,87	24,60	24,33	24,06	23,79	0,001	5
1,048	1,060	1,072	1,084	1,091	1,097	1,103	1,109	1,116	1,122	0,500	6
1,771	1,767	1,762	1,757	1,754	1,751	1,748	1,744	1,741	1,737	0,250	6
2,937	2,905	2,871	2,836	2,818	2,800	2,781	2,762	2,742	2,722	0,100	6
4,060	4,000	3,938	3,874	3,841	3,808	3,774	3,740	3,705	3,669	0,050	6
5,984	5,876	5,765	5,651	5,593	5,534	5,474	5,413	5,352	5,289	0,020	6
7,874	7,718	7,559	7,396	7,313	7,229	7,143	7,057	6,969	6,880	0,010	6
10,25	10,03	9,814	9,589	9,474	9,358	9,241	9,122	9,001	8,879	0,005	6
14,36	14,04	13,71	13,38	13,21	13,04	12,87	12,69	12,51	12,34	0,002	6
18,41	17,99	17,56	17,12	16,90	16,67	16,44	16,21	15,98	15,75	0,001	6

Tables

TABLE D Fisher's *F* Distribution (*Continued*)

df_2	pr	1	2	3	4	5	6	7	8	9	10
7	0,500	,5057	,7665	,8709	,9262	,9603	,9833	1,000	1,013	1,022	1,030
7	0,250	1,573	1,701	1,717	1,716	1,711	1,706	1,701	1,697	1,693	1,690
7	0,100	3,589	3,257	3,074	2,961	2,883	2,827	2,785	2,752	2,725	2,703
7	0,050	5,591	4,737	4,347	4,120	3,972	3,866	3,787	3,726	3,677	3,637
7	0,020	8,988	7,203	6,454	6,035	5,765	5,576	5,435	5,327	5,241	5,171
7	0,010	12,25	9,547	8,451	7,847	7,460	7,191	6,993	6,840	6,719	6,620
7	0,005	16,24	12,40	10,88	10,05	9,522	9,155	8,885	8,678	8,514	8,380
7	0,002	22,90	17,16	14,93	13,72	12,95	12,42	12,03	11,73	11,50	11,31
7	0,001	29,25	21,69	18,77	17,20	16,21	15,52	15,02	14,63	14,33	14,08
8	0,500	,4990	,7568	,8600	,9146	,9483	,9711	,9876	1,000	1,010	1,018
8	0,250	1,538	1,657	1,668	1,664	1,658	1,651	1,645	1,640	1,635	1,631
8	0,100	3,458	3,113	2,924	2,806	2,726	2,668	2,624	2,589	2,561	2,538
8	0,050	5,318	4,459	4,066	3,838	3,688	3,581	3,500	3,438	3,388	3,347
8	0,020	8,389	6,637	5,901	5,489	5,223	5,036	4,897	4,790	4,705	4,635
8	0,010	11,26	8,649	7,591	7,006	6,632	6,371	6,178	6,029	5,911	5,814
8	0,005	14,69	11,04	9,596	8,805	8,302	7,952	7,694	7,496	7,339	7,211
8	0,002	20,26	14,91	12,84	11,71	11,00	10,50	10,14	9,862	9,642	9,463
8	0,001	25,41	18,49	15,83	14,39	13,48	12,86	12,40	12,05	11,77	11,54
9	0,500	,4938	,7494	,8517	,9058	,9392	,9617	,9781	,9904	1,000	1,008
9	0,250	1,512	1,624	1,632	1,625	1,617	1,609	1,602	1,596	1,591	1,586
9	0,100	3,360	3,006	2,813	2,693	2,611	2,551	2,505	2,469	2,440	2,416
9	0,050	5,117	4,256	3,863	3,633	3,482	3,374	3,293	3,230	3,179	3,137
9	0,020	7,961	6,234	5,510	5,103	4,840	4,655	4,517	4,410	4,325	4,256
9	0,010	10,56	8,022	6,992	6,422	6,057	5,802	5,613	5,467	5,351	5,257
9	0,005	13,61	10,11	8,717	7,956	7,471	7,134	6,885	6,693	6,541	6,417
9	0,002	18,46	13,41	11,44	10,38	9,702	9,234	8,890	8,626	8,416	8,246
9	0,001	22,86	16,39	13,90	12,56	11,71	11,13	10,70	10,37	10,11	9,894
10	0,500	,4897	,7435	,8451	,8988	,9319	,9544	,9705	,9828	,9923	1,000
10	0,250	1,491	1,598	1,603	1,595	1,585	1,576	1,569	1,562	1,556	1,551
10	0,100	3,285	2,924	2,728	2,605	2,522	2,461	2,414	2,377	2,347	2,323
10	0,050	4,965	4,103	3,708	3,478	3,326	3,217	3,135	3,072	3,020	2,978
10	0,020	7,638	5,934	5,218	4,816	4,555	4,371	4,235	4,129	4,044	3,975
10	0,010	10,04	7,559	6,552	5,994	5,636	5,386	5,200	5,057	4,942	4,849
10	0,005	12,83	9,427	8,081	7,343	6,872	6,545	6,302	6,116	5,968	5,847
10	0,002	17,17	12,33	10,45	9,432	8,786	8,338	8,008	7,754	7,553	7,389
10	0,001	21,04	14,91	12,55	11,28	10,48	9,926	9,517	9,204	8,956	8,754
11	0,500	,4864	,7387	,8397	,8932	,9261	,9484	,9645	,9766	,9861	,9937
11	0,250	1,475	1,577	1,580	1,570	1,560	1,550	1,542	1,535	1,528	1,523
11	0,100	3,225	2,860	2,660	2,536	2,451	2,389	2,342	2,304	2,274	2,248
11	0,050	4,844	3,982	3,587	3,357	3,204	3,095	3,012	2,948	2,896	2,854
11	0,020	7,388	5,701	4,993	4,594	4,336	4,153	4,017	3,912	3,828	3,758
11	0,010	9,646	7,206	6,217	5,668	5,316	5,069	4,886	4,744	4,632	4,539
11	0,005	12,23	8,912	7,600	6,881	6,422	6,102	5,865	5,682	5,537	5,418
11	0,002	16,20	11,52	9,714	8,731	8,107	7,674	7,355	7,110	6,915	6,756
11	0,001	19,69	13,81	11,56	10,35	9,578	9,047	8,655	8,355	8,116	7,922

TABLE D   Fisher's *F* Distribution   (*Continued*)

df_1

10	12	15	20	24	30	40	60	120	∞	pr	df_2
1,030	1,042	1,054	1,066	1,072	1,079	1,085	1,091	1,097	1,103	0,500	7
1,690	1,684	1,678	1,671	1,667	1,663	1,659	1,655	1,650	1,645	0,250	7
2,703	2,668	2,632	2,595	2,575	2,555	2,535	2,514	2,493	2,471	0,100	7
3,637	3,575	3,511	3,445	3,410	3,376	3,340	3,304	3,267	3,230	0,050	7
5,171	5,064	4,953	4,839	4,781	4,722	4,662	4,601	4,538	4,475	0,020	7
6,620	6,469	6,314	6,155	6,074	5,992	5,908	5,824	5,737	5,650	0,010	7
8,380	8,176	7,968	7,754	7,645	7,534	7,422	7,309	7,193	7,076	0,005	7
11,31	11,01	10,72	10,41	10,25	10,10	9,938	9,776	9,612	9,446	0,002	7
14,08	13,71	13,32	12,93	12,73	12,53	12,33	12,12	11,91	11,70	0,001	7
1,018	1,029	1,041	1,053	1,059	1,065	1,071	1,077	1,083	1,089	0,500	8
1,631	1,624	1,617	1,609	1,604	1,600	1,595	1,589	1,584	1,578	0,250	8
2,538	2,502	2,464	2,425	2,404	2,383	2,361	2,339	2,316	2,293	0,100	8
3,347	3,284	3,218	3,150	3,115	3,079	3,043	3,005	2,967	2,928	0,050	8
4,635	4,528	4,417	4,304	4,245	4,186	4,125	4,063	4,000	3,936	0,020	8
5,814	5,667	5,515	5,359	5,279	5,198	5,116	5,032	4,946	4,859	0,010	8
7,211	7,015	6,814	6,608	6,503	6,396	6,288	6,177	6,065	5,951	0,005	8
9,463	9,189	8,909	8,622	8,476	8,328	8,177	8,024	7,869	7,711	0,002	8
11,54	11,19	10,84	10,48	10,30	10,11	9,919	9,727	9,532	9,334	0,001	8
1,008	1,019	1,031	1,043	1,049	1,055	1,061	1,067	1,073	1,079	0,500	9
1,586	1,579	1,570	1,561	1,556	1,551	1,545	1,539	1,533	1,526	0,250	9
2,416	2,379	2,340	2,298	2,277	2,255	2,232	2,208	2,184	2,159	0,100	9
3,137	3,073	3,006	2,936	2,900	2,864	2,826	2,787	2,748	2,707	0,050	9
4,256	4,149	4,039	3,925	3,866	3,806	3,745	3,683	3,619	3,554	0,020	9
5,257	5,111	4,962	4,808	4,729	4,649	4,567	4,483	4,398	4,311	0,010	9
6,417	6,227	6,032	5,832	5,729	5,625	5,519	5,410	5,300	5,188	0,005	9
8,246	7,985	7,718	7,444	7,304	7,162	7,018	6,871	6,721	6,568	0,002	9
9,894	9,570	9,238	8,898	8,724	8,548	8,369	8,187	8,001	7,813	0,001	9
1,000	1,012	1,023	1,035	1,041	1,047	1,053	1,059	1,064	1,070	0,500	10
1,551	1,543	1,534	1,523	1,518	1,512	1,506	1,499	1,492	1,484	0,250	10
2,323	2,284	2,244	2,201	2,178	2,155	2,132	2,107	2,082	2,055	0,100	10
2,978	2,913	2,845	2,774	2,737	2,700	2,661	2,621	2,580	2,538	0,050	10
3,975	3,868	3,758	3,644	3,585	3,525	3,463	3,400	3,335	3,269	0,020	10
4,849	4,706	4,558	4,405	4,327	4,247	4,165	4,082	3,996	3,909	0,010	10
5,847	5,661	5,471	5,274	5,173	5,071	4,966	4,859	4,750	4,639	0,005	10
7,389	7,138	6,881	6,616	6,481	6,343	6,203	6,060	5,914	5,765	0,002	10
8,754	8,445	8,129	7,804	7,638	7,469	7,297	7,122	6,944	6,763	0,001	10
,9937	1,005	1,017	1,028	1,034	1,040	1,046	1,052	1,058	1,064	0,500	11
1,523	1,514	1,504	1,493	1,487	1,481	1,474	1,466	1,459	1,450	0,250	11
2,248	2,209	2,167	2,123	2,100	2,076	2,052	2,026	2,000	1,972	0,100	11
2,854	2,788	2,719	2,646	2,609	2,570	2,531	2,490	2,448	2,404	0,050	11
3,758	3,652	3,542	3,427	3,367	3,307	3,245	3,181	3,116	3,048	0,020	11
4,539	4,397	4,251	4,099	4,021	3,941	3,860	3,776	3,690	3,602	0,010	11
5,418	5,236	5,049	4,855	4,756	4,654	4,551	4,445	4,337	4,226	0,005	11
6,756	6,513	6,263	6,005	5,873	5,739	5,602	5,462	5,319	5,173	0,002	11
7,922	7,626	7,321	7,008	6,847	6,684	6,518	6,348	6,175	5,998	0,001	11

TABLE D Fisher's F Distribution (*Continued*)

df_2	pr	1	2	3	4	5	6	7	8	9	10
12	0,500	,4837	,7348	,8353	,8885	,9212	,9434	,9594	,9715	,9810	,9886
12	0,250	1,461	1,560	1,561	1,550	1,539	1,529	1,520	1,512	1,505	1,500
12	0,100	3,177	2,807	2,606	2,480	2,394	2,331	2,283	2,245	2,214	2,188
12	0,050	4,747	3,885	3,490	3,259	3,106	2,996	2,913	2,849	2,796	2,753
12	0,020	7,188	5,516	4,814	4,419	4,162	3,980	3,845	3,740	3,656	3,587
12	0,010	9,330	6,927	5,953	5,412	5,064	4,821	4,640	4,499	4,388	4,296
12	0,005	11,75	8,510	7,226	6,521	6,071	5,757	5,525	5,345	5,202	5,085
12	0,002	15,44	10,90	9,146	8,192	7,586	7,165	6,855	6,616	6,426	6,271
12	0,001	18,64	12,97	10,80	9,633	8,892	8,379	8,001	7,710	7,480	7,292
13	0,500	,4814	,7315	,8316	,8845	,9172	,9393	,9552	,9672	,9767	,9842
13	0,250	1,450	1,545	1,545	1,534	1,521	1,511	1,501	1,493	1,486	1,480
13	0,100	3,136	2,763	2,560	2,434	2,347	2,283	2,234	2,195	2,164	2,138
13	0,050	4,667	3,806	3,411	3,179	3,025	2,915	2,832	2,767	2,714	2,671
13	0,020	7,024	5,366	4,669	4,276	4,020	3,840	3,705	3,600	3,516	3,447
13	0,010	9,074	6,701	5,739	5,205	4,862	4,620	4,441	4,302	4,191	4,100
13	0,005	11,37	8,186	6,926	6,233	5,791	5,482	5,253	5,076	4,935	4,820
13	0,002	14,84	10,41	8,696	7,765	7,174	6,763	6,460	6,226	6,040	5,889
13	0,001	17,82	12,31	10,21	9,073	8,354	7,856	7,489	7,206	6,982	6,799
14	0,500	,4794	,7286	,8284	,8812	,9137	,9357	,9516	,9636	,9730	,9805
14	0,250	1,440	1,533	1,532	1,519	1,507	1,495	1,485	1,477	1,470	1,463
14	0,100	3,102	2,726	2,522	2,395	2,307	2,243	2,193	2,154	2,122	2,095
14	0,050	4,600	3,739	3,344	3,112	2,958	2,848	2,764	2,699	2,646	2,602
14	0,020	6,888	5,241	4,549	4,158	3,904	3,724	3,589	3,485	3,401	3,332
14	0,010	8,862	6,515	5,564	5,035	4,695	4,456	4,278	4,140	4,030	3,939
14	0,005	11,06	7,922	6,680	5,998	5,562	5,257	5,031	4,857	4,717	4,603
14	0,002	14,34	10,01	8,332	7,420	6,841	6,438	6,141	5,912	5,729	5,580
14	0,001	17,14	11,78	9,729	8,622	7,922	7,436	7,077	6,802	6,583	6,404
15	0,500	,4778	,7262	,8257	,8783	,9107	,9327	,9485	,9605	,9698	,9773
15	0,250	1,432	1,523	1,520	1,507	1,494	1,482	1,472	1,463	1,456	1,449
15	0,100	3,073	2,695	2,490	2,361	2,273	2,208	2,158	2,119	2,086	2,059
15	0,050	4,543	3,682	3,287	3,056	2,901	2,790	2,707	2,641	2,588	2,544
15	0,020	6,773	5,135	4,447	4,058	3,805	3,626	3,492	3,387	3,303	3,235
15	0,010	8,683	6,359	5,417	4,893	4,556	4,318	4,142	4,004	3,895	3,805
15	0,005	10,80	7,701	6,476	5,803	5,372	5,071	4,847	4,674	4,536	4,424
15	0,002	13,93	9,676	8,030	7,135	6,566	6,171	5,878	5,652	5,473	5,326
15	0,001	16,59	11,34	9,335	8,253	7,567	7,092	6,741	6,471	6,256	6,081
16	0,500	,4763	,7241	,8233	,8758	,9081	,9300	,9458	,9577	,9670	,9745
16	0,250	1,425	1,514	1,510	1,497	1,483	1,471	1,460	1,451	1,443	1,437
16	0,100	3,048	2,668	2,462	2,333	2,244	2,178	2,128	2,088	2,055	2,028
16	0,050	4,494	3,634	3,239	3,007	2,852	2,741	2,657	2,591	2,538	2,494
16	0,020	6,674	5,046	4,361	3,974	3,721	3,543	3,409	3,304	3,221	3,152
16	0,010	8,531	6,226	5,292	4,773	4,437	4,202	4,026	3,890	3,780	3,691
16	0,005	10,58	7,514	6,303	5,638	5,212	4,913	4,692	4,521	4,384	4,272
16	0,002	13,59	9,396	7,777	6,896	6,336	5,946	5,658	5,435	5,258	5,113
16	0,001	16,12	10,97	9,006	7,944	7,272	6,805	6,460	6,195	5,984	5,812

TABLE D Fisher's *F* Distribution (*Continued*)

df₁ → df_1

10	12	15	20	24	30	40	60	120	∞	pr	df₂
,9886	1,000	1,012	1,023	1,029	1,035	1,041	1,046	1,052	1,058	0,500	12
1,500	1,490	1,480	1,468	1,461	1,454	1,447	1,439	1,431	1,422	0,250	12
2,188	2,147	2,105	2,060	2,036	2,011	1,986	1,960	1,932	1,904	0,100	12
2,753	2,687	2,617	2,544	2,505	2,466	2,426	2,384	2,341	2,296	0,050	12
3,587	3,480	3,370	3,254	3,195	3,134	3,071	3,007	2,940	2,872	0,020	12
4,296	4,155	4,010	3,858	3,780	3,701	3,619	3,535	3,449	3,361	0,010	12
5,085	4,906	4,721	4,530	4,431	4,331	4,228	4,123	4,015	3,904	0,005	12
6,271	6,034	5,790	5,538	5,408	5,277	5,142	5,004	4,863	4,719	0,002	12
7,292	7,005	6,709	6,405	6,249	6,090	5,928	5,762	5,593	5,420	0,001	12
,9842	,9956	1,007	1,019	1,024	1,030	1,036	1,042	1,048	1,054	0,500	13
1,480	1,470	1,459	1,447	1,440	1,432	1,425	1,416	1,408	1,398	0,250	13
2,138	2,097	2,053	2,007	1,983	1,958	1,931	1,904	1,876	1,846	0,100	13
2,671	2,604	2,533	2,459	2,420	2,380	2,339	2,297	2,252	2,206	0,050	13
3,447	3,341	3,230	3,114	3,054	2,993	2,930	2,865	2,798	2,728	0,020	13
4,100	3,960	3,815	3,665	3,587	3,507	3,425	3,341	3,255	3,165	0,010	13
4,820	4,643	4,460	4,270	4,173	4,073	3,970	3,866	3,758	3,647	0,005	13
5,889	5,656	5,417	5,169	5,042	4,912	4,779	4,643	4,504	4,360	0,002	13
6,799	6,519	6,231	5,934	5,781	5,626	5,467	5,305	5,138	4,967	0,001	13
,9805	,9919	1,003	1,015	1,020	1,026	1,032	1,038	1,044	1,050	0,500	14
1,463	1,453	1,441	1,428	1,421	1,414	1,405	1,397	1,387	1,377	0,250	14
2,095	2,054	2,010	1,962	1,938	1,912	1,885	1,857	1,828	1,797	0,100	14
2,602	2,534	2,463	2,388	2,349	2,308	2,266	2,223	2,178	2,131	0,050	14
3,332	3,225	3,114	2,998	2,938	2,876	2,812	2,747	2,679	2,608	0,020	14
3,939	3,800	3,656	3,505	3,427	3,348	3,266	3,181	3,094	3,004	0,010	14
4,603	4,428	4,247	4,059	3,961	3,862	3,760	3,655	3,547	3,436	0,005	14
5,580	5,352	5,116	4,872	4,746	4,618	4,486	4,351	4,213	4,070	0,002	14
6,404	6,130	5,848	5,557	5,407	5,254	5,098	4,938	4,773	4,604	0,001	14
,9773	,9886	1,000	1,011	1,017	1,023	1,029	1,034	1,040	1,046	0,500	15
1,449	1,438	1,426	1,413	1,405	1,397	1,389	1,380	1,370	1,359	0,250	15
2,059	2,017	1,972	1,924	1,899	1,873	1,845	1,817	1,787	1,755	0,100	15
2,544	2,475	2,403	2,328	2,288	2,247	2,204	2,160	2,114	2,066	0,050	15
3,235	3,128	3,017	2,900	2,840	2,777	2,713	2,647	2,578	2,506	0,020	15
3,805	3,666	3,522	3,372	3,294	3,214	3,132	3,047	2,959	2,868	0,010	15
4,424	4,250	4,070	3,883	3,786	3,687	3,585	3,480	3,372	3,260	0,005	15
5,326	5,101	4,868	4,627	4,502	4,375	4,245	4,111	3,973	3,831	0,002	15
6,081	5,812	5,535	5,248	5,101	4,950	4,796	4,638	4,475	4,307	0,001	15
,9745	,9858	,9972	1,009	1,014	1,020	1,026	1,032	1,037	1,043	0,500	16
1,437	1,426	1,413	1,399	1,391	1,383	1,374	1,365	1,354	1,343	0,250	16
2,028	1,985	1,940	1,891	1,866	1,839	1,811	1,782	1,751	1,718	0,100	16
2,494	2,425	2,352	2,276	2,235	2,194	2,151	2,106	2,059	2,010	0,050	16
3,152	3,045	2,934	2,817	2,756	2,693	2,628	2,561	2,492	2,419	0,020	16
3,691	3,553	3,409	3,259	3,181	3,101	3,018	2,933	2,845	2,753	0,010	16
4,272	4,099	3,920	3,734	3,638	3,539	3,437	3,332	3,224	3,112	0,005	16
5,113	4,890	4,660	4,422	4,298	4,172	4,043	3,909	3,772	3,630	0,002	16
5,812	5,547	5,274	4,992	4,846	4,697	4,545	4,388	4,226	4,059	0,001	16

TABLE D   Fisher's *F* Distribution   (*Continued*)

df₂	pr	1	2	3	4	5	6	7	8	9	10
17	0,500	,4750	,7222	,8212	,8736	,9058	,9277	,9434	,9553	,9646	,9721
17	0,250	1,419	1,506	1,502	1,487	1,473	,461	1,450	1,441	1,433	1,426
17	0,100	3,026	2,645	2,437	2,308	2,218	2,152	2,102	2,061	2,028	2,001
17	0,050	4,451	3,592	3,197	2,965	2,810	2,699	2,614	2,548	2,494	2,450
17	0,020	6,589	4,968	4,286	3,901	3,649	3,471	3,337	3,233	3,149	3,080
17	0,010	8,400	6,112	5,185	4,669	4,336	4,102	3,927	3,791	3,682	3,593
17	0,005	10,38	7,354	6,156	5,497	5,075	4,779	4,559	4,389	4,254	4,142
17	0,002	13,29	9,158	7,561	6,693	6,140	5,755	5,471	5,251	5,076	4,932
17	0,001	15,72	10,66	8,727	7,683	7,022	6,562	6,223	5,962	5,754	5,584
18	0,500	,4738	,7205	,8194	,8716	,9038	,9256	,9413	,9532	,9625	,9699
18	0,250	1,413	1,499	1,494	1,479	1,464	1,452	1,441	1,431	1,423	1,416
18	0,100	3,007	2,624	2,416	2,286	2,196	2,130	2,079	2,038	2,005	1,977
18	0,050	4,414	3,555	3,160	2,928	2,773	2,661	2,577	2,510	2,456	2,412
18	0,020	6,515	4,900	4,221	3,837	3,586	3,408	3,275	3,171	3,087	3,018
18	0,010	8,285	6,013	5,092	4,579	4,248	4,015	3,841	3,705	3,597	3,508
18	0,005	10,22	7,215	6,028	5,375	4,956	4,663	4,445	4,276	4,141	4,030
18	0,002	13,04	8,953	7,376	6,518	5,972	5,592	5,310	5,092	4,919	4,777
18	0,001	15,38	10,39	8,487	7,459	6,808	6,355	6,021	5,763	5,558	5,390
19	0,500	,4728	,7191	,8177	,8699	,9020	,9238	,9394	,9513	,9606	,9680
19	0,250	1,408	1,493	1,487	1,472	1,457	1,444	1,432	1,423	1,414	1,407
19	0,100	2,990	2,606	2,397	2,266	2,176	2,109	2,058	2,017	1,984	1,956
19	0,050	4,381	3,522	3,127	2,895	2,740	2,628	2,544	2,477	2,423	2,378
19	0,020	6,449	4,840	4,164	3,781	3,531	3,353	3,220	3,116	3,032	2,963
19	0,010	8,185	5,926	5,010	4,500	4,171	3,939	3,765	3,631	3,523	3,434
19	0,005	10,07	7,093	5,916	5,268	4,853	4,561	4,345	4,177	4,043	3,933
19	0,002	12,81	8,774	7,214	6,366	5,826	5,449	5,170	4,955	4,783	4,643
19	0,001	15,08	10,16	8,280	7,265	6,622	6,175	5,845	5,590	5,388	5,222
20	0,500	,4719	,7177	,8162	,8683	,9004	,9221	,9378	,9496	,9588	,9663
20	0,250	1,404	1,487	1,481	1,465	1,450	1,437	1,425	1,415	1,407	1,399
20	0,100	2,975	2,589	2,380	2,249	2,158	2,091	2,040	1,999	1,965	1,937
20	0,050	4,351	3,493	3,098	2,866	2,711	2,599	2,514	2,447	2,393	2,348
20	0,020	6,391	4,788	4,113	3,731	3,482	3,304	3,171	3,067	2,984	2,915
20	0,010	8,096	5,849	4,938	4,431	4,103	3,871	3,699	3,564	3,457	3,368
20	0,005	9,944	6,986	5,818	5,174	4,762	4,472	4,257	4,090	3,956	3,847
20	0,002	12,62	8,616	7,073	6,233	5,698	5,325	5,048	4,835	4,664	4,525
20	0,001	14,82	9,953	8,098	7,096	6,461	6,019	5,692	5,440	5,239	5,075
21	0,500	,4711	,7165	,8149	,8669	,8989	,9206	,9362	,9481	,9573	,9647
21	0,250	1,400	1,482	1,475	1,459	1,444	1,430	1,419	1,409	1,400	1,392
21	0,100	2,961	2,575	2,365	2,233	2,142	2,075	2,023	1,982	1,948	1,920
21	0,050	4,325	3,467	3,072	2,840	2,685	2,573	2,488	2,420	2,366	2,321
21	0,020	6,339	4,740	4,068	3,687	3,438	3,261	3,128	3,024	2,941	2,872
21	0,010	8,017	5,780	4,874	4,369	4,042	3,812	3,640	3,506	3,398	3,310
21	0,005	9,830	6,891	5,730	5,091	4,681	4,393	4,179	4,013	3,880	3,771
21	0,002	12,44	8,477	6,947	6,115	5,585	5,215	4,940	4,728	4,559	4,421
21	0,001	14,59	9,772	7,938	6,947	6,318	5,881	5,557	5,308	5,109	4,946

TABLE D Fisher's F Distribution (*Continued*)

df_1 10	12	15	20	24	30	40	60	120	∞	pr	df_2
.9721	.9833	.9947	1,006	1,012	1,017	1,023	1,029	1,035	1,041	0,500	17
1,426	1,414	1,401	1,387	1,379	1,370	1,361	1,351	1,341	1,329	0,250	17
2,001	1,958	1,912	1,862	1,836	1,809	1,781	1,751	1,719	1,686	0,100	17
2,450	2,381	2,308	2,230	2,190	2,148	2,104	2,058	2,011	1,960	0,050	17
3,080	2,974	2,862	2,745	2,683	2,620	2,555	2,487	2,417	2,343	0,020	17
3,593	3,455	3,312	3,162	3,084	3,003	2,920	2,835	2,746	2,653	0,010	17
4,142	3,971	3,793	3,607	3,511	3,412	3,311	3,206	3,097	2,984	0,005	17
4,932	4,712	4,484	4,248	4,125	4,000	3,871	3,738	3,601	3,459	0,002	17
5,584	5,324	5,054	4,775	4,631	4,484	4,332	4,177	4,016	3,850	0,001	17
.9699	.9812	.9924	1,004	1,009	1,015	1,021	1,027	1,032	1,038	0,500	18
1,416	1,404	1,391	1,376	1,368	1,359	1,350	1,340	1,328	1,316	0,250	18
1,977	1,933	1,887	1,837	1,810	1,783	1,754	1,723	1,691	1,657	0,100	18
2,412	2,342	2,269	2,191	2,150	2,107	2,063	2,017	1,968	1,917	0,050	18
3,018	2,911	2,799	2,682	2,620	2,557	2,491	2,423	2,351	2,277	0,020	18
3,508	3,371	3,227	3,077	2,999	2,919	2,835	2,749	2,660	2,566	0,010	18
4,030	3,860	3,683	3,498	3,402	3,303	3,201	3,096	2,987	2,873	0,005	18
4,777	4,559	4,333	4,098	3,976	3,852	3,723	3,591	3,454	3,311	0,002	18
5,390	5,132	4,866	4,590	4,447	4,301	4,151	3,996	3,836	3,670	0,001	18
.9680	.9792	.9905	1,002	1,007	1,013	1,019	1,025	1,030	1,036	0,500	19
1,407	1,395	1,382	1,367	1,358	1,349	1,339	1,329	1,317	1,305	0,250	19
1,956	1,912	1,865	1,814	1,787	1,759	1,730	1,699	1,666	1,631	0,100	19
2,378	2,308	2,234	2,156	2,114	2,071	2,026	1,980	1,930	1,878	0,050	19
2,963	2,856	2,744	2,626	2,564	2,501	2,434	2,366	2,294	2,218	0,020	19
3,434	3,297	3,153	3,003	2,925	2,844	2,761	2,674	2,584	2,489	0,010	19
3,933	3,763	3,587	3,402	3,306	3,208	3,106	3,000	2,891	2,776	0,005	19
4,643	4,426	4,202	3,968	3,847	3,723	3,595	3,463	3,326	3,183	0,002	19
5,222	4,967	4,704	4,430	4,288	4,143	3,994	3,840	3,680	3,514	0,001	19
.9663	.9775	.9887	1,000	1,006	1,011	1,017	1,023	1,029	1,034	0,500	20
1,399	1,387	1,374	1,358	1,349	1,340	1,330	1,319	1,307	1,294	0,250	20
1,937	1,892	1,845	1,794	1,767	1,738	1,708	1,677	1,643	1,607	0,100	20
2,348	2,278	2,203	2,124	2,082	2,039	1,994	1,946	1,896	1,843	0,050	20
2,915	2,808	2,695	2,577	2,515	2,451	2,384	2,315	2,242	2,165	0,020	20
3,368	3,231	3,088	2,938	2,859	2,778	2,695	2,608	2,517	2,421	0,010	20
3,847	3,678	3,502	3,318	3,222	3,123	3,022	2,916	2,806	2,690	0,005	20
4,525	4,310	4,087	3,855	3,734	3,610	3,483	3,351	3,214	3,070	0,002	20
5,075	4,823	4,562	4,290	4,149	4,005	3,856	3,703	3,544	3,378	0,001	20
.9647	.9759	.9871	.9984	1,004	1,010	1,015	1,021	1,027	1,033	0,500	21
1,392	1,380	1,366	1,350	1,341	1,332	1,322	1,311	1,298	1,285	0,250	21
1,920	1,875	1,827	1,776	1,748	1,719	1,689	1,657	1,623	1,586	0,100	21
2,321	2,250	2,176	2,096	2,054	2,010	1,965	1,916	1,866	1,812	0,050	21
2,872	2,764	2,652	2,533	2,471	2,406	2,339	2,269	2,196	2,118	0,020	21
3,310	3,173	3,030	2,880	2,801	2,720	2,636	2,548	2,457	2,360	0,010	21
3,771	3,602	3,427	3,243	3,147	3,049	2,947	2,841	2,730	2,614	0,005	21
4,421	4,207	3,986	3,754	3,634	3,511	3,383	3,251	3,114	2,970	0,002	21
4,946	4,696	4,437	4,167	4,027	3,884	3,736	3,583	3,424	3,257	0,001	21

TABLE D   Fisher's *F* Distribution   (*Continued*)

df_2	pr	1	2	3	4	5	6	7	8	9	10
22	0,500	,4703	,7155	,8137	,8656	,8976	,9192	,9349	,9467	,9559	,9633
22	0,250	1,396	1,477	1,470	1,454	1,438	1,424	1,413	1,402	1,394	1,386
22	0,100	2,949	2,561	2,351	2,219	2,128	2,061	2,008	1,967	1,933	1,904
22	0,050	4,301	3,443	3,049	2,817	2,661	2,549	2,464	2,397	2,342	2,297
22	0,020	6,292	4,698	4,028	3,647	3,399	3,222	3,089	2,985	2,902	2,833
22	0,010	7,945	5,719	4,817	4,313	3,988	3,758	3,587	3,453	3,346	3,258
22	0,005	9,727	6,806	5,652	5,017	4,609	4,322	4,109	3,944	3,812	3,703
22	0,002	12,29	8,353	6,836	6,010	5,484	5,117	4,844	4,634	4,466	4,328
22	0,001	14,38	9,612	7,796	6,814	6,191	5,758	5,438	5,190	4,993	4,832
23	0,500	,4696	,7145	,8125	,8644	,8964	,9180	,9336	,9454	,9546	,9620
23	0,250	1,393	1,473	1,466	1,449	1,433	1,419	1,407	1,397	1,388	1,380
23	0,100	2,937	2,549	2,339	2,207	2,115	2,047	1,995	1,953	1,919	1,890
23	0,050	4,279	3,422	3,028	2,796	2,640	2,528	2,442	2,375	2,320	2,275
23	0,020	6,249	4,660	3,991	3,611	3,363	3,187	3,054	2,950	2,867	2,798
23	0,010	7,881	5,664	4,765	4,264	3,939	3,710	3,539	3,406	3,299	3,211
23	0,005	9,635	6,730	5,582	4,950	4,544	4,259	4,047	3,882	3,750	3,642
23	0,002	12,15	8,242	6,736	5,916	5,394	5,029	4,759	4,549	4,382	4,245
23	0,001	14,20	9,469	7,669	6,696	6,078	5,649	5,331	5,085	4,890	4,730
24	0,500	,4690	,7136	,8115	,8633	,8953	,9169	,9325	,9442	,9534	,9608
24	0,250	1,390	1,470	1,462	1,445	1,428	1,414	1,402	1,392	1,383	1,375
24	0,100	2,927	2,538	2,327	2,195	2,103	2,035	1,983	1,941	1,906	1,877
24	0,050	4,260	3,403	3,009	2,776	2,621	2,508	2,423	2,355	2,300	2,255
24	0,020	6,211	4,625	3,958	3,579	3,331	3,155	3,022	2,919	2,835	2,766
24	0,010	7,823	5,614	4,718	4,218	3,895	3,667	3,496	3,363	3,256	3,168
24	0,005	9,551	6,661	5,519	4,890	4,486	4,202	3,991	3,826	3,695	3,587
24	0,002	12,02	8,142	6,646	5,832	5,313	4,950	4,681	4,473	4,307	4,171
24	0,001	14,03	9,339	7,554	6,589	5,977	5,550	5,235	4,991	4,797	4,638
25	0,500	,4684	,7127	,8106	,8624	,8942	,9158	,9314	,9432	,9523	,9597
25	0,250	1,387	1,466	1,458	1,441	1,424	1,410	1,398	1,387	1,378	1,370
25	0,100	2,918	2,528	2,317	2,184	2,092	2,024	1,971	1,929	1,895	1,866
25	0,050	4,242	3,385	2,991	2,759	2,603	2,490	2,405	2,337	2,282	2,236
25	0,020	6,176	4,593	3,928	3,549	3,302	3,126	2,993	2,890	2,806	2,737
25	0,010	7,770	5,568	4,675	4,177	3,855	3,627	3,457	3,324	3,217	3,129
25	0,005	9,475	6,598	5,462	4,835	4,433	4,150	3,939	3,776	3,645	3,537
25	0,002	11,90	8,051	6,564	5,755	5,239	4,879	4,611	4,404	4,239	4,103
25	0,001	13,88	9,223	7,451	6,493	5,885	5,462	5,148	4,906	4,713	4,555
26	0,500	,4679	,7120	,8097	,8615	,8933	,9149	,9304	,9422	,9513	,9587
26	0,250	1,384	1,463	1,454	1,437	1,420	1,406	1,393	1,383	1,374	1,366
26	0,100	2,909	2,519	2,307	2,174	2,082	2,014	1,961	1,919	1,884	1,855
26	0,050	4,225	3,369	2,975	2,743	2,587	2,474	2,388	2,321	2,265	2,220
26	0,020	6,144	4,564	3,900	3,522	3,275	3,099	2,967	2,863	2,780	2,711
26	0,010	7,721	5,526	4,637	4,140	3,818	3,591	3,421	3,288	3,182	3,094
26	0,005	9,406	6,541	5,409	4,785	4,384	4,103	3,893	3,730	3,599	3,492
26	0,002	11,80	7,968	6,490	5,686	5,172	4,814	4,548	4,342	4,177	4,042
26	0,001	13,74	9,116	7,357	6,406	5,802	5,381	5,070	4,829	4,637	4,480

TABLE D Fisher's *F* Distribution (*Continued*)

10	12	15	20	24	30	40	60	120	∞	pr	df_2
,9633	,9744	,9857	,9969	1,003	1,008	1,014	1,020	1,025	1,031	0,500	22
1,386	1,374	1,359	1,343	1,334	1,324	1,314	1,303	1,290	1,276	0,250	22
1,904	1,859	1,811	1,759	1,731	1,702	1,671	1,639	1,604	1,567	0,100	22
2,297	2,226	2,151	2,071	2,028	1,984	1,938	1,889	1,838	1,783	0,050	22
2,833	2,725	2,613	2,494	2,431	2,366	2,299	2,228	2,154	2,075	0,020	22
3,258	3,121	2,978	2,827	2,749	2,667	2,583	2,495	2,403	2,305	0,010	22
3,703	3,535	3,360	3,176	3,081	2,982	2,880	2,774	2,663	2,546	0,005	22
4,328	4,116	3,895	3,665	3,545	3,422	3,295	3,163	3,025	2,881	0,002	22
4,832	4,583	4,326	4,058	3,919	3,776	3,629	3,476	3,317	3,151	0,001	22
,9620	,9731	,9843	,9956	1,001	1,007	1,013	1,018	1,024	1,030	0,500	23
1,380	1,368	1,353	1,337	1,327	1,318	1,307	1,295	1,282	1,268	0,250	23
1,890	1,845	1,796	1,744	1,716	1,686	1,655	1,622	1,587	1,549	0,100	23
2,275	2,204	2,128	2,048	2,005	1,961	1,914	1,865	1,813	1,757	0,050	23
2,798	2,690	2,578	2,458	2,395	2,330	2,262	2,191	2,116	2,037	0,020	23
3,211	3,074	2,931	2,781	2,702	2,620	2,536	2,447	2,354	2,256	0,010	23
3,642	3,475	3,300	3,116	3,021	2,922	2,820	2,713	2,602	2,484	0,005	23
4,245	4,034	3,815	3,585	3,466	3,343	3,216	3,084	2,946	2,801	0,002	23
4,730	4,483	4,227	3,961	3,822	3,680	3,533	3,380	3,222	3,055	0,001	23
,9608	,9719	,9831	,9944	1,000	1,006	1,011	1,017	1,023	1,028	0,500	24
1,375	1,362	1,347	1,331	1,321	1,311	1,300	1,289	1,275	1,261	0,250	24
1,877	1,832	1,783	1,730	1,702	1,672	1,641	1,607	1,571	1,533	0,100	24
2,255	2,183	2,108	2,027	1,984	1,939	1,892	1,842	1,790	1,733	0,050	24
2,766	2,659	2,545	2,426	2,363	2,297	2,229	2,157	2,082	2,001	0,020	24
3,168	3,032	2,889	2,738	2,659	2,577	2,492	2,403	2,310	2,211	0,010	24
3,587	3,420	3,246	3,062	2,967	2,868	2,765	2,658	2,546	2,428	0,005	24
4,171	3,961	3,742	3,513	3,394	3,271	3,144	3,012	2,874	2,729	0,002	24
4,638	4,393	4,139	3,873	3,735	3,593	3,447	3,295	3,136	2,969	0,001	24
,9597	,9708	,9820	,9932	,9989	1,005	1,010	1,016	1,022	1,027	0,500	25
1,370	1,357	1,342	1,325	1,316	1,306	1,294	1,282	1,269	1,254	0,250	25
1,866	1,820	1,771	1,718	1,689	1,659	1,627	1,593	1,557	1,518	0,100	25
2,236	2,165	2,089	2,007	1,964	1,919	1,872	1,822	1,768	1,711	0,050	25
2,737	2,629	2,516	2,396	2,333	2,267	2,199	2,127	2,050	1,969	0,020	25
3,129	2,993	2,850	2,699	2,620	2,538	2,453	2,364	2,270	2,169	0,010	25
3,537	3,370	3,196	3,013	2,918	2,819	2,716	2,609	2,496	2,377	0,005	25
4,103	3,894	3,676	3,448	3,329	3,207	3,080	2,948	2,809	2,663	0,002	25
4,555	4,312	4,059	3,794	3,657	3,516	3,369	3,217	3,058	2,890	0,001	25
,9587	,9698	,9810	,9922	,9978	1,003	1,009	1,015	1,020	1,026	0,500	26
1,366	1,352	1,337	1,320	1,311	1,300	1,289	1,277	1,263	1,247	0,250	26
1,855	1,809	1,760	1,706	1,677	1,647	1,615	1,581	1,544	1,504	0,100	26
2,220	2,148	2,072	1,990	1,946	1,901	1,853	1,803	1,749	1,691	0,050	26
2,711	2,603	2,490	2,369	2,306	2,240	2,171	2,098	2,021	1,939	0,020	26
3,094	2,958	2,815	2,664	2,585	2,503	2,417	2,327	2,233	2,131	0,010	26
3,492	3,325	3,151	2,969	2,873	2,774	2,671	2,563	2,450	2,330	0,005	26
4,042	3,834	3,617	3,389	3,270	3,148	3,021	2,889	2,750	2,603	0,002	26
4,480	4,238	3,986	3,723	3,586	3,445	3,299	3,147	2,988	2,819	0,001	26

TABLE D   Fisher's F Distribution   (*Continued*)

df_2	pr	1	2	3	4	5	6	7	8	9	10
27	0,500	,4674	,7113	,8089	,8606	,8924	,9140	,9295	,9413	,9504	,9578
27	0,250	1,382	1,460	1,451	1,433	1,417	1,402	1,390	1,379	1,370	1,361
27	0,100	2,901	2,511	2,299	2,165	2,073	2,005	1,952	1,909	1,874	1,845
27	0,050	4,210	3,354	2,960	2,728	2,572	2,459	2,373	2,305	2,250	2,204
27	0,020	6,114	4,538	3,874	3,498	3,251	3,075	2,943	2,839	2,755	2,686
27	0,010	7,677	5,488	4,601	4,106	3,785	3,558	3,388	3,256	3,149	3,062
27	0,005	9,342	6,489	5,361	4,740	4,340	4,059	3,850	3,687	3,557	3,450
27	0,002	11,70	7,892	6,423	5,622	5,111	4,755	4,490	4,285	4,121	3,986
27	0,001	13,61	9,019	7,272	6,326	5,726	5,308	4,998	4,759	4,568	4,412
28	0,500	,4670	,7106	,8082	,8598	,8916	,9132	,9287	,9404	,9496	,9569
28	0,250	1,380	1,457	1,448	1,430	1,413	1,399	1,386	1,375	1,366	1,358
28	0,100	2,894	2,503	2,291	2,157	2,064	1,996	1,943	1,900	1,865	1,836
28	0,050	4,196	3,340	2,947	2,714	2,558	2,445	2,359	2,291	2,236	2,190
28	0,020	6,087	4,513	3,851	3,475	3,228	3,052	2,920	2,817	2,733	2,664
28	0,010	7,636	5,453	4,568	4,074	3,754	3,528	3,358	3,226	3,120	3,032
28	0,005	9,284	6,440	5,317	4,698	4,300	4,020	3,811	3,649	3,519	3,412
28	0,002	11,62	7,823	6,360	5,564	5,056	4,700	4,437	4,232	4,069	3,935
28	0,001	13,50	8,931	7,193	6,253	5,657	5,241	4,933	4,695	4,505	4,349
29	0,500	,4665	,7100	,8075	,8591	,8909	,9124	,9279	,9396	,9488	,9561
29	0,250	1,378	1,455	1,445	1,427	1,410	1,395	1,383	1,372	1,362	1,354
29	0,100	2,887	2,495	2,283	2,149	2,057	1,988	1,935	1,892	1,857	1,827
29	0,050	4,183	3,328	2,934	2,701	2,545	2,432	2,346	2,278	2,223	2,177
29	0,020	6,062	4,491	3,829	3,453	3,207	3,032	2,899	2,796	2,712	2,643
29	0,010	7,598	5,420	4,538	4,045	3,725	3,499	3,330	3,198	3,092	3,005
29	0,005	9,230	6,396	5,276	4,659	4,262	3,983	3,775	3,613	3,483	3,377
29	0,002	11,53	7,759	6,303	5,510	5,004	4,650	4,388	4,184	4,021	3,888
29	0,001	13,39	8,849	7,121	6,186	5,593	5,179	4,873	4,636	4,447	4,292
30	0,500	,4662	,7094	,8069	,8584	,8902	,9117	,9272	,9389	,9480	,9554
30	0,250	1,376	1,452	1,443	1,424	1,407	1,392	1,380	1,369	1,359	1,351
30	0,100	2,881	2,489	2,276	2,142	2,049	1,980	1,927	1,884	1,849	1,819
30	0,050	4,171	3,316	2,922	2,690	2,534	2,421	2,334	2,266	2,211	2,165
30	0,020	6,038	4,470	3,809	3,434	3,188	3,012	2,880	2,777	2,693	2,624
30	0,010	7,562	5,390	4,510	4,018	3,699	3,473	3,305	3,173	3,067	2,979
30	0,005	9,180	6,355	5,239	4,623	4,228	3,949	3,742	3,580	3,450	3,344
30	0,002	11,46	7,700	6,250	5,461	4,957	4,604	4,343	4,140	3,978	3,844
30	0,001	13,29	8,773	7,054	6,125	5,534	5,122	4,817	4,581	4,393	4,239
40	0,500	,4633	,7053	,8023	,8536	,8852	,9065	,9220	,9336	,9427	,9500
40	0,250	1,363	1,435	1,424	1,404	1,386	1,371	1,357	1,345	1,335	1,327
40	0,100	2,835	2,440	2,226	2,091	1,997	1,927	1,873	1,829	1,793	1,763
40	0,050	4,085	3,232	2,839	2,606	2,449	2,336	2,249	2,180	2,124	2,077
40	0,020	5,872	4,321	3,667	3,295	3,051	2,877	2,745	2,641	2,558	2,488
40	0,010	7,314	5,179	4,313	3,828	3,514	3,291	3,124	2,993	2,888	2,801
40	0,005	8,828	6,066	4,976	4,374	3,986	3,713	3,509	3,350	3,222	3,117
40	0,002	10,94	7,288	5,883	5,117	4,628	4,285	4,030	3,832	3,674	3,543
40	0,001	12,61	8,251	6,595	5,698	5,128	4,731	4,436	4,207	4,024	3,874

TABLE D   Fisher's *F* Distribution   (*Continued*)

df$_1$

10	12	15	20	24	30	40	60	120	∞	pr	df$_2$
,9578	,9689	,9800	,9912	,9969	1,003	1,008	1,014	1,020	1,025	0,500	27
1,361	1,348	1,333	1,316	1,306	1,295	1,284	1,271	1,257	1,241	0,250	27
1,845	1,799	1,749	1,695	1,666	1,636	1,603	1,569	1,531	1,491	0,100	27
2,204	2,132	2,056	1,974	1,930	1,884	1,836	1,785	1,731	1,672	0,050	27
2,686	2,579	2,465	2,344	2,281	2,214	2,145	2,072	1,995	1,911	0,020	27
3,062	2,926	2,783	2,632	2,552	2,470	2,384	2,294	2,198	2,097	0,010	27
3,450	3,284	3,110	2,928	2,832	2,733	2,630	2,522	2,408	2,287	0,005	27
3,986	3,778	3,562	3,335	3,217	3,094	2,967	2,835	2,696	2,548	0,002	27
4,412	4,171	3,920	3,658	3,521	3,380	3,234	3,082	2,923	2,754	0,001	27
,9569	,9680	,9792	,9904	,9960	1,002	1,007	1,013	1,019	1,024	0,500	28
1,358	1,344	1,329	1,311	1,301	1,291	1,279	1,266	1,252	1,236	0,250	28
1,836	1,790	1,740	1,685	1,656	1,625	1,593	1,558	1,520	1,478	0,100	28
2,190	2,118	2,041	1,959	1,915	1,869	1,820	1,769	1,714	1,654	0,050	28
2,664	2,556	2,442	2,321	2,258	2,191	2,121	2,048	1,970	1,886	0,020	28
3,032	2,896	2,753	2,602	2,522	2,440	2,354	2,263	2,167	2,064	0,010	28
3,412	3,246	3,073	2,890	2,794	2,695	2,592	2,483	2,369	2,247	0,005	28
3,935	3,728	3,512	3,286	3,167	3,045	2,918	2,785	2,646	2,497	0,002	28
4,349	4,109	3,859	3,598	3,462	3,321	3,176	3,024	2,864	2,695	0,001	28
,9561	,9672	,9784	,9895	,9952	1,001	1,006	1,012	1,018	1,023	0,500	29
1,354	1,340	1,325	1,307	1,297	1,286	1,275	1,262	1,247	1,231	0,250	29
1,827	1,781	1,731	1,676	1,647	1,616	1,583	1,547	1,509	1,467	0,100	29
2,177	2,104	2,027	1,945	1,901	1,854	1,806	1,754	1,698	1,638	0,050	29
2,643	2,535	2,421	2,300	2,236	2,169	2,099	2,026	1,947	1,862	0,020	29
3,005	2,868	2,726	2,574	2,495	2,412	2,325	2,234	2,138	2,034	0,010	29
3,377	3,211	3,038	2,855	2,759	2,660	2,557	2,448	2,333	2,210	0,005	29
3,888	3,681	3,466	3,240	3,122	3,000	2,873	2,740	2,600	2,451	0,002	29
4,292	4,053	3,804	3,543	3,407	3,267	3,121	2,970	2,810	2,640	0,001	29
,9554	,9665	,9776	,9888	,9944	1,000	1,006	1,011	1,017	1,023	0,500	30
1,351	1,337	1,321	1,303	1,293	1,282	1,270	1,257	1,242	1,226	0,250	30
1,819	1,773	1,722	1,667	1,638	1,606	1,573	1,538	1,499	1,456	0,100	30
2,165	2,092	2,015	1,932	1,887	1,841	1,792	1,740	1,683	1,622	0,050	30
2,624	2,516	2,402	2,281	2,216	2,149	2,079	2,005	1,926	1,840	0,020	30
2,979	2,843	2,700	2,549	2,469	2,386	2,299	2,208	2,111	2,006	0,010	30
3,344	3,179	3,006	2,823	2,727	2,628	2,524	2,415	2,300	2,176	0,005	30
3,844	3,639	3,424	3,198	3,080	2,958	2,831	2,698	2,557	2,407	0,002	30
4,239	4,001	3,753	3,493	3,357	3,217	3,072	2,920	2,760	2,589	0,001	30
,9500	,9610	,9721	,9832	,9888	,9944	1,000	1,006	1,011	1,017	0,500	40
1,327	1,312	1,295	1,276	1,265	1,253	1,240	1,225	1,208	1,188	0,250	40
1,763	1,715	1,662	1,605	1,574	1,541	1,506	1,467	1,425	1,377	0,100	40
2,077	2,003	1,924	1,839	1,793	1,744	1,693	1,637	1,577	1,509	0,050	40
2,488	2,380	2,265	2,141	2,075	2,006	1,933	1,856	1,771	1,678	0,020	40
2,801	2,665	2,522	2,369	2,288	2,203	2,114	2,019	1,917	1,805	0,010	40
3,117	2,953	2,781	2,598	2,502	2,401	2,296	2,184	2,064	1,932	0,005	40
3,543	3,342	3,130	2,907	2,789	2,667	2,539	2,404	2,259	2,102	0,002	40
3,874	3,642	3,400	3,145	3,011	2,872	2,727	2,574	2,410	2,233	0,001	40

TABLE D   Fisher's *F* Distribution   (*Continued*)

df$_2$	pr	1	2	3	4	5	6	7	8	9	10
50	0,500	,4616	,7028	,7995	,8507	,8822	,9035	,9189	,9305	,9395	,9468
50	0,250	1,355	1,425	1,413	1,393	1,374	1,358	1,344	1,332	1,321	1,312
50	0,100	2,809	2,412	2,197	2,061	1,966	1,895	1,841	1,796	1,760	1,729
50	0,050	4,034	3,183	2,790	2,557	2,400	2,286	2,199	2,130	2,073	2,026
50	0,020	5,776	4,235	3,585	3,215	2,972	2,798	2,667	2,563	2,479	2,410
50	0,010	7,171	5,057	4,199	3,720	3,408	3,186	3,020	2,890	2,785	2,698
50	0,005	8,626	5,902	4,826	4,232	3,849	3,579	3,376	3,219	3,092	2,988
50	0,002	10,64	7,055	5,676	4,923	4,442	4,105	3,854	3,659	3,503	3,374
50	0,001	12,22	7,956	6,336	5,459	4,901	4,512	4,222	3,998	3,818	3,671
60	0,500	,4605	,7012	,7977	,8487	,8802	,9014	,9168	,9284	,9374	,9447
60	0,250	1,349	1,419	1,405	1,385	1,366	1,349	1,335	1,323	1,312	1,303
60	0,100	2,791	2,393	2,177	2,041	1,946	1,875	1,819	1,775	1,738	1,707
60	0,050	4,001	3,150	2,758	2,525	2,368	2,254	2,167	2,097	2,040	1,993
60	0,020	5,713	4,179	3,532	3,163	2,921	2,747	2,616	2,512	2,428	2,359
60	0,010	7,077	4,977	4,126	3,649	3,339	3,119	2,953	2,823	2,718	2,632
60	0,005	8,495	5,795	4,729	4,140	3,760	3,492	3,291	3,134	3,008	2,904
60	0,002	10,44	6,905	5,542	4,799	4,324	3,990	3,742	3,548	3,393	3,266
60	0,001	11,97	7,768	6,171	5,307	4,757	4,372	4,086	3,865	3,687	3,541
80	0,500	,4591	,6992	,7954	,8463	,8777	,8989	,9142	,9258	,9348	,9421
80	0,250	1,343	1,411	1,396	1,375	1,355	1,338	1,324	1,311	1,300	1,291
80	0,100	2,769	2,370	2,154	2,016	1,921	1,849	1,793	1,748	1,711	1,680
80	0,050	3,960	3,111	2,719	2,486	2,329	2,214	2,126	2,056	1,999	1,951
80	0,020	5,635	4,110	3,467	3,100	2,858	2,685	2,553	2,450	2,366	2,296
80	0,010	6,963	4,881	4,036	3,563	3,255	3,036	2,871	2,742	2,637	2,551
80	0,005	8,335	5,665	4,611	4,029	3,652	3,387	3,188	3,032	2,907	2,803
80	0,002	10,21	6,723	5,381	4,649	4,180	3,851	3,606	3,415	3,261	3,135
80	0,001	11,67	7,540	5,972	5,123	4,582	4,204	3,923	3,705	3,530	3,386
120	0,500	,4577	,6972	,7932	,8439	,8752	,8964	,9116	,9232	,9322	,9394
120	0,250	1,336	1,402	1,387	1,365	1,345	1,328	1,313	1,300	1,289	1,279
120	0,100	2,748	2,347	2,130	1,992	1,896	1,824	1,767	1,722	1,684	1,652
120	0,050	3,920	3,072	2,680	2,447	2,290	2,175	2,087	2,016	1,959	1,910
120	0,020	5,559	4,042	3,403	3,037	2,796	2,623	2,492	2,389	2,305	2,235
120	0,010	6,851	4,787	3,949	3,480	3,174	2,956	2,792	2,663	2,559	2,472
120	0,005	8,179	5,539	4,497	3,921	3,548	3,285	3,087	2,933	2,808	2,705
120	0,002	9,983	6,548	5,226	4,504	4,042	3,717	3,475	3,286	3,134	3,009
120	0,001	11,38	7,321	5,781	4,947	4,416	4,044	3,767	3,552	3,379	3,237
∞	0,500	,4549	,6931	,7887	,8392	,8703	,8914	,9065	,9180	,9270	,9342
∞	0,250	1,323	1,386	1,369	1,346	1,325	1,307	1,291	1,277	1,265	1,255
∞	0,100	2,706	2,303	2,084	1,945	1,847	1,774	1,717	1,670	1,632	1,599
∞	0,050	3,841	2,996	2,605	2,372	2,214	2,099	2,010	1,938	1,880	1,831
∞	0,020	5,412	3,912	3,279	2,917	2,678	2,506	2,375	2,271	2,187	2,116
∞	0,010	6,635	4,605	3,782	3,319	3,017	2,802	2,639	2,511	2,407	2,321
∞	0,005	7,879	5,298	4,279	3,715	3,350	3,091	2,897	2,744	2,621	2,519
∞	0,002	9,550	6,215	4,932	4,231	3,781	3,465	3,229	3,044	2,895	2,772
∞	0,001	10,83	6,908	5,422	4,617	4,103	3,743	3,475	3,266	3,097	2,959

TABLE D Fisher's *F* Distribution (*Continued*)

df_1

10	12	15	20	24	30	40	60	120	∞	pr	df_2
,9468	,9578	,9688	,9799	,9855	,9911	,9966	1,002	1,008	1,013	0,500	50
1,312	1,297	1,280	1,259	1,248	1,235	1,221	1,205	1,186	1,164	0,250	50
1,729	1,680	1,627	1,568	1,536	1,502	1,465	1,424	1,379	1,327	0,100	50
2,026	1,952	1,871	1,784	1,737	1,687	1,634	1,576	1,511	1,438	0,050	50
2,410	2,301	2,185	2,060	1,993	1,923	1,848	1,767	1,679	1,579	0,020	50
2,698	2,563	2,419	2,265	2,183	2,098	2,007	1,909	1,803	1,683	0,010	50
2,988	2,825	2,653	2,470	2,373	2,272	2,164	2,050	1,925	1,786	0,005	50
3,374	3,174	2,965	2,742	2,625	2,503	2,374	2,236	2,088	1,923	0,002	50
3,671	3,443	3,204	2,951	2,817	2,679	2,533	2,378	2,211	2,026	0,001	50
,9447	,9557	,9667	,9777	,9833	,9888	,9944	1,000	1,006	1,011	0,500	60
1,303	1,287	1,269	1,248	1,236	1,223	1,208	1,191	1,172	1,147	0,250	60
1,707	1,657	1,603	1,543	1,511	1,476	1,437	1,395	1,348	1,291	0,100	60
1,993	1,917	1,836	1,748	1,700	1,649	1,594	1,534	1,467	1,389	0,050	60
2,359	2,249	2,133	2,007	1,939	1,868	1,791	1,709	1,617	1,511	0,020	60
2,632	2,496	2,352	2,198	2,115	2,028	1,936	1,836	1,726	1,601	0,010	60
2,904	2,742	2,570	2,387	2,290	2,187	2,079	1,962	1,834	1,689	0,005	60
3,266	3,067	2,859	2,637	2,520	2,397	2,267	2,127	1,975	1,804	0,002	60
3,541	3,315	3,078	2,827	2,694	2,555	2,409	2,252	2,082	1,890	0,001	60
,9421	,9530	,9640	,9750	,9805	,9861	,9916	,9972	1,003	1,008	0,500	80
1,291	1,275	1,256	1,234	1,222	1,208	1,192	1,174	1,152	1,124	0,250	80
1,680	1,629	1,574	1,513	1,479	1,443	1,403	1,358	1,307	1,245	0,100	80
1,951	1,875	1,793	1,703	1,654	1,602	1,545	1,482	1,411	1,325	0,050	80
2,296	2,186	2,069	1,941	1,873	1,800	1,721	1,635	1,538	1,423	0,020	80
2,551	2,415	2,271	2,115	2,032	1,944	1,849	1,746	1,630	1,494	0,010	80
2,803	2,641	2,470	2,286	2,188	2,084	1,974	1,854	1,720	1,563	0,005	80
3,135	2,938	2,731	2,509	2,392	2,268	2,136	1,994	1,837	1,653	0,002	80
3,386	3,162	2,927	2,677	2,545	2,406	2,258	2,099	1,924	1,720	0,001	80
,9394	,9503	,9613	,9723	,9778	,9833	,9889	,9944	1,000	1,006	0,500	120
1,279	1,262	1,243	1,220	1,207	1,192	1,175	1,156	1,131	1,099	0,250	120
1,652	1,601	1,545	1,482	1,447	1,409	1,368	1,320	1,265	1,193	0,100	120
1,910	1,834	1,751	1,659	1,608	1,554	1,495	1,429	1,352	1,254	0,050	120
2,235	2,124	2,006	1,877	1,807	1,732	1,651	1,561	1,458	1,328	0,020	120
2,472	2,336	2,192	2,035	1,950	1,860	1,763	1,656	1,533	1,381	0,010	120
2,705	2,544	2,373	2,188	2,089	1,984	1,871	1,747	1,606	1,431	0,005	120
3,009	2,814	2,608	2,387	2,268	2,144	2,010	1,864	1,698	1,496	0,002	120
3,237	3,016	2,783	2,534	2,402	2,262	2,113	1,950	1,767	1,543	0,001	120
,9342	,9450	,9559	,9669	,9724	,9779	,9834	,9889	,9945	1,000	0,500	∞
1,255	1,237	1,216	1,191	1,177	1,160	1,140	1,116	1,084	1,000	0,250	∞
1,599	1,546	1,487	1,421	1,383	1,342	1,295	1,240	1,169	1,000	0,100	∞
1,831	1,752	1,666	1,571	1,517	1,459	1,394	1,318	1,221	1,000	0,050	∞
2,116	2,005	1,884	1,751	1,678	1,599	1,511	1,410	1,283	1,000	0,020	∞
2,321	2,185	2,039	1,878	1,791	1,696	1,592	1,473	1,325	1,000	0,010	∞
2,519	2,358	2,187	2,000	1,898	1,789	1,669	1,533	1,364	1,000	0,005	∞
2,772	2,580	2,375	2,154	2,034	1,906	1,765	1,607	1,412	1,000	0,002	∞
2,959	2,742	2,513	2,266	2,132	1,990	1,835	1,660	1,447	1,000	0,001	∞

TABLE E   Wilcoxon's Rank Sum

Probability that Wilcoxon's rank sum minus the expected value
will be equal to or larger than W

					n = 1						
W	m=1	m=3	m=5	m=7	m=9	m=11	m=13	m=15	m=17	m=19	W
0,5	,500	,500	,500	,500	,500	,500	,500	,500	,500	,500	0,5
1,5	,000	,250	,333	,375	,400	,417	,429	,437	,444	,450	1,5
2,5		,000	,167	,250	,300	,333	,357	,375	,389	,400	2,5
3,5			,000	,125	,200	,250	,286	,312	,333	,350	3,5
4,5				,000	,100	,167	,214	,250	,278	,300	4,5
5,5					,000	,083	,143	,187	,222	,250	5,5
6,5						,000	,071	,125	,167	,200	6,5
7,5							,000	,062	,111	,150	7,5
8,5								,000	,056	,100	8,5
9,5									,000	,050	9,5
10,5										,000	10,5

					n = 1						
W	m=2	m=4	m=6	m=8	m=10	m=12	m=14	m=16	m=18	m=20	W
1	,333	,400	,429	,444	,455	,462	,467	,471	,474	,476	1
2	,000	,200	,286	,333	,364	,385	,400	,412	,421	,429	2
3		,000	,143	,222	,273	,308	,333	,353	,368	,381	3
4			,000	,111	,182	,231	,267	,294	,316	,333	4
5				,000	,091	,154	,200	,235	,263	,286	5
6					,000	,077	,133	,176	,211	,238	6
7						,000	,067	,118	,158	,190	7
8							,000	,059	,105	,143	8
9								,000	,053	,095	9
10									,000	,048	10
11										,000	11

					n = 2						
W	m=1	m=2	m=3	m=4	m=5	m=6	m=7	m=8	m=9	m=10	W
1	,333	,333	,400	,400	,429	,429	,444	,444	,455	,455	1
2	,000	,167	,200	,267	,286	,321	,333	,356	,364	,379	2
3		,000	,100	,133	,190	,214	,250	,267	,291	,303	3
4			,000	,067	,095	,143	,167	,200	,218	,242	4
5				,000	,048	,071	,111	,133	,164	,182	5
6					,000	,036	,056	,089	,109	,136	6
7						,000	,028	,044	,073	,091	7
8							,000	,022	,036	,061	8
9								,000	,018	,030	9
10									,000	,015	10
11										,000	11

TABLE E Wilcoxon's Rank Sum   (*Continued*)

Probability that Wilcoxon's rank sum minus the expected value
will be equal to or larger than W

| | | | | | n = | 2 | | | | | |
W	m=11	m=12	m=13	m=14	m=15	m=16	m=17	m=18	m=19	m=20	W
1	,462	,462	,467	,467	,471	,471	,474	,474	,476	,476	1
2	,385	,396	,400	,408	,412	,418	,421	,426	,429	,433	2
3	,321	,330	,343	,350	,360	,366	,374	,379	,386	,390	3
4	,256	,275	,286	,300	,309	,320	,327	,337	,343	,351	4
5	,205	,220	,238	,250	,265	,275	,287	,295	,305	,312	5
6	,154	,176	,190	,208	,221	,235	,246	,258	,267	,277	6
7	,115	,132	,152	,167	,184	,196	,211	,221	,233	,242	7
8	,077	,099	,114	,133	,147	,163	,175	,189	,200	,212	8
9	,051	,066	,086	,100	,118	,131	,146	,158	,171	,182	9
10	,026	,044	,057	,075	,088	,105	,117	,132	,143	,156	10
11	,013	,022	,038	,050	,066	,078	,094	,105	,119	,130	11
12	,000	,011	,019	,033	,044	,059	,070	,084	,095	,108	12
13		,000	,010	,017	,029	,039	,053	,063	,076	,087	13
14			,000	,008	,015	,026	,035	,047	,057	,069	14
15				,000	,007	,013	,023	,032	,043	,052	15
16					,000	,007	,012	,021	,029	,039	16
17						,000	,006	,011	,019	,026	17
18							,000	,005	,010	,017	18
19								,000	,005	,009	19
20									,000	,004	20
21										,000	21

TABLE E Wilcoxon's Rank Sum (*Continued*)

Probability that Wilcoxon's rank sum minus the expected value
will be equal to or larger than W

n = 3

W	m=1	m=3	m=5	m=7	m=9	m=11	m=13	m=15	m=17	m=19	W
0,5	,500	,500	,500	,500	,500	,500	,500	,500	,500	,500	0,5
1,5	,250	,350	,393	,417	,432	,442	,450	,456	,461	,464	1,5
2,5	,000	,200	,286	,333	,364	,385	,400	,412	,421	,429	2,5
3,5		,100	,196	,258	,300	,330	,352	,369	,382	,394	3,5
4,5		,050	,125	,192	,241	,277	,305	,327	,345	,359	4,5
5,5		,000	,071	,133	,186	,228	,261	,287	,308	,325	5,5
6,5			,036	,092	,141	,184	,220	,249	,273	,293	6,5
7,5			,018	,058	,105	,146	,182	,213	,239	,262	7,5
8,5			,000	,033	,073	,113	,148	,180	,208	,232	8,5
9,5				,017	,050	,085	,120	,151	,179	,204	9,5
10,5				,008	,032	,063	,095	,125	,153	,178	10,5
11,5				,000	,018	,044	,073	,102	,129	,154	11,5
12,5					,009	,030	,055	,082	,108	,132	12,5
13,5					,005	,019	,041	,065	,089	,113	13,5
14,5					,000	,011	,029	,050	,073	,095	14,5
15,5						,005	,020	,038	,059	,080	15,5
16,5						,003	,013	,028	,046	,066	16,5
17,5						,000	,007	,020	,036	,054	17,5
18,5							,004	,013	,027	,044	18,5
19,5							,002	,009	,020	,034	19,5
20,5							,000	,005	,014	,027	20,5
21,5								,002	,010	,020	21,5
22,5								,001	,006	,015	22,5
23,5								,000	,004	,010	23,5
24,5									,002	,007	24,5
25,5									,001	,005	25,5
26,5									,000	,003	26,5
27,5										,001	27,5
28,5										,001	28,5
29,5										,000	29,5

n = 3

W	m=2	m=4	m=6	m=8	m=10	m=12	m=14	m=16	m=18	m=20	W
1	,400	,429	,452	,461	,469	,473	,476	,479	,481	,483	1
2	,200	,314	,357	,388	,406	,420	,429	,438	,444	,449	2
3	,100	,200	,274	,315	,346	,367	,384	,396	,407	,415	3
4	,000	,114	,190	,248	,287	,316	,338	,356	,370	,382	4
5		,057	,131	,188	,234	,268	,296	,317	,335	,349	5
6		,029	,083	,139	,185	,224	,254	,280	,300	,317	6
7		,000	,048	,097	,143	,182	,216	,244	,267	,286	7
8			,024	,067	,108	,147	,181	,211	,235	,257	8
9			,012	,042	,080	,116	,150	,180	,206	,229	9
10			,000	,024	,056	,090	,122	,152	,178	,202	10
11				,012	,038	,068	,099	,127	,153	,177	11

TABLE E Wilcoxon's Rank Sum (*Continued*)

Probability that Wilcoxon's rank sum minus the expected value
will be equal to or larger than W

					n =	3					
W	m=2	m=4	m=6	m=8	m=10	m=12	m=14	m=16	m=18	m=20	W
12				,006	,024	,051	,078	,105	,131	,155	12
13				,000	,014	,035	,060	,086	,111	,134	13
14					,007	,024	,046	,069	,092	,115	14
15					,003	,015	,034	,055	,077	,098	15
16					,000	,009	,024	,042	,062	,083	16
17						,004	,016	,032	,050	,069	17
18						,002	,010	,024	,040	,058	18
19						,000	,006	,017	,031	,047	19
20							,003	,011	,023	,038	20
21							,001	,007	,017	,030	21
22							,000	,004	,012	,023	22
23								,002	,008	,018	23
24								,001	,005	,013	24
25								,000	,003	,009	25
26									,002	,006	26
27									,001	,004	27
28									,000	,002	28
29										,001	29
30										,001	30
31										,000	31

					n =	4					
W	m=1	m=2	m=3	m=4	m=5	m=6	m=7	m=8	m=9	m=10	W
1	,400	,400	,429	,443	,452	,457	,464	,467	,470	,473	1
2	,200	,267	,314	,343	,365	,381	,394	,404	,413	,420	2
3	,000	,133	,200	,243	,278	,305	,324	,341	,355	,367	3
4		,067	,114	,171	,206	,238	,264	,285	,302	,318	4
5		,000	,057	,100	,143	,176	,206	,230	,252	,270	5
6			,029	,057	,095	,129	,158	,184	,207	,227	6
7			,000	,029	,056	,086	,115	,141	,165	,187	7
8				,014	,032	,057	,082	,107	,130	,152	8
9				,000	,016	,033	,055	,077	,099	,120	9
10					,008	,019	,036	,055	,074	,094	10
11					,000	,010	,021	,036	,053	,071	11
12						,005	,012	,024	,038	,053	12
13						,000	,006	,014	,025	,038	13
14							,003	,008	,017	,027	14
15							,000	,004	,010	,018	15
16								,002	,006	,012	16
17								,000	,003	,007	17
18									,001	,004	18
19									,000	,002	19
20										,001	20
21										,000	21

TABLE E Wilcoxon's Rank Sum *(Continued)*

Probability that Wilcoxon's rank sum minus the expected value
will be equal to or larger than W

					n = 4						
W	m=11	m=12	m=13	m=14	m=15	m=16	m=17	m=18	m=19	m=20	W
1	.475	.476	.478	.479	.481	.482	.483	.484	.484	.485	1
2	.426	.431	.435	.439	.443	.446	.449	.451	.453	.455	2
3	.377	.385	.392	.399	.405	.410	.415	.419	.422	.426	3
4	.330	.342	.352	.360	.368	.375	.381	.387	.392	.397	4
5	.286	.299	.312	.323	.332	.341	.349	.356	.363	.368	5
6	.245	.260	.274	.287	.298	.308	.318	.326	.334	.341	6
7	.206	.223	.239	.253	.265	.277	.287	.297	.306	.314	7
8	.171	.190	.206	.221	.235	.247	.258	.269	.279	.288	8
9	.140	.158	.175	.191	.205	.219	.231	.242	.253	.262	9
10	.113	.131	.148	.164	.179	.192	.205	.217	.228	.239	10
11	.089	.106	.123	.139	.154	.168	.181	.193	.205	.216	11
12	.069	.085	.101	.116	.131	.145	.158	.171	.183	.194	12
13	.052	.066	.082	.096	.110	.124	.138	.150	.162	.174	13
14	.039	.052	.065	.079	.092	.106	.119	.131	.143	.155	14
15	.028	.039	.051	.063	.076	.089	.101	.113	.125	.137	15
16	.020	.029	.039	.051	.062	.074	.086	.098	.109	.120	16
17	.013	.021	.030	.040	.050	.061	.072	.083	.094	.105	17
18	.009	.015	.022	.031	.040	.050	.060	.070	.081	.091	18
19	.005	.010	.016	.023	.031	.040	.049	.059	.069	.079	19
20	.003	.007	.011	.017	.024	.032	.040	.049	.058	.067	20
21	.001	.004	.008	.012	.018	.025	.032	.040	.049	.057	21
22	.001	.002	.005	.009	.014	.019	.026	.033	.041	.048	22
23	.000	.001	.003	.006	.010	.015	.020	.027	.033	.041	23
24		.001	.002	.004	.007	.011	.016	.021	.027	.034	24
25		.000	.001	.002	.005	.008	.012	.017	.022	.028	25
26			.000	.001	.003	.006	.009	.013	.018	.023	26
27				.001	.002	.004	.006	.010	.014	.018	27
28				.000	.001	.002	.005	.007	.011	.015	28
29					.001	.001	.003	.005	.008	.011	29
30					.000	.001	.002	.004	.006	.009	30
31						.000	.001	.002	.004	.007	31
32							.001	.002	.003	.005	32
33							.000	.001	.002	.004	33
34								.001	.001	.003	34
35								.000	.001	.002	35
36									.000	.001	36
37										.001	37
38										.000	38

TABLE E   Wilcoxon's Rank Sum   (*Continued*)
Probability that Wilcoxon's rank sum minus the expected value
will be equal to or larger than W

W	m=1	m=3	m=5	m=7	m=9	m=11	m=13	m=15	m=17	m=19	W
					n = 5						
0,5	,500	,500	,500	,500	,500	,500	,500	,500	,500	,500	0,5
1,5	,333	,393	,421	,438	,449	,457	,462	,466	,470	,473	1,5
2,5	,167	,286	,345	,378	,399	,413	,424	,433	,440	,445	2,5
3,5	,000	,196	,274	,319	,350	,371	,387	,400	,410	,418	3,5
4,5		,125	,210	,265	,303	,331	,351	,368	,381	,392	4,5
5,5		,071	,155	,216	,259	,292	,317	,336	,352	,365	5,5
6,5		,036	,111	,172	,219	,255	,283	,306	,324	,340	6,5
7,5		,018	,075	,134	,182	,220	,251	,277	,297	,315	7,5
8,5		,000	,048	,101	,149	,189	,222	,249	,271	,290	8,5
9,5			,028	,074	,120	,160	,194	,222	,246	,267	9,5
10,5			,016	,053	,095	,134	,168	,197	,223	,245	10,5
11,5			,008	,037	,073	,111	,144	,174	,200	,223	11,5
12,5			,004	,024	,056	,090	,123	,153	,179	,203	12,5
13,5			,000	,015	,041	,073	,104	,133	,160	,183	13,5
14,5				,009	,030	,057	,087	,115	,141	,165	14,5
15,5				,005	,021	,045	,072	,099	,124	,148	15,5
16,5				,003	,014	,034	,059	,084	,109	,132	16,5
17,5				,001	,009	,026	,047	,071	,095	,118	17,5
18,5				,000	,006	,019	,038	,059	,082	,104	18,5
19,5					,003	,014	,030	,049	,070	,091	19,5
20,5					,002	,010	,023	,040	,060	,080	20,5
21,5					,001	,007	,018	,033	,051	,070	21,5
22,5					,000	,004	,013	,026	,042	,060	22,5
23,5						,003	,010	,021	,035	,052	23,5
24,5						,002	,007	,016	,029	,044	24,5
25,5						,001	,005	,013	,024	,038	25,5
26,5						,000	,003	,010	,019	,032	26,5
27,5							,002	,007	,015	,026	27,5
28,5							,001	,005	,012	,022	28,5
29,5							,001	,004	,010	,018	29,5
30,5							,000	,003	,007	,015	30,5
31,5								,002	,006	,012	31,5
32,5								,001	,004	,010	32,5
33,5								,001	,003	,008	33,5
34,5								,000	,002	,006	34,5
35,5									,002	,005	35,5
36,5									,001	,004	36,5
37,5									,001	,003	37,5
38,5									,000	,002	38,5
39,5										,001	39,5
40,5										,001	40,5
41,5										,001	41,5
42,5										,000	42,5

TABLE E Wilcoxon's Rank Sum (*Continued*)

Probability that Wilcoxon's rank sum minus the expected value
will be equal to or larger than W

					n =	5				
W	m=2	m=4	m=6	m=8	m=10	m=12	m=14	m=16	m=18	m=20	W
1	,429	,452	,465	,472	,477	,480	,482	,484	,486	,487	1
2	,286	,365	,396	,416	,430	,439	,447	,452	,457	,461	2
3	,190	,278	,331	,362	,384	,399	,412	,421	,428	,435	3
4	,095	,206	,268	,311	,339	,361	,377	,390	,400	,409	4
5	,048	,143	,214	,262	,297	,323	,343	,359	,373	,384	5
6	,000	,095	,165	,218	,257	,287	,311	,330	,346	,359	6
7		,056	,123	,177	,220	,253	,280	,301	,319	,334	7
8		,032	,089	,142	,185	,221	,250	,274	,294	,311	8
9		,016	,063	,111	,155	,191	,222	,247	,269	,287	9
10		,008	,041	,085	,127	,164	,196	,223	,245	,265	10
11		,000	,026	,064	,103	,139	,171	,199	,223	,244	11
12			,015	,047	,082	,117	,149	,177	,202	,223	12
13			,009	,033	,065	,097	,128	,156	,182	,204	13
14			,004	,023	,050	,080	,110	,137	,163	,185	14
15			,002	,015	,038	,065	,093	,120	,145	,168	15
16			,000	,009	,028	,052	,078	,104	,129	,151	16
17				,005	,020	,041	,065	,089	,113	,136	17
18				,003	,014	,032	,053	,077	,099	,122	18
19				,002	,010	,024	,044	,065	,087	,108	19
20				,001	,006	,018	,035	,055	,075	,096	20
21				,000	,004	,013	,028	,046	,065	,085	21
22					,002	,010	,022	,038	,055	,074	22
23					,001	,007	,017	,031	,047	,065	23
24					,001	,005	,013	,025	,040	,056	24
25					,000	,003	,010	,020	,033	,048	25
26						,002	,007	,016	,028	,042	26
27						,001	,005	,012	,023	,035	27
28						,001	,004	,010	,019	,030	28
29						,000	,002	,007	,015	,025	29
30							,002	,006	,012	,021	30
31							,001	,004	,010	,018	31
32							,001	,003	,008	,014	32
33							,000	,002	,006	,012	33
34								,001	,004	,010	34
35								,001	,003	,008	35
36								,001	,002	,006	36
37								,000	,002	,005	37
38									,001	,004	38

TABLE E   Wilcoxon's Rank Sum   (*Continued*)

```
Probability that Wilcoxon's rank sum minus the expected value
              will be equal to or larger than W
```

					n = 5					
W	m=2	m=4	m=6	m=8	m=10	m=12	m=14	m=16	m=18	m=20	W
39									,001	,003	39
40									,001	,002	40
41									,000	,002	41
42										,001	42
43										,001	43
44										,001	44
45										,000	45

					n = 6					
W	m=1	m=2	m=3	m=4	m=5	m=6	m=7	m=8	m=9	m=10	W
1	,429	,429	,452	,457	,465	,469	,473	,475	,477	,479	1
2	,286	,321	,357	,381	,396	,409	,418	,426	,432	,437	2
3	,143	,214	,274	,305	,331	,350	,365	,377	,388	,396	3
4	,000	,143	,190	,238	,268	,294	,314	,331	,344	,356	4
5		,071	,131	,176	,214	,242	,267	,286	,303	,318	5
6		,036	,083	,129	,165	,197	,223	,245	,264	,281	6
7		,000	,048	,086	,123	,155	,183	,207	,228	,246	7
8			,024	,057	,089	,120	,147	,172	,194	,214	8
9			,012	,033	,063	,090	,117	,141	,164	,184	9
10			,000	,019	,041	,066	,090	,114	,136	,157	10
11				,010	,026	,047	,069	,091	,112	,132	11
12				,005	,015	,032	,051	,071	,091	,110	12
13				,000	,009	,021	,037	,054	,072	,090	13
14					,004	,013	,026	,041	,057	,074	14
15					,002	,008	,017	,030	,044	,059	15
16					,000	,004	,011	,021	,033	,047	16
17						,002	,007	,015	,025	,036	17
18						,001	,004	,010	,018	,028	18
19						,000	,002	,006	,013	,021	19
20							,001	,004	,009	,016	20
21							,001	,002	,006	,011	21
22							,000	,001	,004	,008	22
23								,001	,002	,005	23
24								,000	,001	,004	24
25									,001	,002	25
26									,000	,001	26
27										,001	27
28										,000	28

TABLE E Wilcoxon's Rank Sum *(Continued)*

Probability that Wilcoxon's rank sum minus the expected value
will be equal to or larger than W

W	m=11	m=12	m=13	m=14	m=15	m=16	m=17	m=18	m=19	m=20	W
					n = 6						
1	,481	,482	,483	,484	,485	,486	,486	,487	,488	,488	1
2	,442	,446	,449	,452	,455	,457	,459	,461	,463	,465	2
3	,404	,410	,416	,421	,425	,429	,432	,436	,439	,441	3
4	,366	,375	,383	,390	,395	,401	,406	,410	,414	,418	4
5	,330	,341	,351	,359	,367	,373	,380	,385	,390	,395	5
6	,295	,308	,319	,329	,338	,347	,354	,361	,367	,372	6
7	,262	,277	,289	,301	,311	,320	,329	,336	,344	,350	7
8	,231	,247	,261	,273	,285	,295	,304	,313	,321	,328	8
9	,202	,219	,234	,247	,259	,271	,281	,290	,299	,307	9
10	,175	,192	,208	,222	,235	,247	,258	,268	,278	,286	10
11	,151	,168	,184	,199	,212	,225	,236	,247	,257	,266	11
12	,128	,145	,161	,177	,190	,204	,216	,227	,237	,247	12
13	,108	,125	,141	,156	,170	,183	,196	,207	,218	,229	13
14	,090	,106	,122	,137	,151	,165	,177	,189	,200	,211	14
15	,074	,090	,105	,120	,134	,147	,160	,172	,183	,194	15
16	,061	,075	,090	,104	,117	,131	,143	,155	,167	,178	16
17	,049	,062	,076	,089	,103	,115	,128	,140	,151	,162	17
18	,039	,051	,064	,076	,089	,102	,114	,125	,137	,148	18
19	,031	,042	,053	,065	,077	,089	,101	,112	,123	,134	19
20	,024	,033	,044	,055	,066	,077	,089	,100	,111	,121	20
21	,018	,026	,036	,046	,056	,067	,078	,088	,099	,109	21
22	,014	,021	,029	,038	,047	,057	,068	,078	,088	,098	22
23	,010	,016	,023	,031	,040	,049	,059	,068	,078	,088	23
24	,007	,012	,018	,025	,033	,042	,050	,060	,069	,078	24
25	,005	,009	,014	,020	,027	,035	,043	,052	,061	,069	25
26	,004	,007	,011	,016	,022	,029	,037	,045	,053	,061	26
27	,002	,005	,008	,013	,018	,024	,031	,038	,046	,054	27
28	,002	,003	,006	,010	,015	,020	,026	,033	,040	,047	28
29	,001	,002	,005	,008	,012	,016	,022	,028	,034	,041	29
30	,001	,002	,003	,006	,009	,013	,018	,024	,030	,036	30
31	,000	,001	,002	,004	,007	,011	,015	,020	,025	,031	31
32		,001	,002	,003	,005	,008	,012	,017	,021	,027	32
33		,000	,001	,002	,004	,007	,010	,014	,018	,023	33
34			,001	,002	,003	,005	,008	,011	,015	,020	34
35			,000	,001	,002	,004	,006	,009	,013	,017	35
36				,001	,002	,003	,005	,007	,010	,014	36
37				,000	,001	,002	,004	,006	,009	,012	37
38					,001	,002	,003	,005	,007	,010	38
39					,001	,001	,002	,004	,006	,008	39

TABLE E Wilcoxon's Rank Sum (*Continued*)

Probability that Wilcoxon's rank sum minus the expected value
will be equal to or larger than W

W	m=11	m=12	m=13	m=14	m=15	m=16	m=17	m=18	m=19	m=20	W
					n =	6					
40					,000	,001	,002	,003	,005	,007	40
41						,001	,001	,002	,004	,005	41
42						,000	,001	,002	,003	,004	42
43							,001	,001	,002	,003	43
44							,000	,001	,002	,003	44
45								,001	,001	,002	45
46								,000	,001	,002	46
47									,001	,001	47
48									,001	,001	48
49									,000	,001	49
50										,001	50

W	m=1	m=3	m=5	m=7	m=9	m=11	m=13	m=15	m=17	m=19	W
					n =	7					
0,5	,500	,500	,500	,500	,500	,500	,500	,500	,500	,500	0,5
1,5	,375	,417	,438	,451	,459	,465	,469	,473	,475	,478	1,5
2,5	,250	,333	,378	,402	,419	,430	,439	,445	,451	,455	2,5
3,5	,125	,258	,319	,355	,379	,396	,408	,418	,426	,433	3,5
4,5	,000	,192	,265	,310	,340	,362	,379	,392	,402	,411	4,5
5,5		,133	,216	,267	,303	,329	,350	,365	,378	,389	5,5
6,5		,092	,172	,228	,268	,298	,321	,340	,355	,367	6,5
7,5		,058	,134	,191	,235	,268	,294	,315	,332	,346	7,5
8,5		,033	,101	,159	,204	,239	,268	,291	,310	,326	8,5
9,5		,017	,074	,130	,176	,213	,243	,267	,288	,305	9,5
10,5		,008	,053	,104	,150	,187	,219	,245	,267	,286	10,5
11,5		,000	,037	,082	,126	,164	,196	,224	,247	,267	11,5
12,5			,024	,064	,105	,143	,175	,203	,228	,248	12,5
13,5			,015	,049	,087	,123	,156	,184	,209	,231	13,5
14,5			,009	,036	,071	,105	,137	,166	,191	,214	14,5
15,5			,005	,027	,057	,090	,121	,149	,175	,197	15,5
16,5			,003	,019	,045	,075	,105	,133	,159	,182	16,5
17,5			,001	,013	,036	,063	,091	,119	,144	,167	17,5
18,5			,000	,009	,027	,052	,079	,105	,130	,153	18,5
19,5				,006	,021	,043	,067	,093	,117	,140	19,5

TABLE E Wilcoxon's Rank Sum *(Continued)*

Probability that Wilcoxon's rank sum minus the expected value
will be equal to or larger than W

					n = 7						
W	m=1	m=3	m=5	m=7	m=9	m=11	m=13	m=15	m=17	m=19	W
20,5				,003	,016	,035	,057	,081	,105	,127	20,5
21,5				,002	,011	,028	,048	,071	,093	,115	21,5
22,5				,001	,008	,022	,041	,061	,083	,104	22,5
23,5				,001	,006	,017	,034	,053	,074	,094	23,5
24,5				,000	,004	,013	,028	,046	,065	,085	24,5
25,5					,003	,010	,023	,039	,057	,076	25,5
26,5					,002	,008	,018	,033	,050	,068	26,5
27,5					,001	,006	,015	,028	,043	,060	27,5
28,5					,001	,004	,012	,023	,037	,053	28,5
29,5					,000	,003	,009	,019	,032	,047	29,5
30,5						,002	,007	,016	,028	,041	30,5
31,5						,001	,006	,013	,024	,036	31,5
32,5						,001	,004	,011	,020	,032	32,5
33,5						,001	,003	,009	,017	,028	33,5
34,5						,000	,002	,007	,014	,024	34,5
35,5							,002	,005	,012	,021	35,5
36,5							,001	,004	,010	,018	36,5
37,5							,001	,003	,008	,015	37,5
38,5							,001	,003	,007	,013	38,5
39,5							,000	,002	,005	,011	39,5
40,5								,001	,004	,009	40,5
41,5								,001	,003	,008	41,5
42,5								,001	,003	,006	42,5
43,5								,001	,002	,005	43,5
44,5								,000	,002	,004	44,5
45,5									,001	,003	45,5
46,5									,001	,003	46,5
47,5									,001	,002	47,5
48,5									,001	,002	48,5
49,5									,000	,001	49,5
50,5										,001	50,5
51,5										,001	51,5
52,5										,001	52,5
53,5										,000	53,5

TABLE E Wilcoxon's Rank Sum *(Continued)*

Probability that Wilcoxon's rank sum minus the expected value
will be equal to or larger than W

					n =	7					
W	m=2	m=4	m=6	m=8	m=10	m=12	m=14	m=16	m=18	m=20	W
1	.444	.464	.473	.478	.481	.484	.486	.487	.488	.489	1
2	.333	.394	.418	.433	.443	.451	.457	.461	.465	.468	2
3	.250	.324	.365	.389	.406	.418	.428	.435	.441	.446	3
4	.167	.264	.314	.347	.370	.387	.400	.410	.418	.425	4
5	.111	.206	.267	.306	.335	.355	.372	.385	.395	.404	5
6	.056	.158	.223	.268	.300	.325	.344	.360	.373	.383	6
7	.028	.115	.183	.232	.268	.296	.318	.336	.350	.363	7
8	.000	.082	.147	.198	.237	.268	.292	.312	.329	.343	8
9		.055	.117	.168	.209	.241	.268	.289	.307	.323	9
10		.036	.090	.140	.182	.216	.244	.267	.287	.304	10
11		.021	.069	.116	.157	.192	.221	.246	.267	.285	11
12		.012	.051	.095	.135	.170	.200	.226	.248	.267	12
13		.006	.037	.076	.115	.150	.180	.206	.229	.249	13
14		.003	.026	.060	.097	.131	.161	.188	.211	.232	14
15		.000	.017	.047	.081	.113	.144	.171	.194	.216	15
16			.011	.036	.067	.098	.127	.154	.178	.200	16
17			.007	.027	.054	.084	.112	.139	.163	.185	17
18			.004	.020	.044	.071	.098	.124	.149	.171	18
19			.002	.014	.035	.060	.086	.111	.135	.157	19
20			.001	.010	.028	.050	.074	.099	.122	.144	20
21			.001	.007	.022	.042	.064	.087	.110	.132	21
22			.000	.005	.017	.034	.055	.077	.099	.120	22
23				.003	.012	.028	.047	.068	.089	.109	23
24				.002	.009	.022	.040	.059	.079	.099	24
25				.001	.007	.018	.033	.051	.070	.090	25
26				.001	.005	.014	.028	.044	.062	.081	26
27				.000	.003	.011	.023	.038	.055	.073	27
28					.002	.009	.019	.033	.048	.065	28
29					.002	.006	.015	.028	.042	.058	29
30					.001	.005	.012	.023	.037	.052	30
31					.001	.004	.010	.020	.032	.046	31
32					.000	.003	.008	.016	.028	.041	32
33						.002	.006	.014	.024	.036	33
34						.001	.005	.011	.020	.031	34
35						.001	.004	.009	.017	.027	35
36						.001	.003	.007	.015	.024	36
37						.000	.002	.006	.012	.021	37
38							.002	.005	.010	.018	38
39							.001	.004	.009	.015	39
40							.001	.003	.007	.013	40

TABLE E   Wilcoxon's Rank Sum   (*Continued*)

Probability that Wilcoxon's rank sum minus the expected value
will be equal to or larger than W

					n = 7					
W	m=2	m=4	m=6	m=8	m=10	m=12	m=14	m=16	m=18	m=20	W
41							,001	,002	,006	,011	41
42							,000	,002	,005	,010	42
43								,001	,004	,008	43
44								,001	,003	,007	44
45								,001	,002	,006	45
46								,001	,002	,005	46
47								,000	,002	,004	47
48									,001	,003	48
49									,001	,003	49
50									,001	,002	50
51									,001	,002	51
52									,000	,001	52
53										,001	53
54										,001	54
55										,001	55
56										,000	56

					n = 8					
W	m=1	m=2	m=3	m=4	m=5	m=6	m=7	m=8	m=9	m=10	W
1	,444	,444	,461	,467	,472	,475	,478	,480	,481	,483	1
2	,333	,356	,388	,404	,416	,426	,433	,439	,444	,448	2
3	,222	,267	,315	,341	,362	,377	,389	,399	,407	,414	3
4	,111	,200	,248	,285	,311	,331	,347	,360	,371	,381	4
5	,000	,133	,188	,230	,262	,286	,306	,323	,336	,348	5
6		,089	,139	,184	,218	,245	,268	,287	,303	,317	6
7		,044	,097	,141	,177	,207	,232	,253	,271	,286	7
8		,022	,067	,107	,142	,172	,198	,221	,240	,257	8
9		,000	,042	,077	,111	,141	,168	,191	,212	,230	9
10			,024	,055	,085	,114	,140	,164	,185	,204	10
11			,012	,036	,064	,091	,116	,139	,161	,180	11
12			,006	,024	,047	,071	,095	,117	,138	,158	12
13			,000	,014	,033	,054	,076	,097	,118	,137	13
14				,008	,023	,041	,060	,080	,100	,118	14
15				,004	,015	,030	,047	,065	,084	,102	15
16				,002	,009	,021	,036	,052	,069	,086	16
17				,000	,005	,015	,027	,041	,057	,073	17
18					,003	,010	,020	,032	,046	,061	18
19					,002	,006	,014	,025	,037	,051	19
20					,001	,004	,010	,019	,030	,042	20

TABLE E   Wilcoxon's Rank Sum   (*Continued*)

Probability that Wilcoxon's rank sum minus the expected value
will be equal to or larger than W

					n = 8					
W	m=1	m=2	m=3	m=4	m=5	m=6	m=7	m=8	m=9	m=10	W
21					,000	,002	,007	,014	,023	,034	21
22						,001	,005	,010	,018	,027	22
23						,001	,003	,007	,014	,022	23
24						,000	,002	,005	,010	,017	24
25							,001	,003	,008	,013	25
26							,001	,002	,006	,010	26
27							,000	,001	,004	,008	27
28								,001	,003	,006	28
29								,001	,002	,004	29
30								,000	,001	,003	30
31									,001	,002	31
32									,000	,002	32
33										,001	33
34										,001	34
35										,000	35

					n = 8					
W	m=11	m=12	m=13	m=14	m=15	m=16	m=17	m=18	m=19	m=20	W
1	,484	,485	,486	,487	,487	,488	,489	,489	,490	,490	1
2	,452	,455	,458	,460	,462	,464	,466	,468	,469	,470	2
3	,420	,425	,430	,434	,437	,440	,443	,446	,448	,450	3
4	,389	,396	,402	,408	,413	,417	,421	,425	,428	,431	4
5	,358	,367	,375	,382	,388	,394	,399	,403	,408	,411	5
6	,329	,339	,348	,357	,364	,371	,377	,382	,387	,392	6
7	,300	,312	,323	,332	,341	,348	,355	,362	,368	,373	7
8	,272	,286	,298	,308	,318	,326	,334	,342	,348	,354	8
9	,246	,260	,273	,285	,295	,305	,314	,322	,329	,336	9
10	,221	,236	,250	,263	,274	,284	,294	,302	,310	,318	10
11	,198	,213	,228	,241	,253	,264	,274	,284	,292	,300	11
12	,176	,192	,207	,221	,233	,245	,255	,265	,275	,283	12
13	,155	,172	,187	,201	,214	,226	,237	,248	,257	,266	13
14	,136	,153	,168	,183	,196	,208	,220	,231	,241	,250	14
15	,119	,135	,151	,165	,179	,191	,203	,214	,225	,234	15
16	,103	,119	,134	,149	,162	,175	,187	,199	,209	,219	16
17	,089	,104	,119	,133	,147	,160	,172	,184	,194	,205	17
18	,076	,091	,105	,119	,133	,145	,158	,169	,180	,191	18
19	,064	,078	,092	,106	,119	,132	,144	,156	,167	,177	19
20	,054	,067	,081	,094	,107	,119	,131	,143	,154	,164	20
21	,045	,058	,070	,083	,095	,107	,119	,130	,141	,152	21

TABLE E Wilcoxon's Rank Sum (*Continued*)

Probability that Wilcoxon's rank sum minus the expected value
will be equal to or larger than W

					n = 8						
W	m=11	m=12	m=13	m=14	m=15	m=16	m=17	m=18	m=19	m=20	W
22	,038	,049	,061	,073	,084	,096	,108	,119	,130	,140	22
23	,031	,041	,052	,063	,075	,086	,097	,108	,119	,129	23
24	,025	,035	,045	,055	,066	,077	,087	,098	,109	,119	24
25	,020	,029	,038	,048	,058	,068	,078	,089	,099	,109	25
26	,016	,024	,032	,041	,050	,060	,070	,080	,090	,100	26
27	,013	,019	,027	,035	,044	,053	,062	,072	,081	,091	27
28	,010	,016	,022	,030	,038	,046	,055	,064	,073	,083	28
29	,008	,013	,018	,025	,032	,040	,049	,057	,066	,075	29
30	,006	,010	,015	,021	,028	,035	,043	,051	,059	,068	30
31	,005	,008	,012	,018	,024	,030	,038	,045	,053	,061	31
32	,003	,006	,010	,015	,020	,026	,033	,040	,047	,055	32
33	,002	,005	,008	,012	,017	,022	,029	,035	,042	,050	33
34	,002	,004	,006	,010	,014	,019	,025	,031	,037	,044	34
35	,001	,003	,005	,008	,012	,016	,021	,027	,033	,040	35
36	,001	,002	,004	,006	,010	,014	,018	,023	,029	,035	36
37	,001	,001	,003	,005	,008	,011	,016	,020	,026	,031	37
38	,000	,001	,002	,004	,006	,010	,013	,018	,022	,028	38
39		,001	,002	,003	,005	,008	,011	,015	,020	,024	39
40		,001	,001	,002	,004	,007	,009	,013	,017	,021	40
41		,000	,001	,002	,003	,005	,008	,011	,015	,019	41
42			,001	,001	,003	,004	,007	,009	,013	,016	42
43			,000	,001	,002	,004	,005	,008	,011	,014	43
44				,001	,002	,003	,005	,007	,009	,012	44
45				,001	,001	,002	,004	,006	,008	,011	45
46				,000	,001	,002	,003	,005	,007	,009	46
47					,001	,001	,002	,004	,006	,008	47
48					,001	,001	,002	,003	,005	,007	48
49					,000	,001	,002	,003	,004	,006	49
50						,001	,001	,002	,003	,005	50
51						,000	,001	,002	,003	,004	51
52							,001	,001	,002	,003	52
53							,001	,001	,002	,003	53
54							,000	,001	,001	,002	54
55								,001	,001	,002	55
56								,001	,001	,002	56
57								,000	,001	,001	57
58									,001	,001	58
59									,000	,001	59
60										,001	60
61										,001	61
62										,000	62

TABLE E Wilcoxon's Rank Sum (*Continued*)

Probability that Wilcoxon's rank sum minus the expected value
will be equal to or larger than W

W	m=1	m=3	m=5	m=7	m=9	m=11	m=13	m=15	m=17	m=19	W
					n = 9						
0,5	,500	,500	,500	,500	,500	,500	,500	,500	,500	,500	0,5
1,5	,400	,432	,449	,459	,466	,470	,474	,477	,479	,481	1,5
2,5	,300	,364	,399	,419	,432	,441	,448	,454	,458	,462	2,5
3,5	,200	,300	,350	,379	,398	,412	,422	,430	,437	,442	3,5
4,5	,100	,241	,303	,340	,365	,383	,397	,408	,416	,423	4,5
5,5	,000	,186	,259	,303	,333	,355	,372	,385	,396	,405	5,5
6,5		,141	,219	,268	,302	,328	,347	,363	,376	,386	6,5
7,5		,105	,182	,235	,273	,301	,323	,341	,356	,368	7,5
8,5		,073	,149	,204	,245	,276	,300	,320	,336	,350	8,5
9,5		,050	,120	,176	,218	,251	,278	,299	,317	,332	9,5
10,5		,032	,095	,150	,193	,228	,256	,279	,298	,315	10,5
11,5		,018	,073	,126	,170	,206	,235	,260	,280	,298	11,5
12,5		,009	,056	,105	,149	,185	,216	,241	,263	,281	12,5
13,5		,005	,041	,087	,129	,166	,197	,223	,245	,265	13,5
14,5		,000	,030	,071	,111	,147	,179	,206	,229	,249	14,5
15,5			,021	,057	,095	,130	,162	,189	,213	,234	15,5
16,5			,014	,045	,081	,115	,146	,174	,198	,219	16,5
17,5			,009	,036	,068	,101	,131	,159	,183	,205	17,5
18,5			,006	,027	,057	,088	,117	,145	,169	,191	18,5
19,5			,003	,021	,047	,076	,105	,132	,156	,178	19,5
20,5			,002	,016	,039	,065	,093	,119	,144	,166	20,5
21,5			,001	,011	,031	,056	,082	,108	,132	,154	21,5
22,5			,000	,008	,025	,048	,072	,097	,120	,143	22,5
23,5				,006	,020	,040	,063	,087	,110	,132	23,5
24,5				,004	,016	,034	,055	,078	,100	,121	24,5
25,5				,003	,012	,028	,048	,069	,091	,112	25,5
26,5				,002	,009	,023	,041	,061	,082	,103	26,5
27,5				,001	,007	,019	,035	,054	,074	,094	27,5
28,5				,001	,005	,016	,030	,048	,067	,086	28,5
29,5				,000	,004	,013	,026	,042	,060	,078	29,5
30,5					,003	,010	,022	,037	,053	,071	30,5
31,5					,002	,008	,018	,032	,048	,065	31,5
32,5					,001	,006	,015	,028	,042	,058	32,5

TABLE E   Wilcoxon's Rank Sum   (*Continued*)

Probability that Wilcoxon's rank sum minus the expected value
will be equal to or larger than W

| | | | | | n = 9 | | | | | |
W	m=1	m=3	m=5	m=7	m=9	m=11	m=13	m=15	m=17	m=19	W
33.5					.001	.005	.013	.024	.037	.053	33.5
34.5					.001	.004	.010	.020	.033	.047	34.5
35.5					.000	.003	.008	.017	.029	.043	35.5
36.5						.002	.007	.015	.026	.038	36.5
37.5						.002	.006	.013	.022	.034	37.5
38.5						.001	.004	.011	.019	.030	38.5
39.5						.001	.004	.009	.017	.027	39.5
40.5						.001	.003	.007	.015	.024	40.5
41.5						.000	.002	.006	.012	.021	41.5
42.5							.002	.005	.011	.019	42.5
43.5							.001	.004	.009	.016	43.5
44.5							.001	.003	.008	.014	44.5
45.5							.001	.003	.007	.012	45.5
46.5							.001	.002	.006	.011	46.5
47.5							.000	.002	.005	.009	47.5
48.5								.001	.004	.008	48.5
49.5								.001	.003	.007	49.5
50.5								.001	.003	.006	50.5
51.5								.001	.002	.005	51.5
52.5								.000	.002	.004	52.5
53.5									.001	.004	53.5
54.5									.001	.003	54.5
55.5									.001	.003	55.5
56.5									.001	.002	56.5
57.5									.001	.002	57.5
58.5									.000	.001	58.5
59.5										.001	59.5
60.5										.001	60.5
61.5										.001	61.5
62.5										.001	62.5
63.5										.001	63.5
64.5										.000	64.5

| | | | | | n = 9 | | | | | |
W	m=2	m=4	m=6	m=8	m=10	m=12	m=14	m=16	m=18	m=20	W
1	.455	.470	.477	.481	.484	.486	.488	.489	.490	.491	1
2	.364	.413	.432	.444	.452	.459	.463	.467	.470	.472	2
3	.291	.355	.388	.407	.421	.431	.439	.445	.450	.454	3
4	.218	.302	.344	.371	.390	.404	.415	.423	.430	.436	4
5	.164	.252	.303	.336	.360	.377	.391	.401	.410	.418	5

TABLE E Wilcoxon's Rank Sum (*Continued*)

Probability that Wilcoxon's rank sum minus the expected value
will be equal to or larger than W

					n =	9					
W	m=2	m=4	m=6	m=8	m=10	m=12	m=14	m=16	m=18	m=20	W
6	,109	,207	,264	,303	,330	,351	,367	,380	,391	,400	6
7	,073	,165	,228	,271	,302	,326	,344	,359	,371	,382	7
8	,036	,130	,194	,240	,274	,301	,322	,339	,353	,364	8
9	,018	,099	,164	,212	,248	,277	,300	,318	,334	,347	9
10	,000	,074	,136	,185	,223	,254	,279	,299	,316	,330	10
11		,053	,112	,161	,200	,232	,258	,280	,298	,313	11
12		,038	,091	,138	,178	,211	,238	,261	,281	,297	12
13		,025	,072	,118	,158	,191	,219	,243	,264	,281	13
14		,017	,057	,100	,139	,173	,201	,226	,247	,266	14
15		,010	,044	,084	,121	,155	,184	,210	,232	,251	15
16		,006	,033	,069	,106	,139	,168	,194	,216	,236	16
17		,003	,025	,057	,091	,123	,153	,179	,202	,222	17
18		,001	,018	,046	,078	,109	,138	,164	,188	,208	18
19		,000	,013	,037	,067	,097	,125	,151	,174	,195	19
20			,009	,030	,056	,085	,112	,138	,161	,182	20
21			,006	,023	,047	,074	,101	,126	,149	,170	21
22			,004	,018	,039	,064	,090	,114	,137	,159	22
23			,002	,014	,033	,056	,080	,104	,126	,147	23
24			,001	,010	,027	,048	,071	,094	,116	,137	24
25			,001	,008	,022	,041	,062	,084	,106	,127	25
26			,000	,006	,017	,035	,055	,076	,097	,117	26
27				,004	,014	,029	,048	,068	,088	,108	27
28				,003	,011	,025	,042	,061	,080	,099	28
29				,002	,009	,020	,036	,054	,073	,091	29
30				,001	,007	,017	,031	,048	,066	,084	30
31				,001	,005	,014	,027	,042	,059	,077	31
32				,000	,004	,011	,023	,037	,053	,070	32
33					,003	,009	,019	,032	,048	,064	33
34					,002	,007	,016	,028	,042	,058	34
35					,001	,006	,014	,025	,038	,052	35
36					,001	,005	,012	,021	,034	,047	36
37					,001	,004	,010	,018	,030	,043	37
38					,000	,003	,008	,016	,026	,038	38
39						,002	,006	,014	,023	,034	39
40						,002	,005	,012	,020	,031	40
41						,001	,004	,010	,018	,028	41
42						,001	,003	,008	,015	,025	42

TABLE E Wilcoxon's Rank Sum   (*Continued*)

Probability that Wilcoxon's rank sum minus the expected value
will be equal to or larger than W

n = 9

W	m=2	m=4	m=6	m=8	m=10	m=12	m=14	m=16	m=18	m=20	W
43						,001	,003	,007	,013	,022	43
44						,000	,002	,006	,012	,019	44
45							,002	,005	,010	,017	45
46							,001	,004	,009	,015	46
47							,001	,003	,007	,013	47
48							,001	,003	,006	,012	48
49							,001	,002	,005	,010	49
50							,000	,002	,004	,009	50
51								,001	,004	,008	51
52								,001	,003	,007	52
53								,001	,003	,006	53
54								,001	,002	,005	54
55								,001	,002	,004	55
56								,000	,001	,004	56
57									,001	,003	57
58									,001	,003	58
59									,001	,002	59
60									,001	,002	60
61									,000	,001	61
62										,001	62
63										,001	63
64										,001	64
65										,001	65
66										,001	66
67										,000	67

n = 10

W	m=1	m=2	m=3	m=4	m=5	m=6	m=7	m=8	m=9	m=10	W
1	,455	,455	,469	,473	,477	,479	,481	,483	,484	,485	1
2	,364	,379	,406	,420	,430	,437	,443	,448	,452	,456	2
3	,273	,303	,346	,367	,384	,396	,406	,414	,421	,427	3
4	,182	,242	,287	,318	,339	,356	,370	,381	,390	,398	4
5	,091	,182	,234	,270	,297	,318	,335	,348	,360	,370	5
6	,000	,136	,185	,227	,257	,281	,300	,317	,330	,342	6
7		,091	,143	,187	,220	,246	,268	,286	,302	,315	7
8		,061	,108	,152	,185	,214	,237	,257	,274	,289	8
9		,030	,080	,120	,155	,184	,209	,230	,248	,264	9
10		,015	,056	,094	,127	,157	,182	,204	,223	,241	10
11		,000	,038	,071	,103	,132	,157	,180	,200	,218	11

TABLE E Wilcoxon's Rank Sum *(Continued)*

Probability that Wilcoxon's rank sum minus the expected value
will be equal to or larger than W

					n = 10					
W	m=1	m=2	m=3	m=4	m=5	m=6	m=7	m=8	m=9	m=10	W
12			,024	,053	,082	,110	,135	,158	,178	,197	12
13			,014	,038	,065	,090	,115	,137	,158	,176	13
14			,007	,027	,050	,074	,097	,118	,139	,157	14
15			,003	,018	,038	,059	,081	,102	,121	,140	15
16			,000	,012	,028	,047	,067	,086	,106	,124	16
17				,007	,020	,036	,054	,073	,091	,109	17
18				,004	,014	,028	,044	,061	,078	,095	18
19				,002	,010	,021	,035	,051	,067	,083	19
20				,001	,006	,016	,028	,042	,056	,072	20
21				,000	,004	,011	,022	,034	,047	,062	21
22					,002	,008	,017	,027	,039	,053	22
23					,001	,005	,012	,022	,033	,045	23
24					,001	,004	,009	,017	,027	,038	24
25					,000	,002	,007	,013	,022	,032	25
26						,001	,005	,010	,017	,026	26
27						,001	,003	,008	,014	,022	27
28						,000	,002	,006	,011	,018	28
29							,002	,004	,009	,014	29
30							,001	,003	,007	,012	30
31							,001	,002	,005	,009	31
32							,000	,002	,004	,007	32
33								,001	,003	,006	33
34								,001	,002	,004	34
35								,000	,001	,003	35
36									,001	,003	36
37									,001	,002	37
38									,000	,001	38
39										,001	39
40										,001	40
41										,001	41
42										,000	42

					n = 10					
W	m=11	m=12	m=13	m=14	m=15	m=16	m=17	m=18	m=19	m=20	W
1	,486	,487	,488	,489	,489	,490	,490	,491	,491	,491	1
2	,459	,461	,464	,466	,467	,469	,471	,472	,473	,474	2
3	,432	,436	,440	,443	,446	,449	,451	,453	,455	,457	3
4	,405	,411	,416	,420	,424	,428	,432	,435	,437	,440	4
5	,378	,386	,392	,398	,403	,408	,412	,416	,420	,423	5

TABLE E   Wilcoxon's Rank Sum   (*Continued*)

```
Probability that Wilcoxon's rank sum minus the expected value
             will be equal to or larger than W
```

					n = 10						
W	m=11	m=12	m=13	m=14	m=15	m=16	m=17	m=18	m=19	m=20	W
6	,352	,361	,369	,376	,382	,388	,393	,398	,402	,406	6
7	,327	,337	,346	,354	,362	,368	,374	,380	,385	,389	7
8	,302	,314	,324	,333	,341	,349	,356	,362	,368	,373	8
9	,279	,291	,302	,313	,322	,330	,337	,344	,351	,357	9
10	,256	,269	,281	,292	,302	,311	,320	,327	,334	,341	10
11	,234	,248	,261	,273	,284	,293	,302	,310	,318	,325	11
12	,213	,228	,242	,254	,265	,276	,285	,294	,302	,309	12
13	,193	,209	,223	,236	,248	,258	,268	,278	,286	,294	13
14	,175	,190	,205	,218	,231	,242	,252	,262	,271	,279	14
15	,157	,173	,188	,202	,214	,226	,237	,247	,256	,265	15
16	,141	,157	,172	,186	,198	,210	,222	,232	,242	,251	16
17	,126	,141	,156	,170	,183	,196	,207	,218	,228	,237	17
18	,112	,127	,142	,156	,169	,182	,193	,204	,214	,224	18
19	,099	,114	,128	,142	,156	,168	,180	,191	,201	,211	19
20	,087	,101	,116	,130	,143	,155	,167	,178	,189	,199	20
21	,076	,090	,104	,117	,130	,143	,155	,166	,177	,187	21
22	,066	,080	,093	,106	,119	,131	,143	,154	,165	,175	22
23	,057	,070	,083	,096	,108	,120	,132	,143	,154	,164	23
24	,049	,061	,074	,086	,098	,110	,121	,133	,143	,153	24
25	,042	,054	,065	,077	,089	,100	,112	,122	,133	,143	25
26	,036	,047	,058	,069	,080	,091	,102	,113	,123	,134	26
27	,031	,040	,051	,061	,072	,083	,093	,104	,114	,124	27
28	,026	,035	,044	,054	,064	,075	,085	,095	,106	,115	28
29	,021	,030	,038	,048	,058	,067	,077	,087	,097	,107	29
30	,018	,025	,033	,042	,051	,061	,070	,080	,090	,099	30
31	,015	,021	,029	,037	,045	,054	,064	,073	,082	,091	31
32	,012	,018	,025	,032	,040	,049	,057	,066	,075	,084	32
33	,010	,015	,021	,028	,035	,043	,052	,060	,069	,078	33
34	,008	,012	,018	,024	,031	,039	,046	,055	,063	,071	34
35	,006	,010	,015	,021	,027	,034	,042	,049	,057	,065	35
36	,005	,008	,013	,018	,024	,030	,037	,044	,052	,060	36
37	,004	,007	,011	,015	,021	,027	,033	,040	,047	,055	37
38	,003	,006	,009	,013	,018	,023	,029	,036	,043	,050	38
39	,002	,004	,007	,011	,015	,020	,026	,032	,039	,045	39
40	,002	,004	,006	,009	,013	,018	,023	,029	,035	,041	40

TABLE E   Wilcoxon's Rank Sum   (*Continued*)

Probability that Wilcoxon's rank sum minus the expected value
will be equal to or larger than W

					n = 10						
W	m=11	m=12	m=13	m=14	m=15	m=16	m=17	m=18	m=19	m=20	W
41	.001	.003	.005	.008	.011	.015	.020	.025	.031	.037	41
42	.001	.002	.004	.006	.010	.013	.018	.023	.028	.034	42
43	.001	.002	.003	.005	.008	.012	.016	.020	.025	.030	43
44	.001	.001	.003	.004	.007	.010	.014	.018	.022	.027	44
45	.000	.001	.002	.004	.006	.008	.012	.016	.020	.025	45
46		.001	.002	.003	.005	.007	.010	.014	.018	.022	46
47		.001	.001	.002	.004	.006	.009	.012	.016	.020	47
48		.000	.001	.002	.003	.005	.008	.010	.014	.017	48
49			.001	.002	.003	.004	.006	.009	.012	.016	49
50			.001	.001	.002	.004	.005	.008	.011	.014	50
51			.000	.001	.002	.003	.005	.007	.009	.012	51
52				.001	.001	.003	.004	.006	.008	.011	52
53				.001	.001	.002	.003	.005	.007	.009	53
54				.000	.001	.002	.003	.004	.006	.008	54
55					.001	.001	.002	.004	.005	.007	55
56					.001	.001	.002	.003	.005	.006	56
57					.000	.001	.002	.003	.004	.006	57
58						.001	.001	.002	.003	.005	58
59						.001	.001	.002	.003	.004	59
60						.000	.001	.002	.002	.004	60
61							.001	.001	.002	.003	61
62							.001	.001	.002	.003	62
63							.000	.001	.001	.002	63
64								.001	.001	.002	64
65								.001	.001	.002	65
66								.000	.001	.001	66
67									.001	.001	67
68									.001	.001	68
69									.000	.001	69
70										.001	70
71										.001	71
72										.000	72

					n = 11					
W	m=1	m=3	m=5	m=7	m=9	m=11	m=13	m=15	m=17	m=19	W
0,5	.500	.500	.500	.500	.500	.500	.500	.500	.500	.500	0,5
1,5	.417	.442	.457	.465	.470	.474	.477	.480	.482	.483	1,5
2,5	.333	.385	.413	.430	.441	.449	.455	.459	.463	.466	2,5
3,5	.250	.330	.371	.396	.412	.423	.432	.439	.445	.449	3,5
4,5	.167	.277	.331	.362	.383	.398	.410	.419	.427	.433	4,5
5,5	.083	.228	.292	.329	.355	.374	.388	.399	.409	.416	5,5
6,5	.000	.184	.255	.298	.328	.350	.366	.380	.391	.400	6,5

TABLE E   Wilcoxon's Rank Sum   (*Continued*)

Probability that Wilcoxon's rank sum minus the expected value
will be equal to or larger than W

					n = 11					
W	m=1	m=3	m=5	m=7	m=9	m=11	m=13	m=15	m=17	m=19	W
7.5		.146	.220	.268	.301	.326	.345	.361	.373	.383	7.5
8.5		.113	.189	.239	.276	.303	.324	.342	.356	.367	8.5
9.5		.085	.160	.213	.251	.281	.304	.323	.339	.352	9.5
10.5		.063	.134	.187	.228	.260	.285	.305	.322	.336	10.5
11.5		.044	.111	.164	.206	.239	.265	.287	.305	.321	11.5
12.5		.030	.090	.143	.185	.219	.247	.270	.289	.306	12.5
13.5		.019	.073	.123	.166	.200	.229	.253	.274	.291	13.5
14.5		.011	.057	.105	.147	.183	.212	.237	.258	.276	14.5
15.5		.005	.045	.090	.130	.166	.196	.222	.243	.262	15.5
16.5		.003	.034	.075	.115	.150	.180	.206	.229	.249	16.5
17.5		.000	.026	.063	.101	.135	.166	.192	.215	.235	17.5
18.5			.019	.052	.088	.121	.152	.178	.202	.222	18.5
19.5			.014	.043	.076	.108	.138	.165	.189	.210	19.5
20.5			.010	.035	.065	.097	.126	.153	.176	.198	20.5
21.5			.007	.028	.056	.086	.114	.141	.165	.186	21.5
22.5			.004	.022	.048	.076	.103	.129	.153	.175	22.5
23.5			.003	.017	.040	.066	.093	.119	.142	.164	23.5
24.5			.002	.013	.034	.058	.084	.109	.132	.154	24.5
25.5			.001	.010	.028	.051	.075	.099	.122	.144	25.5
26.5			.000	.008	.023	.044	.067	.090	.113	.134	26.5
27.5				.006	.019	.038	.060	.082	.104	.125	27.5
28.5				.004	.016	.033	.053	.074	.096	.116	28.5
29.5				.003	.013	.028	.047	.067	.088	.108	29.5
30.5				.002	.010	.024	.041	.060	.080	.100	30.5
31.5				.001	.008	.020	.036	.054	.073	.093	31.5
32.5				.001	.006	.017	.031	.049	.067	.086	32.5
33.5				.001	.005	.014	.027	.043	.061	.079	33.5
34.5				.000	.004	.012	.024	.039	.055	.073	34.5
35.5					.003	.010	.020	.034	.050	.067	35.5
36.5					.002	.008	.018	.030	.045	.061	36.5
37.5					.002	.006	.015	.027	.041	.056	37.5
38.5					.001	.005	.013	.024	.037	.051	38.5
39.5					.001	.004	.011	.021	.033	.047	39.5
40.5					.001	.003	.009	.018	.030	.043	40.5
41.5					.000	.003	.008	.016	.026	.039	41.5
42.5						.002	.006	.014	.023	.035	42.5
43.5						.002	.005	.012	.021	.032	43.5
44.5						.001	.004	.010	.018	.029	44.5
45.5						.001	.004	.009	.016	.026	45.5
46.5						.001	.003	.007	.014	.023	46.5
47.5						.001	.002	.006	.013	.021	47.5
48.5						.000	.002	.005	.011	.019	48.5
49.5							.002	.005	.010	.017	49.5
50.5							.001	.004	.008	.015	50.5
51.5							.001	.003	.007	.013	51.5

TABLE E Wilcoxon's Rank Sum (*Continued*)

Probability that Wilcoxon's rank sum minus the expected value
will be equal to or larger than W

W	m=1	m=3	m=5	m=7	m=9	m=11	m=13	m=15	m=17	m=19	W
					n = 11						
52,5							,001	,003	,006	,012	52,5
53,5							,001	,002	,005	,010	53,5
54,5							,000	,002	,005	,009	54,5
55,5								,002	,004	,008	55,5
56,5								,001	,003	,007	56,5
57,5								,001	,003	,006	57,5
58,5								,001	,002	,005	58,5
59,5								,001	,002	,005	59,5
60,5								,001	,002	,004	60,5
61,5								,000	,001	,004	61,5
62,5									,001	,003	62,5
63,5									,001	,003	63,5
64,5									,001	,002	64,5
65,5									,001	,002	65,5
66,5									,001	,002	66,5
67,5									,000	,001	67,5
68,5										,001	68,5
69,5										,001	69,5
70,5										,001	70,5
71,5										,001	71,5
72,5										,001	72,5
73,5										,000	73,5

W	m=2	m=4	m=6	m=8	m=10	m=12	m=14	m=16	m=18	m=20	W
					n = 11						
1	,462	,475	,481	,484	,486	,488	,489	,490	,491	,492	1
2	,385	,426	,442	,452	,459	,464	,468	,471	,474	,476	2
3	,321	,377	,404	,420	,432	,440	,447	,452	,456	,460	3
4	,256	,330	,366	,389	,405	,416	,425	,433	,439	,443	4
5	,205	,286	,330	,358	,378	,393	,404	,414	,421	,427	5
6	,154	,245	,295	,329	,352	,370	,384	,395	,404	,412	6
7	,115	,206	,262	,300	,327	,347	,363	,376	,387	,396	7
8	,077	,171	,231	,272	,302	,325	,343	,358	,370	,380	8
9	,051	,140	,202	,246	,279	,304	,324	,340	,354	,365	9

TABLE E Wilcoxon's Rank Sum (*Continued*)

Probability that Wilcoxon's rank sum minus the expected value
will be equal to or larger than W

n = 11

W	m=2	m=4	m=6	m=8	m=10	m=12	m=14	m=16	m=18	m=20	W
10	.026	.113	.175	.221	.256	.283	.305	.322	.337	.350	10
11	.013	.089	.151	.198	.234	.263	.286	.305	.321	.335	11
12	.000	.069	.128	.176	.213	.243	.268	.288	.305	.320	12
13		.052	.108	.155	.193	.225	.250	.272	.290	.306	13
14		.039	.090	.136	.175	.207	.233	.256	.275	.291	14
15		.028	.074	.119	.157	.190	.217	.240	.260	.278	15
16		.020	.061	.103	.141	.173	.201	.225	.246	.264	16
17		.013	.049	.089	.126	.158	.187	.211	.232	.251	17
18		.009	.039	.076	.112	.144	.172	.197	.219	.238	18
19		.005	.031	.064	.099	.130	.159	.184	.206	.226	19
20		.003	.024	.054	.087	.118	.146	.171	.193	.213	20
21		.001	.018	.045	.076	.106	.134	.159	.181	.202	21
22		.001	.014	.038	.066	.095	.122	.147	.170	.190	22
23		.000	.010	.031	.057	.085	.111	.136	.159	.179	23
24			.007	.025	.049	.075	.101	.125	.148	.169	24
25			.005	.020	.042	.067	.092	.116	.138	.159	25
26			.004	.016	.036	.059	.083	.106	.128	.149	26
27			.002	.013	.031	.052	.075	.097	.119	.139	27
28			.002	.010	.026	.045	.067	.089	.110	.130	28
29			.001	.008	.021	.040	.060	.081	.102	.122	29
30			.001	.006	.018	.034	.054	.074	.094	.114	30
31			.000	.005	.015	.030	.048	.067	.087	.106	31
32				.003	.012	.026	.042	.061	.080	.098	32
33				.002	.010	.022	.037	.055	.073	.091	33
34				.002	.008	.019	.033	.049	.067	.085	34
35				.001	.006	.016	.029	.044	.061	.078	35
36				.001	.005	.013	.025	.040	.056	.072	36
37				.001	.004	.011	.022	.036	.051	.067	37
38				.000	.003	.009	.019	.032	.046	.061	38
39					.002	.008	.017	.028	.042	.056	39
40					.002	.006	.014	.025	.038	.052	40
41					.001	.005	.012	.022	.034	.047	41
42					.001	.004	.011	.020	.031	.043	42
43					.001	.003	.009	.017	.028	.040	43
44					.001	.003	.008	.015	.025	.036	44
45					.000	.002	.006	.013	.022	.033	45
46						.002	.005	.011	.020	.030	46
47						.001	.005	.010	.018	.027	47
48						.001	.004	.009	.016	.024	48

TABLE E   Wilcoxon's Rank Sum   (*Continued*)

Probability that Wilcoxon's rank sum minus the expected value
will be equal to or larger than W

n = 11

W	m=2	m=4	m=6	m=8	m=10	m=12	m=14	m=16	m=18	m=20	W
49						.001	.003	.007	.014	.022	49
50						.001	.003	.006	.012	.020	50
51						.000	.002	.005	.011	.018	51
52							.002	.005	.009	.016	52
53							.001	.004	.008	.014	53
54							.001	.003	.007	.013	54
55							.001	.003	.006	.011	55
56							.001	.002	.005	.010	56
57							.001	.002	.005	.009	57
58							.000	.002	.004	.008	58
59								.001	.003	.007	59
60								.001	.003	.006	60
61								.001	.003	.005	61
62								.001	.002	.005	62
63								.001	.002	.004	63
64								.000	.002	.004	64
65									.001	.003	65
66									.001	.003	66
67									.001	.002	67
68									.001	.002	68
69									.001	.002	69
70									.001	.001	70
71									.000	.001	71
72										.001	72
73										.001	73
74										.001	74
75										.001	75
76										.001	76
77										.000	77

n = 12

W	m=1	m=2	m=3	m=4	m=5	m=6	m=7	m=8	m=9	m=10	W
1	.462	.462	.473	.476	.480	.482	.484	.485	.486	.487	1
2	.385	.396	.420	.431	.439	.446	.451	.455	.459	.461	2
3	.308	.330	.367	.385	.399	.410	.418	.425	.431	.436	3
4	.231	.275	.316	.342	.361	.375	.387	.396	.404	.411	4
5	.154	.220	.268	.299	.323	.341	.355	.367	.377	.386	5
6	.077	.176	.224	.260	.287	.308	.325	.339	.351	.361	6
7	.000	.132	.182	.223	.253	.277	.296	.312	.326	.337	7

TABLE E   Wilcoxon's Rank Sum   (*Continued*)

Probability that Wilcoxon's rank sum minus the expected value
will be equal to or larger than W

						n = 12					
W	m=1	m=2	m=3	m=4	m=5	m=6	m=7	m=8	m=9	m=10	W
8		,099	,147	,190	,221	,247	,268	,286	,301	,314	8
9		,066	,116	,158	,191	,219	,241	,260	,277	,291	9
10		,044	,090	,131	,164	,192	,216	,236	,254	,269	10
11		,022	,068	,106	,139	,168	,192	,213	,232	,248	11
12		,011	,051	,085	,117	,145	,170	,192	,211	,228	12
13		,000	,035	,066	,097	,125	,150	,172	,191	,209	13
14			,024	,052	,080	,106	,131	,153	,173	,190	14
15			,015	,039	,065	,090	,113	,135	,155	,173	15
16			,009	,029	,052	,075	,098	,119	,139	,157	16
17			,004	,021	,041	,062	,084	,104	,123	,141	17
18			,002	,015	,032	,051	,071	,091	,109	,127	18
19			,000	,010	,024	,042	,060	,078	,097	,114	19
20				,007	,018	,033	,050	,067	,085	,101	20
21				,004	,013	,026	,042	,058	,074	,090	21
22				,002	,010	,021	,034	,049	,064	,080	22
23				,001	,007	,016	,028	,041	,056	,070	23
24				,001	,005	,012	,022	,035	,048	,061	24
25				,000	,003	,009	,018	,029	,041	,054	25
26					,002	,007	,014	,024	,035	,047	26
27					,001	,005	,011	,019	,029	,040	27
28					,001	,003	,009	,016	,025	,035	28
29					,000	,002	,006	,013	,020	,030	29
30						,002	,005	,010	,017	,025	30
31						,001	,004	,008	,014	,021	31
32						,001	,003	,006	,011	,018	32
33						,000	,002	,005	,009	,015	33
34							,001	,004	,007	,012	34
35							,001	,003	,006	,010	35
36							,001	,002	,005	,008	36
37							,000	,001	,004	,007	37
38								,001	,003	,006	38
39								,001	,002	,004	39
40								,001	,002	,004	40
41								,000	,001	,003	41
42									,001	,002	42
43									,001	,002	43
44									,000	,001	44
45										,001	45
46										,001	46
47										,001	47
48										,000	48

TABLE E   Wilcoxon's Rank Sum   (*Continued*)

Probability that Wilcoxon's rank sum minus the expected value
will be equal to or larger than W

					n = 12						
W	m=11	m=12	m=13	m=14	m=15	m=16	m=17	m=18	m=19	m=20	W
1	.488	.489	.489	.490	.490	.491	.491	.492	.492	.492	1
2	.464	.466	.468	.470	.471	.473	.474	.475	.476	.477	2
3	.440	.444	.447	.450	.452	.455	.457	.459	.460	.462	3
4	.416	.421	.426	.430	.433	.437	.439	.442	.444	.447	4
5	.393	.399	.405	.410	.415	.419	.422	.426	.429	.432	5
6	.370	.378	.384	.391	.396	.401	.405	.409	.413	.417	6
7	.347	.356	.364	.371	.378	.383	.388	.393	.398	.402	7
8	.325	.335	.344	.352	.359	.366	.372	.377	.382	.387	8
9	.304	.315	.325	.334	.342	.349	.356	.362	.367	.372	9
10	.283	.295	.306	.315	.324	.332	.339	.346	.352	.358	10
11	.263	.276	.287	.298	.307	.316	.324	.331	.337	.344	11
12	.243	.257	.269	.280	.290	.300	.308	.316	.323	.330	12
13	.225	.239	.252	.263	.274	.284	.293	.301	.309	.316	13
14	.207	.221	.235	.247	.258	.268	.278	.287	.295	.302	14
15	.190	.205	.219	.231	.243	.254	.263	.273	.281	.289	15
16	.173	.189	.203	.216	.228	.239	.249	.259	.268	.276	16
17	.158	.174	.188	.201	.214	.225	.236	.245	.255	.263	17
18	.144	.159	.174	.187	.200	.211	.222	.232	.242	.251	18
19	.130	.146	.160	.174	.186	.198	.209	.220	.229	.239	19
20	.118	.133	.147	.161	.174	.186	.197	.207	.217	.227	20
21	.106	.121	.135	.149	.161	.174	.185	.196	.206	.215	21
22	.095	.109	.124	.137	.150	.162	.173	.184	.194	.204	22
23	.085	.099	.113	.126	.139	.151	.162	.173	.184	.193	23
24	.075	.089	.103	.116	.128	.140	.152	.163	.173	.183	24
25	.067	.080	.093	.106	.118	.130	.141	.152	.163	.173	25
26	.059	.072	.084	.097	.109	.120	.132	.143	.153	.163	26
27	.052	.064	.076	.088	.100	.111	.122	.133	.144	.154	27
28	.045	.057	.068	.080	.091	.103	.114	.124	.135	.145	28
29	.040	.050	.061	.072	.084	.095	.105	.116	.126	.136	29
30	.034	.044	.055	.065	.076	.087	.097	.108	.118	.128	30
31	.030	.039	.049	.059	.069	.080	.090	.100	.110	.120	31
32	.026	.034	.043	.053	.063	.073	.083	.093	.102	.112	32
33	.022	.030	.038	.048	.057	.066	.076	.086	.095	.105	33
34	.019	.026	.034	.042	.051	.061	.070	.079	.088	.098	34
35	.016	.022	.030	.038	.046	.055	.064	.073	.082	.091	35
36	.013	.019	.026	.034	.042	.050	.059	.067	.076	.085	36
37	.011	.017	.023	.030	.037	.045	.053	.062	.070	.079	37
38	.009	.014	.020	.026	.033	.041	.049	.057	.065	.073	38
39	.008	.012	.017	.023	.030	.037	.044	.052	.060	.068	39
40	.006	.010	.015	.020	.026	.033	.040	.047	.055	.063	40
41	.005	.009	.013	.018	.023	.030	.036	.043	.051	.058	41

Tables

TABLE E Wilcoxon's Rank Sum   (*Continued*)

Probability that Wilcoxon's rank sum minus the expected value
will be equal to or larger than W

W	m=11	m=12	m=13	m=14	m=15	m=16	m=17	m=18	m=19	m=20	W
					n = 12						
42	,004	,007	,011	,016	,021	,026	,033	,039	,046	,053	42
43	,003	,006	,009	,013	,018	,024	,030	,036	,042	,049	43
44	,003	,005	,008	,012	,016	,021	,027	,032	,039	,045	44
45	,002	,004	,007	,010	,014	,019	,024	,029	,035	,042	45
46	,002	,003	,006	,009	,012	,017	,021	,027	,032	,038	46
47	,001	,003	,005	,007	,011	,015	,019	,024	,029	,035	47
48	,001	,002	,004	,006	,009	,013	,017	,022	,027	,032	48
49	,001	,002	,003	,005	,008	,011	,015	,019	,024	,029	49
50	,001	,001	,003	,005	,007	,010	,013	,017	,022	,027	50
51	,000	,001	,002	,004	,006	,009	,012	,015	,020	,024	51
52		,001	,002	,003	,005	,008	,010	,014	,018	,022	52
53		,001	,001	,003	,004	,007	,009	,012	,016	,020	53
54		,001	,001	,002	,004	,006	,008	,011	,014	,018	54
55		,000	,001	,002	,003	,005	,007	,010	,013	,016	55
56			,001	,002	,003	,004	,006	,009	,011	,015	56
57			,001	,001	,002	,004	,005	,008	,010	,013	57
58			,000	,001	,002	,003	,005	,007	,009	,012	58
59				,001	,002	,003	,004	,006	,008	,011	59
60				,001	,001	,002	,003	,005	,007	,009	60
61				,001	,001	,002	,003	,004	,006	,008	61
62				,000	,001	,002	,003	,004	,006	,008	62
63					,001	,001	,002	,003	,005	,007	63
64					,001	,001	,002	,003	,004	,006	64
65					,000	,001	,002	,002	,004	,005	65
66						,001	,001	,002	,003	,005	66
67						,001	,001	,002	,003	,004	67
68						,001	,001	,002	,002	,004	68
69						,000	,001	,001	,002	,003	69
70							,001	,001	,002	,003	70
71							,001	,001	,002	,002	71
72							,000	,001	,001	,002	72
73								,001	,001	,002	73
74								,001	,001	,002	74
75								,000	,001	,001	75
76									,001	,001	76
77									,001	,001	77
78									,001	,001	78
79									,000	,001	79
80										,001	80
81										,001	81
82										,000	82

TABLE E   Wilcoxon's Rank Sum   (*Continued*)

Probability that Wilcoxon's rank sum minus the expected value
will be equal to or larger than W

W	m=1	m=3	m=5	m=7	m=9	m=11	m=13	m=15	m=17	m=19	W
					n = 13						
0.5	.500	.500	.500	.500	.500	.500	.500	.500	.500	.500	0.5
1.5	.429	.450	.462	.469	.474	.477	.480	.482	.484	.485	1.5
2.5	.357	.400	.424	.439	.448	.455	.460	.464	.467	.470	2.5
3.5	.286	.352	.387	.408	.422	.432	.440	.446	.451	.455	3.5
4.5	.214	.305	.351	.379	.397	.410	.420	.428	.435	.440	4.5
5.5	.143	.261	.317	.350	.372	.388	.401	.410	.418	.425	5.5
6.5	.071	.220	.283	.321	.347	.366	.381	.393	.402	.410	6.5
7.5	.000	.182	.251	.294	.323	.345	.362	.375	.386	.396	7.5
8.5		.148	.222	.268	.300	.324	.343	.358	.371	.381	8.5
9.5		.120	.194	.243	.278	.304	.325	.342	.355	.367	9.5
10.5		.095	.168	.219	.256	.285	.307	.325	.340	.353	10.5
11.5		.073	.144	.196	.235	.265	.289	.309	.325	.339	11.5
12.5		.055	.123	.175	.216	.247	.272	.293	.310	.325	12.5
13.5		.041	.104	.156	.197	.229	.256	.278	.296	.311	13.5
14.5		.029	.087	.137	.179	.212	.240	.262	.282	.298	14.5
15.5		.020	.072	.121	.162	.196	.224	.248	.268	.285	15.5
16.5		.013	.059	.105	.146	.180	.209	.234	.254	.272	16.5
17.5		.007	.047	.091	.131	.166	.195	.220	.241	.260	17.5
18.5		.004	.038	.079	.117	.152	.181	.207	.229	.248	18.5
19.5		.002	.030	.067	.105	.138	.168	.194	.216	.236	19.5
20.5		.000	.023	.057	.093	.126	.155	.181	.204	.224	20.5
21.5			.018	.048	.082	.114	.143	.169	.193	.213	21.5
22.5			.013	.041	.072	.103	.132	.158	.181	.202	22.5
23.5			.010	.034	.063	.093	.121	.147	.171	.191	23.5
24.5			.007	.028	.055	.084	.111	.137	.160	.181	24.5
25.5			.005	.023	.048	.075	.102	.127	.150	.171	25.5
26.5			.003	.018	.041	.067	.093	.118	.141	.162	26.5
27.5			.002	.015	.035	.060	.084	.109	.131	.152	27.5
28.5			.001	.012	.030	.053	.077	.100	.123	.144	28.5
29.5			.001	.009	.026	.047	.069	.092	.114	.135	29.5
30.5			.000	.007	.022	.041	.063	.085	.106	.127	30.5
31.5				.006	.018	.036	.056	.078	.099	.119	31.5
32.5				.004	.015	.031	.051	.071	.092	.112	32.5
33.5				.003	.013	.027	.045	.065	.085	.104	33.5
34.5				.002	.010	.024	.041	.059	.078	.098	34.5

TABLE E   Wilcoxon's Rank Sum   (*Continued*)

Probability that Wilcoxon's rank sum minus the expected value
will be equal to or larger than W

n = 13

W	m=1	m=3	m=5	m=7	m=9	m=11	m=13	m=15	m=17	m=19	W
35.5				,002	,008	,020	,036	,054	,072	,091	35.5
36.5				,001	,007	,018	,032	,049	,067	,085	36.5
37.5				,001	,006	,015	,028	,044	,061	,079	37.5
38.5				,001	,004	,013	,025	,040	,056	,073	38.5
39.5				,000	,004	,011	,022	,036	,052	,068	39.5
40.5					,003	,009	,019	,032	,047	,063	40.5
41.5					,002	,008	,017	,029	,043	,058	41.5
42.5					,002	,006	,015	,026	,039	,054	42.5
43.5					,001	,005	,013	,023	,036	,050	43.5
44.5					,001	,004	,011	,021	,032	,046	44.5
45.5					,001	,004	,010	,018	,029	,042	45.5
46.5					,001	,003	,008	,016	,027	,039	46.5
47.5					,000	,002	,007	,014	,024	,035	47.5
48.5						,002	,006	,013	,022	,032	48.5
49.5						,002	,005	,011	,019	,030	49.5
50.5						,001	,004	,010	,017	,027	50.5
51.5						,001	,004	,009	,016	,025	51.5
52.5						,001	,003	,007	,014	,022	52.5
53.5						,001	,003	,006	,012	,020	53.5
54.5						,000	,002	,006	,011	,018	54.5
55.5							,002	,005	,010	,017	55.5
56.5							,001	,004	,009	,015	56.5
57.5							,001	,004	,008	,014	57.5
58.5							,001	,003	,007	,012	58.5
59.5							,001	,003	,006	,011	59.5
60.5							,001	,002	,005	,010	60.5
61.5							,000	,002	,005	,009	61.5
62.5								,002	,004	,008	62.5
63.5								,001	,003	,007	63.5
64.5								,001	,003	,006	64.5
65.5								,001	,003	,006	65.5
66.5								,001	,002	,005	66.5
67.5								,001	,002	,004	67.5
68.5								,001	,002	,004	68.5
69.5								,000	,001	,003	69.5
70.5									,001	,003	70.5

TABLE E   Wilcoxon's Rank Sum   (*Continued*)

```
Probability that Wilcoxon's rank sum minus the expected value
               will be equal to or larger than W
```

n = 13

W	m=1	m=3	m=5	m=7	m=9	m=11	m=13	m=15	m=17	m=19	W
71.5									.001	.003	71.5
72.5									.001	.002	72.5
73.5									.001	.002	73.5
74.5									.001	.002	74.5
75.5									.001	.001	75.5
76.5									.000	.001	76.5
77.5										.001	77.5
78.5										.001	78.5
79.5										.001	79.5
80.5										.001	80.5
81.5										.001	81.5
82.5										.001	82.5
83.5										.000	83.5

n = 13

W	m=2	m=4	m=6	m=8	m=10	m=12	m=14	m=16	m=18	m=20	W
1	.467	.478	.483	.486	.488	.489	.490	.491	.492	.493	1
2	.400	.435	.449	.458	.464	.468	.471	.474	.476	.478	2
3	.343	.392	.416	.430	.440	.447	.453	.457	.461	.464	3
4	.286	.352	.383	.402	.416	.426	.434	.440	.445	.449	4
5	.238	.312	.351	.375	.392	.405	.415	.423	.430	.435	5
6	.190	.274	.319	.348	.369	.384	.396	.406	.414	.421	6
7	.152	.239	.289	.323	.346	.364	.378	.390	.399	.407	7
8	.114	.206	.261	.298	.324	.344	.360	.373	.384	.393	8
9	.086	.175	.234	.273	.302	.325	.342	.357	.369	.379	9
10	.057	.148	.208	.250	.281	.306	.325	.341	.354	.365	10
11	.038	.123	.184	.228	.261	.287	.308	.325	.339	.351	11
12	.019	.101	.161	.207	.242	.269	.291	.310	.325	.338	12
13	.010	.082	.141	.187	.223	.252	.275	.295	.311	.325	13
14	.000	.065	.122	.168	.205	.235	.259	.280	.297	.312	14
15		.051	.105	.151	.188	.219	.244	.265	.283	.299	15
16		.039	.090	.134	.172	.203	.229	.251	.270	.287	16
17		.030	.076	.119	.156	.188	.215	.238	.257	.274	17
18		.022	.064	.105	.142	.174	.201	.224	.245	.262	18
19		.016	.053	.092	.128	.160	.188	.212	.232	.250	19
20		.011	.044	.081	.116	.147	.175	.199	.220	.239	20
21		.008	.036	.070	.104	.135	.163	.187	.209	.228	21
22		.005	.029	.061	.093	.124	.151	.176	.197	.217	22
23		.003	.023	.052	.083	.113	.140	.165	.187	.206	23

TABLE E Wilcoxon's Rank Sum (*Continued*)

Probability that Wilcoxon's rank sum minus the expected value
will be equal to or larger than W

					n = 13					
W	m=2	m=4	m=6	m=8	m=10	m=12	m=14	m=16	m=18	m=20	W
24		,002	,018	,045	,074	,103	,129	,154	,176	,196	24
25		,001	,014	,038	,065	,093	,119	,144	,166	,186	25
26		,000	,011	,032	,058	,084	,110	,134	,156	,176	26
27			,008	,027	,051	,076	,101	,125	,147	,167	27
28			,006	,022	,044	,068	,093	,116	,138	,158	28
29			,005	,018	,038	,061	,085	,107	,129	,149	29
30			,003	,015	,033	,055	,077	,100	,121	,141	30
31			,002	,012	,029	,049	,070	,092	,113	,133	31
32			,002	,010	,025	,043	,064	,085	,105	,125	32
33			,001	,008	,021	,038	,058	,078	,098	,117	33
34			,001	,006	,018	,034	,052	,072	,091	,110	34
35			,000	,005	,015	,030	,047	,066	,085	,103	35
36				,004	,013	,026	,042	,060	,079	,097	36
37				,003	,011	,023	,038	,055	,073	,091	37
38				,002	,009	,020	,034	,050	,067	,085	38
39				,002	,007	,017	,030	,046	,062	,079	39
40				,001	,006	,015	,027	,042	,057	,074	40
41				,001	,005	,013	,024	,038	,053	,069	41
42				,001	,004	,011	,021	,034	,049	,064	42
43				,000	,003	,009	,019	,031	,045	,059	43
44					,003	,008	,017	,028	,041	,055	44
45					,002	,007	,015	,025	,037	,051	45
46					,002	,006	,013	,022	,034	,047	46
47					,001	,005	,011	,020	,031	,043	47
48					,001	,004	,010	,018	,028	,040	48
49					,001	,003	,008	,016	,026	,037	49
50					,001	,003	,007	,014	,023	,034	50
51					,000	,002	,006	,013	,021	,031	51
52						,002	,005	,011	,019	,028	52
53						,001	,005	,010	,017	,026	53
54						,001	,004	,009	,015	,024	54
55						,001	,003	,008	,014	,022	55
56						,001	,003	,007	,012	,020	56
57						,001	,002	,006	,011	,018	57
58						,000	,002	,005	,010	,016	58
59							,002	,004	,009	,015	59

TABLE E   Wilcoxon's Rank Sum   (*Continued*)

Probability that Wilcoxon's rank sum minus the expected value
will be equal to or larger than W

n = 13

W	m=2	m=4	m=6	m=8	m=10	m=12	m=14	m=16	m=18	m=20	W
60							,001	,004	,008	,013	60
61							,001	,003	,007	,012	61
62							,001	,003	,006	,011	62
63							,001	,002	,005	,010	63
64							,001	,002	,005	,009	64
65							,001	,002	,004	,008	65
66							,000	,001	,004	,007	66
67								,001	,003	,006	67
68								,001	,003	,006	68
69								,001	,002	,005	69
70								,001	,002	,005	70
71								,001	,002	,004	71
72								,001	,002	,004	72
73								,000	,001	,003	73
74									,001	,003	74
75									,001	,002	75
76									,001	,002	76
77									,001	,002	77
78									,001	,002	78
79									,001	,001	79
80									,000	,001	80
81										,001	81
82										,001	82
83										,001	83
84										,001	84
85										,001	85
86										,001	86
87										,000	87

n = 14

W	m=1	m=2	m=3	m=4	m=5	m=6	m=7	m=8	m=9	m=10	W
1	,467	,467	,476	,479	,482	,484	,486	,487	,488	,489	1
2	,400	,408	,429	,439	,447	,452	,457	,460	,463	,466	2
3	,333	,350	,384	,399	,412	,421	,428	,434	,439	,443	3
4	,267	,300	,338	,360	,377	,390	,400	,408	,415	,420	4
5	,200	,250	,296	,323	,343	,359	,372	,382	,391	,398	5
6	,133	,208	,254	,287	,311	,329	,344	,357	,367	,376	6
7	,067	,167	,216	,253	,280	,301	,318	,332	,344	,354	7

Probability that Wilcoxon's rank sum minus the expected value will be equal to or larger than W

					$n = 14$						
W	m=1	m=2	m=3	m=4	m=5	m=6	m=7	m=8	m=9	m=10	W
8	,000	,133	,181	,221	,250	,273	,292	,308	,322	,333	8
9		,100	,150	,191	,222	,247	,268	,285	,300	,313	9
10		,075	,122	,164	,196	,222	,244	,263	,279	,292	10
11		,050	,099	,139	,171	,199	,221	,241	,258	,273	11
12		,033	,078	,116	,149	,177	,200	,221	,238	,254	12
13		,017	,060	,096	,128	,156	,180	,201	,219	,236	13
14		,008	,046	,079	,110	,137	,161	,183	,201	,218	14
15		,000	,034	,063	,093	,120	,144	,165	,184	,202	15
16			,024	,051	,078	,104	,127	,149	,168	,186	16
17			,016	,040	,065	,089	,112	,133	,153	,170	17
18			,010	,031	,053	,076	,098	,119	,138	,156	18
19			,006	,023	,044	,065	,086	,106	,125	,142	19
20			,003	,017	,035	,055	,074	,094	,112	,130	20
21			,001	,012	,028	,046	,064	,083	,101	,117	21
22			,000	,009	,022	,038	,055	,073	,090	,106	22
23				,006	,017	,031	,047	,063	,080	,096	23
24				,004	,013	,025	,040	,055	,071	,086	24
25				,002	,010	,020	,033	,048	,062	,077	25
26				,001	,007	,016	,028	,041	,055	,069	26
27				,001	,005	,013	,023	,035	,048	,061	27
28				,000	,004	,010	,019	,030	,042	,054	28
29					,002	,008	,015	,025	,036	,048	29
30					,002	,006	,012	,021	,031	,042	30
31					,001	,004	,010	,018	,027	,037	31
32					,001	,003	,008	,015	,023	,032	32
33					,000	,002	,006	,012	,019	,028	33
34						,002	,005	,010	,016	,024	34
35						,001	,004	,008	,014	,021	35
36						,001	,003	,006	,012	,018	36
37						,000	,002	,005	,010	,015	37
38							,002	,004	,008	,013	38
39							,001	,003	,006	,011	39
40							,001	,002	,005	,009	40
41							,001	,002	,004	,008	41
42							,000	,001	,003	,006	42
43								,001	,003	,005	43
44								,001	,002	,004	44
45								,001	,002	,004	45
46								,000	,001	,003	46
47									,001	,002	47
48									,001	,002	48
49									,001	,002	49
50									,000	,001	50
51										,001	51
52										,001	52
53										,001	53
54										,000	54

TABLE E   Wilcoxon's Rank Sum   (*Continued*)

Probability that Wilcoxon's rank sum minus the expected value
will be equal to or larger than W

W	m=11	m=12	m=13	m=14	m=15	m=16	m=17	m=18	m=19	m=20	W
					n = 14						
1	,489	,490	,490	,491	,491	,492	,492	,493	,493	,493	1
2	,468	,470	,471	,473	,474	,476	,477	,478	,479	,479	2
3	,447	,450	,453	,455	,457	,459	,461	,463	,464	,466	3
4	,425	,430	,434	,437	,440	,443	,446	,448	,450	,452	4
5	,404	,410	,415	,419	,423	,427	,430	,433	,436	,438	5
6	,384	,391	,396	,402	,407	,411	,415	,418	,422	,425	6
7	,363	,371	,378	,384	,390	,395	,400	,404	,408	,411	7
8	,343	,352	,360	,367	,374	,379	,385	,389	,394	,398	8
9	,324	,334	,342	,350	,357	,364	,370	,375	,380	,385	9
10	,305	,315	,325	,334	,341	,349	,355	,361	,366	,371	10
11	,286	,298	,308	,317	,326	,334	,341	,347	,353	,358	11
12	,268	,280	,291	,301	,310	,319	,326	,333	,340	,346	12
13	,250	,263	,275	,286	,295	,304	,312	,320	,327	,333	13
14	,233	,247	,259	,270	,281	,290	,299	,306	,314	,321	14
15	,217	,231	,244	,256	,266	,276	,285	,293	,301	,308	15
16	,201	,216	,229	,241	,252	,262	,272	,281	,289	,296	16
17	,187	,201	,215	,227	,239	,249	,259	,268	,276	,284	17
18	,172	,187	,201	,214	,225	,236	,246	,256	,264	,273	18
19	,159	,174	,188	,201	,213	,224	,234	,244	,253	,261	19
20	,146	,161	,175	,188	,200	,212	,222	,232	,241	,250	20
21	,134	,149	,163	,176	,188	,200	,211	,221	,230	,239	21
22	,122	,137	,151	,164	,177	,188	,199	,210	,219	,229	22
23	,111	,126	,140	,153	,166	,177	,189	,199	,209	,218	23
24	,101	,116	,129	,143	,155	,167	,178	,189	,199	,208	24
25	,092	,106	,119	,132	,145	,157	,168	,179	,189	,198	25
26	,083	,097	,110	,123	,135	,147	,158	,169	,179	,189	26
27	,075	,088	,101	,114	,126	,138	,149	,159	,170	,179	27
28	,067	,080	,093	,105	,117	,129	,140	,150	,161	,170	28
29	,060	,072	,085	,097	,109	,120	,131	,142	,152	,162	29
30	,054	,065	,077	,089	,101	,112	,123	,133	,143	,153	30
31	,048	,059	,070	,082	,093	,104	,115	,125	,135	,145	31
32	,042	,053	,064	,075	,086	,097	,107	,118	,128	,137	32
33	,037	,048	,058	,069	,079	,090	,100	,110	,120	,130	33
34	,033	,042	,052	,063	,073	,083	,093	,103	,113	,122	34
35	,029	,038	,047	,057	,067	,077	,087	,096	,106	,115	35
36	,025	,034	,042	,052	,061	,071	,080	,090	,099	,109	36
37	,022	,030	,038	,047	,056	,065	,075	,084	,093	,102	37
38	,019	,026	,034	,042	,051	,060	,069	,078	,087	,096	38
39	,017	,023	,030	,038	,047	,055	,064	,073	,081	,090	39
40	,014	,020	,027	,035	,042	,051	,059	,067	,076	,085	40
41	,012	,018	,024	,031	,038	,046	,054	,063	,071	,079	41
42	,011	,016	,021	,028	,035	,042	,050	,058	,066	,074	42
43	,009	,013	,019	,025	,031	,039	,046	,054	,061	,069	43

TABLE E   Wilcoxon's Rank Sum   (*Continued*)

Probability that Wilcoxon's rank sum minus the expected value
will be equal to or larger than W

n = 14

W	m=11	m=12	m=13	m=14	m=15	m=16	m=17	m=18	m=19	m=20	W
44	,008	,012	,017	,022	,028	,035	,042	,049	,057	,065	44
45	,006	,010	,015	,020	,026	,032	,039	,046	,053	,060	45
46	,005	,009	,013	,018	,023	,029	,035	,042	,049	,056	46
47	,005	,007	,011	,016	,021	,026	,032	,039	,045	,052	47
48	,004	,006	,010	,014	,018	,024	,029	,035	,042	,048	48
49	,003	,005	,008	,012	,016	,021	,027	,032	,039	,045	49
50	,003	,005	,007	,011	,015	,019	,024	,030	,036	,042	50
51	,002	,004	,006	,009	,013	,017	,022	,027	,033	,039	51
52	,002	,003	,005	,008	,011	,015	,020	,025	,030	,036	52
53	,001	,003	,005	,007	,010	,014	,018	,023	,028	,033	53
54	,001	,002	,004	,006	,009	,012	,016	,020	,025	,030	54
55	,001	,002	,003	,005	,008	,011	,015	,019	,023	,028	55
56	,001	,002	,003	,005	,007	,010	,013	,017	,021	,026	56
57	,001	,001	,002	,004	,006	,009	,012	,015	,019	,024	57
58	,000	,001	,002	,003	,005	,008	,010	,014	,017	,022	58
59		,001	,002	,003	,005	,007	,009	,012	,016	,020	59
60		,001	,001	,002	,004	,006	,008	,011	,014	,018	60
61		,001	,001	,002	,003	,005	,007	,010	,013	,016	61
62		,000	,001	,002	,003	,005	,007	,009	,012	,015	62
63			,001	,001	,003	,004	,006	,008	,011	,014	63
64			,001	,001	,002	,003	,005	,007	,010	,012	64
65			,001	,001	,002	,003	,004	,006	,009	,011	65
66			,000	,001	,002	,003	,004	,006	,008	,010	66
67				,001	,001	,002	,003	,005	,007	,009	67
68				,001	,001	,002	,003	,004	,006	,008	68
69				,000	,001	,002	,003	,004	,006	,008	69
70					,001	,001	,002	,003	,005	,007	70
71					,001	,001	,002	,003	,004	,006	71
72					,001	,001	,002	,003	,004	,005	72
73					,000	,001	,001	,002	,003	,005	73
74						,001	,001	,002	,003	,004	74
75						,001	,001	,002	,003	,004	75
76						,001	,001	,002	,002	,003	76
77						,000	,001	,001	,002	,003	77
78							,001	,001	,002	,003	78
79							,001	,001	,002	,002	79

TABLE E Wilcoxon's Rank Sum (*Continued*)

Probability that Wilcoxon's rank sum minus the expected value will be equal to or larger than W

n = 14

W	m=11	m=12	m=13	m=14	m=15	m=16	m=17	m=18	m=19	m=20	W
80							,000	,001	,001	,002	80
81								,001	,001	,002	81
82								,001	,001	,002	82
83								,001	,001	,001	83
84								,000	,001	,001	84
85									,001	,001	85
86									,001	,001	86
87									,001	,001	87
88									,000	,001	88
89										,001	89
90										,001	90
91										,000	91

n = 15

W	m=1	m=3	m=5	m=7	m=9	m=11	m=13	m=15	m=17	m=19	W
0,5	,500	,500	,500	,500	,500	,500	,500	,500	,500	,500	0,5
1,5	,437	,456	,466	,473	,477	,480	,482	,484	,485	,486	1,5
2,5	,375	,412	,433	,445	,454	,459	,464	,467	,470	,473	2,5
3,5	,312	,369	,400	,418	,430	,439	,446	,451	,456	,459	3,5
4,5	,250	,327	,368	,392	,408	,419	,428	,435	,441	,446	4,5
5,5	,187	,287	,336	,365	,385	,399	,410	,419	,426	,432	5,5
6,5	,125	,249	,306	,340	,363	,380	,393	,403	,412	,419	6,5
7,5	,062	,213	,277	,315	,341	,361	,375	,387	,397	,405	7,5
8,5	,000	,180	,249	,291	,320	,342	,358	,372	,383	,392	8,5
9,5		,151	,222	,267	,299	,323	,342	,356	,369	,379	9,5
10,5		,125	,197	,245	,279	,305	,325	,341	,355	,366	10,5
11,5		,102	,174	,224	,260	,287	,309	,326	,341	,353	11,5
12,5		,082	,153	,203	,241	,270	,293	,312	,327	,341	12,5
13,5		,065	,133	,184	,223	,253	,278	,297	,314	,328	13,5
14,5		,050	,115	,166	,206	,237	,262	,283	,301	,316	14,5
15,5		,038	,099	,149	,189	,222	,248	,270	,288	,304	15,5
16,5		,028	,084	,133	,174	,206	,234	,256	,275	,292	16,5
17,5		,020	,071	,119	,159	,192	,220	,243	,263	,280	17,5
18,5		,013	,059	,105	,145	,178	,207	,230	,251	,269	18,5
19,5		,009	,049	,093	,132	,165	,194	,218	,239	,257	19,5
20,5		,005	,040	,081	,119	,153	,181	,206	,228	,246	20,5
21,5		,002	,033	,071	,108	,141	,169	,195	,217	,236	21,5
22,5		,001	,026	,061	,097	,129	,158	,183	,206	,225	22,5
23,5		,000	,021	,053	,087	,119	,147	,173	,195	,215	23,5

TABLE E   Wilcoxon's Rank Sum   (*Continued*)

Probability that Wilcoxon's rank sum minus the expected value
will be equal to or larger than W

					n = 15					
W	m=1	m=3	m=5	m=7	m=9	m=11	m=13	m=15	m=17	m=19	W
24.5			.016	.046	.078	.109	.137	.162	.185	.205	24.5
25.5			.013	.039	.069	.099	.127	.152	.175	.195	25.5
26.5			.010	.033	.061	.090	.118	.143	.165	.186	26.5
27.5			.007	.028	.054	.082	.109	.134	.156	.177	27.5
28.5			.005	.023	.048	.074	.100	.125	.147	.168	28.5
29.5			.004	.019	.042	.067	.092	.116	.139	.159	29.5
30.5			.003	.016	.037	.060	.085	.108	.131	.151	30.5
31.5			.002	.013	.032	.054	.078	.101	.123	.143	31.5
32.5			.001	.011	.028	.049	.071	.094	.115	.135	32.5
33.5			.001	.009	.024	.043	.065	.087	.108	.128	33.5
34.5			.000	.007	.020	.039	.059	.080	.101	.121	34.5
35.5				.005	.017	.034	.054	.074	.094	.114	35.5
36.5				.004	.015	.030	.049	.068	.088	.107	36.5
37.5				.003	.013	.027	.044	.063	.082	.101	37.5
38.5				.003	.011	.024	.040	.058	.077	.095	38.5
39.5				.002	.009	.021	.036	.053	.071	.089	39.5
40.5				.001	.007	.018	.032	.049	.066	.084	40.5
41.5				.001	.006	.016	.029	.045	.061	.078	41.5
42.5				.001	.005	.014	.026	.041	.057	.073	42.5
43.5				.001	.004	.012	.023	.037	.053	.069	43.5
44.5				.000	.003	.010	.021	.034	.048	.064	44.5
45.5					.003	.009	.018	.031	.045	.060	45.5
46.5					.002	.007	.016	.028	.041	.056	46.5
47.5					.002	.006	.014	.025	.038	.052	47.5
48.5					.001	.005	.013	.023	.035	.048	48.5
49.5					.001	.005	.011	.020	.032	.045	49.5
50.5					.001	.004	.010	.018	.029	.041	50.5
51.5					.001	.003	.009	.016	.027	.038	51.5
52.5					.000	.003	.007	.015	.024	.035	52.5
53.5						.002	.006	.013	.022	.033	53.5
54.5						.002	.006	.012	.020	.030	54.5
55.5						.002	.005	.010	.018	.028	55.5
56.5						.001	.004	.009	.017	.026	56.5
57.5						.001	.004	.008	.015	.024	57.5
58.5						.001	.003	.007	.014	.022	58.5
59.5						.001	.003	.006	.012	.020	59.5
60.5						.001	.002	.006	.011	.018	60.5
61.5						.000	.002	.005	.010	.017	61.5
62.5							.002	.004	.009	.015	62.5
63.5							.001	.004	.008	.014	63.5
64.5							.001	.003	.007	.013	64.5
65.5							.001	.003	.006	.011	65.5

TABLE E   Wilcoxon's Rank Sum   (*Continued*)

Probability that Wilcoxon's rank sum minus the expected value
will be equal to or larger than W

				n = 15						
W	m=1	m=3	m=5	m=7	m=9	m=11	m=13	m=15	m=17	m=19	W
66.5							.001	.002	.006	.010	66.5
67.5							.001	.002	.005	.009	67.5
68.5							.001	.002	.004	.008	68.5
69.5							.000	.002	.004	.008	69.5
70.5								.001	.003	.007	70.5
71.5								.001	.003	.006	71.5
72.5								.001	.003	.006	72.5
73.5								.001	.002	.005	73.5
74.5								.001	.002	.004	74.5
75.5								.001	.002	.004	75.5
76.5								.000	.002	.004	76.5
77.5									.001	.003	77.5
78.5									.001	.003	78.5
79.5									.001	.003	79.5
80.5									.001	.002	80.5
81.5									.001	.002	81.5
82.5									.001	.002	82.5
83.5									.001	.002	83.5
84.5									.000	.001	84.5
85.5										.001	85.5
86.5										.001	86.5
87.5										.001	87.5
88.5										.001	88.5
89.5										.001	89.5
90.5										.001	90.5
91.5										.001	91.5
92.5										.000	92.5

TABLE E Wilcoxon's Rank Sum (*Continued*)

Probability that Wilcoxon's rank sum minus the expected value
will be equal to or larger than W

					n = 15					
W	m=2	m=4	m=6	m=8	m=10	m=12	m=14	m=16	m=18	m=20	W
1	,471	,481	,485	,487	,489	,490	,491	,492	,493	,493	1
2	,412	,443	,455	,462	,467	,471	,474	,477	,479	,480	2
3	,360	,405	,425	,437	,446	,452	,457	,461	,464	,467	3
4	,309	,368	,395	,413	,424	,433	,440	,446	,450	,454	4
5	,265	,332	,367	,388	,403	,415	,423	,430	,436	,441	5
6	,221	,298	,338	,364	,382	,396	,407	,415	,422	,428	6
7	,184	,265	,311	,341	,362	,378	,390	,400	,408	,415	7
8	,147	,235	,285	,318	,341	,359	,374	,385	,395	,403	8
9	,118	,205	,259	,295	,322	,342	,357	,370	,381	,390	9
10	,088	,179	,235	,274	,302	,324	,341	,356	,367	,377	10
11	,066	,154	,212	,253	,284	,307	,326	,341	,354	,365	11
12	,044	,131	,190	,233	,265	,290	,310	,327	,341	,353	12
13	,029	,110	,170	,214	,248	,274	,295	,313	,328	,340	13
14	,015	,092	,151	,196	,231	,258	,281	,299	,315	,328	14
15	,007	,076	,134	,179	,214	,243	,266	,286	,302	,317	15
16	,000	,062	,117	,162	,198	,228	,252	,273	,290	,305	16
17		,050	,103	,147	,183	,214	,239	,260	,278	,293	17
18		,040	,089	,133	,169	,200	,225	,247	,266	,282	18
19		,031	,077	,119	,156	,186	,213	,235	,254	,271	19
20		,024	,066	,107	,143	,174	,200	,223	,243	,260	20
21		,018	,056	,095	,130	,161	,188	,212	,232	,250	21
22		,014	,047	,084	,119	,150	,177	,200	,221	,239	22
23		,010	,040	,075	,108	,139	,166	,190	,211	,229	23
24		,007	,033	,066	,098	,128	,155	,179	,200	,219	24
25		,005	,027	,058	,089	,118	,145	,169	,190	,210	25
26		,003	,022	,050	,080	,109	,135	,159	,181	,200	26
27		,002	,018	,044	,072	,100	,126	,150	,171	,191	27
28		,001	,015	,038	,064	,091	,117	,141	,162	,182	28
29		,001	,012	,032	,058	,084	,109	,132	,154	,173	29
30		,000	,009	,028	,051	,076	,101	,124	,145	,165	30
31			,007	,024	,045	,069	,093	,116	,137	,157	31
32			,005	,020	,040	,063	,086	,108	,129	,149	32
33			,004	,017	,035	,057	,079	,101	,122	,141	33
34			,003	,014	,031	,051	,073	,094	,115	,134	34
35			,002	,012	,027	,046	,067	,088	,108	,127	35
36			,002	,010	,024	,042	,061	,081	,101	,120	36
37			,001	,008	,021	,037	,056	,075	,095	,113	37
38			,001	,006	,018	,033	,051	,070	,089	,107	38
39			,001	,005	,015	,030	,047	,065	,083	,101	39
40			,000	,004	,013	,026	,042	,060	,078	,095	40
41				,003	,011	,023	,038	,055	,072	,090	41
42				,003	,010	,021	,035	,051	,067	,084	42
43				,002	,008	,018	,031	,047	,063	,079	43
44				,002	,007	,016	,028	,043	,058	,074	44
45				,001	,006	,014	,026	,039	,054	,070	45
46				,001	,005	,012	,023	,036	,050	,065	46

TABLE E Wilcoxon's Rank Sum (*Continued*)

Probability that Wilcoxon's rank sum minus the expected value
will be equal to or larger than W

					n = 15						
W	m=2	m=4	m=6	m=8	m=10	m=12	m=14	m=16	m=18	m=20	W
47				,001	,004	,011	,021	,033	,046	,061	47
48				,001	,003	,009	,018	,030	,043	,057	48
49				,000	,003	,008	,016	,027	,040	,053	49
50					,002	,007	,015	,025	,037	,050	50
51					,002	,006	,013	,022	,034	,046	51
52					,001	,005	,011	,020	,031	,043	52
53					,001	,004	,010	,018	,028	,040	53
54					,001	,004	,009	,017	,026	,037	54
55					,001	,003	,008	,015	,024	,034	55
56					,001	,003	,007	,013	,022	,032	56
57					,000	,002	,006	,012	,020	,029	57
58						,002	,005	,011	,018	,027	58
59						,002	,005	,010	,017	,025	59
60						,001	,004	,009	,015	,023	60
61						,001	,003	,008	,014	,021	61
62						,001	,003	,007	,012	,020	62
63						,001	,003	,006	,011	,018	63
64						,001	,002	,005	,010	,017	64
65						,000	,002	,005	,009	,015	65
66							,002	,004	,008	,014	66
67							,001	,004	,007	,013	67
68							,001	,003	,007	,012	68
69							,001	,003	,006	,010	69
70							,001	,002	,005	,010	70
71							,001	,002	,005	,009	71
72							,001	,002	,004	,008	72
73							,000	,002	,004	,007	73
74								,001	,003	,006	74
75								,001	,003	,006	75
76								,001	,003	,005	76
77								,001	,002	,005	77
78								,001	,002	,004	78
79								,001	,002	,004	79
80								,001	,002	,003	80
81								,000	,001	,003	81

TABLE E   Wilcoxon's Rank Sum   (*Continued*)

```
Probability that Wilcoxon's rank sum minus the expected value
            will be equal to or larger than W
```

					n = 15					
W	m=2	m=4	m=6	m=8	m=10	m=12	m=14	m=16	m=18	m=20	W
82									,001	,003	82
83									,001	,002	83
84									,001	,002	84
85									,001	,002	85
86									,001	,002	86
87									,001	,002	87
88									,001	,001	88
89									,000	,001	89
90										,001	90
91										,001	91
92										,001	92
93										,001	93
94										,001	94
95										,001	95
96										,000	96

					n = 16					
W	m=1	m=2	m=3	m=4	m=5	m=6	m=7	m=8	m=9	m=10	W
1	,471	,471	,479	,482	,484	,486	,487	,488	,489	,490	1
2	,412	,418	,438	,446	,452	,457	,461	,464	,467	,469	2
3	,353	,366	,396	,410	,421	,429	,435	,440	,445	,449	3
4	,294	,320	,356	,375	,390	,401	,410	,417	,423	,428	4
5	,235	,275	,317	,341	,359	,373	,385	,394	,401	,408	5
6	,176	,235	,280	,308	,330	,347	,360	,371	,380	,388	6
7	,118	,196	,244	,277	,301	,320	,336	,348	,359	,368	7
8	,059	,163	,211	,247	,274	,295	,312	,326	,339	,349	8
9	,000	,131	,180	,219	,247	,271	,289	,305	,318	,330	9
10		,105	,152	,192	,223	,247	,267	,284	,299	,311	10
11		,078	,127	,168	,199	,225	,246	,264	,280	,293	11
12		,059	,105	,145	,177	,204	,226	,245	,261	,276	12
13		,039	,086	,124	,156	,183	,206	,226	,243	,258	13
14		,026	,069	,106	,137	,165	,188	,208	,226	,242	14
15		,013	,055	,089	,120	,147	,171	,191	,210	,226	15
16		,007	,042	,074	,104	,131	,154	,175	,194	,210	16
17		,000	,032	,061	,089	,115	,139	,160	,179	,196	17
18			,024	,050	,077	,102	,124	,145	,164	,182	18
19			,017	,040	,065	,089	,111	,132	,151	,168	19
20			,011	,032	,055	,077	,099	,119	,138	,155	20
21			,007	,025	,046	,067	,087	,107	,126	,143	21
22			,004	,019	,038	,057	,077	,096	,114	,131	22
23			,002	,015	,031	,049	,068	,086	,104	,120	23

TABLE E   Wilcoxon's Rank Sum   (*Continued*)

Probability that Wilcoxon's rank sum minus the expected value
will be equal to or larger than W

W	m=1	m=2	m=3	m=4	m=5	m=6	m=7	m=8	m=9	m=10	W
					n = 16						
24			,001	,011	,025	,042	,059	,077	,094	,110	24
25			,000	,008	,020	,035	,051	,068	,084	,100	25
26				,006	,016	,029	,044	,060	,076	,091	26
27				,004	,012	,024	,038	,053	,068	,083	27
28				,002	,010	,020	,033	,046	,061	,075	28
29				,001	,007	,016	,028	,040	,054	,067	29
30				,001	,006	,013	,023	,035	,048	,061	30
31				,000	,004	,011	,020	,030	,042	,054	31
32					,003	,008	,016	,026	,037	,049	32
33					,002	,007	,014	,022	,032	,043	33
34					,001	,005	,011	,019	,028	,039	34
35					,001	,004	,009	,016	,025	,034	35
36					,001	,003	,007	,014	,021	,030	36
37					,000	,002	,006	,011	,018	,027	37
38						,002	,005	,010	,016	,023	38
39						,001	,004	,008	,014	,020	39
40						,001	,003	,007	,012	,018	40
41						,001	,002	,005	,010	,015	41
42						,000	,002	,004	,008	,013	42
43							,001	,004	,007	,012	43
44							,001	,003	,006	,010	44
45							,001	,002	,005	,008	45
46							,001	,002	,004	,007	46
47							,000	,001	,003	,006	47
48								,001	,003	,005	48
49								,001	,002	,004	49
50								,001	,002	,004	50
51								,000	,001	,003	51
52									,001	,003	52
53									,001	,002	53
54									,001	,002	54
55									,001	,001	55
56									,000	,001	56
57										,001	57
58										,001	58
59										,001	59
60										,000	60

TABLE E Wilcoxon's Rank Sum (*Continued*)

Probability that Wilcoxon's rank sum minus the expected value
will be equal to or larger than W

n = 16

W	m=11	m=12	m=13	m=14	m=15	m=16	m=17	m=18	m=19	m=20	W
1	,490	,491	,491	,492	,492	,493	,493	,493	,493	,494	1
2	,471	,473	,474	,476	,477	,478	,479	,480	,480	,481	2
3	,452	,455	,457	,459	,461	,463	,465	,466	,467	,469	3
4	,433	,437	,440	,443	,446	,448	,451	,453	,454	,456	4
5	,414	,419	,423	,427	,430	,434	,436	,439	,442	,444	5
6	,395	,401	,406	,411	,415	,419	,423	,426	,429	,431	6
7	,376	,383	,390	,395	,400	,405	,409	,412	,416	,419	7
8	,358	,366	,373	,379	,385	,390	,395	,399	,403	,407	8
9	,340	,349	,357	,364	,370	,376	,381	,386	,391	,395	9
10	,322	,332	,341	,349	,356	,362	,368	,373	,378	,383	10
11	,305	,316	,325	,334	,341	,348	,354	,360	,366	,371	11
12	,288	,300	,310	,319	,327	,334	,341	,348	,353	,359	12
13	,272	,284	,295	,304	,313	,321	,328	,335	,341	,347	13
14	,256	,268	,280	,290	,299	,308	,316	,323	,329	,336	14
15	,240	,254	,265	,276	,286	,295	,303	,311	,318	,324	15
16	,225	,239	,251	,262	,273	,282	,291	,299	,306	,313	16
17	,211	,225	,238	,249	,260	,270	,279	,287	,295	,302	17
18	,197	,211	,224	,236	,247	,257	,267	,275	,283	,291	18
19	,184	,198	,212	,224	,235	,245	,255	,264	,272	,280	19
20	,171	,186	,199	,212	,223	,234	,244	,253	,262	,270	20
21	,159	,174	,187	,200	,212	,223	,233	,242	,251	,259	21
22	,147	,162	,176	,188	,200	,212	,222	,232	,241	,249	22
23	,136	,151	,165	,177	,190	,201	,211	,221	,231	,239	23
24	,125	,140	,154	,167	,179	,190	,201	,211	,221	,230	24
25	,116	,130	,144	,157	,169	,180	,191	,201	,211	,220	25
26	,106	,120	,134	,147	,159	,171	,182	,192	,202	,211	26
27	,097	,111	,125	,138	,150	,161	,172	,183	,192	,202	27
28	,089	,103	,116	,129	,141	,152	,163	,174	,184	,193	28
29	,081	,095	,107	,120	,132	,144	,155	,165	,175	,184	29
30	,074	,087	,100	,112	,124	,135	,146	,157	,167	,176	30
31	,067	,080	,092	,104	,116	,127	,138	,148	,158	,168	31
32	,061	,073	,085	,097	,108	,119	,130	,141	,151	,160	32
33	,055	,066	,078	,090	,101	,112	,123	,133	,143	,152	33
34	,049	,061	,072	,083	,094	,105	,115	,126	,136	,145	34
35	,044	,055	,066	,077	,088	,098	,109	,119	,128	,138	35
36	,040	,050	,060	,071	,081	,092	,102	,112	,122	,131	36
37	,036	,045	,055	,065	,075	,086	,096	,105	,115	,124	37
38	,032	,041	,050	,060	,070	,080	,090	,099	,109	,118	38
39	,028	,037	,046	,055	,065	,074	,084	,093	,103	,112	39
40	,025	,033	,042	,051	,060	,069	,078	,088	,097	,106	40
41	,022	,030	,038	,046	,055	,064	,073	,082	,091	,100	41
42	,020	,026	,034	,042	,051	,059	,068	,077	,086	,094	42
43	,017	,024	,031	,039	,047	,055	,063	,072	,081	,089	43

TABLE E Wilcoxon's Rank Sum   (*Continued*)

```
Probability that Wilcoxon's rank sum minus the expected value
           will be equal to or larger than W
```

W	m=11	m=12	m=13	m=14	m=15	m=16	m=17	m=18	m=19	m=20	W
					n = 16						
44	,015	,021	,028	,035	,043	,051	,059	,067	,076	,084	44
45	,013	,019	,025	,032	,039	,047	,055	,063	,071	,079	45
46	,011	,017	,022	,029	,036	,043	,051	,059	,067	,075	46
47	,010	,015	,020	,026	,033	,040	,047	,055	,062	,070	47
48	,009	,013	,018	,024	,030	,037	,044	,051	,058	,066	48
49	,007	,011	,016	,021	,027	,034	,040	,047	,054	,062	49
50	,006	,010	,014	,019	,025	,031	,037	,044	,051	,058	50
51	,005	,009	,013	,017	,022	,028	,034	,041	,047	,054	51
52	,005	,008	,011	,015	,020	,026	,032	,038	,044	,051	52
53	,004	,007	,010	,014	,018	,023	,029	,035	,041	,047	53
54	,003	,006	,009	,012	,017	,021	,027	,032	,038	,044	54
55	,003	,005	,008	,011	,015	,019	,024	,030	,035	,041	55
56	,002	,004	,007	,010	,013	,018	,022	,027	,033	,039	56
57	,002	,004	,006	,009	,012	,016	,020	,025	,030	,036	57
58	,002	,003	,005	,008	,011	,014	,019	,023	,028	,033	58
59	,001	,003	,004	,007	,010	,013	,017	,021	,026	,031	59
60	,001	,002	,004	,006	,009	,012	,015	,019	,024	,029	60
61	,001	,002	,003	,005	,008	,011	,014	,018	,022	,027	61
62	,001	,002	,003	,005	,007	,009	,013	,016	,020	,025	62
63	,001	,001	,002	,004	,006	,008	,011	,015	,019	,023	63
64	,000	,001	,002	,003	,005	,008	,010	,014	,017	,021	64
65		,001	,002	,003	,005	,007	,009	,012	,016	,019	65
66		,001	,001	,003	,004	,006	,008	,011	,014	,018	66
67		,001	,001	,002	,004	,005	,008	,010	,013	,017	67
68		,001	,001	,002	,003	,005	,007	,009	,012	,015	68
69		,000	,001	,002	,003	,004	,006	,008	,011	,014	69
70			,001	,001	,002	,004	,005	,008	,010	,013	70
71			,001	,001	,002	,003	,005	,007	,009	,012	71
72			,001	,001	,002	,003	,004	,006	,008	,011	72
73			,000	,001	,002	,003	,004	,006	,007	,010	73
74				,001	,001	,002	,003	,005	,007	,009	74
75				,001	,001	,002	,003	,004	,006	,008	75
76				,001	,001	,002	,003	,004	,006	,007	76
77				,000	,001	,001	,002	,004	,005	,007	77
78					,001	,001	,002	,003	,005	,006	78
79					,001	,001	,002	,003	,004	,006	79
80					,001	,001	,002	,002	,004	,005	80
81					,000	,001	,001	,002	,003	,005	81
82						,001	,001	,002	,003	,004	82
83						,001	,001	,002	,003	,004	83
84						,001	,001	,002	,002	,003	84

TABLE E   Wilcoxon's Rank Sum   (*Continued*)

Probability that Wilcoxon's rank sum minus the expected value
will be equal to or larger than W

					n = 16					
W	m=11	m=12	m=13	m=14	m=15	m=16	m=17	m=18	m=19	m=20	W
85						,000	,001	,001	,002	,003	85
86							,001	,001	,002	,003	86
87							,001	,001	,002	,002	87
88							,001	,001	,001	,002	88
89							,000	,001	,001	,002	89
90								,001	,001	,002	90
91								,001	,001	,002	91
92								,001	,001	,001	92
93								,000	,001	,001	93
94									,001	,001	94
95									,001	,001	95
96									,001	,001	96
97									,000	,001	97
98										,001	98
99										,001	99
100										,001	100
101										,000	101

					n = 17					
W	m=1	m=3	m=5	m=7	m=9	m=11	m=13	m=15	m=17	m=19	W
0,5	,500	,500	,500	,500	,500	,500	,500	,500	,500	,500	0,5
1,5	,444	,461	,470	,475	,479	,482	,484	,485	,486	,488	1,5
2,5	,389	,421	,440	,451	,458	,463	,467	,470	,473	,475	2,5
3,5	,333	,382	,410	,426	,437	,445	,451	,456	,459	,463	3,5
4,5	,278	,345	,381	,402	,416	,427	,435	,441	,446	,450	4,5
5,5	,222	,308	,352	,378	,396	,409	,418	,426	,433	,438	5,5
6,5	,167	,273	,324	,355	,376	,391	,402	,412	,419	,426	6,5
7,5	,111	,239	,297	,332	,356	,373	,386	,397	,406	,413	7,5
8,5	,056	,208	,271	,310	,336	,356	,371	,383	,393	,401	8,5
9,5	,000	,179	,246	,288	,317	,339	,355	,369	,380	,389	9,5
10,5		,153	,223	,267	,298	,322	,340	,355	,367	,377	10,5
11,5		,129	,200	,247	,280	,305	,325	,341	,354	,365	11,5
12,5		,108	,179	,228	,263	,289	,310	,327	,342	,354	12,5
13,5		,089	,160	,209	,245	,274	,296	,314	,329	,342	13,5
14,5		,073	,141	,191	,229	,258	,282	,301	,317	,331	14,5
15,5		,059	,124	,175	,213	,243	,268	,288	,305	,319	15,5
16,5		,046	,109	,159	,198	,229	,254	,275	,293	,308	16,5
17,5		,036	,095	,144	,183	,215	,241	,263	,281	,297	17,5
18,5		,027	,082	,130	,169	,202	,229	,251	,270	,286	18,5

TABLE E   Wilcoxon's Rank Sum   (*Continued*)

Probability that Wilcoxon's rank sum minus the expected value
will be equal to or larger than W

n = 17

W	m=1	m=3	m=5	m=7	m=9	m=11	m=13	m=15	m=17	m=19	W
19.5		.020	.070	.117	.156	.189	.216	.239	.259	.276	19.5
20.5		.014	.060	.105	.144	.176	.204	.228	.248	.265	20.5
21.5		.010	.051	.093	.132	.165	.193	.217	.237	.255	21.5
22.5		.006	.042	.083	.120	.153	.181	.206	.227	.245	22.5
23.5		.004	.035	.074	.110	.142	.171	.195	.217	.235	23.5
24.5		.002	.029	.065	.100	.132	.160	.185	.207	.226	24.5
25.5		.001	.024	.057	.091	.122	.150	.175	.197	.216	25.5
26.5		.000	.019	.050	.082	.113	.141	.165	.188	.207	26.5
27.5			.015	.043	.074	.104	.131	.156	.179	.198	27.5
28.5			.012	.037	.067	.096	.123	.147	.170	.190	28.5
29.5			.010	.032	.060	.088	.114	.139	.161	.181	29.5
30.5			.007	.028	.053	.080	.106	.131	.153	.173	30.5
31.5			.006	.024	.048	.073	.099	.123	.145	.165	31.5
32.5			.004	.020	.042	.067	.092	.115	.137	.157	32.5
33.5			.003	.017	.037	.061	.085	.108	.130	.150	33.5
34.5			.002	.014	.033	.055	.078	.101	.123	.143	34.5
35.5			.002	.012	.029	.050	.072	.094	.116	.136	35.5
36.5			.001	.010	.026	.045	.067	.088	.109	.129	36.5
37.5			.001	.008	.022	.041	.061	.082	.103	.122	37.5
38.5			.000	.007	.019	.037	.056	.077	.097	.116	38.5
39.5				.005	.017	.033	.052	.071	.091	.110	39.5
40.5				.004	.015	.030	.047	.066	.085	.104	40.5
41.5				.003	.012	.026	.043	.061	.080	.098	41.5
42.5				.003	.011	.023	.039	.057	.075	.093	42.5
43.5				.002	.009	.021	.036	.053	.070	.088	43.5
44.5				.002	.008	.018	.032	.048	.065	.083	44.5
45.5				.001	.007	.016	.029	.045	.061	.078	45.5
46.5				.001	.006	.014	.027	.041	.057	.073	46.5
47.5				.001	.005	.013	.024	.038	.053	.069	47.5
48.5				.001	.004	.011	.022	.035	.049	.065	48.5
49.5				.000	.003	.010	.019	.032	.046	.061	49.5
50.5					.003	.008	.017	.029	.043	.057	50.5
51.5					.002	.007	.016	.027	.039	.053	51.5
52.5					.002	.006	.014	.024	.036	.050	52.5
53.5					.001	.005	.012	.022	.034	.047	53.5
54.5					.001	.005	.011	.020	.031	.044	54.5
55.5					.001	.004	.010	.018	.029	.041	55.5
56.5					.001	.003	.009	.017	.026	.038	56.5
57.5					.001	.003	.008	.015	.024	.035	57.5
58.5					.000	.002	.007	.014	.022	.033	58.5
59.5						.002	.006	.012	.021	.030	59.5

TABLE E   Wilcoxon's Rank Sum   (*Continued*)

Probability that Wilcoxon's rank sum minus the expected value
will be equal to or larger than W

					n = 17					
W	m=1	m=3	m=5	m=7	m=9	m=11	m=13	m=15	m=17	m=19	W
60.5						,002	,005	,011	,019	,028	60.5
61.5						,001	,005	,010	,017	,026	61.5
62.5						,001	,004	,009	,016	,024	62.5
63.5						,001	,003	,008	,014	,022	63.5
64.5						,001	,003	,007	,013	,021	64.5
65.5						,001	,003	,006	,012	,019	65.5
66.5						,001	,002	,006	,011	,018	66.5
67.5						,000	,002	,005	,010	,016	67.5
68.5							,002	,004	,009	,015	68.5
69.5							,001	,004	,008	,014	69.5
70.5							,001	,003	,007	,013	70.5
71.5							,001	,003	,007	,012	71.5
72.5							,001	,003	,006	,011	72.5
73.5							,001	,002	,005	,010	73.5
74.5							,001	,002	,005	,009	74.5
75.5							,001	,002	,004	,008	75.5
76.5							,000	,002	,004	,007	76.5
77.5								,001	,003	,007	77.5
78.5								,001	,003	,006	78.5
79.5								,001	,003	,006	79.5
80.5								,001	,002	,005	80.5
81.5								,001	,002	,005	81.5
82.5								,001	,002	,004	82.5
83.5								,001	,002	,004	83.5
84.5								,000	,001	,003	84.5
85.5									,001	,003	85.5
86.5									,001	,003	86.5
87.5									,001	,002	87.5
88.5									,001	,002	88.5
89.5									,001	,002	89.5
90.5									,001	,002	90.5
91.5									,001	,002	91.5
92.5									,001	,001	92.5
93.5									,000	,001	93.5
94.5										,001	94.5
95.5										,001	95.5
96.5										,001	96.5
97.5										,001	97.5
98.5										,001	98.5
99.5										,001	99.5
100.5										,001	100.5
101.5										,000	101.5

TABLE E   Wilcoxon's Rank Sum   (*Continued*)

Probability that Wilcoxon's rank sum minus the expected value
will be equal to or larger than W

W	m=2	m=4	m=6	m=8	m=10	m=12	m=14	m=16	m=18	m=20	W
					n = 17						
1	.474	.483	.486	.489	.490	.491	.492	.493	.494	.494	1
2	.421	.449	.459	.466	.471	.474	.477	.479	.481	.482	2
3	.374	.415	.432	.443	.451	.457	.461	.465	.468	.470	3
4	.327	.381	.406	.421	.432	.439	.446	.451	.455	.458	4
5	.287	.349	.380	.399	.412	.422	.430	.436	.442	.446	5
6	.246	.318	.354	.377	.393	.405	.415	.423	.429	.434	6
7	.211	.287	.329	.355	.374	.388	.400	.409	.416	.422	7
8	.175	.258	.304	.334	.356	.372	.385	.395	.403	.411	8
9	.146	.231	.281	.314	.337	.356	.370	.381	.391	.399	9
10	.117	.205	.258	.294	.320	.339	.355	.368	.378	.387	10
11	.094	.181	.236	.274	.302	.324	.341	.354	.366	.376	11
12	.070	.158	.216	.255	.285	.308	.326	.341	.354	.365	12
13	.053	.138	.196	.237	.268	.293	.312	.328	.342	.353	13
14	.035	.119	.177	.220	.252	.278	.299	.316	.330	.342	14
15	.023	.101	.160	.203	.237	.263	.285	.303	.318	.331	15
16	.012	.086	.143	.187	.222	.249	.272	.291	.307	.320	16
17	.006	.072	.128	.172	.207	.236	.259	.279	.295	.310	17
18	.000	.060	.114	.158	.193	.222	.246	.267	.284	.299	18
19		.049	.101	.144	.180	.209	.234	.255	.273	.289	19
20		.040	.089	.131	.167	.197	.222	.244	.262	.278	20
21		.032	.078	.119	.155	.185	.211	.233	.252	.268	21
22		.026	.068	.108	.143	.173	.199	.222	.241	.258	22
23		.020	.059	.097	.132	.162	.189	.211	.231	.249	23
24		.016	.050	.087	.121	.152	.178	.201	.221	.239	24
25		.012	.043	.078	.112	.141	.168	.191	.212	.230	25
26		.009	.037	.070	.102	.132	.158	.182	.202	.221	26
27		.006	.031	.062	.093	.122	.149	.172	.193	.212	27
28		.005	.026	.055	.085	.114	.140	.163	.184	.203	28
29		.003	.022	.049	.077	.105	.131	.155	.176	.195	29
30		.002	.018	.043	.070	.097	.123	.146	.167	.187	30
31		.001	.015	.038	.064	.090	.115	.138	.159	.178	31
32		.001	.012	.033	.057	.083	.107	.130	.151	.171	32
33		.000	.010	.029	.052	.076	.100	.123	.144	.163	33
34			.008	.025	.046	.070	.093	.115	.136	.156	34
35			.006	.021	.042	.064	.087	.109	.129	.148	35
36			.005	.018	.037	.059	.080	.102	.122	.142	36
37			.004	.016	.033	.053	.075	.096	.116	.135	37
38			.003	.013	.029	.049	.069	.090	.109	.128	38
39			.002	.011	.026	.044	.064	.084	.103	.122	39
40			.002	.009	.023	.040	.059	.078	.097	.116	40
41			.001	.008	.020	.036	.054	.073	.092	.110	41

TABLE E Wilcoxon's Rank Sum (*Continued*)

Probability that Wilcoxon's rank sum minus the expected value
will be equal to or larger than W

n = 17

W	m=2	m=4	m=6	m=8	m=10	m=12	m=14	m=16	m=18	m=20	W
42			,001	,007	,018	,033	,050	,068	,086	,104	42
43			,001	,005	,016	,030	,046	,063	,081	,099	43
44			,000	,005	,014	,027	,042	,059	,076	,094	44
45				,004	,012	,024	,039	,055	,072	,089	45
46				,003	,010	,021	,035	,051	,067	,084	46
47				,002	,009	,019	,032	,047	,063	,079	47
48				,002	,008	,017	,029	,044	,059	,075	48
49				,002	,006	,015	,027	,040	,055	,070	49
50				,001	,005	,013	,024	,037	,051	,066	50
51				,001	,005	,012	,022	,034	,048	,062	51
52				,001	,004	,010	,020	,032	,045	,059	52
53				,001	,003	,009	,018	,029	,042	,055	53
54				,000	,003	,008	,016	,027	,039	,052	54
55					,002	,007	,015	,024	,036	,049	55
56					,002	,006	,013	,022	,033	,045	56
57					,002	,005	,012	,020	,031	,043	57
58					,001	,005	,010	,019	,029	,040	58
59					,001	,004	,009	,017	,026	,037	59
60					,001	,003	,008	,015	,024	,035	60
61					,001	,003	,007	,014	,022	,032	61
62					,001	,003	,007	,013	,021	,030	62
63					,000	,002	,006	,011	,019	,028	63
64						,002	,005	,010	,017	,026	64
65						,002	,004	,009	,016	,024	65
66						,001	,004	,008	,015	,022	66
67						,001	,003	,008	,013	,021	67
68						,001	,003	,007	,012	,019	68
69						,001	,003	,006	,011	,018	69
70						,001	,002	,005	,010	,016	70
71						,001	,002	,005	,009	,015	71
72						,000	,002	,004	,008	,014	72
73							,001	,004	,008	,013	73
74							,001	,003	,007	,012	74
75							,001	,003	,006	,011	75
76							,001	,003	,006	,010	76
77							,001	,002	,005	,009	77
78							,001	,002	,005	,008	78
79							,001	,002	,004	,008	79
80							,000	,002	,004	,007	80
81								,001	,003	,006	81
82								,001	,003	,006	82

TABLE E   Wilcoxon's Rank Sum   (*Continued*)

Probability that Wilcoxon's rank sum minus the expected value
will be equal to or larger than W

					n = 17					
W	m=2	m=4	m=6	m=8	m=10	m=12	m=14	m=16	m=18	m=20	W
83								,001	,003	,005	83
84								,001	,002	,005	84
85								,001	,002	,004	85
86								,001	,002	,004	86
87								,001	,002	,004	87
88								,001	,002	,003	88
89								,000	,001	,003	89
90									,001	,003	90
91									,001	,002	91
92									,001	,002	92
93									,001	,002	93
94									,001	,002	94
95									,001	,002	95
96									,001	,001	96
97									,000	,001	97
98										,001	98
99										,001	99
100										,001	100
101										,001	101
102										,001	102
103										,001	103
104										,001	104
105										,000	105

					n = 18					
W	m=1	m=2	m=3	m=4	m=5	m=6	m=7	m=8	m=9	m=10	W
1	,474	,474	,481	,484	,486	,487	,488	,489	,490	,491	1
2	,421	,426	,444	,451	,457	,461	,465	,468	,470	,472	2
3	,368	,379	,407	,419	,428	,436	,441	,446	,450	,453	3
4	,316	,337	,370	,387	,400	,410	,418	,425	,430	,435	4
5	,263	,295	,335	,356	,373	,385	,395	,403	,410	,416	5
6	,211	,258	,300	,326	,346	,361	,373	,382	,391	,398	6
7	,158	,221	,267	,297	,319	,336	,350	,362	,371	,380	7
8	,105	,189	,235	,269	,294	,313	,329	,342	,353	,362	8
9	,053	,158	,206	,242	,269	,290	,307	,322	,334	,344	9
10	,000	,132	,178	,217	,245	,268	,287	,302	,316	,327	10
11		,105	,153	,193	,223	,247	,267	,284	,298	,310	11
12		,084	,131	,171	,202	,227	,248	,265	,281	,294	12

TABLE E Wilcoxon's Rank Sum   (*Continued*)

Probability that Wilcoxon's rank sum minus the expected value
will be equal to or larger than W

n = 18

W	m=1	m=2	m=3	m=4	m=5	m=6	m=7	m=8	m=9	m=10	W
13		,063	,111	,150	,182	,207	,229	,248	,264	,278	13
14		,047	,092	,131	,163	,189	,211	,231	,247	,262	14
15		,032	,077	,113	,145	,172	,194	,214	,232	,247	15
16		,021	,062	,098	,129	,155	,178	,199	,216	,232	16
17		,011	,050	,083	,113	,140	,163	,184	,202	,218	17
18		,005	,040	,070	,099	,125	,149	,169	,188	,204	18
19		,000	,031	,059	,087	,112	,135	,156	,174	,191	19
20			,023	,049	,075	,100	,122	,143	,161	,178	20
21			,017	,040	,065	,088	,110	,130	,149	,166	21
22			,012	,033	,055	,078	,099	,119	,137	,154	22
23			,008	,027	,047	,068	,089	,108	,126	,143	23
24			,005	,021	,040	,060	,079	,098	,116	,133	24
25			,003	,017	,033	,052	,070	,089	,106	,122	25
26			,002	,013	,028	,045	,062	,080	,097	,113	26
27			,001	,010	,023	,038	,055	,072	,088	,104	27
28			,000	,007	,019	,033	,048	,064	,080	,095	28
29				,005	,015	,028	,042	,057	,073	,087	29
30				,004	,012	,024	,037	,051	,066	,080	30
31				,002	,010	,020	,032	,045	,059	,073	31
32				,002	,008	,017	,028	,040	,053	,066	32
33				,001	,006	,014	,024	,035	,048	,060	33
34				,001	,004	,011	,020	,031	,042	,055	34
35				,000	,003	,009	,017	,027	,038	,049	35
36					,002	,007	,015	,023	,034	,044	36
37					,002	,006	,012	,020	,030	,040	37
38					,001	,005	,010	,018	,026	,036	38
39					,001	,004	,009	,015	,023	,032	39
40					,001	,003	,007	,013	,020	,029	40
41					,000	,002	,006	,011	,018	,025	41
42						,002	,005	,009	,015	,023	42
43						,001	,004	,008	,013	,020	43
44						,001	,003	,007	,012	,018	44
45						,001	,002	,006	,010	,016	45
46						,000	,002	,005	,009	,014	46
47							,002	,004	,007	,012	47
48							,001	,003	,006	,010	48
49							,001	,003	,005	,009	49
50							,001	,002	,004	,008	50
51							,001	,002	,004	,007	51
52							,000	,001	,003	,006	52
53								,001	,003	,005	53
54								,001	,002	,004	54

TABLE E   Wilcoxon's Rank Sum   (*Continued*)

Probability that Wilcoxon's rank sum minus the expected value
will be equal to or larger than W

					n = 18						
W	m=1	m=2	m=3	m=4	m=5	m=6	m=7	m=8	m=9	m=10	W
55								,001	,002	,004	55
56								,001	,001	,003	56
57								,000	,001	,003	57
58									,001	,002	58
59									,001	,002	59
60									,001	,002	60
61									,000	,001	61
62										,001	62
63										,001	63
64										,001	64
65										,001	65
66										,000	66

					n = 18						
W	m=11	m=12	m=13	m=14	m=15	m=16	m=17	m=18	m=19	m=20	W
1	,491	,492	,492	,493	,493	,493	,494	,494	,494	,494	1
2	,474	,475	,476	,478	,479	,480	,481	,481	,482	,483	2
3	,456	,459	,461	,463	,464	,466	,468	,469	,470	,471	3
4	,439	,442	,445	,448	,450	,453	,455	,456	,458	,460	4
5	,421	,426	,430	,433	,436	,439	,442	,444	,446	,448	5
6	,404	,409	,414	,418	,422	,426	,429	,432	,434	,437	6
7	,387	,393	,399	,404	,408	,412	,416	,420	,423	,426	7
8	,370	,377	,384	,389	,395	,399	,403	,407	,411	,414	8
9	,354	,362	,369	,375	,381	,386	,391	,395	,399	,403	9
10	,337	,346	,354	,361	,367	,373	,378	,383	,388	,392	10
11	,321	,331	,339	,347	,354	,360	,366	,371	,376	,381	11
12	,305	,316	,325	,333	,341	,348	,354	,360	,365	,370	12
13	,290	,301	,311	,320	,328	,335	,342	,348	,354	,359	13
14	,275	,287	,297	,306	,315	,323	,330	,337	,343	,348	14
15	,260	,273	,283	,293	,302	,311	,318	,325	,332	,338	15
16	,246	,259	,270	,281	,290	,299	,307	,314	,321	,327	16
17	,232	,245	,257	,268	,278	,287	,295	,303	,310	,317	17
18	,219	,232	,245	,256	,266	,275	,284	,292	,300	,306	18
19	,206	,220	,232	,244	,254	,264	,273	,281	,289	,296	19
20	,193	,207	,220	,232	,243	,253	,262	,271	,279	,286	20
21	,181	,196	,209	,221	,232	,242	,252	,261	,269	,277	21
22	,170	,184	,197	,210	,221	,232	,241	,251	,259	,267	22
23	,159	,173	,187	,199	,211	,221	,231	,241	,249	,258	23
24	,148	,163	,176	,189	,200	,211	,221	,231	,240	,248	24

TABLE E Wilcoxon's Rank Sum (*Continued*)

Probability that Wilcoxon's rank sum minus the expected value
will be equal to or larger than W

n = 18

W	m=11	m=12	m=13	m=14	m=15	m=16	m=17	m=18	m=19	m=20	W
25	,138	,152	,166	,179	,190	,201	,212	,221	,231	,239	25
26	,128	,143	,156	,169	,181	,192	,202	,212	,221	,230	26
27	,119	,133	,147	,159	,171	,183	,193	,203	,213	,221	27
28	,110	,124	,138	,150	,162	,174	,184	,194	,204	,213	28
29	,102	,116	,129	,142	,154	,165	,176	,186	,195	,205	29
30	,094	,108	,121	,133	,145	,157	,167	,178	,187	,196	30
31	,087	,100	,113	,125	,137	,148	,159	,169	,179	,188	31
32	,080	,093	,105	,118	,129	,141	,151	,162	,171	,181	32
33	,073	,086	,098	,110	,122	,133	,144	,154	,164	,173	33
34	,067	,079	,091	,103	,115	,126	,136	,147	,156	,166	34
35	,061	,073	,085	,096	,108	,119	,129	,139	,149	,159	35
36	,056	,067	,079	,090	,101	,112	,122	,132	,142	,152	36
37	,051	,062	,073	,084	,095	,105	,116	,126	,135	,145	37
38	,046	,057	,067	,078	,089	,099	,109	,119	,129	,138	38
39	,042	,052	,062	,073	,083	,093	,103	,113	,123	,132	39
40	,038	,047	,057	,067	,078	,088	,097	,107	,117	,126	40
41	,034	,043	,053	,063	,072	,082	,092	,101	,111	,120	41
42	,031	,039	,049	,058	,067	,077	,086	,096	,105	,114	42
43	,028	,036	,045	,054	,063	,072	,081	,090	,100	,108	43
44	,025	,032	,041	,049	,058	,067	,076	,085	,094	,103	44
45	,022	,029	,037	,046	,054	,063	,072	,081	,089	,098	45
46	,020	,027	,034	,042	,050	,059	,067	,076	,084	,093	46
47	,018	,024	,031	,039	,046	,055	,063	,071	,080	,088	47
48	,016	,022	,028	,035	,043	,051	,059	,067	,075	,083	48
49	,014	,019	,026	,032	,040	,047	,055	,063	,071	,079	49
50	,012	,017	,023	,030	,037	,044	,051	,059	,067	,075	50
51	,011	,015	,021	,027	,034	,041	,048	,055	,063	,071	51
52	,009	,014	,019	,025	,031	,038	,045	,052	,059	,067	52
53	,008	,012	,017	,023	,028	,035	,042	,049	,056	,063	53
54	,007	,011	,015	,020	,026	,032	,039	,045	,052	,059	54
55	,006	,010	,014	,019	,024	,030	,036	,042	,049	,056	55
56	,005	,009	,012	,017	,022	,027	,033	,039	,046	,053	56
57	,005	,008	,011	,015	,020	,025	,031	,037	,043	,049	57
58	,004	,007	,010	,014	,018	,023	,029	,034	,040	,046	58
59	,003	,006	,009	,012	,017	,021	,026	,032	,038	,044	59
60	,003	,005	,008	,011	,015	,019	,024	,030	,035	,041	60
61	,003	,004	,007	,010	,014	,018	,022	,027	,033	,038	61
62	,002	,004	,006	,009	,012	,016	,021	,025	,031	,036	62
63	,002	,003	,005	,008	,011	,015	,019	,024	,028	,034	63

TABLE E   Wilcoxon's Rank Sum   (*Continued*)

Probability that Wilcoxon's rank sum minus the expected value
will be equal to or larger than W

n = 18

W	m=11	m=12	m=13	m=14	m=15	m=16	m=17	m=18	m=19	m=20	W
64	,002	,003	,005	,007	,010	,014	,017	,022	,026	,031	64
65	,001	,002	,004	,006	,009	,012	,016	,020	,025	,029	65
66	,001	,002	,004	,006	,008	,011	,015	,019	,023	,027	66
67	,001	,002	,003	,005	,007	,010	,013	,017	,021	,026	67
68	,001	,002	,003	,004	,007	,009	,012	,016	,020	,024	68
69	,001	,001	,002	,004	,006	,008	,011	,014	,018	,022	69
70	,001	,001	,002	,003	,005	,008	,010	,013	,017	,021	70
71	,000	,001	,002	,003	,005	,007	,009	,012	,015	,019	71
72		,001	,002	,003	,004	,006	,008	,011	,014	,018	72
73		,001	,001	,002	,004	,006	,008	,010	,013	,016	73
74		,001	,001	,002	,003	,005	,007	,009	,012	,015	74
75		,000	,001	,002	,003	,004	,006	,009	,011	,014	75
76			,001	,002	,003	,004	,006	,008	,010	,013	76
77			,001	,001	,002	,004	,005	,007	,009	,012	77
78			,001	,001	,002	,003	,005	,006	,009	,011	78
79			,001	,001	,002	,003	,004	,006	,008	,010	79
80			,000	,001	,002	,002	,004	,005	,007	,009	80
81				,001	,001	,002	,003	,005	,007	,009	81
82				,001	,001	,002	,003	,004	,006	,008	82
83				,001	,001	,002	,003	,004	,005	,007	83
84				,000	,001	,002	,002	,004	,005	,007	84
85					,001	,001	,002	,003	,005	,006	85
86					,001	,001	,002	,003	,004	,006	86
87					,001	,001	,002	,003	,004	,005	87
88					,001	,001	,002	,002	,003	,005	88
89					,000	,001	,001	,002	,003	,004	89
90						,001	,001	,002	,003	,004	90
91						,001	,001	,002	,002	,004	91
92						,001	,001	,001	,002	,003	92
93						,000	,001	,001	,002	,003	93
94							,001	,001	,002	,003	94
95							,001	,001	,002	,002	95
96							,001	,001	,001	,002	96
97							,000	,001	,001	,002	97
98								,001	,001	,002	98
99								,001	,001	,002	99
100								,001	,001	,001	100
101								,000	,001	,001	101
102									,001	,001	102
103									,001	,001	103

TABLE E   Wilcoxon's Rank Sum   (*Continued*)

Probability that Wilcoxon's rank sum minus the expected value
will be equal to or larger than W

n = 18

W	m=11	m=12	m=13	m=14	m=15	m=16	m=17	m=18	m=19	m=20	W
104									,001	,001	104
105									,001	,001	105
106									,000	,001	106
107										,001	107
108										,001	108
109										,001	109
110										,000	110

n = 19

W	m=1	m=3	m=5	m=7	m=9	m=11	m=13	m=15	m=17	m=19	W
0,5	,500	,500	,500	,500	,500	,500	,500	,500	,500	,500	0,5
1,5	,450	,464	,473	,478	,481	,483	,485	,486	,488	,488	1,5
2,5	,400	,429	,445	,455	,462	,466	,470	,473	,475	,477	2,5
3,5	,350	,394	,418	,433	,442	,449	,455	,459	,463	,466	3,5
4,5	,300	,359	,392	,411	,423	,433	,440	,446	,450	,454	4,5
5,5	,250	,325	,365	,389	,405	,416	,425	,432	,438	,443	5,5
6,5	,200	,293	,340	,367	,386	,400	,410	,419	,426	,431	6,5
7,5	,150	,262	,315	,346	,368	,383	,396	,405	,413	,420	7,5
8,5	,100	,232	,290	,326	,350	,367	,381	,392	,401	,409	8,5
9,5	,050	,204	,267	,305	,332	,352	,367	,379	,389	,398	9,5
10,5	,000	,178	,245	,286	,315	,336	,353	,366	,377	,386	10,5
11,5		,154	,223	,267	,298	,321	,339	,353	,365	,375	11,5
12,5		,132	,203	,248	,281	,306	,325	,341	,354	,365	12,5
13,5		,113	,183	,231	,265	,291	,311	,328	,342	,354	13,5
14,5		,095	,165	,214	,249	,276	,298	,316	,331	,343	14,5
15,5		,080	,148	,197	,234	,262	,285	,304	,319	,333	15,5
16,5		,066	,132	,182	,219	,249	,272	,292	,308	,322	16,5
17,5		,054	,118	,167	,205	,235	,260	,280	,297	,312	17,5
18,5		,044	,104	,153	,191	,222	,248	,269	,286	,302	18,5
19,5		,034	,091	,140	,178	,210	,236	,257	,276	,292	19,5
20,5		,027	,080	,127	,166	,198	,224	,246	,265	,282	20,5
21,5		,020	,070	,115	,154	,186	,213	,236	,255	,272	21,5
22,5		,015	,060	,104	,143	,175	,202	,225	,245	,263	22,5
23,5		,010	,052	,094	,132	,164	,191	,215	,235	,253	23,5
24,5		,007	,044	,085	,121	,154	,181	,205	,226	,244	24,5
25,5		,005	,038	,076	,112	,144	,171	,195	,216	,235	25,5
26,5		,003	,032	,068	,103	,134	,162	,186	,207	,226	26,5
27,5		,001	,026	,060	,094	,125	,152	,177	,198	,217	27,5
28,5		,001	,022	,053	,086	,116	,144	,168	,190	,209	28,5

TABLE E Wilcoxon's Rank Sum (*Continued*)

Probability that Wilcoxon's rank sum minus the expected value
will be equal to or larger than W

W	m=1	m=3	m=5	m=7	m=9	m=11	m=13	m=15	m=17	m=19	W
					n = 19						
29.5		.000	.018	.047	.078	.108	.135	.159	.181	.201	29.5
30.5			.015	.041	.071	.100	.127	.151	.173	.193	30.5
31.5			.012	.036	.065	.093	.119	.143	.165	.185	31.5
32.5			.010	.032	.058	.086	.112	.135	.157	.177	32.5
33.5			.008	.028	.053	.079	.104	.128	.150	.170	33.5
34.5			.006	.024	.047	.073	.098	.121	.143	.162	34.5
35.5			.005	.021	.043	.067	.091	.114	.136	.155	35.5
36.5			.004	.018	.038	.061	.085	.107	.129	.148	36.5
37.5			.003	.015	.034	.056	.079	.101	.122	.142	37.5
38.5			.002	.013	.030	.051	.073	.095	.116	.135	38.5
39.5			.001	.011	.027	.047	.068	.089	.110	.129	39.5
40.5			.001	.009	.024	.043	.063	.084	.104	.123	40.5
41.5			.001	.008	.021	.039	.058	.078	.098	.117	41.5
42.5			.000	.006	.019	.035	.054	.073	.093	.111	42.5
43.5				.005	.016	.032	.050	.069	.088	.106	43.5
44.5				.004	.014	.029	.046	.064	.083	.101	44.5
45.5				.003	.012	.026	.042	.060	.078	.096	45.5
46.5				.003	.011	.023	.039	.056	.073	.091	46.5
47.5				.002	.009	.021	.035	.052	.069	.086	47.5
48.5				.002	.008	.019	.032	.048	.065	.081	48.5
49.5				.001	.007	.017	.030	.045	.061	.077	49.5
50.5				.001	.006	.015	.027	.041	.057	.073	50.5
51.5				.001	.005	.013	.025	.038	.053	.069	51.5
52.5				.001	.004	.012	.022	.035	.050	.065	52.5
53.5				.000	.004	.010	.020	.033	.047	.061	53.5
54.5					.003	.009	.018	.030	.044	.058	54.5
55.5					.003	.008	.017	.028	.041	.054	55.5
56.5					.002	.007	.015	.026	.038	.051	56.5
57.5					.002	.006	.014	.024	.035	.048	57.5
58.5					.001	.005	.012	.022	.033	.045	58.5
59.5					.001	.005	.011	.020	.030	.043	59.5
60.5					.001	.004	.010	.018	.028	.040	60.5
61.5					.001	.004	.009	.017	.026	.037	61.5
62.5					.001	.003	.008	.015	.024	.035	62.5
63.5					.001	.003	.007	.014	.022	.033	63.5
64.5					.000	.002	.006	.013	.021	.031	64.5
65.5						.002	.006	.011	.019	.029	65.5
66.5						.002	.005	.010	.018	.027	66.5
67.5						.001	.004	.009	.016	.025	67.5
68.5						.001	.004	.008	.015	.023	68.5
69.5						.001	.003	.008	.014	.022	69.5

TABLE E Wilcoxon's Rank Sum *(Continued)*

Probability that Wilcoxon's rank sum minus the expected value
will be equal to or larger than W

						n = 19					
W	m=1	m=3	m=5	m=7	m=9	m=11	m=13	m=15	m=17	m=19	W
70,5						,001	,003	,007	,013	,020	70,5
71,5						,001	,003	,006	,012	,019	71,5
72,5						,001	,002	,006	,011	,017	72,5
73,5						,000	,002	,005	,010	,016	73,5
74,5							,002	,004	,009	,015	74,5
75,5							,001	,004	,008	,014	75,5
76,5							,001	,004	,007	,013	76,5
77,5							,001	,003	,007	,012	77,5
78,5							,001	,003	,006	,011	78,5
79,5							,001	,003	,006	,010	79,5
80,5							,001	,002	,005	,009	80,5
81,5							,001	,002	,005	,008	81,5
82,5							,001	,002	,004	,008	82,5
83,5							,000	,002	,004	,007	83,5
84,5								,001	,003	,006	84,5
85,5								,001	,003	,006	85,5
86,5								,001	,003	,005	86,5
87,5								,001	,002	,005	87,5
88,5								,001	,002	,005	88,5
89,5								,001	,002	,004	89,5
90,5								,001	,002	,004	90,5
91,5								,001	,002	,003	91,5
92,5								,000	,001	,003	92,5
93,5									,001	,003	93,5
94,5									,001	,003	94,5
95,5									,001	,002	95,5
96,5									,001	,002	96,5
97,5									,001	,002	97,5
98,5									,001	,002	98,5
99,5									,001	,002	99,5
100,5									,001	,001	100,5
101,5									,000	,001	101,5
102,5										,001	102,5
103,5										,001	103,5
104,5										,001	104,5
105,5										,001	105,5
106,5										,001	106,5
107,5										,001	107,5
108,5										,001	108,5
109,5										,001	109,5
110,5										,000	110,5

TABLE E Wilcoxon's Rank Sum *(Continued)*

Probability that Wilcoxon's rank sum minus the expected value
will be equal to or larger than W

W	m=2	m=4	m=6	m=8	m=10	m=12	m=14	m=16	m=18	m=20	W
1	.476	.484	.488	.490	.491	.492	.493	.493	.494	.494	1
2	.429	.453	.463	.469	.473	.476	.479	.480	.482	.483	2
3	.386	.422	.439	.448	.455	.460	.464	.467	.470	.472	3
4	.343	.392	.414	.428	.437	.444	.450	.454	.458	.461	4
5	.305	.363	.390	.408	.420	.429	.436	.442	.446	.450	5
6	.267	.334	.367	.387	.402	.413	.422	.429	.434	.439	6
7	.233	.306	.344	.368	.385	.398	.408	.416	.423	.428	7
8	.200	.279	.321	.348	.368	.382	.394	.403	.411	.417	8
9	.171	.253	.299	.329	.351	.367	.380	.391	.399	.407	9
10	.143	.228	.278	.310	.334	.352	.366	.378	.388	.396	10
11	.119	.205	.257	.292	.318	.337	.353	.366	.376	.385	11
12	.095	.183	.237	.275	.302	.323	.340	.353	.365	.375	12
13	.076	.162	.218	.257	.286	.309	.327	.341	.354	.364	13
14	.057	.143	.200	.241	.271	.295	.314	.329	.343	.354	14
15	.043	.125	.183	.225	.256	.281	.301	.318	.332	.344	15
16	.029	.109	.167	.209	.242	.268	.289	.306	.321	.333	16
17	.019	.094	.151	.194	.228	.255	.276	.295	.310	.323	17
18	.010	.081	.137	.180	.214	.242	.264	.283	.300	.313	18
19	.005	.069	.123	.167	.201	.229	.253	.272	.289	.304	19
20	.000	.058	.111	.154	.189	.217	.241	.262	.279	.294	20
21		.049	.099	.141	.177	.206	.230	.251	.269	.284	21
22		.041	.088	.130	.165	.194	.219	.241	.259	.275	22
23		.033	.078	.119	.154	.184	.209	.231	.249	.266	23
24		.027	.069	.109	.143	.173	.199	.221	.240	.257	24
25		.022	.061	.099	.133	.163	.189	.211	.231	.248	25
26		.018	.053	.090	.123	.153	.179	.202	.221	.239	26
27		.014	.046	.081	.114	.144	.170	.192	.213	.230	27
28		.011	.040	.073	.106	.135	.161	.184	.204	.222	28
29		.008	.034	.066	.097	.126	.152	.175	.195	.214	29
30		.006	.030	.059	.090	.118	.143	.167	.187	.206	30
31		.004	.025	.053	.082	.110	.135	.158	.179	.198	31
32		.003	.021	.047	.075	.102	.128	.151	.171	.190	32
33		.002	.018	.042	.069	.095	.120	.143	.164	.183	33
34		.001	.015	.037	.063	.088	.113	.136	.156	.175	34
35		.001	.013	.033	.057	.082	.106	.128	.149	.168	35
36		.000	.010	.029	.052	.076	.099	.122	.142	.161	36
37			.009	.026	.047	.070	.093	.115	.135	.154	37
38			.007	.022	.043	.065	.087	.109	.129	.148	38
39			.006	.020	.039	.060	.081	.103	.123	.141	39
40			.005	.017	.035	.055	.076	.097	.117	.135	40
41			.004	.015	.031	.051	.071	.091	.111	.129	41

TABLE E   Wilcoxon's Rank Sum   (*Continued*)

Probability that Wilcoxon's rank sum minus the expected value
will be equal to or larger than W

						n = 19					
W	m=2	m=4	m=6	m=8	m=10	m=12	m=14	m=16	m=18	m=20	W
42			,003	,013	,028	,046	,066	,086	,105	,123	42
43			,002	,011	,025	,042	,061	,081	,100	,118	43
44			,002	,009	,022	,039	,057	,076	,094	,112	44
45			,001	,008	,020	,035	,053	,071	,089	,107	45
46			,001	,007	,018	,032	,049	,067	,084	,102	46
47			,001	,006	,016	,029	,045	,062	,080	,097	47
48			,001	,005	,014	,027	,042	,058	,075	,092	48
49			,000	,004	,012	,024	,039	,054	,071	,087	49
50				,003	,011	,022	,036	,051	,067	,083	50
51				,003	,009	,020	,033	,047	,063	,079	51
52				,002	,008	,018	,030	,044	,059	,075	52
53				,002	,007	,016	,028	,041	,056	,071	53
54				,001	,006	,014	,025	,038	,052	,067	54
55				,001	,005	,013	,023	,035	,049	,063	55
56				,001	,005	,011	,021	,033	,046	,060	56
57				,001	,004	,010	,019	,030	,043	,057	57
58				,001	,003	,009	,017	,028	,040	,053	58
59				,000	,003	,008	,016	,026	,038	,050	59
60					,002	,007	,014	,024	,035	,047	60
61					,002	,006	,013	,022	,033	,045	61
62					,002	,006	,012	,020	,031	,042	62
63					,001	,005	,011	,019	,028	,039	63
64					,001	,004	,010	,017	,026	,037	64
65					,001	,004	,009	,016	,025	,035	65
66					,001	,003	,008	,014	,023	,033	66
67					,001	,003	,007	,013	,021	,031	67
68					,001	,002	,006	,012	,020	,029	68
69					,000	,002	,006	,011	,018	,027	69
70						,002	,005	,010	,017	,025	70
71						,002	,004	,009	,015	,023	71
72						,001	,004	,008	,014	,022	72
73						,001	,003	,007	,013	,020	73
74						,001	,003	,007	,012	,019	74
75						,001	,003	,006	,011	,018	75
76						,001	,002	,006	,010	,016	76
77						,001	,002	,005	,009	,015	77
78						,001	,002	,005	,009	,014	78
79						,000	,002	,004	,008	,013	79
80							,001	,004	,007	,012	80
81							,001	,003	,007	,011	81
82							,001	,003	,006	,010	82

TABLE E   Wilcoxon's Rank Sum   (*Continued*)

Probability that Wilcoxon's rank sum minus the expected value
will be equal to or larger than W

W	m=2	m=4	m=6	m=8	m=10	m=12	m=14	m=16	m=18	m=20	W
								n = 19			
83							,001	,003	,005	,010	83
84							,001	,002	,005	,009	84
85							,001	,002	,005	,008	85
86							,001	,002	,004	,008	86
87							,001	,002	,004	,007	87
88							,000	,001	,003	,006	88
89								,001	,003	,006	89
90								,001	,003	,005	90
91								,001	,002	,005	91
92								,001	,002	,005	92
93								,001	,002	,004	93
94								,001	,002	,004	94
95								,001	,002	,003	95
96								,001	,001	,003	96
97								,000	,001	,003	97
98									,001	,003	98
99									,001	,002	99
100									,001	,002	100
101									,001	,002	101
102									,001	,002	102
103									,001	,002	103
104									,001	,001	104
105									,001	,001	105
106									,000	,001	106
107										,001	107
108										,001	108
109										,001	109
110										,001	110
111										,001	111
112										,001	112
113										,001	113
114										,000	114

W	m=1	m=2	m=3	m=4	m=5	m=6	m=7	m=8	m=9	m=10	W
						n = 20					
1	,476	,476	,483	,485	,487	,488	,489	,490	,491	,491	1
2	,429	,433	,449	,455	,461	,465	,468	,470	,472	,474	2
3	,381	,390	,415	,426	,435	,441	,446	,450	,454	,457	3
4	,333	,351	,382	,397	,409	,418	,425	,431	,436	,440	4

TABLE E Wilcoxon's Rank Sum (*Continued*)

Probability that Wilcoxon's rank sum minus the expected value
will be equal to or larger than W

					n = 20						
W	m=1	m=2	m=3	m=4	m=5	m=6	m=7	m=8	m=9	m=10	W
5	,286	,312	,349	,368	,384	,395	,404	,411	,418	,423	5
6	,238	,277	,317	,341	,359	,372	,383	,392	,400	,406	6
7	,190	,242	,286	,314	,334	,350	,363	,373	,382	,389	7
8	,143	,212	,257	,288	,311	,328	,343	,354	,364	,373	8
9	,095	,182	,229	,262	,287	,307	,323	,336	,347	,357	9
10	,048	,156	,202	,239	,265	,286	,304	,318	,330	,341	10
11	,000	,130	,177	,216	,244	,266	,285	,300	,313	,325	11
12		,108	,155	,194	,223	,247	,267	,283	,297	,309	12
13		,087	,134	,174	,204	,229	,249	,266	,281	,294	13
14		,069	,115	,155	,185	,211	,232	,250	,266	,279	14
15		,052	,098	,137	,168	,194	,216	,234	,251	,265	15
16		,039	,083	,120	,151	,178	,200	,219	,236	,251	16
17		,026	,069	,105	,136	,162	,185	,205	,222	,237	17
18		,017	,058	,091	,122	,148	,171	,191	,208	,224	18
19		,009	,047	,079	,108	,134	,157	,177	,195	,211	19
20		,004	,038	,067	,096	,121	,144	,164	,182	,199	20
21		,000	,030	,057	,085	,109	,132	,152	,170	,187	21
22			,023	,048	,074	,098	,120	,140	,159	,175	22
23			,018	,041	,065	,088	,109	,129	,147	,164	23
24			,013	,034	,056	,078	,099	,119	,137	,153	24
25			,009	,028	,048	,069	,090	,109	,127	,143	25
26			,006	,023	,042	,061	,081	,100	,117	,134	26
27			,004	,018	,035	,054	,073	,091	,108	,124	27
28			,002	,015	,030	,047	,065	,083	,099	,115	28
29			,001	,011	,025	,041	,058	,075	,091	,107	29
30			,001	,009	,021	,036	,052	,068	,084	,099	30
31			,000	,007	,018	,031	,046	,061	,077	,091	31
32				,005	,014	,027	,041	,055	,070	,084	32
33				,004	,012	,023	,036	,050	,064	,078	33
34				,003	,010	,020	,031	,044	,058	,071	34
35				,002	,008	,017	,027	,040	,052	,065	35
36				,001	,006	,014	,024	,035	,047	,060	36
37				,001	,005	,012	,021	,031	,043	,055	37
38				,000	,004	,010	,018	,028	,038	,050	38
39					,003	,008	,015	,024	,034	,045	39
40					,002	,007	,013	,021	,031	,041	40
41					,002	,005	,011	,019	,028	,037	41
42					,001	,004	,010	,016	,025	,034	42
43					,001	,003	,008	,014	,022	,030	43
44					,001	,003	,007	,012	,019	,027	44
45					,000	,002	,006	,011	,017	,025	45

TABLE E   Wilcoxon's Rank Sum   (*Continued*)

Probability that Wilcoxon's rank sum minus the expected value
will be equal to or larger than W

					n = 20					
W	m=1	m=2	m=3	m=4	m=5	m=6	m=7	m=8	m=9	m=10	W
46						,002	,005	,009	,015	,022	46
47						,001	,004	,008	,013	,020	47
48						,001	,003	,007	,012	,017	48
49						,001	,003	,006	,010	,016	49
50						,001	,002	,005	,009	,014	50
51						,000	,002	,004	,008	,012	51
52							,001	,003	,007	,011	52
53							,001	,003	,006	,009	53
54							,001	,002	,005	,008	54
55							,001	,002	,004	,007	55
56							,000	,002	,004	,006	56
57								,001	,003	,006	57
58								,001	,003	,005	58
59								,001	,002	,004	59
60								,001	,002	,004	60
61								,001	,001	,003	61
62								,000	,001	,003	62
63									,001	,002	63
64									,001	,002	64
65									,001	,002	65
66									,001	,001	66
67									,000	,001	67
68										,001	68
69										,001	69
70										,001	70
71										,001	71
72										,000	72

					n = 20					
W	m=11	m=12	m=13	m=14	m=15	m=16	m=17	m=18	m=19	m=20	W
1	,492	,492	,493	,493	,493	,494	,494	,494	,494	,495	1
2	,476	,477	,478	,479	,480	,481	,482	,483	,483	,484	2
3	,460	,462	,464	,466	,467	,469	,470	,471	,472	,473	3
4	,443	,447	,449	,452	,454	,456	,458	,460	,461	,463	4
5	,427	,432	,435	,438	,441	,444	,446	,448	,450	,452	5
6	,412	,417	,421	,425	,428	,431	,434	,437	,439	,442	6
7	,396	,402	,407	,411	,415	,419	,422	,426	,428	,431	7
8	,380	,387	,393	,398	,403	,407	,411	,414	,417	,421	8
9	,365	,372	,379	,385	,390	,395	,399	,403	,407	,410	9

TABLE E   Wilcoxon's Rank Sum   (*Continued*)

Probability that Wilcoxon's rank sum minus the expected value will be equal to or larger than W

W	m=11	m=12	m=13	m=14	m=15	m=16	m=17	m=18	m=19	m=20	W
					n = 20						
10	,350	,358	,365	,371	,377	,383	,387	,392	,396	,400	10
11	,335	,344	,351	,358	,365	,371	,376	,381	,385	,389	11
12	,320	,330	,338	,346	,353	,359	,365	,370	,375	,379	12
13	,306	,316	,325	,333	,340	,347	,353	,359	,364	,369	13
14	,291	,302	,312	,321	,328	,336	,342	,348	,354	,359	14
15	,278	,289	,299	,308	,317	,324	,331	,338	,344	,349	15
16	,264	,276	,287	,296	,305	,313	,320	,327	,333	,339	16
17	,251	,263	,274	,284	,293	,302	,310	,317	,323	,329	17
18	,238	,251	,262	,273	,282	,291	,299	,306	,313	,320	18
19	,226	,239	,250	,261	,271	,280	,289	,296	,304	,310	19
20	,213	,227	,239	,250	,260	,270	,278	,286	,294	,301	20
21	,202	,215	,228	,239	,250	,259	,268	,277	,284	,292	21
22	,190	,204	,217	,229	,239	,249	,258	,267	,275	,282	22
23	,179	,193	,206	,218	,229	,239	,249	,258	,266	,273	23
24	,169	,183	,196	,208	,219	,230	,239	,248	,257	,265	24
25	,159	,173	,186	,198	,210	,220	,230	,239	,248	,256	25
26	,149	,163	,176	,189	,200	,211	,221	,230	,239	,247	26
27	,139	,154	,167	,179	,191	,202	,212	,221	,230	,239	27
28	,130	,145	,158	,170	,182	,193	,203	,213	,222	,231	28
29	,122	,136	,149	,162	,173	,184	,195	,205	,214	,222	29
30	,114	,128	,141	,153	,165	,176	,187	,196	,206	,215	30
31	,106	,120	,133	,145	,157	,168	,178	,188	,198	,207	31
32	,098	,112	,125	,137	,149	,160	,171	,181	,190	,199	32
33	,091	,105	,117	,130	,141	,152	,163	,173	,183	,192	33
34	,085	,098	,110	,122	,134	,145	,156	,166	,175	,184	34
35	,078	,091	,103	,115	,127	,138	,148	,159	,168	,177	35
36	,072	,085	,097	,109	,120	,131	,142	,152	,161	,170	36
37	,067	,079	,091	,102	,113	,124	,135	,145	,154	,164	37
38	,061	,073	,085	,096	,107	,118	,128	,138	,148	,157	38
39	,056	,068	,079	,090	,101	,112	,122	,132	,141	,151	39
40	,052	,063	,074	,085	,095	,106	,116	,126	,135	,144	40
41	,047	,058	,069	,079	,090	,100	,110	,120	,129	,138	41
42	,043	,053	,064	,074	,084	,094	,104	,114	,123	,132	42
43	,040	,049	,059	,069	,079	,089	,099	,108	,118	,127	43
44	,036	,045	,055	,065	,074	,084	,094	,103	,112	,121	44
45	,033	,042	,051	,060	,070	,079	,089	,098	,107	,116	45
46	,030	,038	,047	,056	,065	,075	,084	,093	,102	,111	46
47	,027	,035	,043	,052	,061	,070	,079	,088	,097	,105	47
48	,024	,032	,040	,048	,057	,066	,075	,083	,092	,101	48
49	,022	,029	,037	,045	,053	,062	,070	,079	,087	,096	49
50	,020	,027	,034	,042	,050	,058	,066	,075	,083	,091	50

TABLE E   Wilcoxon's Rank Sum   (*Continued*)

Probability that Wilcoxon's rank sum minus the expected value
will be equal to or larger than W

					n = 20						
W	m=11	m=12	m=13	m=14	m=15	m=16	m=17	m=18	m=19	m=20	W
51	,018	,024	,031	,039	,046	,054	,062	,071	,079	,087	51
52	,016	,022	,028	,036	,043	,051	,059	,067	,075	,083	52
53	,014	,020	,026	,033	,040	,047	,055	,063	,071	,079	53
54	,013	,018	,024	,030	,037	,044	,052	,059	,067	,075	54
55	,011	,016	,022	,028	,034	,041	,049	,056	,063	,071	55
56	,010	,015	,020	,026	,032	,039	,045	,053	,060	,067	56
57	,009	,013	,018	,024	,029	,036	,043	,049	,057	,064	57
58	,008	,012	,016	,022	,027	,033	,040	,046	,053	,060	58
59	,007	,011	,015	,020	,025	,031	,037	,044	,050	,057	59
60	,006	,009	,013	,018	,023	,029	,035	,041	,047	,054	60
61	,005	,008	,012	,016	,021	,027	,032	,038	,045	,051	61
62	,005	,008	,011	,015	,020	,025	,030	,036	,042	,048	62
63	,004	,007	,010	,014	,018	,023	,028	,034	,039	,046	63
64	,004	,006	,009	,012	,017	,021	,026	,031	,037	,043	64
65	,003	,005	,008	,011	,015	,019	,024	,029	,035	,040	65
66	,003	,005	,007	,010	,014	,018	,022	,027	,033	,038	66
67	,002	,004	,006	,009	,013	,017	,021	,026	,031	,036	67
68	,002	,004	,006	,008	,012	,015	,019	,024	,029	,034	68
69	,002	,003	,005	,008	,010	,014	,018	,022	,027	,032	69
70	,001	,003	,005	,007	,010	,013	,016	,021	,025	,030	70
71	,001	,002	,004	,006	,009	,012	,015	,019	,023	,028	71
72	,001	,002	,004	,005	,008	,011	,014	,018	,022	,026	72
73	,001	,002	,003	,005	,007	,010	,013	,016	,020	,025	73
74	,001	,002	,003	,004	,006	,009	,012	,015	,019	,023	74
75	,001	,001	,002	,004	,006	,008	,011	,014	,018	,021	75
76	,001	,001	,002	,003	,005	,007	,010	,013	,016	,020	76
77	,000	,001	,002	,003	,005	,007	,009	,012	,015	,019	77
78		,001	,002	,003	,004	,006	,008	,011	,014	,018	78
79		,001	,001	,002	,004	,006	,008	,010	,013	,016	79
80		,001	,001	,002	,003	,005	,007	,009	,012	,015	80
81		,001	,001	,002	,003	,005	,006	,009	,011	,014	81
82		,000	,001	,002	,003	,004	,006	,008	,010	,013	82
83			,001	,001	,002	,004	,005	,007	,010	,012	83
84			,001	,001	,002	,003	,005	,007	,009	,011	84
85			,001	,001	,002	,003	,004	,006	,008	,011	85
86			,001	,001	,002	,003	,004	,006	,008	,010	86
87			,000	,001	,002	,002	,004	,005	,007	,009	87
88				,001	,001	,002	,003	,005	,006	,008	88
89				,001	,001	,002	,003	,004	,006	,008	89
90				,001	,001	,002	,003	,004	,005	,007	90

TABLE E   Wilcoxon's Rank Sum   (*Continued*)

Probability that Wilcoxon's rank sum minus the expected value
will be equal to or larger than W

					n = 20						
W	m=11	m=12	m=13	m=14	m=15	m=16	m=17	m=18	m=19	m=20	W
91				,000	,001	,002	,002	,004	,005	,007	91
92					,001	,001	,002	,003	,005	,006	92
93					,001	,001	,002	,003	,004	,006	93
94					,001	,001	,002	,003	,004	,005	94
95					,001	,001	,002	,002	,003	,005	95
96					,000	,001	,001	,002	,003	,004	96
97						,001	,001	,002	,003	,004	97
98						,001	,001	,002	,003	,004	98
99						,001	,001	,002	,002	,003	99
100						,001	,001	,001	,002	,003	100
101						,000	,001	,001	,002	,003	101
102							,001	,001	,002	,003	102
103							,001	,001	,002	,002	103
104							,001	,001	,001	,002	104
105							,000	,001	,001	,002	105
106								,001	,001	,002	106
107								,001	,001	,002	107
108								,001	,001	,001	108
109								,001	,001	,001	109
110								,000	,001	,001	110
111									,001	,001	111
112									,001	,001	112
113									,001	,001	113
114									,000	,001	114
115										,001	115
116										,001	116
117										,001	117
118										,001	118
119										,000	119

TABLE F Wilcoxon's Signed Rank

Probability that Wilcoxon's signed rank minus the
expected value will be equal to or larger than W

W	n=1	n=2	W	n=3	n=4	W	n=5	n=6	W	n=7	n=8	W	n=9	n=10
0.5	.500	.500	1	.375	.437	0.5	.500	.500	1	.469	.473	0.5	.500	.500
1.5	.000	.250	2	.250	.312	1.5	.406	.422	2	.406	.422	1.5	.455	.461
2.5		.000	3	.125	.187	2.5	.312	.344	3	.344	.371	2.5	.410	.423
			4	.000	.125	3.5	.219	.281	4	.289	.320	3.5	.367	.385
			5		.062	4.5	.156	.219	5	.234	.273	4.5	.326	.348
			6		.000	5.5	.094	.156	6	.187	.230	5.5	.285	.312
						6.5	.062	.109	7	.148	.191	6.5	.248	.278
						7.5	.031	.078	8	.109	.156	7.5	.213	.246
						8.5	.000	.047	9	.078	.125	8.5	.180	.216
						9.5		.031	10	.055	.098	9.5	.150	.187
						10.5		.016	11	.039	.074	10.5	.125	.161
						11.5		.000	12	.023	.055	11.5	.102	.138
									13	.016	.039	12.5	.082	.116
									14	.008	.027	13.5	.064	.097
									15	.000	.020	14.5	.049	.080
									16		.012	15.5	.037	.065
									17		.008	16.5	.027	.053
									18		.004	17.5	.020	.042
									19		.000	18.5	.014	.032

W	n=11	n=12	W	n=13	n=14	W	n=15	n=16	W	n=17	n=18	W	n=19	n=20
1	.483	.485	0.5	.500	.500	1	.489	.490	0.5	.500	.500	1	.492	.493
2	.449	.455	1.5	.473	.476	2	.467	.470	1.5	.482	.483	2	.476	.478
3	.416	.425	2.5	.446	.452	3	.445	.450	2.5	.463	.466	3	.461	.464
4	.382	.396	3.5	.420	.428	4	.423	.430	3.5	.445	.449	4	.445	.449
5	.350	.367	4.5	.393	.404	5	.402	.410	4.5	.427	.433	5	.430	.435
6	.319	.339	5.5	.368	.380	6	.381	.391	5.5	.409	.416	6	.414	.420
7	.289	.311	6.5	.342	.357	7	.360	.372	6.5	.391	.399	7	.399	.406
8	.260	.285	7.5	.318	.335	8	.339	.353	7.5	.373	.383	8	.384	.392
9	.232	.259	8.5	.294	.313	9	.319	.334	8.5	.356	.367	9	.369	.378
10	.207	.235	9.5	.271	.292	10	.300	.316	9.5	.339	.351	10	.354	.364
11	.183	.212	10.5	.249	.271	11	.281	.298	10.5	.322	.335	11	.340	.351
12	.160	.190	11.5	.227	.251	12	.262	.281	11.5	.306	.320	12	.325	.337
13	.139	.170	12.5	.207	.232	13	.244	.264	12.5	.290	.305	13	.311	.324
14	.120	.151	13.5	.188	.213	14	.227	.248	13.5	.274	.290	14	.297	.311
15	.103	.133	14.5	.170	.195	15	.211	.232	14.5	.259	.275	15	.284	.298
16	.087	.117	15.5	.153	.179	16	.195	.217	15.5	.244	.261	16	.271	.285
17	.074	.102	16.5	.137	.163	17	.180	.202	16.5	.229	.248	17	.258	.273
18	.062	.088	17.5	.122	.148	18	.165	.188	17.5	.215	.234	18	.245	.261
19	.051	.076	18.5	.108	.134	19	.151	.174	18.5	.202	.221	19	.233	.249
20	.042	.065	19.5	.095	.121	20	.138	.161	19.5	.189	.209	20	.221	.237
21	.034	.055	20.5	.084	.108	21	.126	.149	20.5	.176	.196	21	.209	.226
22	.027	.046	21.5	.073	.097	22	.115	.137	21.5	.164	.185	22	.198	.215
23	.021	.039	22.5	.064	.086	23	.104	.126	22.5	.153	.173	23	.187	.205
24	.016	.032	23.5	.055	.077	24	.094	.116	23.5	.142	.162	24	.176	.194
25	.012	.026	24.5	.047	.068	25	.084	.106	24.5	.132	.152	25	.166	.184
26	.009	.021	25.5	.040	.059	26	.076	.096	25.5	.122	.142	26	.156	.174
27	.007	.017	26.5	.034	.052	27	.068	.088	26.5	.112	.132	27	.147	.165
28	.005	.013	27.5	.029	.045	28	.060	.080	27.5	.103	.123	28	.138	.156
29	.003	.010	28.5	.024	.039	29	.053	.072	28.5	.095	.114	29	.129	.147
30	.002	.008	29.5	.020	.034	30	.047	.065	29.5	.087	.106	30	.121	.139
31	.001	.006	30.5	.016	.029	31	.042	.058	30.5	.080	.098	31	.113	.131
32	.001	.005	31.5	.013	.025	32	.036	.052	31.5	.073	.091	32	.105	.123
33	.000	.003	32.5	.011	.021	33	.032	.047	32.5	.066	.084	33	.098	.115
34		.002	33.5	.009	.018	34	.028	.042	33.5	.060	.077	34	.091	.108
35		.002	34.5	.007	.015	35	.024	.037	34.5	.054	.071	35	.084	.101
36		.001	35.5	.005	.012	36	.021	.033	35.5	.049	.065	36	.078	.095
37		.001	36.5	.004	.010	37	.018	.029	36.5	.044	.059	37	.072	.088
38		.000	37.5	.003	.008	38	.015	.025	37.5	.040	.054	38	.067	.082
			38.5	.002	.007	39	.013	.022	38.5	.036	.049	39	.062	.077

TABLE F Wilcoxon's Signed Rank (*Continued*)

Probability that Wilcoxon's signed rank minus the
expected value will be equal to or larger than W

W	n=11	n=12	W	n=13	n=14	W	n=15	n=16	W	n=17	n=18	W	n=19	n=20
			39.5	.002	.005	40	.011	.019	39.5	.032	.045	40	.057	.071
			40.5	.001	.004	41	.009	.017	40.5	.028	.041	41	.052	.066
			41.5	.001	.003	42	.008	.014	41.5	.025	.037	42	.048	.062
			42.5	.001	.003	43	.006	.012	42.5	.022	.033	43	.044	.057
			43.5	.000	.002	44	.005	.011	43.5	.020	.030	44	.040	.053
			44.5		.002	45	.004	.009	44.5	.017	.027	45	.036	.049
			45.5		.001	46	.003	.008	45.5	.015	.024	46	.033	.045
			46.5		.001	47	.003	.007	46.5	.013	.022	47	.030	.041
			47.5		.001	48	.002	.005	47.5	.012	.019	48	.027	.038
			48.5		.000	49	.002	.005	48.5	.010	.017	49	.025	.035
						50	.001	.004	49.5	.009	.015	50	.022	.032
						51	.001	.003	50.5	.007	.013	51	.020	.029
						52	.001	.003	51.5	.006	.012	52	.018	.027
						53	.001	.002	52.5	.005	.010	53	.016	.024
						54	.000	.002	53.5	.005	.009	54	.014	.022
						55		.001	54.5	.004	.008	55	.013	.020
						56		.001	55.5	.003	.007	56	.011	.018
						57		.001	56.5	.003	.006	57	.010	.016
						58		.001	57.5	.002	.005	58	.009	.015
						59		.001	58.5	.002	.004	59	.008	.013
						60		.000	59.5	.002	.004	60	.007	.012
									60.5	.001	.003	61	.006	.011
									61.5	.001	.003	62	.005	.010
									62.5	.001	.002	63	.005	.009
									63.5	.001	.002	64	.004	.008
									64.5	.001	.002	65	.004	.007
									65.5	.000	.001	66	.003	.006
									66.5		.001	67	.003	.005
									67.5		.001	68	.002	.005
									68.5		.001	69	.002	.004
									69.5		.001	70	.002	.004
									70.5		.001	71	.001	.003
									71.5		.000	72	.001	.003
												73	.001	.002
												76	.001	.002
												77	.000	.001
												83		.001
												84		.000

W	n=21	n=22	W	n=23	n=24	W	n=25	n=26	W	n=27	n=28	W	n=29	n=30
0.5	.500	.500	1	.494	.494	0.5	.500	.500	1	.495	.496	0.5	.500	.500
1.5	.486	.487	2	.482	.483	1.5	.489	.490	2	.486	.487	1.5	.492	.492
2.5	.473	.475	3	.470	.472	2.5	.479	.480	3	.476	.478	2.5	.483	.484
3.5	.459	.462	4	.458	.461	3.5	.468	.470	4	.467	.469	3.5	.475	.476
4.5	.446	.449	5	.447	.450	4.5	.458	.460	5	.458	.460	4.5	.466	.468
5.5	.432	.437	6	.435	.439	5.5	.447	.450	6	.448	.451	5.5	.458	.460
6.5	.419	.424	7	.423	.428	6.5	.437	.440	7	.439	.442	6.5	.449	.452
7.5	.406	.412	8	.411	.417	7.5	.427	.431	8	.430	.433	7.5	.441	.444
8.5	.393	.400	9	.400	.406	8.5	.416	.421	9	.420	.424	8.5	.432	.436
9.5	.380	.387	10	.388	.395	9.5	.406	.411	10	.411	.416	9.5	.424	.428
10.5	.367	.375	11	.377	.384	10.5	.396	.401	11	.402	.407	10.5	.416	.420
11.5	.354	.363	12	.366	.373	11.5	.386	.392	12	.393	.398	11.5	.407	.412
12.5	.341	.351	13	.355	.363	12.5	.375	.382	13	.384	.390	12.5	.399	.404
13.5	.329	.339	14	.343	.352	13.5	.365	.373	14	.375	.381	13.5	.391	.396
14.5	.317	.328	15	.332	.342	14.5	.356	.363	15	.366	.373	14.5	.383	.388
15.5	.305	.316	16	.322	.332	15.5	.346	.354	16	.357	.364	15.5	.375	.381
16.5	.293	.305	17	.311	.322	16.5	.336	.345	17	.348	.356	16.5	.367	.373
17.5	.281	.294	18	.301	.312	17.5	.326	.335	18	.339	.347	17.5	.359	.365
18.5	.270	.283	19	.290	.302	18.5	.317	.326	19	.331	.339	18.5	.351	.358
19.5	.258	.272	20	.280	.292	19.5	.308	.317	20	.322	.331	19.5	.343	.350

TABLE F Wilcoxon's Signed Rank (*Continued*)

Probability that Wilcoxon's signed rank minus the
expected value will be equal to or larger than W

W	n=21	n=22	W	n=23	n=24	W	n=25	n=26	W	n=27	n=28	W	n=29	n=30
20.5	.247	.262	21	.270	.282	20.5	.298	.309	21	.314	.323	20.5	.335	.343
21.5	.237	.251	22	.260	.273	21.5	.289	.300	22	.305	.315	21.5	.327	.335
22.5	.226	.241	23	.250	.264	22.5	.280	.291	23	.297	.307	22.5	.320	.328
23.5	.216	.231	24	.241	.254	23.5	.271	.283	24	.289	.299	23.5	.312	.320
24.5	.206	.222	25	.232	.245	24.5	.262	.274	25	.281	.291	24.5	.304	.313
25.5	.196	.212	26	.223	.237	25.5	.254	.266	26	.273	.284	25.5	.297	.306
26.5	.187	.203	27	.214	.228	26.5	.245	.258	27	.265	.276	26.5	.290	.299
27.5	.178	.194	28	.205	.219	27.5	.237	.250	28	.257	.268	27.5	.282	.292
28.5	.169	.185	29	.197	.211	28.5	.229	.242	29	.250	.261	28.5	.275	.285
29.5	.160	.177	30	.188	.203	29.5	.221	.234	30	.242	.254	29.5	.268	.278
30.5	.152	.168	31	.180	.195	30.5	.213	.226	31	.235	.247	30.5	.261	.271
31.5	.144	.160	32	.172	.187	31.5	.205	.219	32	.228	.240	31.5	.254	.265
32.5	.136	.153	33	.165	.180	32.5	.198	.211	33	.221	.233	32.5	.247	.258
33.5	.129	.145	34	.157	.172	33.5	.190	.204	34	.214	.226	33.5	.241	.251
34.5	.121	.138	35	.150	.165	34.5	.183	.197	35	.207	.219	34.5	.234	.245
35.5	.114	.131	36	.143	.158	35.5	.176	.190	36	.200	.212	35.5	.227	.239
36.5	.108	.124	37	.136	.151	36.5	.169	.183	37	.193	.206	36.5	.221	.232
37.5	.101	.117	38	.130	.145	37.5	.163	.177	38	.187	.199	37.5	.215	.226
38.5	.095	.111	39	.123	.138	38.5	.156	.170	39	.180	.193	38.5	.208	.220
39.5	.089	.105	40	.117	.132	39.5	.150	.164	40	.174	.187	39.5	.202	.214
40.5	.084	.099	41	.111	.126	40.5	.144	.158	41	.168	.181	40.5	.196	.208
41.5	.079	.093	42	.106	.120	41.5	.138	.152	42	.162	.175	41.5	.190	.202
42.5	.073	.088	43	.100	.115	42.5	.132	.146	43	.156	.169	42.5	.185	.197
43.5	.069	.083	44	.095	.109	43.5	.126	.140	44	.151	.164	43.5	.179	.191
44.5	.064	.078	45	.090	.104	44.5	.121	.134	45	.145	.158	44.5	.173	.185
45.5	.060	.073	46	.085	.099	45.5	.115	.129	46	.140	.153	45.5	.168	.180
46.5	.056	.069	47	.080	.094	46.5	.110	.124	47	.134	.147	46.5	.163	.175
47.5	.052	.064	48	.076	.089	47.5	.105	.118	48	.129	.142	47.5	.157	.169
48.5	.048	.060	49	.071	.084	48.5	.100	.113	49	.124	.137	48.5	.152	.164
49.5	.044	.056	50	.067	.080	49.5	.095	.109	50	.119	.132	49.5	.147	.159
50.5	.041	.053	51	.063	.076	50.5	.091	.104	51	.115	.127	50.5	.142	.154
51.5	.038	.049	52	.059	.072	51.5	.086	.099	52	.110	.123	51.5	.137	.149
52.5	.035	.046	53	.056	.068	52.5	.082	.095	53	.105	.118	52.5	.133	.145
53.5	.032	.043	54	.052	.064	53.5	.078	.091	54	.101	.113	53.5	.128	.140
54.5	.030	.040	55	.049	.060	54.5	.074	.087	55	.097	.109	54.5	.124	.136
55.5	.027	.037	56	.046	.057	55.5	.070	.083	56	.093	.105	55.5	.119	.131
56.5	.025	.034	57	.043	.054	56.5	.067	.079	57	.089	.101	56.5	.115	.127
57.5	.023	.032	58	.040	.051	57.5	.063	.075	58	.085	.097	57.5	.111	.122
58.5	.021	.029	59	.037	.048	58.5	.060	.071	59	.081	.093	58.5	.107	.118
59.5	.019	.027	60	.035	.045	59.5	.057	.068	60	.078	.089	59.5	.103	.114
60.5	.018	.025	61	.033	.042	60.5	.054	.065	61	.074	.085	60.5	.099	.110
61.5	.016	.023	62	.030	.039	61.5	.051	.061	62	.071	.082	61.5	.095	.106
62.5	.015	.021	63	.028	.037	62.5	.048	.058	63	.067	.078	62.5	.091	.103
63.5	.013	.020	64	.026	.035	63.5	.045	.055	64	.064	.075	63.5	.088	.099
64.5	.012	.018	65	.024	.032	64.5	.043	.052	65	.061	.072	64.5	.084	.095
65.5	.011	.016	66	.022	.030	65.5	.040	.050	66	.058	.069	65.5	.081	.092
66.5	.010	.015	67	.021	.028	66.5	.038	.047	67	.056	.066	66.5	.078	.089
67.5	.009	.014	68	.019	.026	67.5	.035	.045	68	.053	.063	67.5	.075	.085
68.5	.008	.013	69	.018	.025	68.5	.033	.042	69	.050	.060	68.5	.072	.082
69.5	.007	.011	70	.016	.023	69.5	.031	.040	70	.048	.057	69.5	.069	.079
70.5	.006	.010	71	.015	.021	70.5	.029	.038	71	.045	.055	70.5	.066	.076
71.5	.006	.009	72	.014	.020	71.5	.028	.035	72	.043	.052	71.5	.063	.073
72.5	.005	.009	73	.013	.018	72.5	.026	.033	73	.041	.050	72.5	.060	.070
73.5	.005	.008	74	.012	.017	73.5	.024	.032	74	.039	.047	73.5	.058	.067
74.5	.004	.007	75	.011	.016	74.5	.023	.030	75	.037	.045	74.5	.055	.065
75.5	.004	.006	76	.010	.015	75.5	.021	.028	76	.035	.043	75.5	.053	.062
76.5	.003	.006	77	.009	.013	76.5	.020	.026	77	.033	.041	76.5	.050	.060
77.5	.003	.005	78	.008	.012	77.5	.018	.025	78	.031	.039	77.5	.048	.057
78.5	.002	.005	79	.007	.011	78.5	.017	.023	79	.029	.037	78.5	.046	.055
79.5	.002	.004	80	.007	.011	79.5	.016	.022	80	.028	.035	79.5	.044	.052
80.5	.002	.004	81	.006	.010	80.5	.015	.020	81	.026	.033	80.5	.042	.050
81.5	.002	.003	82	.006	.009	81.5	.014	.019	82	.025	.031	81.5	.040	.048

TABLE F Wilcoxon's Signed Rank (Continued)

Probability that Wilcoxon's signed rank minus the
expected value will be equal to or larger than W

W	n=21	n=22	W	n=23	n=24	W	n=25	n=26	W	n=27	n=28	W	n=29	n=30
82.5	.001	.003	83	.005	.008	82.5	.013	.018	83	.023	.030	82.5	.038	.046
83.5	.001	.003	84	.005	.008	83.5	.012	.017	84	.022	.028	83.5	.036	.044
84.5	.001	.002	85	.004	.007	84.5	.011	.016	85	.020	.027	84.5	.034	.042
85.5	.001	.002	86	.004	.006	85.5	.010	.015	86	.019	.025	85.5	.033	.040
86.5	.001	.002	87	.003	.006	86.5	.009	.014	87	.018	.024	86.5	.031	.038
87.5	.001	.002	88	.003	.005	87.5	.009	.013	88	.017	.023	87.5	.030	.037
88.5	.001	.001	89	.003	.005	88.5	.008	.012	89	.016	.021	88.5	.028	.035
89.5	.001	.001	90	.002	.004	89.5	.007	.011	90	.015	.020	89.5	.027	.033
90.5	.000	.001	91	.002	.004	90.5	.007	.010	91	.014	.019	90.5	.025	.032
91.5		.001	92	.002	.004	91.5	.006	.009	92	.013	.018	91.5	.024	.030
92.5		.001	93	.002	.003	92.5	.006	.009	93	.012	.017	92.5	.023	.029
93.5		.001	94	.002	.003	93.5	.005	.008	94	.011	.016	93.5	.022	.027
94.5		.001	95	.001	.003	94.5	.005	.008	95	.011	.015	94.5	.020	.026
95.5		.001	96	.001	.002	95.5	.004	.007	96	.010	.014	95.5	.019	.025
96.5		.000	97	.001	.002	96.5	.004	.006	97	.009	.013	96.5	.018	.024
			98	.001	.002	97.5	.004	.006	98	.009	.012	97.5	.017	.022
			99	.001	.002	98.5	.003	.006	99	.008	.012	98.5	.016	.021
			100	.001	.002	99.5	.003	.005	100	.008	.011	99.5	.015	.020
			101	.001	.001	100.5	.003	.005	101	.007	.010	100.5	.015	.019
			102	.001	.001	101.5	.003	.004	102	.007	.010	101.5	.014	.018
			103	.000	.001	102.5	.002	.004	103	.006	.009	102.5	.013	.017
			104		.001	103.5	.002	.004	104	.006	.008	103.5	.012	.016
			105		.001	104.5	.002	.003	105	.005	.008	104.5	.011	.015
			106		.001	105.5	.002	.003	106	.005	.007	105.5	.011	.015
			107		.001	106.5	.002	.003	107	.004	.007	106.5	.010	.014
			108		.001	107.5	.001	.003	108	.004	.006	107.5	.010	.013
			109		.001	108.5	.001	.002	109	.004	.006	108.5	.009	.012
			110		.000	109.5	.001	.002	110	.003	.006	109.5	.008	.012
						110.5	.001	.002	111	.003	.005	110.5	.008	.011
						111.5	.001	.002	112	.003	.005	111.5	.007	.010
						112.5	.001	.002	113	.003	.004	112.5	.007	.010
						113.5	.001	.001	114	.003	.004	113.5	.006	.009
						114.5	.001	.001	115	.002	.004	114.5	.006	.009
						115.5	.001	.001	116	.002	.004	115.5	.006	.008
						116.5	.001	.001	117	.002	.003	116.5	.005	.008
						117.5	.000	.001	118	.002	.003	117.5	.005	.007
						118.5		.001	119	.002	.003	118.5	.005	.007
						119.5		.001	120	.001	.003	119.5	.004	.006
						120.5		.001	121	.001	.002	120.5	.004	.006
						121.5		.001	122	.001	.002	121.5	.004	.006
						122.5		.001	123	.001	.002	122.5	.003	.005
						123.5		.001	124	.001	.002	123.5	.003	.005
						124.5		.000	125	.001	.002	124.5	.003	.005
									126	.001	.001	125.5	.003	.004
									127	.001	.001	126.5	.003	.004
									128	.001	.001	127.5	.002	.004
									129	.001	.001	128.5	.002	.004
									130	.001	.001	129.5	.002	.003
									131	.001	.001	130.5	.002	.003
									132	.000	.001	131.5	.002	.003
									133		.001	132.5	.002	.003
									134		.001	133.5	.001	.003
									135		.001	134.5	.001	.002
									138		.001	137.5	.001	.002
									139		.000	138.5	.001	.002
												139.5	.001	.002
												140.5	.001	.001
												145.5	.001	.001
												146.5	.000	.001
												153.5		.001
												154.5		.000

TABLE G   The Sign Test

The first part (pages T.96 and T.97) is used when the least-frequent attribute is counted and gives upper limits on counts for particular probabilities.

The second part (pages T.98 and T.99) is used when the most-frequent attribute is counted and gives lower limits on counts for particular probabilities.

TABLE G The Sign Test

Column headings are probabilities of an equal or smaller count than the table entry

n	.001	.005	.010	.025	.050	.100	.250
4	0	0	0	0	0	0	0
5	0	0	0	0	0	0	1
6	0	0	0	0	0	0	1
7	0	0	0	0	0	1	2
8	0	0	0	0	1	1	2
9	0	0	0	1	1	2	2
10	0	0	0	1	1	2	3
11	0	0	1	1	2	2	3
12	0	1	1	2	2	3	3
13	0	1	1	2	3	3	4
14	1	1	2	2	3	4	4
15	1	2	2	3	3	4	5
16	1	2	2	3	4	4	5
17	1	2	3	4	4	5	6
18	2	3	3	4	5	5	6
19	2	3	4	4	5	6	7
20	2	3	4	5	5	6	7
21	3	4	4	5	6	7	8
22	3	4	5	5	6	7	8
23	3	4	5	6	7	7	9
24	4	5	5	6	7	8	9
25	4	5	6	7	7	8	10
26	4	6	6	7	8	9	10
27	5	6	7	7	8	9	11
28	5	6	7	8	9	10	11
29	5	7	7	8	9	10	12
30	6	7	8	9	10	10	12

n	.001	.005	.010	.025	.050	.100	.250
70	21	23	24	26	27	29	31
72	22	24	25	27	28	30	32
74	23	25	26	28	29	30	33
76	24	26	27	28	30	31	34
78	24	27	28	29	31	32	35
80	25	28	29	30	32	33	36
82	26	28	30	31	33	34	37
84	27	29	30	32	33	35	38
86	28	30	31	33	34	36	39
88	29	31	32	34	35	37	40
90	29	32	33	35	36	38	41
92	30	33	34	36	37	39	42
94	31	34	35	37	38	40	43
96	32	34	36	37	39	41	44
98	33	35	37	38	40	42	45
100	34	36	37	39	41	43	46
110	38	41	42	44	45	47	50
120	42	45	46	48	50	52	55
130	46	49	51	53	55	57	60
140	51	54	55	57	59	61	65
150	55	58	60	62	64	66	70
160	60	63	64	67	69	71	75
170	64	67	69	71	73	76	80
180	68	72	73	76	78	80	84
190	73	76	78	81	83	85	89
200	77	81	83	85	87	90	94
210	82	85	87	90	92	95	99

31	6	7	8	9	10	11	13
32	6	8	8	9	10	11	13
33	7	8	9	10	11	12	14
34	7	9	9	10	11	12	14
35	8	9	10	11	12	13	15
36	8	9	10	11	12	13	15
37	8	10	10	12	13	14	15
38	9	10	11	12	13	14	16
39	9	11	11	12	13	15	16
40	9	11	12	13	14	15	17
41	10	11	12	13	14	15	17
42	10	12	13	14	15	16	18
43	11	12	13	14	15	16	18
44	11	13	13	15	16	17	19
45	11	13	14	15	16	17	19
46	12	13	14	15	16	18	20
47	12	14	15	16	17	18	20
48	12	14	15	16	17	19	21
49	13	15	15	17	18	19	21
50	13	15	16	17	18	19	22
52	14	16	17	18	19	20	23
54	15	17	18	19	20	21	24
56	16	17	18	20	21	22	24
58	16	18	19	21	22	23	25
60	17	19	20	21	23	24	26
62	18	20	21	22	24	25	27
64	19	21	22	23	24	26	28
66	20	22	23	24	25	27	29
68	20	22	23	25	26	28	30

220	86	90	92	94	97	99	104
230	91	95	96	99	102	104	109
240	95	99	101	104	106	109	114
250	100	104	106	109	111	114	119
260	104	108	110	113	116	119	124
270	109	113	115	118	120	123	128
280	113	117	120	123	125	128	133
290	118	122	124	127	130	133	138
300	122	127	129	132	135	138	143
310	127	131	134	137	140	143	148
320	131	136	138	141	144	148	153
330	136	141	143	146	149	152	158
340	141	145	148	151	154	157	163
350	145	150	152	156	159	162	168
360	150	155	157	160	163	167	173
370	154	159	162	165	168	172	178
380	159	164	166	170	173	177	182
390	164	169	171	175	178	181	187
400	168	173	176	179	183	186	192
410	173	178	180	184	187	191	197
420	177	183	185	189	192	196	202
430	182	187	190	194	197	201	207
440	187	192	195	198	202	206	212
450	191	197	199	203	207	210	217
460	196	201	204	208	211	215	222
470	201	206	209	213	216	220	227
480	205	211	214	218	221	225	232
490	210	216	218	222	226	230	237
500	214	220	223	227	231	235	241

TABLE G The Sign Test (Continued)

Column headings are probabilities of an equal or greater count than the table entry

n	.250	.100	.050	.025	.010	.005	.001
4	4	4	4	4	4	4	4
5	4	5	5	5	5	5	5
6	5	6	6	6	6	6	6
7	5	6	7	7	7	7	7
8	6	7	7	8	8	8	8
9	7	7	8	8	9	9	9
10	7	8	9	9	10	10	10
11	8	9	9	10	10	11	11
12	8	9	10	10	11	11	12
13	9	10	10	11	12	12	13
14	9	10	11	12	12	13	13
15	10	11	12	12	13	13	14
16	10	12	12	13	14	14	15
17	11	12	13	13	14	15	16
18	11	13	13	14	15	15	16
19	12	13	14	15	15	16	17
20	13	14	15	15	16	17	18
21	13	14	15	16	17	17	18
22	14	15	16	17	17	18	19
23	14	16	16	17	18	19	20
24	15	16	17	18	19	19	20
25	15	17	18	18	19	20	21
26	16	17	18	19	20	20	22
27	16	18	19	20	20	21	22
28	17	18	19	20	21	22	23
29	17	19	20	21	22	22	24
30	18	20	20	21	22	23	24

n	.250	.100	.050	.025	.010	.005	.001
70	39	41	43	44	46	47	49
72	40	42	44	45	47	48	50
74	41	44	45	46	48	49	51
76	42	45	46	48	49	50	52
78	43	46	47	49	50	51	54
80	44	47	48	50	51	52	55
82	45	48	49	51	52	54	56
84	46	49	51	52	54	55	57
86	47	50	52	53	55	56	58
88	48	51	53	54	56	57	59
90	49	52	54	55	57	58	61
92	50	53	55	56	58	59	62
94	51	54	56	57	59	60	63
96	52	55	57	59	60	62	64
98	53	56	58	60	61	63	65
100	54	57	59	61	63	64	66
110	60	63	65	66	68	69	72
120	65	68	70	72	74	75	78
130	70	73	75	77	79	81	84
140	75	79	81	83	85	86	89
150	80	84	86	88	90	92	95
160	85	89	91	93	96	97	100
170	90	94	97	99	101	103	106
180	96	100	102	104	107	108	112
190	101	105	107	109	112	114	117
200	106	110	113	115	117	119	123
210	111	115	118	120	123	125	128

134	130	128	126	123	121	116	220
139	135	134	131	128	126	121	230
145	141	139	136	134	131	126	240
150	146	144	141	139	136	131	250
156	152	150	147	144	141	136	260
161	157	155	152	150	147	142	270
167	163	160	157	155	152	147	280
172	168	166	163	160	157	152	290
178	173	171	168	165	162	157	300
183	179	176	173	170	167	162	310
189	184	182	179	176	172	167	320
194	189	187	184	181	178	172	330
199	195	192	189	186	183	177	340
205	200	198	194	191	188	182	350
210	205	203	200	197	193	187	360
216	211	208	205	202	198	192	370
221	216	214	210	207	203	198	380
226	221	219	215	212	209	203	390
232	227	224	221	217	214	208	400
237	232	230	226	223	219	213	410
243	237	235	231	228	224	218	420
248	243	240	236	233	229	223	430
253	248	245	242	238	234	228	440
259	253	251	247	243	240	233	450
264	259	256	252	249	245	238	460
269	264	261	257	254	250	243	470
275	269	266	262	259	255	248	480
280	274	272	268	264	260	253	490
286	280	277	273	269	265	259	500

31	18	20	21	22	23	24	25
32	19	21	22	23	24	24	26
33	19	21	22	23	24	25	26
34	20	22	23	24	25	25	27
35	20	22	23	24	25	26	27
36	21	23	24	25	26	27	28
37	22	23	25	25	27	27	29
38	22	24	25	26	27	28	29
39	23	24	26	27	28	28	30
40	23	25	26	27	28	29	31
41	24	26	27	28	29	30	31
42	24	26	27	28	29	30	32
43	25	27	28	29	30	31	32
44	25	27	28	29	31	31	33
45	26	28	29	30	31	32	34
46	26	28	30	31	32	33	34
47	27	29	30	31	32	33	35
48	27	29	31	32	33	34	36
49	28	30	31	32	34	34	36
50	28	31	32	33	34	35	37
52	29	32	33	34	35	36	38
54	30	33	34	35	36	37	39
56	32	34	35	36	38	39	40
58	33	35	36	37	39	40	42
60	34	36	37	39	40	41	43
62	35	37	38	40	41	42	44
64	36	38	40	41	42	43	45
66	37	39	41	42	43	44	46
68	38	40	42	43	45	46	48

TABLE H The Binomial Distribution

n	c	p=.01	p=.05	p=.10	p=.15	p=.20	p=.25	p=.30	p=.35	p=.40	p=.45	p=.50
2	0	1,000	1,000	1,000	1,000	1,000	1,000	1,000	1,000	1,000	1,000	1,000
2	1	,0199	,0975	,1900	,2775	,3600	,4375	,5100	,5775	,6400	,6975	,7500
2	2	,0001	,0025	,0100	,0225	,0400	,0625	,0900	,1225	,1600	,2025	,2500

n	c	p=.01	p=.05	p=.10	p=.15	p=.20	p=.25	p=.30	p=.35	p=.40	p=.45	p=.50
3	0	1,000	1,000	1,000	1,000	1,000	1,000	1,000	1,000	1,000	1,000	1,000
3	1	,0297	,1426	,2710	,3859	,4880	,5781	,6570	,7254	,7840	,8336	,8750
3	2	,0003	,0073	,0280	,0608	,1040	,1563	,2160	,2818	,3520	,4253	,5000
3	3	,0000	,0001	,0010	,0034	,0080	,0156	,0270	,0429	,0640	,0911	,1250

n	c	p=.01	p=.05	p=.10	p=.15	p=.20	p=.25	p=.30	p=.35	p=.40	p=.45	p=.50
4	0	1,000	1,000	1,000	1,000	1,000	1,000	1,000	1,000	1,000	1,000	1,000
4	1	,0394	,1855	,3439	,4780	,5904	,6836	,7599	,8215	,8704	,9085	,9375
4	2	,0006	,0140	,0523	,1095	,1808	,2617	,3483	,4370	,5248	,6090	,6875
4	3	,0000	,0005	,0037	,0120	,0272	,0508	,0837	,1265	,1792	,2415	,3125
4	4	,0000	,0000	,0001	,0005	,0016	,0039	,0081	,0150	,0256	,0410	,0625

n	c	p=.01	p=.05	p=.10	p=.15	p=.20	p=.25	p=.30	p=.35	p=.40	p=.45	p=.50
5	0	1,000	1,000	1,000	1,000	1,000	1,000	1,000	1,000	1,000	1,000	1,000
5	1	,0490	,2262	,4095	,5563	,6723	,7627	,8319	,8840	,9222	,9497	,9688
5	2	,0010	,0226	,0815	,1648	,2627	,3672	,4718	,5716	,6630	,7438	,8125
5	3	,0000	,0012	,0086	,0266	,0579	,1035	,1631	,2352	,3174	,4069	,5000
5	4	,0000	,0000	,0005	,0022	,0067	,0156	,0308	,0540	,0870	,1312	,1875
5	5	,0000	,0000	,0000	,0001	,0003	,0010	,0024	,0053	,0102	,0185	,0313

n	c	p=.01	p=.05	p=.10	p=.15	p=.20	p=.25	p=.30	p=.35	p=.40	p=.45	p=.50
6	0	1,000	1,000	1,000	1,000	1,000	1,000	1,000	1,000	1,000	1,000	1,000
6	1	,0585	,2649	,4686	,6229	,7379	,8220	,8824	,9246	,9533	,9723	,9844
6	2	,0015	,0328	,1143	,2235	,3446	,4661	,5798	,6809	,7667	,8364	,8906
6	3	,0000	,0022	,0159	,0473	,0989	,1694	,2557	,3529	,4557	,5585	,6563
6	4	,0000	,0001	,0013	,0059	,0170	,0376	,0705	,1174	,1792	,2553	,3438
6	5	,0000	,0000	,0001	,0004	,0016	,0046	,0109	,0223	,0410	,0692	,1094
6	6	,0000	,0000	,0000	,0000	,0001	,0002	,0007	,0018	,0041	,0083	,0156

n	c	p=.01	p=.05	p=.10	p=.15	p=.20	p=.25	p=.30	p=.35	p=.40	p=.45	p=.50
7	0	1,000	1,000	1,000	1,000	1,000	1,000	1,000	1,000	1,000	1,000	1,000
7	1	,0679	,3017	,5217	,6794	,7903	,8665	,9176	,9510	,9720	,9848	,9922
7	2	,0020	,0444	,1497	,2834	,4233	,5551	,6706	,7662	,8414	,8976	,9375
7	3	,0000	,0038	,0257	,0738	,1480	,2436	,3529	,4677	,5801	,6836	,7734
7	4	,0000	,0002	,0027	,0121	,0333	,0706	,1260	,1998	,2898	,3917	,5000
7	5	,0000	,0000	,0002	,0012	,0047	,0129	,0288	,0556	,0963	,1529	,2266
7	6	,0000	,0000	,0000	,0001	,0004	,0013	,0038	,0090	,0188	,0357	,0625
7	7	,0000	,0000	,0000	,0000	,0000	,0001	,0002	,0006	,0016	,0037	,0078

n	c	p=.01	p=.05	p=.10	p=.15	p=.20	p=.25	p=.30	p=.35	p=.40	p=.45	p=.50
8	0	1,000	1,000	1,000	1,000	1,000	1,000	1,000	1,000	1,000	1,000	1,000
8	1	,0773	,3366	,5695	,7275	,8322	,8999	,9424	,9681	,9832	,9916	,9961
8	2	,0027	,0572	,1869	,3428	,4967	,6329	,7447	,8309	,8936	,9368	,9648
8	3	,0001	,0058	,0381	,1052	,2031	,3215	,4482	,5722	,6846	,7799	,8555
8	4	,0000	,0004	,0050	,0214	,0563	,1138	,1941	,2936	,4059	,5230	,6367
8	5	,0000	,0000	,0004	,0029	,0104	,0273	,0580	,1061	,1737	,2604	,3633
8	6	,0000	,0000	,0000	,0002	,0012	,0042	,0113	,0253	,0498	,0885	,1445
8	7	,0000	,0000	,0000	,0000	,0001	,0004	,0013	,0036	,0085	,0181	,0352
8	8	,0000	,0000	,0000	,0000	,0000	,0000	,0001	,0002	,0007	,0017	,0039

TABLE H   The Binomial Distribution   (*Continued*)

n	c	p=.50	p=.55	p=.60	p=.65	p=.70	p=.75	p=.80	p=.85	p=.90	p=.95	p=.99
2	0	1,000	1,000	1,000	1,000	1,000	1,000	1,000	1,000	1,000	1,000	1,000
2	1	.7500	.7975	.8400	.8775	.9100	.9375	.9600	.9775	.9900	.9975	.9999
2	2	.2500	.3025	.3600	.4225	.4900	.5625	.6400	.7225	.8100	.9025	.9801

n	c	p=.50	p=.55	p=.60	p=.65	p=.70	p=.75	p=.80	p=.85	p=.90	p=.95	p=.99
3	0	1,000	1,000	1,000	1,000	1,000	1,000	1,000	1,000	1,000	1,000	1,000
3	1	.8750	.9089	.9360	.9571	.9730	.9844	.9920	.9966	.9990	.9999	1,000
3	2	.5000	.5748	.6480	.7183	.7840	.8438	.8960	.9393	.9720	.9928	.9997
3	3	.1250	.1664	.2160	.2746	.3430	.4219	.5120	.6141	.7290	.8574	.9703

n	c	p=.50	p=.55	p=.60	p=.65	p=.70	p=.75	p=.80	p=.85	p=.90	p=.95	p=.99
4	0	1,000	1,000	1,000	1,000	1,000	1,000	1,000	1,000	1,000	1,000	1,000
4	1	.9375	.9590	.9744	.9850	.9919	.9961	.9984	.9995	.9999	1,000	1,000
4	2	.6875	.7585	.8208	.8735	.9163	.9492	.9728	.9880	.9963	.9995	1,000
4	3	.3125	.3910	.4752	.5630	.6517	.7383	.8192	.8905	.9477	.9860	.9994
4	4	.0625	.0915	.1296	.1785	.2401	.3164	.4096	.5220	.6561	.8145	.9606

n	c	p=.50	p=.55	p=.60	p=.65	p=.70	p=.75	p=.80	p=.85	p=.90	p=.95	p=.99
5	0	1,000	1,000	1,000	1,000	1,000	1,000	1,000	1,000	1,000	1,000	1,000
5	1	.9688	.9815	.9898	.9947	.9976	.9990	.9997	.9999	1,000	1,000	1,000
5	2	.8125	.8688	.9130	.9460	.9692	.9844	.9933	.9978	.9995	1,000	1,000
5	3	.5000	.5931	.6826	.7648	.8369	.8965	.9421	.9734	.9914	.9988	1,000
5	4	.1875	.2562	.3370	.4284	.5282	.6328	.7373	.8352	.9185	.9774	.9990
5	5	.0313	.0503	.0778	.1160	.1681	.2373	.3277	.4437	.5905	.7738	.9510

n	c	p=.50	p=.55	p=.60	p=.65	p=.70	p=.75	p=.80	p=.85	p=.90	p=.95	p=.99
6	0	1,000	1,000	1,000	1,000	1,000	1,000	1,000	1,000	1,000	1,000	1,000
6	1	.9844	.9917	.9959	.9982	.9993	.9998	.9999	1,000	1,000	1,000	1,000
6	2	.8906	.9308	.9590	.9777	.9891	.9954	.9984	.9996	.9999	1,000	1,000
6	3	.6563	.7447	.8208	.8826	.9295	.9624	.9830	.9941	.9987	.9999	1,000
6	4	.3438	.4415	.5443	.6471	.7443	.8306	.9011	.9527	.9842	.9978	1,000
6	5	.1094	.1636	.2333	.3191	.4202	.5339	.6554	.7765	.8857	.9672	.9985
6	6	.0156	.0277	.0467	.0754	.1176	.1780	.2621	.3771	.5314	.7351	.9415

n	c	p=.50	p=.55	p=.60	p=.65	p=.70	p=.75	p=.80	p=.85	p=.90	p=.95	p=.99
7	0	1,000	1,000	1,000	1,000	1,000	1,000	1,000	1,000	1,000	1,000	1,000
7	1	.9922	.9963	.9984	.9994	.9998	.9999	1,000	1,000	1,000	1,000	1,000
7	2	.9375	.9643	.9812	.9910	.9962	.9987	.9996	.9999	1,000	1,000	1,000
7	3	.7734	.8471	.9037	.9444	.9712	.9871	.9953	.9988	.9998	1,000	1,000
7	4	.5000	.6083	.7102	.8002	.8740	.9294	.9667	.9879	.9973	.9998	1,000
7	5	.2266	.3164	.4199	.5323	.6471	.7564	.8520	.9262	.9743	.9962	1,000
7	6	.0625	.1024	.1586	.2338	.3294	.4449	.5767	.7166	.8503	.9556	.9980
7	7	.0078	.0152	.0280	.0490	.0824	.1335	.2097	.3206	.4783	.6983	.9321

n	c	p=.50	p=.55	p=.60	p=.65	p=.70	p=.75	p=.80	p=.85	p=.90	p=.95	p=.99
8	0	1,000	1,000	1,000	1,000	1,000	1,000	1,000	1,000	1,000	1,000	1,000
8	1	.9961	.9983	.9993	.9998	.9999	1,000	1,000	1,000	1,000	1,000	1,000
8	2	.9648	.9819	.9915	.9964	.9987	.9996	.9999	1,000	1,000	1,000	1,000
8	3	.8555	.9115	.9502	.9747	.9887	.9958	.9988	.9998	1,000	1,000	1,000
8	4	.6367	.7396	.8263	.8939	.9420	.9727	.9896	.9971	.9996	1,000	1,000
8	5	.3633	.4770	.5941	.7064	.8059	.8862	.9437	.9786	.9950	.9996	1,000
8	6	.1445	.2201	.3154	.4278	.5518	.6785	.7969	.8948	.9619	.9942	.9999
8	7	.0352	.0632	.1064	.1691	.2553	.3671	.5033	.6572	.8131	.9428	.9973
8	8	.0039	.0084	.0168	.0319	.0576	.1001	.1678	.2725	.4305	.6634	.9227

TABLE H The Binomial Distribution (*Continued*)

n	c	p=,01	p=,05	p=,10	p=,15	p=,20	p=,25	p=,30	p=,35	p=,40	p=,45	p=,50
9	0	1,000	1,000	1,000	1,000	1,000	1,000	1,000	1,000	1,000	1,000	1,000
9	1	,0865	,3698	,6126	,7684	,8658	,9249	,9596	,9793	,9899	,9954	,9980
9	2	,0034	,0712	,2252	,4005	,5638	,6997	,8040	,8789	,9295	,9615	,9805
9	3	,0001	,0084	,0530	,1409	,2618	,3993	,5372	,6627	,7682	,8505	,9102
9	4	,0000	,0006	,0083	,0339	,0856	,1657	,2703	,3911	,5174	,6386	,7461
9	5	,0000	,0000	,0009	,0056	,0196	,0489	,0988	,1717	,2666	,3786	,5000
9	6	,0000	,0000	,0001	,0006	,0031	,0100	,0253	,0536	,0994	,1658	,2539
9	7	,0000	,0000	,0000	,0000	,0003	,0013	,0043	,0112	,0250	,0498	,0898
9	8	,0000	,0000	,0000	,0000	,0000	,0001	,0004	,0014	,0038	,0091	,0195
9	9	,0000	,0000	,0000	,0000	,0000	,0000	,0000	,0001	,0003	,0008	,0020

n	c	p=,01	p=,05	p=,10	p=,15	p=,20	p=,25	p=,30	p=,35	p=,40	p=,45	p=,50
10	0	1,000	1,000	1,000	1,000	1,000	1,000	1,000	1,000	1,000	1,000	1,000
10	1	,0956	,4013	,6513	,8031	,8926	,9437	,9718	,9865	,9940	,9975	,9990
10	2	,0043	,0861	,2639	,4557	,6242	,7560	,8507	,9140	,9536	,9767	,9893
10	3	,0001	,0115	,0702	,1798	,3222	,4744	,6172	,7384	,8327	,9004	,9453
10	4	,0000	,0010	,0128	,0500	,1209	,2241	,3504	,4862	,6177	,7340	,8281
10	5	,0000	,0001	,0016	,0099	,0328	,0781	,1503	,2485	,3669	,4956	,6230
10	6	,0000	,0000	,0001	,0014	,0064	,0197	,0473	,0949	,1662	,2616	,3770
10	7	,0000	,0000	,0000	,0001	,0009	,0035	,0106	,0260	,0548	,1020	,1719
10	8	,0000	,0000	,0000	,0000	,0001	,0004	,0016	,0048	,0123	,0274	,0547
10	9	,0000	,0000	,0000	,0000	,0000	,0000	,0001	,0005	,0017	,0045	,0107
10	10	,0000	,0000	,0000	,0000	,0000	,0000	,0000	,0000	,0001	,0003	,0010

n	c	p=,01	p=,05	p=,10	p=,15	p=,20	p=,25	p=,30	p=,35	p=,40	p=,45	p=,50
11	0	1,000	1,000	1,000	1,000	1,000	1,000	1,000	1,000	1,000	1,000	1,000
11	1	,1047	,4312	,6862	,8327	,9141	,9578	,9802	,9912	,9964	,9986	,9995
11	2	,0052	,1019	,3026	,5078	,6779	,8029	,8870	,9394	,9698	,9861	,9941
11	3	,0002	,0152	,0896	,2212	,3826	,5448	,6873	,7999	,8811	,9348	,9673
11	4	,0000	,0016	,0185	,0694	,1611	,2867	,4304	,5744	,7037	,8089	,8867
11	5	,0000	,0001	,0028	,0159	,0504	,1146	,2103	,3317	,4672	,6029	,7256
11	6	,0000	,0000	,0003	,0027	,0117	,0343	,0782	,1487	,2465	,3669	,5000
11	7	,0000	,0000	,0000	,0003	,0020	,0076	,0216	,0501	,0994	,1738	,2744
11	8	,0000	,0000	,0000	,0000	,0002	,0012	,0043	,0122	,0293	,0610	,1133
11	9	,0000	,0000	,0000	,0000	,0000	,0001	,0006	,0020	,0059	,0148	,0327
11	10	,0000	,0000	,0000	,0000	,0000	,0000	,0000	,0002	,0007	,0022	,0059
11	11	,0000	,0000	,0000	,0000	,0000	,0000	,0000	,0000	,0000	,0002	,0005

n	c	p=,01	p=,05	p=,10	p=,15	p=,20	p=,25	p=,30	p=,35	p=,40	p=,45	p=,50
12	0	1,000	1,000	1,000	1,000	1,000	1,000	1,000	1,000	1,000	1,000	1,000
12	1	,1136	,4596	,7176	,8578	,9313	,9683	,9862	,9943	,9978	,9992	,9998
12	2	,0062	,1184	,3410	,5565	,7251	,8416	,9150	,9576	,9804	,9917	,9968
12	3	,0002	,0196	,1109	,2642	,4417	,6093	,7472	,8487	,9166	,9579	,9807
12	4	,0000	,0022	,0256	,0922	,2054	,3512	,5075	,6533	,7747	,8655	,9270
12	5	,0000	,0002	,0043	,0239	,0726	,1576	,2763	,4167	,5618	,6956	,8062
12	6	,0000	,0000	,0005	,0046	,0194	,0544	,1178	,2127	,3348	,4731	,6128
12	7	,0000	,0000	,0001	,0007	,0039	,0143	,0386	,0846	,1582	,2607	,3872
12	8	,0000	,0000	,0000	,0001	,0006	,0028	,0095	,0255	,0573	,1117	,1938
12	9	,0000	,0000	,0000	,0000	,0001	,0004	,0017	,0056	,0153	,0356	,0730
12	10	,0000	,0000	,0000	,0000	,0000	,0000	,0002	,0008	,0028	,0079	,0193
12	11	,0000	,0000	,0000	,0000	,0000	,0000	,0000	,0001	,0003	,0011	,0032
12	12	,0000	,0000	,0000	,0000	,0000	,0000	,0000	,0000	,0000	,0001	,0002

TABLE H　The Binomial Distribution　(*Continued*)

n	c	p=.50	p=.55	p=.60	p=.65	p=.70	p=.75	p=.80	p=.85	p=.90	p=.95	p=.99
9	0	1,000	1,000	1,000	1,000	1,000	1,000	1,000	1,000	1,000	1,000	1,000
9	1	,9980	,9992	,9997	,9999	1,000	1,000	1,000	1,000	1,000	1,000	1,000
9	2	,9805	,9909	,9962	,9986	,9996	,9999	1,000	1,000	1,000	1,000	1,000
9	3	,9102	,9502	,9750	,9888	,9957	,9987	,9997	1,000	1,000	1,000	1,000
9	4	,7461	,8342	,9006	,9464	,9747	,9900	,9969	,9994	,9999	1,000	1,000
9	5	,5000	,6214	,7334	,8283	,9012	,9511	,9804	,9944	,9991	1,000	1,000
9	6	,2539	,3614	,4826	,6089	,7297	,8343	,9144	,9661	,9917	,9994	1,000
9	7	,0898	,1495	,2318	,3373	,4628	,6007	,7382	,8591	,9470	,9916	,9999
9	8	,0195	,0385	,0705	,1211	,1960	,3003	,4362	,5995	,7748	,9288	,9966
9	9	,0020	,0046	,0101	,0207	,0404	,0751	,1342	,2316	,3874	,6302	,9135

n	c	p=.50	p=.55	p=.60	p=.65	p=.70	p=.75	p=.80	p=.85	p=.90	p=.95	p=.99
10	0	1,000	1,000	1,000	1,000	1,000	1,000	1,000	1,000	1,000	1,000	1,000
10	1	,9990	,9997	,9999	1,000	1,000	1,000	1,000	1,000	1,000	1,000	1,000
10	2	,9893	,9955	,9983	,9995	,9999	1,000	1,000	1,000	1,000	1,000	1,000
10	3	,9453	,9726	,9877	,9952	,9984	,9996	,9999	1,000	1,000	1,000	1,000
10	4	,8281	,8980	,9452	,9740	,9894	,9965	,9991	,9999	1,000	1,000	1,000
10	5	,6230	,7384	,8338	,9051	,9527	,9803	,9936	,9986	,9999	1,000	1,000
10	6	,3770	,5044	,6331	,7515	,8497	,9219	,9672	,9901	,9984	,9999	1,000
10	7	,1719	,2660	,3823	,5138	,6496	,7759	,8791	,9500	,9872	,9990	1,000
10	8	,0547	,0996	,1673	,2616	,3828	,5256	,6778	,8202	,9298	,9885	,9999
10	9	,0107	,0233	,0464	,0860	,1493	,2440	,3758	,5443	,7361	,9139	,9957
10	10	,0010	,0025	,0060	,0135	,0282	,0563	,1074	,1969	,3487	,5987	,9044

n	c	p=.50	p=.55	p=.60	p=.65	p=.70	p=.75	p=.80	p=.85	p=.90	p=.95	p=.99
11	0	1,000	1,000	1,000	1,000	1,000	1,000	1,000	1,000	1,000	1,000	1,000
11	1	,9995	,9998	1,000	1,000	1,000	1,000	1,000	1,000	1,000	1,000	1,000
11	2	,9941	,9978	,9993	,9998	1,000	1,000	1,000	1,000	1,000	1,000	1,000
11	3	,9673	,9852	,9941	,9980	,9994	,9999	1,000	1,000	1,000	1,000	1,000
11	4	,8867	,9390	,9707	,9878	,9957	,9988	,9998	1,000	1,000	1,000	1,000
11	5	,7256	,8262	,9006	,9499	,9784	,9924	,9980	,9997	1,000	1,000	1,000
11	6	,5000	,6331	,7535	,8513	,9218	,9657	,9883	,9973	,9997	1,000	1,000
11	7	,2744	,3971	,5328	,6683	,7897	,8854	,9496	,9841	,9972	,9999	1,000
11	8	,1133	,1911	,2963	,4256	,5696	,7133	,8389	,9306	,9815	,9984	1,000
11	9	,0327	,0652	,1189	,2001	,3127	,4552	,6174	,7788	,9104	,9848	,9998
11	10	,0059	,0139	,0302	,0606	,1130	,1971	,3221	,4922	,6974	,8981	,9948
11	11	,0005	,0014	,0036	,0088	,0198	,0422	,0859	,1673	,3138	,5688	,8953

n	c	p=.50	p=.55	p=.60	p=.65	p=.70	p=.75	p=.80	p=.85	p=.90	p=.95	p=.99
12	0	1,000	1,000	1,000	1,000	1,000	1,000	1,000	1,000	1,000	1,000	1,000
12	1	,9998	,9999	1,000	1,000	1,000	1,000	1,000	1,000	1,000	1,000	1,003
12	2	,9968	,9989	,9997	,9999	1,000	1,000	1,000	1,000	1,000	1,000	1,000
12	3	,9807	,9921	,9972	,9992	,9998	1,000	1,000	1,000	1,000	1,000	1,000
12	4	,9270	,9644	,9847	,9944	,9983	,9996	,9999	1,000	1,000	1,000	1,000
12	5	,8062	,8883	,9427	,9745	,9905	,9972	,9994	,9999	1,000	1,000	1,000
12	6	,6128	,7393	,8418	,9154	,9614	,9857	,9961	,9993	,9999	1,000	1,000
12	7	,3872	,5269	,6652	,7873	,8822	,9456	,9806	,9954	,9995	1,000	1,000
12	8	,1938	,3044	,4382	,5833	,7237	,8424	,9274	,9761	,9957	,9998	1,000
12	9	,0730	,1345	,2253	,3467	,4925	,6488	,7946	,9078	,9744	,9978	1,000
12	10	,0193	,0421	,0834	,1513	,2528	,3907	,5583	,7358	,8891	,9804	,9998
12	11	,0032	,0083	,0196	,0424	,0850	,1584	,2749	,4435	,6590	,8816	,9938
12	12	,0002	,0008	,0022	,0057	,0138	,0317	,0687	,1422	,2824	,5404	,8864

TABLE H   The Binomial Distribution   (*Continued*)

n	c	p=.01	p=.05	p=.10	p=.15	p=.20	p=.25	p=.30	p=.35	p=.40	p=.45	p=.50
13	0	1,000	1,000	1,000	1,000	1,000	1,000	1,000	1,000	1,000	1,000	1,000
13	1	,1225	,4867	,7458	,8791	,9450	,9762	,9903	,9963	,9987	,9996	,9999
13	2	,0072	,1354	,3787	,6017	,7664	,8733	,9363	,9704	,9874	,9951	,9983
13	3	,0003	,0245	,1339	,3080	,4983	,6674	,7975	,8868	,9421	,9731	,9888
13	4	,0000	,0031	,0342	,1180	,2527	,4157	,5794	,7217	,8314	,9071	,9539
13	5	,0000	,0003	,0065	,0342	,0991	,2060	,3457	,4995	,6470	,7721	,8666
13	6	,0000	,0000	,0009	,0075	,0300	,0802	,1654	,2841	,4256	,5732	,7095
13	7	,0000	,0000	,0001	,0013	,0070	,0243	,0624	,1295	,2288	,3563	,5000
13	8	,0000	,0000	,0000	,0002	,0012	,0056	,0182	,0462	,0977	,1788	,2905
13	9	,0000	,0000	,0000	,0000	,0002	,0010	,0040	,0126	,0321	,0698	,1334
13	10	,0000	,0000	,0000	,0000	,0000	,0001	,0007	,0025	,0078	,0203	,0461
13	11	,0000	,0000	,0000	,0000	,0000	,0000	,0001	,0003	,0013	,0041	,0112
13	12	,0000	,0000	,0000	,0000	,0000	,0000	,0000	,0000	,0001	,0005	,0017
13	13	,0000	,0000	,0000	,0000	,0000	,0000	,0000	,0000	,0000	,0000	,0001

n	c	p=.01	p=.05	p=.10	p=.15	p=.20	p=.25	p=.30	p=.35	p=.40	p=.45	p=.50
14	0	1,000	1,000	1,000	1,000	1,000	1,000	1,000	1,000	1,000	1,000	1,000
14	1	,1313	,5123	,7712	,8972	,9560	,9822	,9932	,9976	,9992	,9998	,9999
14	2	,0084	,1530	,4154	,6433	,8021	,8990	,9525	,9795	,9919	,9971	,9991
14	3	,0003	,0301	,1584	,3521	,5519	,7189	,8392	,9161	,9602	,9830	,9935
14	4	,0000	,0042	,0441	,1465	,3018	,4787	,6448	,7795	,8757	,9368	,9713
14	5	,0000	,0004	,0092	,0467	,1298	,2585	,4158	,5773	,7207	,8328	,9102
14	6	,0000	,0000	,0015	,0115	,0439	,1117	,2195	,3595	,5141	,6627	,7880
14	7	,0000	,0000	,0002	,0022	,0116	,0383	,0933	,1836	,3075	,4539	,6047
14	8	,0000	,0000	,0000	,0003	,0024	,0103	,0315	,0753	,1501	,2586	,3953
14	9	,0000	,0000	,0000	,0000	,0004	,0022	,0083	,0243	,0583	,1189	,2120
14	10	,0000	,0000	,0000	,0000	,0000	,0003	,0017	,0060	,0175	,0426	,0898
14	11	,0000	,0000	,0000	,0000	,0000	,0000	,0002	,0011	,0039	,0114	,0287
14	12	,0000	,0000	,0000	,0000	,0000	,0000	,0000	,0001	,0006	,0022	,0065
14	13	,0000	,0000	,0000	,0000	,0000	,0000	,0000	,0000	,0001	,0003	,0009
14	14	,0000	,0000	,0000	,0000	,0000	,0000	,0000	,0000	,0000	,0000	,0001

n	c	p=.01	p=.05	p=.10	p=.15	p=.20	p=.25	p=.30	p=.35	p=.40	p=.45	p=.50
15	0	1,000	1,000	1,000	1,000	1,000	1,000	1,000	1,000	1,000	1,000	1,000
15	1	,1399	,5367	,7941	,9126	,9648	,9866	,9953	,9984	,9995	,9999	1,000
15	2	,0096	,1710	,4510	,6814	,8329	,9198	,9647	,9858	,9948	,9983	,9995
15	3	,0004	,0362	,1841	,3958	,6020	,7639	,8732	,9383	,9729	,9893	,9963
15	4	,0000	,0055	,0556	,1773	,3518	,5387	,7031	,8273	,9095	,9576	,9824
15	5	,0000	,0006	,0127	,0617	,1642	,3135	,4845	,6481	,7827	,8796	,9408
15	6	,0000	,0001	,0022	,0168	,0611	,1484	,2784	,4357	,5968	,7392	,8491
15	7	,0000	,0000	,0003	,0036	,0181	,0566	,1311	,2452	,3902	,5478	,6964
15	8	,0000	,0000	,0000	,0006	,0042	,0173	,0500	,1132	,2131	,3465	,5000
15	9	,0000	,0000	,0000	,0001	,0008	,0042	,0152	,0422	,0950	,1818	,3036
15	10	,0000	,0000	,0000	,0000	,0001	,0008	,0037	,0124	,0338	,0769	,1509
15	11	,0000	,0000	,0000	,0000	,0000	,0001	,0007	,0028	,0093	,0255	,0592
15	12	,0000	,0000	,0000	,0000	,0000	,0000	,0001	,0005	,0019	,0063	,0176
15	13	,0000	,0000	,0000	,0000	,0000	,0000	,0000	,0001	,0003	,0011	,0037
15	14	,0000	,0000	,0000	,0000	,0000	,0000	,0000	,0000	,0000	,0001	,0005
15	15	,0000	,0000	,0000	,0000	,0000	,0000	,0000	,0000	,0000	,0000	,0000

TABLE H   The Binomial Distribution   (*Continued*)

n	c	p=,50	p=,55	p=,60	p=,65	p=,70	p=,75	p=,80	p=,85	p=,90	p=,95	p=,99
13	0	1,000	1,000	1,000	1,000	1,000	1,000	1,000	1,000	1,000	1,000	1,000
13	1	,9999	1,000	1,000	1,000	1,000	1,000	1,000	1,000	1,000	1,000	1,000
13	2	,9983	,9995	,9999	1,000	1,000	1,000	1,000	1,000	1,000	1,000	1,000
13	3	,9888	,9959	,9987	,9997	,9999	1,000	1,000	1,000	1,000	1,000	1,000
13	4	,9539	,9797	,9922	,9975	,9993	,9999	1,000	1,000	1,000	1,000	1,000
13	5	,8666	,9302	,9679	,9874	,9960	,9990	,9998	1,000	1,000	1,000	1,000
13	6	,7095	,8212	,9023	,9538	,9818	,9944	,9988	,9998	1,000	1,000	1,000
13	7	,5000	,6437	,7712	,8705	,9376	,9757	,9930	,9987	,9999	1,000	1,000
13	8	,2905	,4268	,5744	,7159	,8346	,9198	,9700	,9925	,9991	1,000	1,000
13	9	,1334	,2279	,3530	,5005	,6543	,7940	,9009	,9658	,9935	,9997	1,000
13	10	,0461	,0929	,1686	,2783	,4206	,5843	,7473	,8820	,9658	,9969	1,000
13	11	,0112	,0269	,0579	,1132	,2025	,3326	,5017	,6920	,8661	,9755	,9997
13	12	,0017	,0049	,0126	,0296	,0637	,1267	,2336	,3983	,6213	,8646	,9928
13	13	,0001	,0004	,0013	,0037	,0097	,0238	,0550	,1209	,2542	,5133	,8775

n	c	p=,50	p=,55	p=,60	p=,65	p=,70	p=,75	p=,80	p=,85	p=,90	p=,95	p=,99
14	0	1,000	1,000	1,000	1,000	1,000	1,000	1,000	1,000	1,000	1,000	1,000
14	1	,9999	1,000	1,000	1,000	1,000	1,000	1,000	1,000	1,000	1,000	1,000
14	2	,9991	,9997	,9999	1,000	1,000	1,000	1,000	1,000	1,000	1,000	1,000
14	3	,9935	,9978	,9994	,9999	1,000	1,000	1,000	1,000	1,000	1,000	1,000
14	4	,9713	,9886	,9961	,9989	,9998	1,000	1,000	1,000	1,000	1,000	1,000
14	5	,9102	,9574	,9825	,9940	,9983	,9997	1,000	1,000	1,000	1,000	1,000
14	6	,7880	,8811	,9417	,9757	,9917	,9978	,9996	1,000	1,000	1,000	1,000
14	7	,6047	,7414	,8499	,9247	,9685	,9897	,9976	,9997	1,000	1,000	1,000
14	8	,3953	,5461	,6925	,8164	,9067	,9617	,9884	,9978	,9998	1,000	1,000
14	9	,2120	,3373	,4859	,6405	,7805	,8883	,9561	,9885	,9985	1,000	1,000
14	10	,0898	,1672	,2793	,4227	,5842	,7415	,8702	,9533	,9908	,9996	1,000
14	11	,0287	,0632	,1243	,2205	,3552	,5213	,6982	,8535	,9559	,9958	1,000
14	12	,0065	,0170	,0398	,0839	,1608	,2811	,4481	,6479	,8416	,9699	,9997
14	13	,0009	,0029	,0081	,0205	,0475	,1010	,1979	,3567	,5846	,8470	,9916
14	14	,0001	,0002	,0008	,0024	,0068	,0178	,0440	,1028	,2288	,4877	,8687

n	c	p=,50	p=,55	p=,60	p=,65	p=,70	p=,75	p=,80	p=,85	p=,90	p=,95	p=,99
15	0	1,000	1,000	1,000	1,000	1,000	1,000	1,000	1,000	1,000	1,000	1,000
15	1	1,000	1,000	1,000	1,000	1,000	1,000	1,000	1,000	1,000	1,000	1,000
15	2	,9995	,9999	1,000	1,000	1,000	1,000	1,000	1,000	1,000	1,000	1,000
15	3	,9963	,9989	,9997	,9999	1,000	1,000	1,000	1,000	1,000	1,000	1,000
15	4	,9824	,9937	,9981	,9995	,9999	1,000	1,000	1,000	1,000	1,000	1,000
15	5	,9408	,9745	,9907	,9972	,9993	,9999	1,000	1,000	1,000	1,000	1,000
15	6	,8491	,9231	,9662	,9876	,9963	,9992	,9999	1,000	1,000	1,000	1,000
15	7	,6964	,8182	,9050	,9578	,9848	,9958	,9992	,9999	1,000	1,000	1,000
15	8	,5000	,6535	,7869	,8868	,9500	,9827	,9958	,9994	1,000	1,000	1,000
15	9	,3036	,4522	,6098	,7548	,8689	,9434	,9819	,9964	,9997	1,000	1,000
15	10	,1509	,2608	,4032	,5643	,7216	,8516	,9389	,9832	,9978	,9999	1,000
15	11	,0592	,1204	,2173	,3519	,5155	,6865	,8358	,9383	,9873	,9994	1,000
15	12	,0176	,0424	,0905	,1727	,2969	,4613	,6482	,8227	,9444	,9945	1,000
15	13	,0037	,0107	,0271	,0617	,1268	,2361	,3980	,6042	,8159	,9638	,9996
15	14	,0005	,0017	,0052	,0142	,0353	,0802	,1671	,3186	,5490	,8290	,9904
15	15	,0000	,0001	,0005	,0016	,0047	,0134	,0352	,0874	,2059	,4633	,8601

TABLE H The Binomial Distribution (*Continued*)

n	c	p=.01	p=.05	p=.10	p=.15	p=.20	p=.25	p=.30	p=.35	p=.40	p=.45	p=.50
16	0	1,000	1,000	1,000	1,000	1,000	1,000	1,000	1,000	1,000	1,000	1,000
16	1	,1485	,5599	,8147	,9257	,9719	,9900	,9967	,9990	,9997	,9999	1,000
16	2	,0109	,1892	,4853	,7161	,8593	,9365	,9739	,9902	,9967	,9990	,9997
16	3	,0005	,0429	,2108	,4386	,6482	,8029	,9006	,9549	,9817	,9934	,9979
16	4	,0000	,0070	,0684	,2101	,4019	,5950	,7541	,8661	,9349	,9719	,9894
16	5	,0000	,0009	,0170	,0791	,2018	,3698	,5501	,7108	,8334	,9147	,9616
16	6	,0000	,0001	,0033	,0235	,0817	,1897	,3402	,5100	,6712	,8024	,8949
16	7	,0000	,0000	,0005	,0056	,0267	,0796	,1753	,3119	,4728	,6340	,7728
16	8	,0000	,0000	,0001	,0011	,0070	,0271	,0744	,1594	,2839	,4371	,5982
16	9	,0000	,0000	,0000	,0002	,0015	,0075	,0257	,0671	,1423	,2559	,4018
16	10	,0000	,0000	,0000	,0000	,0002	,0016	,0071	,0229	,0583	,1241	,2272
16	11	,0000	,0000	,0000	,0000	,0000	,0003	,0016	,0062	,0191	,0486	,1051
16	12	,0000	,0000	,0000	,0000	,0000	,0000	,0003	,0013	,0049	,0149	,0384
16	13	,0000	,0000	,0000	,0000	,0000	,0000	,0000	,0002	,0009	,0035	,0106
16	14	,0000	,0000	,0000	,0000	,0000	,0000	,0000	,0000	,0001	,0006	,0021
16	15	,0000	,0000	,0000	,0000	,0000	,0000	,0000	,0000	,0000	,0001	,0003
16	16	,0000	,0000	,0000	,0000	,0000	,0000	,0000	,0000	,0000	,0000	,0000

n	c	p=.01	p=.05	p=.10	p=.15	p=.20	p=.25	p=.30	p=.35	p=.40	p=.45	p=.50
17	0	1,000	1,000	1,000	1,000	1,000	1,000	1,000	1,000	1,000	1,000	1,000
17	1	,1571	,5819	,8332	,9369	,9775	,9925	,9977	,9993	,9998	1,000	1,000
17	2	,0123	,2078	,5182	,7475	,8818	,9499	,9807	,9933	,9979	,9994	,9999
17	3	,0006	,0503	,2382	,4802	,6904	,8363	,9226	,9673	,9877	,9959	,9988
17	4	,0000	,0088	,0826	,2444	,4511	,6470	,7981	,8972	,9536	,9816	,9936
17	5	,0000	,0012	,0221	,0987	,2418	,4261	,6113	,7652	,8740	,9404	,9755
17	6	,0000	,0001	,0047	,0319	,1057	,2347	,4032	,5803	,7361	,8529	,9283
17	7	,0000	,0000	,0008	,0083	,0377	,1071	,2248	,3812	,5522	,7098	,8338
17	8	,0000	,0000	,0001	,0017	,0109	,0402	,1046	,2128	,3595	,5257	,6855
17	9	,0000	,0000	,0000	,0003	,0026	,0124	,0403	,0994	,1989	,3374	,5000
17	10	,0000	,0000	,0000	,0000	,0005	,0031	,0127	,0383	,0919	,1834	,3145
17	11	,0000	,0000	,0000	,0000	,0001	,0006	,0032	,0120	,0348	,0826	,1662
17	12	,0000	,0000	,0000	,0000	,0000	,0001	,0007	,0030	,0106	,0301	,0717
17	13	,0000	,0000	,0000	,0000	,0000	,0000	,0001	,0006	,0025	,0086	,0245
17	14	,0000	,0000	,0000	,0000	,0000	,0000	,0000	,0001	,0005	,0019	,0064
17	15	,0000	,0000	,0000	,0000	,0000	,0000	,0000	,0000	,0001	,0003	,0012
17	16	,0000	,0000	,0000	,0000	,0000	,0000	,0000	,0000	,0000	,0000	,0001
17	17	,0000	,0000	,0000	,0000	,0000	,0000	,0000	,0000	,0000	,0000	,0000

n	c	p=.01	p=.05	p=.10	p=.15	p=.20	p=.25	p=.30	p=.35	p=.40	p=.45	p=.50
18	0	1,000	1,000	1,000	1,000	1,000	1,000	1,000	1,000	1,000	1,000	1,000
18	1	,1655	,6028	,8499	,9464	,9820	,9944	,9984	,9996	,9999	1,000	1,000
18	2	,0138	,2265	,5497	,7759	,9009	,9605	,9858	,9954	,9987	,9997	,9999
18	3	,0007	,0581	,2662	,5203	,7287	,8647	,9400	,9764	,9918	,9975	,9993
18	4	,0000	,0109	,0982	,2798	,4990	,6943	,8354	,9217	,9672	,9880	,9962
18	5	,0000	,0015	,0282	,1206	,2836	,4813	,6673	,8114	,9058	,9589	,9846
18	6	,0000	,0002	,0064	,0419	,1329	,2825	,4656	,6450	,7912	,8923	,9519
18	7	,0000	,0000	,0012	,0118	,0513	,1390	,2783	,4509	,6257	,7742	,8811
18	8	,0000	,0000	,0002	,0027	,0163	,0569	,1407	,2717	,4366	,6085	,7597
18	9	,0000	,0000	,0000	,0005	,0043	,0193	,0596	,1391	,2632	,4222	,5927
18	10	,0000	,0000	,0000	,0001	,0009	,0054	,0210	,0597	,1347	,2527	,4073
18	11	,0000	,0000	,0000	,0000	,0002	,0012	,0061	,0212	,0576	,1280	,2403
18	12	,0000	,0000	,0000	,0000	,0000	,0002	,0014	,0062	,0203	,0537	,1189
18	13	,0000	,0000	,0000	,0000	,0000	,0000	,0003	,0014	,0058	,0183	,0481
18	14	,0000	,0000	,0000	,0000	,0000	,0000	,0000	,0003	,0013	,0049	,0154
18	15	,0000	,0000	,0000	,0000	,0000	,0000	,0000	,0000	,0002	,0010	,0038
18	16	,0000	,0000	,0000	,0000	,0000	,0000	,0000	,0000	,0000	,0001	,0007
18	17	,0000	,0000	,0000	,0000	,0000	,0000	,0000	,0000	,0000	,0000	,0001
18	18	,0000	,0000	,0000	,0000	,0000	,0000	,0000	,0000	,0000	,0000	,0000

TABLE H The Binomial Distribution (*Continued*)

n	c	p=.50	p=.55	p=.60	p=.65	p=.70	p=.75	p=.80	p=.85	p=.90	p=.95	p=.99
16	0	1,000	1,000	1,000	1,000	1,000	1,000	1,000	1,000	1,000	1,000	1,000
16	1	1,000	1,000	1,000	1,000	1,000	1,000	1,000	1,000	1,000	1,000	1,000
16	2	.9997	.9999	1,000	1,000	1,000	1,000	1,000	1,000	1,000	1,000	1,000
16	3	.9979	.9994	.9999	1,000	1,000	1,000	1,000	1,000	1,000	1,000	1,000
16	4	.9894	.9965	.9991	.9998	1,000	1,000	1,000	1,000	1,000	1,000	1,000
16	5	.9616	.9851	.9951	.9987	.9997	1,000	1,000	1,000	1,000	1,000	1,000
16	6	.8949	.9514	.9809	.9938	.9984	.9997	1,000	1,000	1,000	1,000	1,000
16	7	.7728	.8759	.9417	.9771	.9929	.9984	.9998	1,000	1,000	1,000	1,000
16	8	.5982	.7441	.8577	.9329	.9743	.9925	.9985	.9998	1,000	1,000	1,000
16	9	.4018	.5629	.7161	.8406	.9256	.9729	.9930	.9989	.9999	1,000	1,000
16	10	.2272	.3660	.5272	.6881	.8247	.9204	.9733	.9944	.9995	1,000	1,000
16	11	.1051	.1976	.3288	.4900	.6598	.8103	.9183	.9765	.9967	.9999	1,000
16	12	.0384	.0853	.1666	.2892	.4499	.6302	.7982	.9209	.9830	.9991	1,000
16	13	.0106	.0281	.0651	.1339	.2459	.4050	.5981	.7899	.9316	.9930	1,000
16	14	.0021	.0066	.0183	.0451	.0994	.1971	.3518	.5614	.7892	.9571	.9995
16	15	.0003	.0010	.0033	.0098	.0261	.0635	.1407	.2839	.5147	.8108	.9891
16	16	.0000	.0001	.0003	.0010	.0033	.0100	.0281	.0743	.1853	.4401	.8515

n	c	p=.50	p=.55	p=.60	p=.65	p=.70	p=.75	p=.80	p=.85	p=.90	p=.95	p=.99
17	0	1,000	1,000	1,000	1,000	1,000	1,000	1,000	1,000	1,000	1,000	1,000
17	1	1,000	1,000	1,000	1,000	1,000	1,000	1,000	1,000	1,000	1,000	1,000
17	2	.9999	1,000	1,000	1,000	1,000	1,000	1,000	1,000	1,000	1,000	1,000
17	3	.9988	.9997	.9999	1,000	1,000	1,000	1,000	1,000	1,000	1,000	1,000
17	4	.9936	.9981	.9995	.9999	1,000	1,000	1,000	1,000	1,000	1,000	1,000
17	5	.9755	.9914	.9975	.9994	.9999	1,000	1,000	1,000	1,000	1,000	1,000
17	6	.9283	.9699	.9894	.9970	.9993	.9999	1,000	1,000	1,000	1,000	1,000
17	7	.8338	.9174	.9652	.9880	.9968	.9994	.9999	1,000	1,000	1,000	1,000
17	8	.6855	.8166	.9081	.9617	.9873	.9969	.9995	1,000	1,000	1,000	1,000
17	9	.5000	.6626	.8011	.9006	.9597	.9876	.9974	.9997	1,000	1,000	1,000
17	10	.3145	.4743	.6405	.7872	.8954	.9598	.9891	.9983	.9999	1,000	1,000
17	11	.1662	.2902	.4478	.6188	.7752	.8929	.9623	.9917	.9992	1,000	1,000
17	12	.0717	.1471	.2639	.4197	.5968	.7653	.8943	.9681	.9953	.9999	1,000
17	13	.0245	.0596	.1260	.2348	.3887	.5739	.7582	.9013	.9779	.9988	1,000
17	14	.0064	.0184	.0464	.1028	.2019	.3530	.5489	.7556	.9174	.9912	1,000
17	15	.0012	.0041	.0123	.0327	.0774	.1637	.3096	.5198	.7618	.9497	.9994
17	16	.0001	.0006	.0021	.0067	.0193	.0501	.1182	.2525	.4818	.7922	.9877
17	17	.0000	.0000	.0002	.0007	.0023	.0075	.0225	.0631	.1668	.4181	.8429

n	c	p=.50	p=.55	p=.60	p=.65	p=.70	p=.75	p=.80	p=.85	p=.90	p=.95	p=.99
18	0	1,000	1,000	1,000	1,000	1,000	1,000	1,000	1,000	1,000	1,000	1,000
18	1	1,000	1,000	1,000	1,000	1,000	1,000	1,000	1,000	1,000	1,000	1,000
18	2	.9999	1,000	1,000	1,000	1,000	1,000	1,000	1,000	1,000	1,000	1,000
18	3	.9993	.9999	1,000	1,000	1,000	1,000	1,000	1,000	1,000	1,000	1,000
18	4	.9962	.9990	.9998	1,000	1,000	1,000	1,000	1,000	1,000	1,000	1,000
18	5	.9846	.9951	.9987	.9997	1,000	1,000	1,000	1,000	1,000	1,000	1,000
18	6	.9519	.9817	.9942	.9986	.9997	1,000	1,000	1,000	1,000	1,000	1,000
18	7	.8811	.9463	.9797	.9938	.9986	.9998	1,000	1,000	1,000	1,000	1,000
18	8	.7597	.8720	.9424	.9788	.9939	.9988	.9998	1,000	1,000	1,000	1,000
18	9	.5927	.7473	.8653	.9403	.9790	.9946	.9991	.9999	1,000	1,000	1,000
18	10	.4073	.5778	.7368	.8609	.9404	.9807	.9957	.9995	1,000	1,000	1,000
18	11	.2403	.3915	.5634	.7283	.8593	.9431	.9837	.9973	.9998	1,000	1,000
18	12	.1189	.2258	.3743	.5491	.7217	.8610	.9487	.9882	.9988	1,000	1,000
18	13	.0481	.1077	.2088	.3550	.5344	.7175	.8671	.9581	.9936	.9998	1,000
18	14	.0154	.0411	.0942	.1886	.3327	.5187	.7164	.8794	.9718	.9985	1,000
18	15	.0038	.0120	.0328	.0783	.1646	.3057	.5010	.7202	.9018	.9891	1,000
18	16	.0007	.0025	.0082	.0236	.0600	.1353	.2713	.4797	.7338	.9419	.9993
18	17	.0001	.0003	.0013	.0046	.0142	.0395	.0991	.2241	.4503	.7735	.9862
18	18	.0000	.0000	.0001	.0004	.0016	.0056	.0180	.0536	.1501	.3972	.8345

TABLE H The Binomial Distribution *(Continued)*

n	c	p=.01	p=.05	p=.10	p=.15	p=.20	p=.25	p=.30	p=.35	p=.40	p=.45	p=.50
19	0	1,000	1,000	1,000	1,000	1,000	1,000	1,000	1,000	1,000	1,000	1,000
19	1	,1738	,6226	,8649	,9544	,9856	,9958	,9989	,9997	,9999	1,000	1,000
19	2	,0153	,2453	,5797	,8015	,9171	,9690	,9896	,9969	,9992	,9998	1,000
19	3	,0009	,0665	,2946	,5587	,7631	,8887	,9538	,9830	,9945	,9985	,9996
19	4	,0000	,0132	,1150	,3159	,5449	,7369	,8668	,9409	,9770	,9923	,9978
19	5	,0000	,0020	,0352	,1444	,3267	,5346	,7178	,8500	,9304	,9720	,9904
19	6	,0000	,0002	,0086	,0537	,1631	,3322	,5261	,7032	,8371	,9223	,9682
19	7	,0000	,0000	,0017	,0163	,0676	,1749	,3345	,5188	,6919	,8273	,9165
19	8	,0000	,0000	,0003	,0041	,0233	,0775	,1820	,3344	,5122	,6831	,8204
19	9	,0000	,0000	,0000	,0008	,0067	,0287	,0839	,1855	,3325	,5060	,6762
19	10	,0000	,0000	,0000	,0001	,0016	,0089	,0326	,0875	,1861	,3290	,5000
19	11	,0000	,0000	,0000	,0000	,0003	,0023	,0105	,0347	,0885	,1841	,3238
19	12	,0000	,0000	,0000	,0000	,0000	,0005	,0028	,0114	,0352	,0871	,1796
19	13	,0000	,0000	,0000	,0000	,0000	,0001	,0006	,0031	,0116	,0342	,0835
19	14	,0000	,0000	,0000	,0000	,0000	,0000	,0001	,0007	,0031	,0109	,0318
19	15	,0000	,0000	,0000	,0000	,0000	,0000	,0000	,0001	,0006	,0028	,0096
19	16	,0000	,0000	,0000	,0000	,0000	,0000	,0000	,0000	,0001	,0005	,0022
19	17	,0000	,0000	,0000	,0000	,0000	,0000	,0000	,0000	,0000	,0001	,0004
19	18	,0000	,0000	,0000	,0000	,0000	,0000	,0000	,0000	,0000	,0000	,0000
19	19	,0000	,0000	,0000	,0000	,0000	,0000	,0000	,0000	,0000	,0000	,0000

n	c	p=.01	p=.05	p=.10	p=.15	p=.20	p=.25	p=.30	p=.35	p=.40	p=.45	p=.50
20	0	1,000	1,000	1,000	1,000	1,000	1,000	1,000	1,000	1,000	1,000	1,000
20	1	,1821	,6415	,8784	,9612	,9885	,9968	,9992	,9998	1,000	1,000	1,000
20	2	,0169	,2642	,6083	,8244	,9308	,9757	,9924	,9979	,9995	,9999	1,000
20	3	,0010	,0755	,3231	,5951	,7939	,9087	,9645	,9879	,9964	,9991	,9998
20	4	,0000	,0159	,1330	,3523	,5886	,7748	,8929	,9556	,9840	,9951	,9987
20	5	,0000	,0026	,0432	,1702	,3704	,5852	,7625	,8818	,9490	,9811	,9941
20	6	,0000	,0003	,0113	,0673	,1958	,3828	,5836	,7546	,8744	,9447	,9793
20	7	,0000	,0000	,0024	,0219	,0867	,2142	,3920	,5834	,7500	,8701	,9423
20	8	,0000	,0000	,0004	,0059	,0321	,1018	,2277	,3990	,5841	,7480	,8684
20	9	,0000	,0000	,0001	,0013	,0100	,0409	,1133	,2376	,4044	,5857	,7483
20	10	,0000	,0000	,0000	,0002	,0026	,0139	,0480	,1218	,2447	,4086	,5881
20	11	,0000	,0000	,0000	,0000	,0006	,0039	,0171	,0532	,1275	,2493	,4119
20	12	,0000	,0000	,0000	,0000	,0001	,0009	,0051	,0196	,0565	,1308	,2517
20	13	,0000	,0000	,0000	,0000	,0000	,0002	,0013	,0060	,0210	,0580	,1316
20	14	,0000	,0000	,0000	,0000	,0000	,0000	,0003	,0015	,0065	,0214	,0577
20	15	,0000	,0000	,0000	,0000	,0000	,0000	,0000	,0003	,0016	,0064	,0207
20	16	,0000	,0000	,0000	,0000	,0000	,0000	,0000	,0000	,0003	,0015	,0059
20	17	,0000	,0000	,0000	,0000	,0000	,0000	,0000	,0000	,0000	,0003	,0013
20	18	,0000	,0000	,0000	,0000	,0000	,0000	,0000	,0000	,0000	,0000	,0002
20	19	,0000	,0000	,0000	,0000	,0000	,0000	,0000	,0000	,0000	,0000	,0000
20	20	,0000	,0000	,0000	,0000	,0000	,0000	,0000	,0000	,0000	,0000	,0000

TABLE H   The Binomial Distribution   (*Continued*)

n	c	p=.50	p=.55	p=.60	p=.65	p=.70	p=.75	p=.80	p=.85	p=.90	p=.95	p=.99
19	0	1.000	1.000	1.000	1.000	1.000	1.000	1.000	1.000	1.000	1.000	1.000
19	1	1.000	1.000	1.000	1.000	1.000	1.000	1.000	1.000	1.000	1.000	1.000
19	2	1.000	1.000	1.000	1.000	1.000	1.000	1.000	1.000	1.000	1.000	1.000
19	3	.9996	.9999	1.000	1.000	1.000	1.000	1.000	1.000	1.000	1.000	1.000
19	4	.9978	.9995	.9999	1.000	1.000	1.000	1.000	1.000	1.000	1.000	1.000
19	5	.9904	.9972	.9994	.9999	1.000	1.000	1.000	1.000	1.000	1.000	1.000
19	6	.9682	.9891	.9969	.9993	.9999	1.000	1.000	1.000	1.000	1.000	1.000
19	7	.9165	.9658	.9884	.9969	.9994	.9999	1.000	1.000	1.000	1.000	1.000
19	8	.8204	.9129	.9648	.9886	.9972	.9995	1.000	1.000	1.000	1.000	1.000
19	9	.6762	.8159	.9115	.9653	.9895	.9977	.9997	1.000	1.000	1.000	1.000
19	10	.5000	.6710	.8139	.9125	.9674	.9911	.9984	.9999	1.000	1.000	1.000
19	11	.3238	.4940	.6675	.8145	.9161	.9713	.9933	.9992	1.000	1.000	1.000
19	12	.1796	.3169	.4878	.6656	.8180	.9225	.9767	.9959	.9997	1.000	1.000
19	13	.0835	.1727	.3081	.4812	.6655	.8251	.9324	.9837	.9983	1.000	1.000
19	14	.0318	.0777	.1629	.2968	.4739	.6678	.8369	.9463	.9914	.9998	1.000
19	15	.0096	.0280	.0696	.1500	.2822	.4654	.6733	.8556	.9648	.9980	1.000
19	16	.0022	.0077	.0230	.0591	.1332	.2631	.4551	.6841	.8850	.9868	1.000
19	17	.0004	.0015	.0055	.0170	.0462	.1113	.2369	.4413	.7054	.9335	.9991
19	18	.0000	.0002	.0008	.0031	.0104	.0310	.0829	.1985	.4203	.7547	.9847
19	19	.0000	.0000	.0001	.0003	.0011	.0042	.0144	.0456	.1351	.3774	.8262

n	c	p=.50	p=.55	p=.60	p=.65	p=.70	p=.75	p=.80	p=.85	p=.90	p=.95	p=.99
20	0	1.000	1.000	1.000	1.000	1.000	1.000	1.000	1.000	1.000	1.000	1.000
20	1	1.000	1.000	1.000	1.000	1.000	1.000	1.000	1.000	1.000	1.000	1.000
20	2	1.000	1.000	1.000	1.000	1.000	1.000	1.000	1.000	1.000	1.000	1.000
20	3	.9998	1.000	1.000	1.000	1.000	1.000	1.000	1.000	1.000	1.000	1.000
20	4	.9987	.9997	1.000	1.000	1.000	1.000	1.000	1.000	1.000	1.000	1.000
20	5	.9941	.9985	.9997	1.000	1.000	1.000	1.000	1.000	1.000	1.000	1.000
20	6	.9793	.9936	.9984	.9997	1.000	1.000	1.000	1.000	1.000	1.000	1.000
20	7	.9423	.9786	.9935	.9985	.9997	1.000	1.000	1.000	1.000	1.000	1.000
20	8	.8684	.9420	.9790	.9940	.9987	.9998	1.000	1.000	1.000	1.000	1.000
20	9	.7483	.8692	.9435	.9804	.9949	.9991	.9999	1.000	1.000	1.000	1.000
20	10	.5881	.7507	.8725	.9468	.9829	.9961	.9994	1.000	1.000	1.000	1.000
20	11	.4119	.5914	.7553	.8782	.9520	.9861	.9974	.9998	1.000	1.000	1.000
20	12	.2517	.4143	.5956	.7624	.8867	.9591	.9900	.9987	.9999	1.000	1.000
20	13	.1316	.2520	.4159	.6010	.7723	.8982	.9679	.9941	.9996	1.000	1.000
20	14	.0577	.1299	.2500	.4166	.6080	.7858	.9133	.9781	.9976	1.000	1.000
20	15	.0207	.0553	.1256	.2454	.4164	.6172	.8042	.9327	.9887	.9997	1.000
20	16	.0059	.0189	.0510	.1182	.2375	.4148	.6296	.8298	.9568	.9974	1.000
20	17	.0013	.0049	.0160	.0444	.1071	.2252	.4114	.6477	.8670	.9841	1.000
20	18	.0002	.0009	.0036	.0121	.0355	.0913	.2061	.4049	.6769	.9245	.9990
20	19	.0000	.0001	.0005	.0021	.0076	.0243	.0692	.1756	.3917	.7358	.9831
20	20	.0000	.0000	.0000	.0002	.0008	.0032	.0115	.0388	.1216	.3585	.8179

TABLE I   The Kruskal-Wallis Test Statistic

$n_1=1$ $n_2=1$ $n_3=2$

KW	prob
.300	1.000
1.800	.833
2.700	.500

$n_1=1$ $n_2=1$ $n_3=3$

KW	prob
.533	1.000
.800	.800
2.133	.700
3.200	.300

$n_1=1$ $n_2=1$ $n_3=4$

KW	prob
.143	1.000
.786	.933
1.000	.800
1.286	.667
2.143	.600
2.500	.467
3.571	.200

$n_1=1$ $n_2=1$ $n_3=5$

KW	prob
2.314	.524
2.829	.333
3.857	.143

$n_1=1$ $n_2=2$ $n_3=2$

KW	prob
2.400	.467
3.000	.333
3.600	.200

$n_1=1$ $n_2=2$ $n_3=3$

KW	prob
3.095	.267
3.524	.200
3.857	.133
4.286	.100

$n_1=1$ $n_2=2$ $n_3=4$

KW	prob
2.893	.267
3.161	.190
3.696	.171
3.750	.133
4.018	.114
4.500	.076
4.821	.057

$n_1=1$ $n_2=2$ $n_3=5$

KW	prob
2.800	.286
2.867	.214
3.133	.202
3.333	.190
3.383	.179
3.783	.131
4.050	.119
4.200	.095
4.450	.071
5.000	.048
5.250	.036

$n_1=1$ $n_2=3$ $n_3=3$

KW	prob
3.143	.243
3.286	.157
4.000	.129
4.571	.100
5.143	.043

$n_1=1$ $n_2=3$ $n_3=4$

KW	prob
2.764	.229
3.000	.221
3.097	.214
3.208	.200
3.222	.157
3.764	.136
3.889	.129
4.056	.093
4.097	.086
4.208	.079
4.764	.071
5.000	.057
5.208	.050
5.389	.036
5.833	.021

$n_1=1$ $n_2=3$ $n_3=5$

KW	prob
2.844	.258
2.951	.218
3.040	.210
3.218	.190
3.271	.183
3.378	.143
3.484	.135
3.804	.131
3.840	.123
4.018	.095
4.284	.083
4.338	.079
4.551	.075
4.711	.056
4.871	.052
4.960	.048
5.404	.044
5.440	.036
5.760	.028
6.044	.020
6.400	.012

$n_1=1$ $n_2=4$ $n_3=4$

KW	prob
2.700	.260
2.967	.235
3.000	.222
3.267	.178
3.367	.171
3.467	.152
3.867	.121
3.900	.108
4.067	.102
4.167	.083
4.267	.070
4.800	.067
4.867	.054
4.967	.048
5.100	.041
5.667	.035
6.000	.029
6.167	.022
6.667	.010

$n_1=1$ $n_2=4$ $n_3=5$

KW	prob
2.896	.251
2.913	.222
2.940	.216
3.000	.208
3.087	.194
3.158	.187
3.240	.183
3.349	.151
3.524	.146
3.595	.138
3.682	.132
3.813	.110
3.960	.102
3.987	.098
4.206	.095
4.222	.087
4.287	.071
4.549	.067
4.636	.063
4.724	.060
4.833	.059
4.860	.056
4.986	.044
5.078	.041
5.160	.038
5.515	.037
5.558	.035
5.596	.033
5.733	.027
5.776	.025
5.858	.024
5.864	.022
5.967	.021
6.431	.019
6.578	.016
6.818	.013
6.840	.011
6.954	.008
7.364	.005

$n_1=1$ $n_2=5$ $n_3=5$

KW	prob
2.909	.242
2.946	.227
3.236	.188
3.346	.168
3.382	.161
3.527	.141
3.600	.132
3.636	.116
3.927	.113
4.036	.105
4.109	.086
4.182	.082
4.400	.076
4.546	.074
4.800	.056
4.909	.053
5.127	.046
5.236	.039
5.636	.033
5.709	.030
5.782	.027
6.000	.022
6.146	.019
6.509	.018
6.546	.015
6.582	.014
6.727	.012
6.836	.011
7.309	.009
7.527	.008
7.746	.005
8.182	.002

$n_1=2$ $n_2=2$ $n_3=2$

KW	prob
3.429	.333
3.714	.200
4.571	.067

$n_1=2$ $n_2=2$ $n_3=3$

KW	prob
3.429	.248
3.607	.238
3.750	.219
3.929	.181

$n_1=2$ $n_2=2$ $n_3=4$

KW	prob
4.464	.105
4.500	.067
4.714	.048
5.357	.029
3.125	.248
3.167	.229
3.458	.210
3.667	.190
4.000	.181
4.125	.152
4.167	.105
4.458	.100
4.500	.090
5.125	.052
5.333	.033
5.500	.024
6.000	.014

$n_1=2$ $n_2=2$ $n_3=5$

KW	prob
3.133	.254
3.240	.238
3.333	.206
3.360	.196
3.573	.185
3.773	.175
3.840	.164
3.973	.159
4.093	.148
4.200	.138
4.293	.122
4.373	.090
4.573	.085
4.800	.063
4.893	.061
5.040	.056
5.160	.034
5.693	.029
6.000	.019
6.133	.013
6.533	.008

$n_1=2$ $n_2=3$ $n_3=3$

KW	prob
3.139	.243
3.222	.221
3.361	.207
3.778	.200
3.806	.179
4.028	.164
4.111	.129
4.250	.121
4.556	.100
4.694	.093
5.000	.075
5.139	.061
5.361	.032
5.556	.025
6.250	.011

TABLE I The Kruskal-Wallis Test Statistic (*Continued*)

$n_1=2$ $n_2=3$ $n_3=4$

KW	prob
3.100	.251
3.111	.238
3.244	.232
3.278	.225
3.300	.216
3.311	.203
3.444	.197
3.478	.190
3.544	.184
3.600	.175
3.811	.168
3.844	.163
3.911	.159
3.978	.156
4.000	.149
4.078	.140
4.200	.137
4.278	.124
4.311	.108
4.378	.105
4.444	.102
4.511	.098
4.544	.086
4.611	.083
4.711	.079
4.811	.076
4.878	.073
4.900	.071
4.978	.059
5.078	.057
5.144	.054
5.378	.052
5.400	.051
5.444	.046
5.500	.040
5.611	.032
5.800	.030
6.000	.024
6.111	.021
6.144	.014
6.300	.011
6.444	.008
7.000	.005

$n_1=2$ $n_2=3$ $n_3=5$

KW	prob
2.978	.252
3.022	.248
3.069	.243
3.167	.237
3.186	.233
3.273	.222
3.331	.211
3.342	.206
3.386	.201
3.414	.193
3.506	.189
3.546	.183
3.604	.175
3.676	.171
3.767	.167
3.778	.159
3.822	.156
3.909	.152

$n_1=2$ $n_2=3$ $n_3=5$

KW	prob
3.942	.146
3.996	.139
4.058	.137
4.069	.132
4.204	.129
4.214	.125
4.233	.122
4.258	.120
4.331	.117
4.378	.113
4.494	.101
4.651	.091
4.694	.089
4.724	.087
4.727	.085
4.814	.071
4.869	.067
4.913	.063
4.942	.062
5.076	.060
5.087	.053
5.106	.052
5.251	.049
5.349	.046
5.513	.044
5.524	.043
5.542	.041
5.727	.037
5.742	.034
5.786	.033
5.804	.033
5.949	.026
6.004	.025
6.033	.024
6.091	.021
6.124	.020
6.294	.017
6.386	.016
6.414	.015
6.818	.012
6.822	.010
6.909	.009
6.949	.006
7.182	.004
7.636	.002

$n_1=2$ $n_2=4$ $n_3=4$

KW	prob
3.054	.239
3.136	.228
3.327	.220
3.354	.210
3.464	.192
3.491	.185
3.682	.180
3.764	.166
3.818	.152
4.009	.142
4.364	.125
4.418	.120
4.446	.103
4.554	.098
4.582	.094
4.691	.080

$n_1=2$ $n_2=4$ $n_3=4$

KW	prob
4.773	.075
4.854	.071
4.991	.065
5.127	.057
5.236	.052
5.454	.046
5.509	.044
5.536	.042
5.646	.039
5.727	.034
5.946	.028
6.082	.025
6.327	.024
6.409	.022
6.546	.020
6.600	.017
6.627	.016
6.873	.011
7.036	.006
7.282	.004
7.854	.002

$n_1=2$ $n_2=4$ $n_3=5$

KW	prob
2.914	.249
2.973	.246
3.023	.237
3.050	.234
3.064	.231
3.118	.226
3.164	.221
3.268	.217
3.314	.214
3.341	.208
3.364	.200
3.414	.197
3.454	.193
3.523	.190
3.564	.187
3.568	.184
3.573	.181
3.618	.178
3.641	.175
3.654	.170
3.700	.164
3.704	.160
3.791	.157
3.800	.151
3.818	.148
3.823	.145
3.864	.143
4.041	.139
4.064	.135
4.073	.133
4.091	.130
4.141	.128
4.154	.126
4.200	.123
4.223	.121
4.250	.119
4.323	.116
4.364	.114
4.368	.112
4.404	.110

$n_1=2$ $n_2=4$ $n_3=5$

KW	prob
4.500	.104
4.518	.101
4.541	.098
4.614	.090
4.664	.088
4.768	.079
4.791	.078
4.800	.076
4.818	.074
4.841	.072
4.868	.071
4.950	.063
5.073	.061
5.154	.059
5.164	.053
5.254	.052
5.268	.051
5.273	.049
5.300	.048
5.314	.046
5.414	.045
5.518	.043
5.523	.042
5.564	.038
5.641	.037
5.664	.036
5.754	.035
5.823	.034
5.891	.032
5.954	.030
5.973	.029
6.004	.026
6.041	.025
6.068	.025
6.118	.024
6.141	.023
6.223	.022
6.368	.021
6.391	.021
6.473	.020
6.504	.020
6.541	.017
6.550	.017
6.564	.016
6.654	.016
6.723	.015
6.904	.014
6.914	.013
7.000	.013
7.018	.012
7.064	.012
7.118	.010
7.204	.009
7.254	.009
7.291	.008
7.450	.007
7.500	.007
7.568	.006
7.573	.005
7.773	.004
7.814	.003
8.018	.002
8.114	.001
8.591	.001

$n_1=2$ $n_2=5$ $n_3=5$

KW	prob
3.023	.243
3.031	.234
3.123	.228
3.146	.218
3.331	.210
3.369	.203
3.392	.198
3.492	.190
3.515	.186
3.577	.181
3.646	.169
3.738	.165
3.769	.163
3.862	.150
3.885	.146
4.015	.136
4.069	.132
4.131	.130
4.138	.127
4.231	.124
4.254	.114
4.438	.106
4.477	.103
4.508	.100
4.623	.097
4.685	.092
4.754	.084
4.808	.081
4.846	.073
4.877	.068
4.992	.066
5.054	.060
5.177	.057
5.238	.054
5.246	.051
5.338	.047
5.546	.045
5.585	.041
5.608	.040
5.615	.039
5.708	.037
5.731	.036
5.792	.032
5.915	.030
5.985	.028
6.077	.027
6.231	.026
6.346	.025
6.354	.021
6.446	.020
6.469	.019
6.654	.017
6.692	.016
6.815	.015
6.838	.014
6.969	.013
7.023	.013
7.185	.012
7.208	.011
7.269	.010
7.338	.010
7.392	.009
7.462	.008
7.577	.007
7.762	.007
7.923	.006

TABLE I The Kruskal-Wallis Test Statistic (*Continued*)

$n_1=2$ $n_2=5$ $n_3=5$

KW	prob
8.008	.006
8.077	.006
8.131	.005
8.169	.003
8.292	.003
8.377	.002
8.562	.002
8.685	.001
8.938	.001
9.423	.000

$n_1=3$ $n_2=3$ $n_3=3$

KW	prob
3.200	.254
3.289	.232
3.467	.196
3.822	.168
4.267	.139
4.356	.132
4.622	.100
5.067	.086
5.422	.071
5.600	.050
5.689	.029
5.956	.025
6.489	.011
7.200	.004

$n_1=3$ $n_2=3$ $n_3=4$

KW	prob
2.954	.253
3.027	.244
3.073	.234
3.109	.220
3.254	.212
3.364	.203
3.391	.196
3.609	.188
3.682	.180
3.754	.178
3.800	.165
3.836	.150
3.973	.143
4.046	.132
4.091	.126
4.273	.123
4.336	.117
4.382	.111
4.564	.106
4.700	.101
4.709	.092
4.818	.085
4.846	.081
5.000	.074
5.064	.070
5.109	.068
5.254	.064
5.436	.062
5.500	.056
5.573	.053
5.727	.050
5.791	.046
5.936	.036
5.982	.034
6.018	.027
6.154	.025
6.300	.023
6.564	.017
6.664	.014
6.709	.013
6.746	.010
7.000	.006
7.318	.004
7.436	.002
8.018	.001

$n_1=3$ $n_2=3$ $n_3=5$

KW	prob
2.970	.242
3.079	.239
3.103	.232
3.333	.218
3.382	.215
3.394	.209
3.442	.196
3.467	.184
3.503	.179
3.576	.173
3.648	.167
3.709	.162
3.879	.156
3.927	.149
4.012	.144
4.048	.139
4.170	.135
4.194	.126
4.242	.122
4.303	.117
4.315	.113
4.412	.109
4.533	.097
4.679	.094
4.776	.090
4.800	.087
4.848	.085
4.861	.082
4.909	.079
5.042	.077
5.079	.069
5.103	.067
5.212	.065
5.261	.062
5.346	.058
5.442	.055
5.503	.053
5.515	.051
5.648	.049
5.770	.047
5.867	.042
6.012	.040
6.061	.033
6.109	.032
6.194	.027
6.303	.026
6.315	.021
6.376	.020
6.533	.019
6.594	.019
6.715	.014
6.776	.013
6.861	.012
6.982	.011
7.079	.009
7.333	.008
7.467	.008
7.503	.006
7.515	.005
7.636	.004
7.879	.003
8.048	.002
8.242	.001
8.727	.001

$n_1=3$ $n_2=4$ $n_3=4$

KW	prob
2.932	.250
2.962	.243
3.076	.230
3.136	.218
3.326	.212
3.386	.207
3.394	.201
3.417	.195
3.477	.190
3.576	.184
3.598	.178
3.659	.173
3.682	.162
3.727	.160
3.803	.154
3.848	.150
3.932	.145
3.962	.140
4.144	.135
4.167	.131
4.212	.129
4.296	.125
4.303	.121
4.326	.116
4.348	.113
4.409	.106
4.477	.102
4.546	.099
4.576	.097
4.598	.093
4.712	.090
4.750	.087
4.894	.084
5.053	.078
5.144	.073
5.182	.068
5.212	.066
5.296	.063
5.303	.061
5.326	.058
5.386	.054
5.500	.052
5.576	.051
5.598	.049
5.667	.047
5.803	.045
5.932	.043
5.962	.041
6.000	.040
6.046	.039
6.053	.035
6.144	.032
6.167	.031
6.182	.030
6.348	.027
6.386	.026
6.394	.025
6.409	.023
6.417	.022
6.546	.021
6.659	.020
6.712	.019
6.727	.018
6.962	.017
7.000	.016
7.053	.014
7.076	.011
7.136	.011
7.144	.010
7.212	.009
7.477	.006
7.598	.004
7.636	.004
7.682	.003
7.848	.003
8.227	.002
8.326	.001
8.909	.001

$n_1=3$ $n_2=4$ $n_3=5$

KW	prob
2.949	.251
2.953	.248
2.964	.240
3.010	.238
3.035	.235
3.087	.232
3.092	.222
3.106	.219
3.137	.216
3.195	.214
3.256	.209
3.260	.206
3.312	.204
3.318	.199
3.353	.197
3.414	.194
3.445	.192
3.462	.190
3.496	.188
3.503	.183
3.506	.181
3.568	.179
3.580	.177
3.599	.173
3.626	.169
3.703	.165
3.722	.163
3.753	.161
3.773	.159
3.785	.156
3.810	.152
3.831	.150
3.865	.148
3.876	.146
3.958	.144
4.015	.140
4.030	.137
4.060	.134
4.122	.132
4.154	.131
4.180	.125
4.195	.124
4.235	.121
4.241	.119
4.276	.117
4.318	.115
4.327	.112
4.368	.110
4.419	.109
4.426	.107
4.487	.106
4.522	.105
4.523	.103
4.549	.099
4.564	.097
4.645	.095
4.676	.093
4.754	.091
4.789	.089
4.810	.088
4.830	.083
4.856	.082
4.881	.081
4.891	.078
4.939	.075
4.953	.074
4.983	.073
5.041	.072
5.045	.071
5.106	.070
5.137	.068
5.158	.067
5.180	.065
5.291	.063
5.308	.062
5.342	.061
5.349	.061
5.353	.059
5.414	.058
5.426	.057
5.549	.054
5.568	.052
5.619	.051
5.631	.050
5.656	.049
5.660	.048
5.677	.047
5.718	.046
5.722	.045
5.753	.044
5.780	.043
5.804	.041
5.814	.040
5.862	.040

TABLE I The Kruskal-Wallis Test Statistic (*Continued*)

$n_1=3\ n_2=4\ n_3=5$

KW	prob
5.876	.039
5.964	.038
6.026	.038
6.030	.037
6.060	.037
6.087	.035
6.164	.035
6.173	.034
6.231	.033
6.265	.032
6.272	.030
6.337	.030
6.368	.029
6.369	.029
6.395	.026
6.410	.025
6.491	.025
6.522	.024
6.542	.023
6.580	.021
6.635	.020
6.676	.020
6.703	.019
6.780	.019
6.785	.018
6.799	.016
6.830	.016
6.891	.015
7.010	.015
7.096	.014
7.106	.014
7.188	.013
7.195	.012
7.272	.012
7.291	.011
7.395	.011
7.445	.010
7.465	.010
7.477	.009
7.523	.007
7.641	.007
7.708	.006
7.887	.006
7.906	.005
8.030	.005
8.060	.004
8.122	.004
8.215	.003
8.256	.003
8.430	.002
8.481	.002
8.503	.001
9.118	.001
9.199	.000
9.692	.000

$n_1=3\ n_2=5\ n_3=5$

KW	prob
2.936	.246
2.963	.241
3.094	.237
3.112	.224
3.121	.220
3.165	.216

$n_1=3\ n_2=5\ n_3=5$

KW	prob
3.191	.208
3.279	.206
3.306	.202
3.429	.195
3.464	.191
3.516	.187
3.622	.173
3.648	.167
3.666	.164
3.745	.161
3.780	.158
3.798	.152
3.807	.147
3.912	.144
3.965	.142
3.991	.139
4.114	.136
4.141	.135
4.150	.132
4.202	.127
4.220	.125
4.255	.117
4.308	.112
4.352	.110
4.378	.107
4.457	.105
4.466	.104
4.536	.102
4.545	.100
4.571	.098
4.694	.094
4.774	.092
4.826	.089
4.835	.088
4.888	.082
4.914	.079
4.941	.077
4.993	.075
5.020	.072
5.064	.070
5.152	.067
5.169	.065
5.222	.065
5.284	.063
5.363	.062
5.407	.059
5.486	.057
5.494	.056
5.521	.055
5.574	.053
5.600	.051
5.626	.051
5.706	.046
5.802	.045
5.837	.042
5.934	.040
5.943	.039
6.022	.038
6.048	.037
6.198	.035
6.207	.034
6.250	.034
6.259	.033
6.286	.031
6.312	.030
6.365	.030

$n_1=3\ n_2=5\ n_3=5$

KW	prob
6.391	.028
6.435	.027
6.488	.025
6.550	.024
6.593	.024
6.655	.022
6.734	.022
6.752	.021
6.866	.019
6.892	.018
6.945	.018
6.963	.017
6.998	.015
7.050	.015
7.121	.014
7.209	.014
7.226	.012
7.306	.012
7.314	.011
7.437	.011
7.543	.010
7.578	.010
7.622	.009
7.736	.009
7.763	.008
7.780	.008
7.859	.007
7.912	.007
8.026	.006
8.106	.006
8.237	.005
8.334	.005
8.545	.004
8.580	.004
8.650	.003
8.659	.003
8.791	.002
9.002	.002
9.055	.001
9.398	.001
9.521	.000
10.550	.000

$n_1=4\ n_2=4\ n_3=4$

KW	prob
2.923	.252
3.038	.234
3.115	.219
3.231	.212
3.500	.197
3.577	.173
3.731	.162
3.846	.151
3.962	.145
4.154	.136
4.192	.131
4.269	.122
4.308	.114
4.500	.104
4.654	.097
4.769	.094
4.885	.086
4.962	.080
5.115	.074

$n_1=4\ n_2=4\ n_3=4$

KW	prob
5.346	.063
5.538	.057
5.654	.055
5.692	.049
5.808	.044
6.000	.040
6.038	.037
6.269	.033
6.500	.030
6.577	.026
6.615	.024
6.731	.021
6.962	.019
7.038	.018
7.269	.016
7.385	.015
7.423	.013
7.538	.011
7.654	.008
7.731	.007
8.000	.005
8.115	.003
8.346	.002
8.654	.001
9.269	.001
9.846	.000

$n_1=4\ n_2=4\ n_3=5$

KW	prob
2.918	.249
2.967	.245
2.987	.240
2.997	.236
3.013	.228
3.086	.224
3.119	.221
3.129	.217
3.168	.214
3.218	.210
3.260	.206
3.297	.202
3.330	.200
3.382	.197
3.432	.190
3.442	.187
3.481	.183
3.511	.180
3.590	.176
3.613	.170
3.630	.167
3.640	.164
3.656	.160
3.696	.157
3.758	.154
3.828	.151
3.910	.146
3.986	.143
3.989	.141
4.025	.139
4.042	.134
4.068	.132
4.075	.130
4.118	.127
4.170	.125

$n_1=4\ n_2=4\ n_3=5$

KW	prob
4.200	.122
4.233	.121
4.253	.119
4.272	.117
4.289	.114
4.332	.112
4.381	.108
4.447	.106
4.497	.104
4.553	.102
4.619	.100
4.668	.098
4.685	.096
4.701	.094
4.711	.092
4.728	.091
4.747	.089
4.760	.088
4.813	.086
4.830	.084
4.833	.082
4.896	.081
4.975	.077
5.014	.076
5.024	.074
5.028	.073
5.090	.071
5.172	.069
5.196	.068
5.225	.066
5.344	.065
5.360	.063
5.370	.062
5.387	.061
5.410	.060
5.440	.059
5.476	.057
5.486	.056
5.489	.056
5.519	.054
5.568	.052
5.571	.051
5.618	.050
5.657	.049
5.687	.048
5.756	.047
5.782	.046
5.815	.045
5.819	.043
5.914	.042
6.003	.042
6.013	.041
6.030	.040
6.096	.039
6.119	.038
6.132	.037
6.201	.036
6.214	.034
6.228	.033
6.267	.032
6.310	.031
6.343	.030
6.382	.029
6.399	.028
6.462	.027
6.544	.027

TABLE I   The Kruskal-Wallis Test Statistic   (*Continued*)

$n_1=4$ $n_2=4$ $n_3=5$		$n_1=4$ $n_2=5$ $n_3=5$		$n_1=4$ $n_2=5$ $n_3=5$		$n_1=4$ $n_2=5$ $n_3=5$		$n_1=5$ $n_2=5$ $n_3=5$	
KW	prob	KW	prob	KW	prob	KW	prob	KW	prob
6,547	,026	3,380	,188	5,643	,050	8,363	,006	6,260	,035
6,597	,026	3,403	,187	5,666	,049	8,371	,005	6,320	,033
6,672	,024	3,471	,184	5,711	,048	8,543	,005	6,480	,032
6,676	,024	3,540	,176	5,780	,048	8,546	,004	6,500	,031
6,804	,023	3,571	,174	5,803	,047	8,771	,004	6,540	,030
6,860	,022	3,586	,170	5,811	,046	8,969	,003	6,620	,028
6,870	,022	3,651	,167	5,871	,045	9,026	,003	6,660	,027
6,887	,021	3,743	,162	5,903	,043	9,071	,002	6,720	,026
6,890	,021	3,746	,160	5,963	,042	9,286	,002	6,740	,025
6,943	,020	3,791	,155	5,983	,042	9,323	,001	6,860	,024
6,953	,020	3,800	,153	5,986	,041	9,926	,001	6,980	,021
6,976	,019	3,846	,151	6,031	,040	9,986	,000	7,020	,020
7,058	,018	3,883	,148	6,086	,040	11,571	,000	7,220	,019
7,075	,017	3,891	,144	6,100	,038			7,260	,018
7,101	,017	3,906	,142	6,123	,037			7,280	,018
7,124	,016	3,926	,140	6,146	,037	$n_1=5$ $n_2=5$ $n_3=5$		7,340	,016
7,190	,016	3,951	,137	6,166	,035	KW	prob	7,440	,015
7,203	,015	3,971	,135	6,211	,035			7,460	,015
7,233	,015	4,043	,133	6,223	,034	2,940	,252	7,580	,014
7,240	,014	4,063	,131	6,283	,034	2,960	,239	7,620	,013
7,418	,014	4,166	,127	6,303	,033	3,020	,231	7,740	,012
7,467	,013	4,200	,124	6,351	,032	3,120	,223	7,760	,012
7,497	,013	4,203	,122	6,406	,031	3,140	,216	7,940	,011
7,503	,012	4,246	,120	6,440	,030	3,260	,208	7,980	,011
7,596	,012	4,271	,118	6,451	,029	3,380	,201	8,000	,009
7,714	,011	4,291	,115	6,486	,029	3,420	,190	8,060	,009
7,744	,011	4,303	,113	6,531	,028	3,440	,184	8,180	,008
7,760	,009	4,363	,111	6,543	,028	3,500	,177	8,240	,008
7,810	,009	4,383	,110	6,603	,027	3,620	,171	8,340	,007
7,833	,008	4,386	,108	6,623	,026	3,660	,165	8,420	,007
7,942	,007	4,486	,106	6,626	,026	3,780	,159	8,540	,006
7,981	,007	4,500	,105	6,671	,025	3,840	,153	8,660	,006
8,047	,006	4,520	,101	6,760	,025	3,860	,150	8,720	,005
8,130	,006	4,523	,099	6,763	,024	3,920	,145	8,820	,005
8,140	,005	4,531	,098	6,771	,024	3,980	,137	8,880	,004
8,189	,005	4,591	,096	6,786	,023	4,020	,132	9,060	,004
8,403	,004	4,611	,095	6,806	,022	4,160	,127	9,140	,003
8,456	,004	4,660	,093	6,831	,022	4,220	,123	9,380	,003
8,525	,003	4,706	,092	6,900	,021	4,340	,118	9,420	,002
8,703	,003	4,806	,089	6,943	,020	4,380	,110	9,620	,002
8,733	,002	4,843	,088	7,000	,019	4,460	,105	9,680	,001
8,868	,002	4,851	,086	7,046	,019	4,500	,102	10,220	,001
8,997	,001	4,866	,084	7,080	,018	4,560	,100	10,260	,000
9,590	,001	4,886	,083	7,171	,018	4,580	,096	12,500	,000
9,613	,000	4,911	,079	7,183	,017	4,740	,092		
10,681	,000	4,943	,078	7,243	,017	4,820	,089		
		4,980	,076	7,266	,016	4,860	,085		
		5,023	,075	7,311	,015	4,880	,084		
$n_1=4$ $n_2=5$ $n_3=5$		5,071	,074	7,426	,015	4,940	,081		
KW	prob	5,126	,073	7,446	,014	5,040	,075		
		5,163	,070	7,491	,014	5,120	,072		
2,886	,250	5,171	,069	7,503	,013	5,180	,070		
2,931	,246	5,186	,068	7,563	,013	5,360	,065		
2,946	,239	5,206	,067	7,586	,012	5,420	,063		
2,966	,236	5,231	,066	7,631	,012	5,460	,060		
2,991	,232	5,263	,064	7,640	,011	5,540	,055		
3,023	,229	5,323	,063	7,720	,011	5,580	,053		
3,083	,224	5,400	,061	7,766	,010	5,660	,051		
3,103	,221	5,446	,059	7,860	,010	5,780	,049		
3,160	,218	5,460	,058	7,903	,009	5,820	,048		
3,240	,215	5,483	,057	8,043	,009	5,840	,046		
3,243	,211	5,491	,056	8,051	,008	6,000	,044		
3,266	,209	5,526	,056	8,143	,008	6,020	,043		
3,286	,203	5,571	,055	8,223	,007	6,080	,040		
3,311	,200	5,583	,052	8,271	,007	6,140	,038		
3,343	,197	5,620	,051	8,280	,006	6,180	,036		

TABLE J Friedman's Chi-Square Statistic

p = 3 n = 2

χ^2	prob
0,000	1,000
1,000	,8333
3,000	,5000
4,000	,1667

p = 3 n = 3

χ^2	prob
0,000	1,000
0,667	,9444
2,000	,5278
2,667	,3611
4,667	,1944
6,000	,0278

p = 3 n = 4

χ^2	prob
0,000	1,000
0,500	,9306
1,500	,6528
2,000	,4306
3,500	,2731
4,500	,1250
6,000	,0694
6,500	,0417
8,000	,0046

p = 3 n = 5

χ^2	prob
0,000	1,000
0,400	,9537
1,200	,6914
1,600	,5216
2,800	,3673
3,600	,1821
4,800	,1242
5,200	,0934
6,400	,0394
7,600	,0239
8,400	,0085
10,400	,0008

p = 3 n = 6

χ^2	prob
0,000	1,000
0,330	,9563
1,000	,7402
1,330	,5705
2,330	,4297
3,000	,2522
4,000	,1840
4,330	,1416
5,330	,0721
6,330	,0521

p = 3 n = 6

χ^2	prob
7,000	,0289
8,330	,0120
9,000	,0081
9,330	,0055
10,330	,0017
12,000	,0001

p = 3 n = 7

χ^2	prob
0,000	1,000
0,286	,9640
0,857	,7682
1,143	,6197
2,000	,4861
2,571	,3046
3,429	,2366
3,714	,1916
4,571	,1118
5,429	,0854
6,000	,0515
7,143	,0272
7,714	,0207
8,000	,0162
8,857	,0084
10,286	,0036
10,571	,0027
11,143	,0012
12,286	,0003
14,000	,0000

p = 3 n = 8

χ^2	prob
0,000	1,000
0,250	,9674
0,750	,7943
1,000	,6543
1,750	,5306
2,250	,3553
3,000	,2851
3,250	,2359
4,000	,1495
4,750	,1197
5,250	,0789
6,250	,0469
6,750	,0375
7,000	,0303
7,750	,0179
9,000	,0099
9,250	,0080
9,750	,0048
10,750	,0024
12,000	,0011
12,250	,0009
13,000	,0003
14,250	,0001
16,000	,0000

p = 3 n = 9

χ^2	prob
0,000	1,000
0,222	,9712
0,667	,8135
0,889	,6854
1,556	,5690
2,000	,3977
2,667	,3285
2,889	,2781
3,556	,1870
4,222	,1540
4,667	,1066
5,556	,0689
6,000	,0570
6,222	,0476
6,889	,0307
8,000	,0190
8,222	,0158
8,667	,0103
9,556	,0061
10,667	,0035
10,889	,0029
11,556	,0013
12,667	,0007
13,556	,0003
14,000	,0002
14,222	,0002
14,889	,0001
16,222	,0000
18,000	,0000

p = 3 n = 10

χ^2	prob
0,000	1,000
0,200	,9737
0,600	,8302
0,800	,7103
1,400	,6013
1,800	,4362
2,400	,3675
2,600	,3159
3,200	,2223
3,800	,1873
4,200	,1352
5,000	,0924
5,400	,0781
5,600	,0665
6,200	,0456
7,200	,0303
7,400	,0257
7,800	,0179
8,600	,0117
9,600	,0073
9,800	,0064
10,400	,0034
11,400	,0022
12,200	,0010
12,600	,0008
12,800	,0007
13,400	,0004

p = 3 n = 10

χ^2	prob
14,600	,0002
15,000	,0001
15,200	,0001
15,800	,0000
20,000	,0000

p = 3 n = 11

χ^2	prob
0,000	1,000
0,182	,9761
0,545	,8438
0,727	,7321
1,273	,6293
1,636	,4698
2,182	,4026
2,364	,3508
2,909	,2557
3,455	,2192
3,818	,1632
4,545	,1165
4,909	,1002
5,091	,0867
5,636	,0622
6,545	,0434
6,727	,0373
7,091	,0273
7,818	,0190
8,727	,0125
8,909	,0112
9,455	,0066
10,364	,0046
11,091	,0025
11,455	,0021
11,636	,0018
12,182	,0011
13,273	,0007
13,636	,0003
13,818	,0003
14,364	,0002
14,727	,0002
15,273	,0001
16,545	,0001
16,909	,0000
22,000	,0000

p = 3 n = 12

χ^2	prob
0,000	1,000
0,167	,9780
0,500	,8556
0,667	,7506
1,167	,6536
1,500	,5000
2,000	,4343
2,167	,3825
2,667	,2870
3,167	,2495

p = 3 n = 12

χ^2	prob
3,500	,1907
4,167	,1406
4,500	,1226
4,667	,1075
5,167	,0798
6,000	,0579
6,167	,0502
6,500	,0382
7,167	,0279
8,000	,0191
8,167	,0174
8,667	,0109
9,500	,0080
10,167	,0046
10,500	,0041
10,667	,0036
11,167	,0024
12,167	,0016
12,500	,0009
12,667	,0008
13,167	,0007
13,500	,0005
14,000	,0003
15,167	,0002
15,500	,0001
16,167	,0001
16,667	,0000
24,000	,0000

p = 3 n = 13

χ^2	prob
0,000	1,000
0,154	,9796
0,462	,8656
0,615	,7669
1,077	,6749
1,385	,5269
1,846	,4630
2,000	,4117
2,462	,3163
2,923	,2781
3,231	,2171
3,846	,1646
4,154	,1451
4,308	,1286
4,769	,0981
5,538	,0734
5,692	,0642
6,000	,0502
6,615	,0380
7,385	,0269
7,538	,0248
8,000	,0163
8,769	,0124
9,385	,0076
9,692	,0068
9,846	,0060
10,308	,0042
11,231	,0030

Donald Owen, HANDBOOK OF STATISTICAL TABLES, c 1962, Addison-Wesley, Reading, Mass. Portions of Table 14.1. Reprinted with permission.

TABLE J Friedman's Chi-Square Statistic (*Continued*)

p = 3, n = 13

χ²	prob
11,538	,0018
11,692	,0016
12,154	,0014
12,462	,0012
12,923	,0008
14,000	,0005
14,308	,0003
14,923	,0002
15,385	,0002
15,846	,0001
16,615	,0001
16,769	,0001
17,077	,0001
17,231	,0000
26,000	,0000

p = 3, n = 14

χ²	prob
16,000	,0001
16,714	,0001
17,286	,0000
28,000	,0000

p = 3, n = 15

χ²	prob
0,000	1,000
0,133	,9822
0,400	,8820
0,533	,7937
0,933	,7107
1,200	,5733
1,600	,5128
1,733	,4628
2,133	,3690
2,533	,3303
2,800	,2668
3,333	,2107
3,600	,1889
3,733	,1704
4,133	,1356
4,800	,1059
4,933	,0957
5,200	,0768
5,733	,0594
6,400	,0470
6,533	,0425
6,933	,0301
7,600	,0222
8,133	,0177
8,400	,0146
8,533	,0114
8,933	,0100
9,733	,0074
10,000	,0049
10,133	,0047
10,533	,0042
10,800	,0035
11,200	,0027
12,133	,0019
12,400	,0011
12,933	,0009
13,333	,0007
13,733	,0006
14,400	,0004
14,533	,0004
14,800	,0003
14,933	,0002
15,600	,0002
16,133	,0001
16,533	,0001
16,933	,0001
17,200	,0001
17,733	,0000
30,000	,0000

p = 3, n = 14

χ²	prob
0,000	1,000
0,143	,9810
0,429	,8744
0,571	,7811
1,000	,6938
1,286	,5513
1,714	,4890
1,857	,4383
2,286	,3436
2,714	,3050
3,000	,2425
3,571	,1880
3,857	,1672
4,000	,1496
4,429	,1168
5,143	,0895
5,286	,0788
5,571	,0632
6,143	,0492
6,857	,0357
7,000	,0332
7,429	,0227
8,143	,0177
8,714	,0114
9,000	,0103
9,143	,0092
9,571	,0068
10,429	,0049
10,714	,0031
10,857	,0029
11,286	,0026
11,571	,0021
12,000	,0016
13,000	,0011
13,286	,0006
13,857	,0005
14,286	,0004
14,714	,0003
15,429	,0002
15,571	,0002
15,857	,0002

p = 4, n = 2

χ²	prob
0,000	1,000
0,600	,9583
1,200	,8333
1,800	,7917
2,400	,6250
3,000	,5417
3,600	,4583
4,200	,3750
4,800	,2083
5,400	,1667
6,000	,0417

p = 4, n = 3

χ²	prob
0,200	1,000
0,600	,9583
1,000	,9097
1,800	,7274
2,200	,6146
2,600	,5243
3,400	,4462
3,800	,3281
4,200	,2934
5,000	,2066
5,400	,1823
5,800	,1615
6,600	,0747
7,000	,0538
7,400	,0260
8,200	,0174
9,000	,0017

p = 4, n = 4

χ²	prob
0,000	1,000
0,300	,9924
0,600	,9303
0,900	,8982
1,200	,7945
1,500	,7534
1,800	,6799
2,100	,6510
2,400	,5279
2,700	,5130
3,000	,4316
3,300	,3904
3,600	,3519
3,900	,3206
4,500	,2374
4,800	,1990
5,100	,1885
5,400	,1587
5,700	,1415
6,000	,1061
6,300	,0934
6,600	,0772
6,900	,0691
7,200	,0580
7,500	,0543
7,800	,0363
8,100	,0347
8,400	,0202
8,700	,0134
9,300	,0112
9,600	,0056
9,900	,0052
10,200	,0020
10,800	,0016
11,100	,0009
12,000	,0001

p = 4, n = 5

χ²	prob
0,120	1,000
0,360	,9743
0,600	,9441
1,080	,8569
1,320	,7690
1,560	,7100
2,040	,6521
2,280	,5631
2,520	,5199
3,000	,4428
3,240	,4055
3,480	,3679
3,960	,3007
4,200	,2664
4,440	,2316
4,920	,2125
5,160	,1618
5,400	,1509
5,880	,1186
6,120	,1017
6,360	,0889
6,840	,0714
7,080	,0667
7,320	,0574
7,800	,0491
8,040	,0331
8,280	,0325
8,760	,0236
9,000	,0210
9,240	,0152
9,720	,0111
9,960	,0092
10,200	,0077
10,680	,0058
10,920	,0028
11,160	,0023
11,640	,0017
11,880	,0014
12,120	,0011
12,600	,0004
12,840	,0004
13,080	,0002
13,560	,0001
14,040	,0000
15,000	,0000

p = 4, n = 6

χ²	prob
0,000	1,000
0,200	,9964
0,400	,9517
0,600	,9376
0,800	,8781
1,000	,8434
1,200	,7970
1,400	,7789
1,600	,6756
1,800	,6656
2,000	,6084
2,200	,5657
2,400	,5408
2,600	,5170
3,000	,4274
3,200	,3849
3,400	,3741
3,600	,3374
3,800	,3210
4,000	,2739
4,200	,2588
4,400	,2321
4,600	,2205
4,800	,1925
5,000	,1897
5,200	,1623
5,400	,1545
5,600	,1273
5,800	,1133
6,200	,1091
6,400	,0877
6,600	,0865
6,800	,0734
7,000	,0667
7,200	,0628
7,400	,0582
7,600	,0428
7,800	,0412
8,000	,0365
8,200	,0334
8,400	,0312
8,600	,0272
8,800	,0211
9,000	,0210
9,400	,0168
9,600	,0150
9,800	,0148
10,000	,0110
10,200	,0097
10,400	,0088
10,600	,0079
10,800	,0063
11,000	,0057
11,400	,0041

TABLE J Friedman's Chi-Square Statistic (*Continued*)

p = 4   n = 6

χ^2	prob
11.600	.0033
11.800	.0029
12.000	.0021
12.200	.0018
12.600	.0014
12.800	.0010
13.000	.0009
13.200	.0008
13.400	.0006
13.600	.0003
13.800	.0003
14.000	.0002
14.600	.0002
14.800	.0001
15.000	.0001
15.200	.0000
18.000	.0000

p = 4   n = 7

χ^2	prob
0.086	1.000
0.257	.9841
0.429	.9636
0.771	.9055
0.943	.8461
1.114	.7954
1.457	.7543
1.629	.6775
1.800	.6521
2.143	.5959
2.314	.5640
2.486	.5330
2.829	.4603
3.000	.4198
3.171	.3781
3.514	.3585
3.686	.3062
3.857	.2999
4.200	.2637
4.371	.2394
4.543	.2156
4.886	.1877
5.057	.1816
5.229	.1634
5.571	.1496
5.743	.1221
5.914	.1182
6.257	.1007
6.429	.0930
6.600	.0810
6.943	.0729
7.114	.0624
7.286	.0582
7.629	.0513
7.800	.0405
7.971	.0371
8.314	.0338
8.486	.0322
8.657	.0297
9.000	.0239
9.171	.0211
9.343	.0184
9.686	.0160
9.857	.0136
10.029	.0128
10.371	.0091
10.543	.0081
10.714	.0080
11.057	.0068
11.229	.0055
11.400	.0039
11.743	.0039
11.914	.0033
12.086	.0031
12.429	.0029
12.600	.0022
12.771	.0020
13.114	.0014
13.286	.0011
13.457	.0011
13.800	.0008
13.971	.0006
14.143	.0005
14.486	.0004
14.657	.0004
14.829	.0002
15.171	.0002
15.343	.0002
15.514	.0001
15.857	.0001
16.029	.0001
16.200	.0001
16.543	.0000
21.000	.0000

p = 4   n = 8

χ^2	prob
0.000	1.000
0.150	.9976
0.300	.9667
0.450	.9567
0.600	.9139
0.750	.8896
0.900	.8530
1.050	.8424
1.200	.7642
1.350	.7539
1.500	.7086
1.650	.6766
1.800	.6595
1.950	.6366
2.250	.5565
2.400	.5094
2.550	.5002
2.700	.4705
2.850	.4528
3.000	.4041
3.150	.3897
3.300	.3645
3.450	.3480
3.750	.3253
3.900	.2969
4.050	.2827
4.200	.2472
4.350	.2305
4.650	.2172
4.800	.1847
4.950	.1819
5.100	.1619
5.250	.1551
5.400	.1526
5.550	.1439
5.700	.1216
5.850	.1195
6.000	.1118
6.150	.1065
6.300	.0982
6.450	.0911
6.750	.0771
7.150	.0671
7.200	.0619
7.350	.0614
7.500	.0521
7.650	.0492
7.800	.0463
7.950	.0428
8.100	.0382
8.250	.0368
8.550	.0310
8.700	.0278
8.850	.0261
9.000	.0227
9.150	.0210
9.450	.0195
9.600	.0155
9.750	.0149
9.900	.0133
10.050	.0128
10.200	.0110
10.350	.0100
10.500	.0092
10.650	.0085
10.800	.0085
10.950	.0085
11.100	.0069
11.250	.0066
11.400	.0056
11.550	.0052
11.850	.0045
12.000	.0038
12.150	.0035
12.300	.0032
12.450	.0028
12.600	.0025
12.750	.0023
12.900	.0018
13.050	.0018
13.200	.0015
13.350	.0015
13.500	.0014
13.650	.0013
13.800	.0010
13.950	.0009
14.250	.0007
14.550	.0007
14.700	.0005
14.850	.0005
15.000	.0005
15.150	.0004
15.300	.0003
15.450	.0003
15.600	.0002
15.750	.0002
15.900	.0002
16.050	.0001
16.200	.0001
16.350	.0001
16.650	.0001
16.800	.0001
16.950	.0001
17.250	.0000
24.000	.0000

TABLE K   Kendall's Tau

t	n=4	n=5
0	.6250	.5917
2	.3750	.4083
4	.1667	.2417
6	.0417	.1167
8		.0417
10		.0083

t	n=6	n=7
1	.5000	.5000
3	.3597	.3863
5	.2347	.2810
7	.1361	.1907
9	.0681	.1194
11	.0278	.0681
13	.0083	.0345
15	.0014	.0151
17		.0054
19		.0014
21		.0002

t	n=8	n=9
0	.5476	.5403
2	.4524	.4597
4	.3598	.3807
6	.2742	.3061
8	.1994	.2384
10	.1375	.1792
12	.0894	.1298
14	.0543	.0901
16	.0305	.0597
18	.0156	.0376
20	.0071	.0223
22	.0028	.0124
24	.0009	.0063
26	.0002	.0029
28	.0000	.0012
30		.0004
32		.0001
34		.0000

t	n=10	n=11
1	.5000	.5000
3	.4309	.4396
5	.3637	.3806
7	.3003	.3240
9	.2422	.2711
11	.1904	.2227
13	.1456	.1794
15	.1082	.1415
17	.0779	.1092
19	.0542	.0823
21	.0363	.0605
23	.0233	.0433
25	.0143	.0301
27	.0083	.0203
29	.0046	.0132
31	.0023	.0083
33	.0011	.0050

t	n=10	n=11
35	.0005	.0029
37	.0002	.0016
39	.0001	.0008
41	.0000	.0004
43		.0002
45		.0001
47		.0000

t	n=12	n=13
0	.5267	.5238
2	.4733	.4762
4	.4203	.4289
6	.3687	.3825
8	.3192	.3377
10	.2726	.2950
12	.2295	.2549
14	.1904	.2177
16	.1554	.1837
18	.1248	.1531
20	.0985	.1259
22	.0763	.1022
24	.0580	.0817
26	.0432	.0644
28	.0314	.0500
30	.0224	.0382
32	.0155	.0286
34	.0105	.0211
36	.0069	.0152
38	.0044	.0108
40	.0027	.0075
42	.0016	.0051
44	.0009	.0033
46	.0005	.0021
48	.0002	.0013
50	.0001	.0008
52	.0001	.0005
54	.0000	.0003
56		.0001
58		.0001
60		.0000

t	n=14	n=15
1	.5000	.5000
3	.4572	.4613
5	.4150	.4229
7	.3736	.3852
9	.3336	.3486
11	.2953	.3132
13	.2591	.2795
15	.2253	.2475
17	.1940	.2176
19	.1654	.1897
21	.1396	.1641
23	.1166	.1408
25	.0963	.1197
27	.0786	.1009
29	.0634	.0843
31	.0505	.0697
33	.0397	.0571
35	.0308	.0463

t	n=14	n=15
37	.0236	.0372
39	.0178	.0295
41	.0132	.0231
43	.0096	.0179
45	.0069	.0137
47	.0049	.0104
49	.0034	.0078
51	.0023	.0057
53	.0015	.0041
55	.0010	.0030
57	.0006	.0021
59	.0004	.0014
61	.0002	.0010
63	.0001	.0006
65	.0001	.0004
67	.0000	.0003
69		.0002
71		.0001
73		.0001
75		.0000

t	n=16	n=17
0	.5177	.5162
2	.4823	.4838
4	.4472	.4516
6	.4124	.4197
8	.3783	.3882
10	.3450	.3575
12	.3129	.3277
14	.2821	.2988
16	.2528	.2712
18	.2251	.2448
20	.1992	.2198
22	.1751	.1964
24	.1529	.1744
26	.1325	.1541
28	.1141	.1353
30	.0975	.1181
32	.0826	.1024
34	.0695	.0883
36	.0580	.0757
38	.0480	.0644
40	.0394	.0544
42	.0321	.0457
44	.0258	.0381
46	.0206	.0315
48	.0163	.0259
50	.0128	.0211
52	.0099	.0170
54	.0076	.0137
56	.0057	.0109
58	.0043	.0086
60	.0032	.0067
62	.0023	.0052
64	.0017	.0040
66	.0012	.0030
68	.0008	.0023
70	.0006	.0017
72	.0004	.0012
74	.0003	.0009
76	.0002	.0006
78	.0001	.0005

t	n=16	n=17
80	.0001	.0003
82	.0000	.0002
84		.0001
86		.0001
88		.0001
90		.0000

t	n=18	n=19
1	.5000	.5000
3	.4703	.4725
5	.4407	.4451
7	.4114	.4180
9	.3826	.3913
11	.3544	.3650
13	.3270	.3394
15	.3005	.3144
17	.2749	.2903
19	.2504	.2670
21	.2271	.2447
23	.2051	.2234
25	.1843	.2031
27	.1648	.1840
29	.1467	.1660
31	.1300	.1492
33	.1145	.1334
35	.1004	.1189
37	.0876	.1055
39	.0760	.0931
41	.0655	.0819
43	.0562	.0716
45	.0479	.0624
47	.0406	.0541
49	.0342	.0466
51	.0287	.0400
53	.0239	.0342
55	.0197	.0290
57	.0162	.0245
59	.0132	.0206
61	.0107	.0172
63	.0086	.0143
65	.0069	.0118
67	.0054	.0097
69	.0043	.0079
71	.0033	.0064
73	.0026	.0052
75	.0020	.0041
77	.0015	.0033
79	.0011	.0026
81	.0008	.0020
83	.0006	.0016
85	.0004	.0012
87	.0003	.0009
89	.0002	.0007
91	.0002	.0005
93	.0001	.0004
95	.0001	.0003
97	.0001	.0002
99	.0000	.0002
101		.0001
103		.0001
105		.0001
107		.0000

t	n=20
0	.5128
2	.4872
4	.4618
6	.4364
8	.4113
10	.3866
12	.3623
14	.3386
16	.3154
18	.2929
20	.2712
22	.2503
24	.2303
26	.2111
28	.1929
30	.1757
32	.1594
34	.1442
36	.1299
38	.1166
40	.1043
42	.0929
44	.0825
46	.0729
48	.0642
50	.0563
52	.0492
54	.0428
56	.0370
58	.0319
60	.0274
62	.0234
64	.0199
66	.0168
68	.0142
70	.0119
72	.0099
74	.0082
76	.0068
78	.0056
80	.0045
82	.0037
84	.0030
86	.0024
88	.0019
90	.0015
92	.0012
94	.0009
96	.0007
98	.0005
100	.0004
102	.0003
104	.0002
106	.0002
108	.0001
110	.0001
112	.0001
114	.0001
116	.0000

TABLE L Spearman's Rho

n=4

s	rho	prob
0	1.000	.0417
2	.8000	.1667
4	.6000	.2083
6	.4000	.3750
8	.2000	.4583
10	.0000	.5417
12	-.2000	.6250
14	-.4000	.7917
16	-.6000	.8333
18	-.8000	.9583
20	-1.000	1.000

n=5

s	rho	prob
0	1.000	.0083
2	.9000	.0417
4	.8000	.0667
6	.7000	.1167
8	.6000	.1750
10	.5000	.2250
12	.4000	.2583
14	.3000	.3417
16	.2000	.3917
18	.1000	.4750
20	.0000	.5250
22	-.1000	.6083
24	-.2000	.6583
26	-.3000	.7417
28	-.4000	.7750
30	-.5000	.8250
32	-.6000	.8833
34	-.7000	.9333
36	-.8000	.9583
38	-.9000	.9917
40	-1.000	1.000

n=6

s	rho	prob
0	1.000	.0014
2	.9429	.0083
4	.8857	.0167
6	.8286	.0292
8	.7714	.0514
10	.7143	.0681
12	.6571	.0875
14	.6000	.1208
16	.5429	.1486
18	.4857	.1778
20	.4286	.2097
22	.3714	.2486
24	.3143	.2819
26	.2571	.3292
28	.2000	.3569
30	.1429	.4014
32	.0857	.4597
34	.0286	.5000
36	-.0286	.5403
38	-.0857	.5986
40	-.1429	.6431
42	-.2000	.6708
44	-.2571	.7181
46	-.3143	.7514
48	-.3714	.7903
50	-.4286	.8222
52	-.4857	.8514
54	-.5429	.8792
56	-.6000	.9125
58	-.6571	.9319
60	-.7143	.9486
62	-.7714	.9708
64	-.8286	.9833
66	-.8857	.9917
68	-.9429	.9986
70	-1.000	1.000

n=7

s	rho	prob
0	1.000	.0002
2	.9643	.0014
4	.9286	.0034
6	.8929	.0062
8	.8571	.0119
10	.8214	.0171
12	.7857	.0240
14	.7500	.0331
16	.7143	.0440
18	.6786	.0548
20	.6429	.0694
22	.6071	.0833
24	.5714	.1000
26	.5357	.1179
28	.5000	.1333
30	.4643	.1512
32	.4286	.1768
34	.3929	.1978
36	.3571	.2222
38	.3214	.2488
40	.2857	.2780
42	.2500	.2974
44	.2143	.3307
46	.1786	.3565
48	.1429	.3913
50	.1071	.4198
52	.0714	.4532
54	.0357	.4818
56	.0000	.5182
58	-.0357	.5468
60	-.0714	.5802
62	-.1071	.6087
64	-.1429	.6435
66	-.1786	.6693
68	-.2143	.7026
70	-.2500	.7220
72	-.2857	.7512
74	-.3214	.7778
76	-.3571	.8022
78	-.3929	.8232
80	-.4286	.8488
82	-.4643	.8667
84	-.5000	.8821
86	-.5357	.9000
88	-.5714	.9167
90	-.6071	.9306
92	-.6429	.9452
94	-.6786	.9560
96	-.7143	.9669
98	-.7500	.9760
100	-.7857	.9829
102	-.8214	.9881
104	-.8571	.9939
106	-.8929	.9966
108	-.9286	.9986
110	-.9643	.9998
112	-1.000	1.000

n=8

s	rho	prob
0	1.000	.0000
2	.9762	.0002
4	.9524	.0006
6	.9286	.0011
8	.9048	.0023
10	.8810	.0036
12	.8571	.0054
14	.8333	.0077
16	.8095	.0109
18	.7857	.0140
20	.7619	.0184
22	.7381	.0229
24	.7143	.0288
26	.6905	.0347
28	.6667	.0415
30	.6429	.0481
32	.6190	.0575
34	.5952	.0662
36	.5714	.0756
38	.5476	.0855
40	.5238	.0983
42	.5000	.1081
44	.4762	.1215
46	.4524	.1337
48	.4286	.1496
50	.4048	.1634
52	.3810	.1799
54	.3571	.1947
56	.3333	.2139
58	.3095	.2309
60	.2857	.2504
62	.2619	.2682
64	.2381	.2911
66	.2143	.3095
68	.1905	.3323
70	.1667	.3517
72	.1429	.3760
74	.1190	.3965
76	.0952	.4201
78	.0714	.4410
80	.0476	.4674
82	.0238	.4884
84	.0000	.5116
86	-.0238	.5326
88	-.0476	.5590
90	-.0714	.5799
92	-.0952	.6035
94	-.1190	.6240
96	-.1429	.6483
98	-.1667	.6677
100	-.1905	.6905
102	-.2143	.7089
104	-.2381	.7318
106	-.2619	.7496
108	-.2857	.7691
110	-.3095	.7861
112	-.3333	.8053
114	-.3571	.8201
116	-.3810	.8366
118	-.4048	.8504
120	-.4286	.8663
122	-.4524	.8785
124	-.4762	.8919
126	-.5000	.9017
128	-.5238	.9145
130	-.5476	.9244
132	-.5714	.9339
134	-.5952	.9425
136	-.6190	.9519
138	-.6429	.9585
140	-.6667	.9653
142	-.6905	.9712
144	-.7143	.9771
146	-.7381	.9816
148	-.7619	.9860
150	-.7857	.9891
152	-.8095	.9923
154	-.8333	.9946
156	-.8571	.9964
158	-.8810	.9977
160	-.9048	.9989
162	-.9286	.9994
164	-.9524	.9998
166	-.9762	1.000
168	-1.000	1.000

n=9

s	rho	prob
0	1.000	.0000
2	.9833	.0000
4	.9667	.0001
6	.9500	.0002
8	.9333	.0004
10	.9167	.0007
12	.9000	.0010
14	.8833	.0015
16	.8667	.0023
18	.8500	.0030
20	.8333	.0041
22	.8167	.0054
24	.8000	.0069
26	.7833	.0086
28	.7667	.0107
30	.7500	.0128
32	.7333	.0156
34	.7167	.0184
36	.7000	.0216
38	.6833	.0252
40	.6667	.0294
42	.6500	.0333
44	.6333	.0380
46	.6167	.0429
48	.6000	.0484

TABLE L Spearman's Rho (Continued)

n=9			n=9			n=10			n=10		
s	rho	prob	s	rho	prob	s	rho	prob	s	rho	prob
50	.5833	.0540	180	−.5000	.9191	64	.6121	.0334	194	−.1758	.6963
52	.5667	.0603	182	−.5167	.9262	66	.6000	.0367	196	−.1879	.7082
54	.5500	.0664	184	−.5333	.9336	68	.5879	.0403	198	−.2000	.7199
56	.5333	.0738	186	−.5500	.9397	70	.5758	.0441	200	−.2121	.7317
58	.5167	.0809	188	−.5667	.9460	72	.5636	.0481	202	−.2242	.7433
60	.5000	.0888	190	−.5833	.9516	74	.5515	.0524	204	−.2364	.7541
62	.4833	.0969	192	−.6000	.9571	76	.5394	.0569	206	−.2485	.7651
64	.4667	.1063	194	−.6167	.9620	78	.5273	.0616	208	−.2606	.7759
66	.4500	.1149	196	−.6333	.9667	80	.5152	.0667	210	−.2727	.7865
68	.4333	.1250	198	−.6500	.9706	82	.5030	.0720	212	−.2848	.7965
70	.4167	.1348	200	−.6667	.9748	84	.4909	.0774	214	−.2970	.8066
72	.4000	.1456	202	−.6833	.9784	86	.4788	.0831	216	−.3091	.8160
74	.3833	.1562	204	−.7000	.9816	88	.4667	.0893	218	−.3212	.8256
76	.3667	.1681	206	−.7167	.9844	90	.4545	.0956	220	−.3333	.8348
78	.3500	.1793	208	−.7333	.9873	92	.4424	.1022	222	−.3455	.8436
80	.3333	.1927	210	−.7500	.9893	94	.4303	.1091	224	−.3576	.8522
82	.3167	.2050	212	−.7667	.9914	96	.4182	.1163	226	−.3697	.8606
84	.3000	.2183	214	−.7833	.9931	98	.4061	.1237	228	−.3818	.8684
86	.2833	.2315	216	−.8000	.9946	100	.3939	.1316	230	−.3939	.8763
88	.2667	.2467	218	−.8167	.9959	102	.3818	.1394	232	−.4061	.8837
90	.2500	.2603	220	−.8333	.9970	104	.3697	.1478	234	−.4182	.8909
92	.2333	.2758	222	−.8500	.9978	106	.3576	.1564	236	−.4303	.8978
94	.2167	.2905	224	−.8667	.9985	108	.3455	.1652	238	−.4424	.9044
96	.2000	.3067	226	−.8833	.9990	110	.3333	.1744	240	−.4545	.9107
98	.1833	.3218	228	−.9000	.9993	112	.3212	.1840	242	−.4667	.9169
100	.1667	.3389	230	−.9167	.9996	114	.3091	.1934	244	−.4788	.9226
102	.1500	.3540	232	−.9333	.9998	116	.2970	.2035	246	−.4909	.9280
104	.1333	.3718	234	−.9500	.9999	118	.2848	.2135	248	−.5030	.9333
106	.1167	.3878	236	−.9667	1.000	120	.2727	.2241	250	−.5152	.9385
108	.1000	.4050	240	−1.000	1.000	122	.2606	.2349	252	−.5273	.9431
110	.0833	.4216				124	.2485	.2459	254	−.5394	.9476
112	.0667	.4401				126	.2364	.2567	256	−.5515	.9519
114	.0500	.4558		n=10		128	.2242	.2683	258	−.5636	.9559
116	.0333	.4742	s	rho	prob	130	.2121	.2801	260	−.5758	.9597
118	.0167	.4908				132	.2000	.2918	262	−.5879	.9633
120	.0000	.5092	0	1.000	.0000	134	.1879	.3037	264	−.6000	.9666
122	−.0167	.5258	6	.9636	.0000	136	.1758	.3161	266	−.6121	.9698
124	−.0333	.5442	8	.9515	.0001	138	.1636	.3284	268	−.6242	.9728
126	−.0500	.5599	10	.9394	.0001	140	.1515	.3410	270	−.6364	.9755
128	−.0667	.5784	12	.9273	.0002	142	.1394	.3536	272	−.6485	.9781
130	−.0833	.5950	14	.9152	.0003	144	.1273	.3665	274	−.6606	.9805
132	−.1000	.6122	16	.9030	.0004	146	.1152	.3795	276	−.6727	.9827
134	−.1167	.6282	18	.8909	.0006	148	.1030	.3925	278	−.6848	.9847
136	−.1333	.6460	20	.8788	.0008	150	.0909	.4056	280	−.6970	.9866
138	−.1500	.6611	22	.8667	.0011	152	.0788	.4191	282	−.7091	.9883
140	−.1667	.6782	24	.8545	.0014	154	.0667	.4326	284	−.7212	.9899
142	−.1833	.6933	26	.8424	.0019	156	.0545	.4458	286	−.7333	.9913
144	−.2000	.7095	28	.8303	.0024	158	.0424	.4592	288	−.7455	.9925
146	−.2167	.7242	30	.8182	.0029	160	.0303	.4730	290	−.7576	.9937
148	−.2333	.7397	32	.8061	.0036	162	.0182	.4865	292	−.7697	.9947
150	−.2500	.7533	34	.7939	.0044	164	.0061	.5000	294	−.7818	.9956
152	−.2667	.7685	36	.7818	.0053	166	−.0061	.5135	296	−.7939	.9964
154	−.2833	.7817	38	.7697	.0063	168	−.0182	.5270	298	−.8061	.9971
156	−.3000	.7950	40	.7576	.0075	170	−.0303	.5408	300	−.8182	.9976
158	−.3167	.8073	42	.7455	.0087	172	−.0424	.5542	302	−.8303	.9982
160	−.3333	.8207	44	.7333	.0101	174	−.0545	.5674	304	−.8424	.9986
162	−.3500	.8319	46	.7212	.0117	176	−.0667	.5809	306	−.8545	.9989
164	−.3667	.8437	48	.7091	.0134	178	−.0788	.5944	308	−.8667	.9992
166	−.3833	.8544	50	.6970	.0153	180	−.0909	.6075	310	−.8788	.9994
168	−.4000	.8652	52	.6848	.0173	182	−.1030	.6205	312	−.8909	.9996
170	−.4167	.8750	54	.6727	.0195	184	−.1152	.6335	314	−.9030	.9997
172	−.4333	.8851	56	.6606	.0219	186	−.1273	.6464	316	−.9152	.9998
174	−.4500	.8937	58	.6485	.0245	188	−.1394	.6590	318	−.9273	.9999
176	−.4667	.9031	60	.6364	.0272	190	−.1515	.6716	320	−.9394	1.000
178	−.4833	.9112	62	.6242	.0302	192	−.1636	.6839	330	−1.000	1.000

TABLE M   Lilliefors' Test Statistic

n	p=.20	p=.15	p=.10	p=.05	p=.01
4	.300	.319	.352	.381	.417
5	.285	.299	.315	.337	.405
6	.265	.277	.294	.319	.364
7	.247	.258	.276	.300	.348
8	.233	.244	.261	.285	.331
9	.223	.233	.249	.271	.311
10	.215	.224	.239	.258	.294
11	.206	.217	.230	.249	.284
12	.199	.212	.223	.242	.275
13	.190	.202	.214	.234	.268
14	.183	.194	.207	.227	.261
15	.177	.187	.201	.220	.257
16	.173	.182	.195	.213	.250
17	.169	.177	.189	.206	.245
18	.166	.173	.184	.200	.239
19	.163	.169	.179	.195	.235
20	.160	.166	.174	.190	.231
25	.142	.147	.158	.173	.200
30	.131	.136	.144	.161	.187
30+	$\dfrac{.736}{\sqrt{n}}$	$\dfrac{.768}{\sqrt{n}}$	$\dfrac{.805}{\sqrt{n}}$	$\dfrac{.886}{\sqrt{n}}$	$\dfrac{1.031}{\sqrt{n}}$

When n is greater than 30, approximate percentiles can be obtained by dividing the factors in the last row by the square root of n.

Hubert W. Lilliefors (1967), Reprinted with permission from the Journal of the American Statistical Association, 806 15th Street NW, Washington, DC 20005

TABLE N1 Coefficients for the Shapiro-Wilk Test

i	n=3	n=4	n=5	n=6	n=7	n=8	n=9	n=10	n=11	n=12	n=13	n=14
1	.7071	.6872	.6646	.6431	.6233	.6052	.5888	.5739	.5601	.5475	.5359	.5251
2	.0000	.1677	.2413	.2806	.3031	.3164	.3244	.3291	.3315	.3325	.3325	.3318
3			.0000	.0875	.1401	.1743	.1976	.2141	.2260	.2347	.2412	.2460
4					.0000	.0561	.0947	.1224	.1429	.1586	.1707	.1802
5							.0000	.0399	.0695	.0922	.1099	.1240
6									.0000	.0303	.0539	.0727
7											.0000	.0240

i	n=15	n=16	n=17	n=18	n=19	n=20	n=21	n=22	n=23	n=24	n=25	n=26
1	.5150	.5056	.4968	.4886	.4808	.4734	.4643	.4590	.4542	.4493	.4450	.4407
2	.3306	.3290	.3273	.3253	.3232	.3211	.3185	.3156	.3126	.3098	.3069	.3043
3	.2495	.2521	.2540	.2553	.2561	.2565	.2578	.2571	.2563	.2554	.2543	.2533
4	.1878	.1939	.1988	.2027	.2059	.2085	.2119	.2131	.2139	.2145	.2148	.2151
5	.1353	.1447	.1524	.1587	.1641	.1686	.1736	.1764	.1787	.1807	.1822	.1836
6	.0880	.1005	.1109	.1197	.1271	.1334	.1399	.1443	.1480	.1512	.1539	.1563
7	.0433	.0593	.0725	.0837	.0932	.1013	.1092	.1150	.1201	.1245	.1283	.1316
8	.0000	.0196	.0359	.0496	.0612	.0711	.0804	.0878	.0941	.0997	.1046	.1089
9			.0000	.0163	.0303	.0422	.0530	.0618	.0696	.0764	.0823	.0876
10					.0000	.0140	.0263	.0368	.0459	.0539	.0610	.0672
11							.0000	.0122	.0228	.0321	.0403	.0476
12									.0000	.0107	.0200	.0284
13											.0000	.0094

i	n=27	n=28	n=29	n=30	n=31	n=32	n=33	n=34	n=35	n=36	n=37	n=38
1	.4366	.4328	.4291	.4254	.4220	.4188	.4156	.4127	.4096	.4068	.4040	.4015
2	.3018	.2992	.2968	.2944	.2921	.2898	.2876	.2854	.2834	.2813	.2794	.2774
3	.2522	.2510	.2499	.2487	.2475	.2463	.2451	.2439	.2427	.2415	.2403	.2391
4	.2152	.2151	.2150	.2148	.2145	.2141	.2137	.2132	.2127	.2121	.2116	.2110
5	.1848	.1857	.1864	.1870	.1874	.1878	.1880	.1882	.1883	.1883	.1883	.1881
6	.1584	.1601	.1616	.1630	.1641	.1651	.1660	.1667	.1673	.1678	.1683	.1686
7	.1346	.1372	.1395	.1415	.1433	.1449	.1463	.1475	.1487	.1496	.1505	.1513
8	.1128	.1162	.1192	.1219	.1243	.1265	.1284	.1301	.1317	.1331	.1344	.1356
9	.0923	.0965	.1002	.1036	.1066	.1093	.1118	.1140	.1160	.1179	.1196	.1211
10	.0728	.0778	.0822	.0862	.0899	.0931	.0961	.0988	.1013	.1036	.1056	.1075
11	.0540	.0598	.0650	.0697	.0739	.0777	.0812	.0844	.0873	.0900	.0924	.0947
12	.0358	.0424	.0483	.0537	.0585	.0629	.0669	.0706	.0739	.0770	.0798	.0824
13	.0178	.0253	.0320	.0381	.0435	.0485	.0530	.0572	.0610	.0645	.0677	.0706
14	.0000	.0084	.0159	.0227	.0289	.0344	.0395	.0441	.0484	.0523	.0559	.0592
15			.0000	.0076	.0144	.0206	.0262	.0314	.0361	.0404	.0444	.0481
16					.0000	.0068	.0131	.0187	.0239	.0287	.0331	.0372
17							.0000	.0062	.0119	.0172	.0220	.0264
18									.0000	.0057	.0110	.0158
19											.0000	.0053

TABLE N1 Coefficients for the Shapiro-Wilk Test *(Continued)*

i	n=39	n=40	n=41	n=42	n=43	n=44	n=45	n=46	n=47	n=48	n=49	n=50
1	.3989	.3964	.3940	.3917	.3894	.3872	.3850	.3830	.3808	.3789	.3770	.3751
2	.2755	.2737	.2719	.2701	.2684	.2667	.2651	.2635	.2620	.2604	.2589	.2574
3	.2380	.2368	.2357	.2345	.2334	.2323	.2313	.2302	.2291	.2281	.2271	.2260
4	.2104	.2098	.2091	.2085	.2078	.2072	.2065	.2058	.2052	.2045	.2038	.2032
5	.1880	.1878	.1876	.1874	.1871	.1868	.1865	.1862	.1859	.1855	.1851	.1847
6	.1689	.1691	.1693	.1694	.1695	.1695	.1695	.1695	.1695	.1693	.1692	.1691
7	.1520	.1526	.1531	.1535	.1539	.1542	.1545	.1548	.1550	.1551	.1553	.1554
8	.1366	.1376	.1384	.1392	.1398	.1405	.1410	.1415	.1420	.1423	.1427	.1430
9	.1225	.1237	.1249	.1259	.1269	.1278	.1286	.1293	.1300	.1306	.1312	.1317
10	.1092	.1108	.1123	.1136	.1149	.1160	.1170	.1180	.1189	.1197	.1205	.1212
11	.0967	.0986	.1004	.1020	.1035	.1049	.1062	.1073	.1085	.1095	.1105	.1113
12	.0848	.0870	.0891	.0909	.0927	.0943	.0959	.0972	.0986	.0998	.1010	.1020
13	.0733	.0759	.0782	.0804	.0824	.0842	.0860	.0876	.0892	.0906	.0919	.0932
14	.0622	.0651	.0677	.0701	.0724	.0745	.0765	.0783	.0801	.0817	.0832	.0846
15	.0515	.0546	.0575	.0602	.0628	.0651	.0673	.0694	.0713	.0731	.0748	.0764
16	.0409	.0444	.0476	.0506	.0534	.0560	.0584	.0607	.0628	.0648	.0667	.0685
17	.0305	.0343	.0379	.0411	.0442	.0471	.0497	.0522	.0546	.0568	.0588	.0608
18	.0203	.0244	.0283	.0318	.0352	.0383	.0412	.0439	.0465	.0489	.0511	.0532
19	.0101	.0146	.0188	.0227	.0263	.0296	.0328	.0357	.0385	.0411	.0436	.0459
20	.0000	.0049	.0094	.0136	.0175	.0211	.0245	.0277	.0307	.0335	.0361	.0386
21			.0000	.0045	.0087	.0126	.0163	.0197	.0229	.0259	.0288	.0314
22					.0000	.0042	.0081	.0118	.0153	.0185	.0215	.0244
23							.0000	.0039	.0076	.0111	.0143	.0174
24									.0000	.0037	.0071	.0104
25											.0000	.0035

E. S. Pearson and H. O. Hartly (1972). Reproduced with permission from Biometrika, University College, Gower Street, London, WC1E 6 BT.

TABLE N2   The Shapiro-Wilk Test Statistic

n	p=.99	p=.98	p=.95	p=.90	p=.50	p=.10	p=.05	p=.02	p=.01
3	1,00	1,00	,999	,998	,959	,789	,767	,756	,753
4	,997	,996	,992	,987	,935	,792	,748	,707	,687
5	,993	,991	,986	,979	,927	,806	,762	,715	,686
6	,989	,986	,981	,974	,927	,826	,788	,743	,713
7	,988	,985	,979	,972	,928	,838	,803	,760	,730
8	,987	,984	,978	,972	,932	,851	,818	,778	,749
9	,986	,984	,978	,972	,935	,859	,829	,791	,764
10	,986	,983	,978	,972	,938	,869	,842	,806	,781
11	,986	,984	,979	,973	,940	,876	,850	,817	,792
12	,986	,984	,979	,973	,943	,883	,859	,828	,805
13	,986	,984	,979	,974	,945	,889	,866	,837	,814
14	,986	,984	,980	,975	,947	,895	,874	,846	,825
15	,987	,984	,980	,975	,950	,901	,881	,855	,835
16	,987	,985	,981	,976	,952	,906	,887	,863	,844
17	,987	,985	,981	,977	,954	,910	,892	,869	,851
18	,988	,986	,982	,978	,956	,914	,897	,874	,858
19	,988	,986	,982	,978	,957	,917	,901	,879	,863
20	,988	,986	,983	,979	,959	,920	,905	,884	,868
21	,989	,987	,983	,980	,960	,923	,908	,888	,873
22	,989	,987	,984	,980	,961	,926	,911	,892	,878
23	,989	,987	,984	,981	,962	,928	,914	,895	,881
24	,989	,987	,984	,981	,963	,930	,916	,898	,884
25	,989	,988	,985	,981	,964	,931	,918	,901	,888
26	,989	,988	,985	,982	,965	,933	,920	,904	,891
27	,990	,988	,985	,982	,965	,935	,923	,906	,894
28	,990	,988	,985	,982	,966	,936	,924	,908	,896
29	,990	,988	,985	,982	,966	,937	,926	,910	,898
30	,990	,988	,985	,983	,967	,939	,927	,912	,900
31	,990	,988	,986	,983	,967	,940	,929	,914	,902
32	,990	,988	,986	,983	,968	,941	,930	,915	,904
33	,990	,989	,986	,983	,968	,942	,931	,917	,906
34	,990	,989	,986	,983	,969	,943	,933	,919	,908
35	,990	,989	,986	,984	,969	,944	,934	,920	,910
36	990	,989	,986	,984	,970	,945	,935	,922	,912
37	,990	,989	,987	,984	,970	,946	,936	,924	,914
38	,990	,989	,987	,984	,971	,947	,938	,925	,916
39	,991	,989	,987	,984	,971	,948	,939	,927	,917
40	,991	,989	,987	,985	,972	,949	,940	,928	,919
41	,991	,989	,987	,985	,972	,950	,941	,929	,920
42	,991	,989	,987	,985	,972	,951	,942	,930	,922
43	,991	,990	,987	,985	,973	,951	,943	,932	,923
44	,991	,990	,987	,985	,973	,952	,944	,933	,924
45	,991	,990	,988	,985	,973	,953	,945	,934	,926
46	,991	,990	,988	,985	,974	,953	,945	,935	,927
47	,991	,990	,988	,985	,974	,954	,946	,936	,928
48	,991	,990	,988	,985	,974	,954	,947	,937	,929
49	,991	,990	,988	,985	,974	,955	,947	,937	,929
50	,991	,990	,988	,985	,974	,955	,947	,938	,930

TABLE O Hart's Test Statistic

The ratio of the mean square successive difference to the sample variance.

n	.001	.01	.05	.95	.99	.999	n	.001	.01	.05	.95	.99	.999
4	.7864	.8341	1.0406	4.2927	4.4992	4.5469	32	1.0245	1.2570	1.4817	2.6473	2.8720	3.1046
5	.5201	.6724	1.0255	3.9745	4.3276	4.4799	33	1.0369	1.2667	1.4885	2.6365	2.8583	3.0882
							34	1.0488	1.2761	1.4951	2.6262	2.8451	3.0725
6	.4361	.6738	1.0682	3.7318	4.1262	4.3639	35	1.0603	1.2852	1.5014	2.6163	2.8324	3.0574
7	.4311	.7163	1.0919	3.5748	3.9504	4.2356							
8	.4612	.7575	1.1228	3.4486	3.8139	4.1102	36	1.0714	1.2940	1.5075	2.6068	2.8202	3.0429
9	.4973	.7574	1.1524	3.3476	3.7025	4.0027	37	1.0822	1.3025	1.5135	2.5977	2.8085	3.0289
10	.5351	.8353	1.1803	3.2642	3.6091	3.9093	38	1.0927	1.3108	1.5193	2.5889	2.7973	3.0154
							39	1.1029	1.3188	1.5249	2.5804	2.7865	3.0024
11	.5717	.8706	1.2062	3.1938	3.5294	3.8283	40	1.1128	1.3266	1.5304	2.5722	2.7760	2.9898
12	.6062	.9033	1.2301	3.1335	3.4603	3.7574							
13	.6390	.9336	1.2521	3.0812	3.3996	3.6944	41	1.1224	1.3342	1.5357	2.5643	2.7658	2.9776
14	.6702	.9618	1.2725	3.0352	3.3458	3.6375	42	1.1317	1.3415	1.5408	2.5567	2.7560	2.9658
15	.6999	.9880	1.2914	2.9943	3.2977	3.5858	43	1.1407	1.3486	1.5458	2.5494	2.7466	2.9545
							44	1.1494	1.3554	1.5506	2.5424	2.7376	2.9436
16	.7281	1.0124	1.3090	2.9577	3.2543	3.5386	45	1.1577	1.3620	1.5552	2.5357	2.7289	2.9332
17	.7548	1.0352	1.3253	2.9247	3.2148	3.4952							
18	.7801	1.0566	1.3405	2.8948	3.1787	3.4552	46	1.1657	1.3684	1.5596	2.5293	2.7205	2.9232
19	.8040	1.0766	1.3547	2.8675	3.1456	3.4182	47	1.1734	1.3745	1.5638	2.5232	2.7125	2.9136
20	.8265	1.0954	1.3680	2.8425	3.1151	3.3840	48	1.1807	1.3802	1.5678	2.5173	2.7049	2.9044
							49	1.1877	1.3856	1.5716	2.5117	2.6977	2.8956
21	.8477	1.1131	1.3805	2.8195	3.0869	3.3523	50	1.1944	1.3907	1.5752	2.5064	2.6908	2.8872
22	.8677	1.1298	1.3923	2.7982	3.0607	3.3228							
23	.8866	1.1456	1.4035	2.7784	3.0362	3.2953	51	1.2010	1.3957	1.5787	2.5013	2.6842	2.8790
24	.9045	1.1606	1.4141	2.7599	3.0133	3.2695	52	1.2075	1.4007	1.5822	2.4963	2.6777	2.8709
25	.9215	1.1748	1.4241	2.7426	2.9919	3.2452	53	1.2139	1.4057	1.5856	2.4914	2.6712	2.8630
							54	1.2202	1.4107	1.5890	2.4866	2.6648	2.8553
26	.9378	1.1883	1.4336	2.7264	2.9718	3.2222	55	1.2264	1.4156	1.5923	2.4819	2.6585	2.8477
27	.9535	1.2012	1.4426	2.7112	2.9528	3.2003							
28	.9687	1.2135	1.4512	2.6969	2.9348	3.1794	56	1.2324	1.4203	1.5955	2.4773	2.6524	2.8403
29	.9835	1.2252	1.4594	2.6834	2.9177	3.1594	57	1.2383	1.4249	1.5987	2.4728	2.6465	2.8331
30	.9978	1.2363	1.4672	2.6707	2.9016	3.1402	58	1.2442	1.4294	1.6019	2.4684	2.6407	2.8260
							59	1.2500	1.4339	1.6051	2.4640	2.6350	2.8190
31	1.0115	1.2469	1.4746	2.6587	2.8864	3.1219	60	1.2558	1.4384	1.6082	2.4596	2.6294	2.8120

B. I. HART (1942) Reprinted with permission from the Institute of Mathematical Statistics, 3401 Investment Boulevard, Hayward, CA 94545.

TABLE P   The Swed-Eisenhart Test Statistic

m = 2

n	.995	.99	.975	.95	.05	.025	.01	.005
2					4	4	4	4
3					5	5	5	5
4					5	5	5	5
5					5	5	5	5
6					5	5	5	5
7					5	5	5	5
8			2		5	5	5	5
9			2		5	5	5	5
10			2		5	5	5	5
11			2		5	5	5	5
12		2	2		5	5	5	5
13		2	2		5	5	5	5
14		2	2		5	5	5	5
15		2	2		5	5	5	5
16		2	2		5	5	5	5
17		2	2		5	5	5	5
18		2	2		5	5	5	5
19	2	2	2		5	5	5	5
20	2	2	2		5	5	5	5

m = 3

n	.995	.99	.975	.95	.05	.025	.01	.005
3					6	6	6	6
4					6	7	7	7
5			2		7	7	7	7
6		2	2		7	7	7	7
7		2	2		7	7	7	7
8		2	2		7	7	7	7
9		2	2	2	7	7	7	7
10		2	2	3	7	7	7	7
11		2	2	3	7	7	7	7
12	2	2	2	3	7	7	7	7
13	2	2	2	3	7	7	7	7
14	2	2	2	3	7	7	7	7
15	2	2	3	3	7	7	7	7
16	2	2	3	3	7	7	7	7
17	2	2	3	3	7	7	7	7
18	2	2	3	3	7	7	7	7
19	2	2	3	3	7	7	7	7
20	2	2	3	3	7	7	7	7

m = 4

n	.995	.99	.975	.95	.05	.025	.01	.005
4				2	7	8	8	8
5			2	2	8	8	8	9
6		2	2	3	8	8	9	9
7		2	2	3	8	9	9	9
8	2	2	3	3	9	9	9	9
9	2	2	3	3	9	9	9	9
10	2	2	3	3	9	9	9	9
11	2	2	3	3	9	9	9	9
12	2	3	3	4	9	9	9	9
13	2	3	3	4	9	9	9	9
14	2	3	3	4	9	9	9	9
15	3	3	3	4	9	9	9	9
16	3	3	4	4	9	9	9	9
17	3	3	4	4	9	9	9	9
18	3	3	4	4	9	9	9	9
19	3	3	4	4	9	9	9	9
20	3	3	4	4	9	9	9	9

m = 5

n	.995	.99	.975	.95	.05	.025	.01	.005
5		2	2	3	8	9	9	10
6	2	2	3	3	9	9	10	10
7	2	2	3	3	9	10	10	11
8	2	2	3	3	10	10	11	11
9	2	3	3	4	10	11	11	11
10	3	3	3	4	10	11	11	11
11	3	3	4	4	11	11	11	11
12	3	3	4	4	11	11	11	11
13	3	3	4	4	11	11	11	11
14	3	3	4	5	11	11	11	11
15	3	4	4	5	11	11	11	11
16	3	4	4	5	11	11	11	11
17	3	4	4	5	11	11	11	11
18	4	4	5	5	11	11	11	11
19	4	4	5	5	11	11	11	11
20	4	4	5	5	11	11	11	11

m = 6

n	.995	.99	.975	.95	.05	.025	.01	.005
6	2	2	3	3	10	10	11	11
7	2	3	3	4	10	11	11	12
8	3	3	3	4	11	11	12	12
9	3	3	4	4	11	12	12	13
10	3	3	4	5	11	12	13	13
11	3	4	4	5	12	12	13	13
12	3	4	4	5	12	12	13	13
13	3	4	5	5	12	13	13	13
14	4	4	5	5	12	13	13	13
15	4	4	5	6	13	13	13	13
16	4	4	5	6	13	13	13	13
17	4	5	5	6	13	13	13	13
18	4	5	5	6	13	13	13	13
19	4	5	6	6	13	13	13	13
20	4	5	6	6	13	13	13	13

m = 7

n	.995	.99	.975	.95	.05	.025	.01	.005
7	3	3	3	4	11	12	12	12
8	3	3	4	4	12	12	13	13
9	3	4	4	5	12	13	13	14
10	3	4	5	5	12	13	14	14
11	4	4	5	5	13	13	14	14
12	4	4	5	6	13	13	14	15
13	4	5	5	6	13	14	15	15
14	4	5	5	6	13	14	15	15
15	4	5	6	6	14	14	15	15
16	5	5	6	6	14	15	15	15
17	5	5	6	7	14	15	15	15
18	5	5	6	7	14	15	15	15
19	5	6	6	7	14	15	15	15
20	5	6	6	7	14	15	15	15

m = 8

n	.995	.99	.975	.95	.05	.025	.01	.005
8	3	4	4	5	12	13	13	14
9	3	4	5	5	13	13	14	14
10	4	4	5	6	13	14	14	15
11	4	5	5	6	14	14	15	15
12	4	5	6	6	14	15	15	16
13	5	5	6	6	14	15	16	16
14	5	5	6	7	15	15	16	16

Frieda S. Swed and Churchhill Eisenhart (1943), Reprinted with permission from
the Institute of Mathematical Statistics, Ohio State University, Columbus, OH 43210

TABLE P   The Swed-Eisenhart Test Statistic   (*Continued*)

m = 8

n	.995	.99	.975	.95	.05	.025	.01	.005
15	5	5	6	7	15	15	16	17
16	5	6	6	7	15	16	16	17
17	5	6	7	7	15	16	17	17
18	6	6	7	8	15	16	17	17
19	6	6	7	8	15	16	17	17
20	6	6	7	8	16	16	17	17

m = 9

n	.995	.99	.975	.95	.05	.025	.01	.005
9	4	4	5	6	13	14	15	15
10	4	5	5	6	14	15	15	16
11	5	5	6	6	14	15	16	16
12	5	5	6	7	15	15	16	17
13	5	6	6	7	15	16	17	17
14	5	6	7	7	16	16	17	17
15	6	6	7	8	16	17	17	18
16	6	6	7	8	16	17	17	18
17	6	7	7	8	16	17	18	18
18	6	7	8	8	17	17	18	19
19	6	7	8	8	17	17	18	19
20	7	7	8	9	17	17	18	19

m = 10

n	.995	.99	.975	.95	.05	.025	.01	.005
10	5	5	6	6	15	15	16	16
11	5	5	6	7	15	16	17	17
12	5	6	7	7	16	16	17	18
13	5	6	7	8	16	17	18	18
14	6	6	7	8	16	17	18	18
15	6	7	7	8	17	17	18	19
16	6	7	8	8	17	18	19	19
17	7	7	8	9	17	18	19	19
18	7	7	8	9	18	18	19	20
19	7	8	8	9	18	19	19	20
20	7	8	9	9	18	19	19	20

m = 11

n	.995	.99	.975	.95	.05	.025	.01	.005
11	5	6	7	7	16	16	17	18
12	6	6	7	8	16	17	18	18
13	6	6	7	8	17	18	18	19
14	6	7	8	8	17	18	19	19
15	7	7	8	9	18	18	19	20
16	7	7	8	9	18	19	20	20
17	7	8	9	9	18	19	20	21
18	7	8	9	10	19	19	20	21
19	8	8	9	10	19	20	21	21
20	8	8	9	10	19	20	21	21

m = 12

n	.995	.99	.975	.95	.05	.025	.01	.005
12	6	7	7	8	17	18	18	19
13	6	7	8	9	17	18	19	20
14	7	7	8	9	18	19	20	20
15	7	8	8	9	18	19	20	21
16	7	8	9	10	19	20	21	21
17	8	8	9	10	19	20	21	21
18	8	8	9	10	20	20	21	22
19	8	9	10	10	20	21	22	22
20	8	9	10	11	20	21	22	22

m = 13

n	.995	.99	.975	.95	.05	.025	.01	.005
13	7	7	8	9	18	19	20	20
14	7	8	9	9	19	19	20	21
15	7	8	9	10	19	20	21	21
16	8	8	9	10	20	20	21	22
17	8	9	10	10	20	21	22	22
18	8	9	10	11	20	21	22	23
19	9	9	10	11	21	22	23	23
20	9	10	10	11	21	22	23	23

m = 14

n	.995	.99	.975	.95	.05	.025	.01	.005
14	7	8	9	10	19	20	21	22
15	8	8	9	10	20	21	22	22
16	8	9	10	11	20	21	22	23
17	8	9	10	11	21	22	23	23
18	9	9	10	11	21	22	23	24
19	9	10	11	12	22	22	23	24
20	9	10	11	12	22	23	24	24

m = 15

n	.995	.99	.975	.95	.05	.025	.01	.005
15	8	9	10	11	20	21	22	23
16	9	9	10	11	21	22	23	23
17	9	10	11	11	21	22	23	24
18	9	10	11	12	22	23	24	24
19	10	10	11	12	22	23	24	25
20	10	11	12	12	23	24	25	25

m = 16

n	.995	.99	.975	.95	.05	.025	.01	.005
16	9	10	11	11	22	22	23	24
17	9	10	11	12	22	23	24	25
18	10	10	11	12	23	24	25	25
19	10	11	12	13	23	24	25	26
20	10	11	12	13	24	24	25	26

m = 17

n	.995	.99	.975	.95	.05	.025	.01	.005
17	10	10	11	12	23	24	25	25
18	10	11	12	13	23	24	25	26
19	10	11	12	13	24	25	26	26
20	11	11	13	13	24	25	26	27

m = 18

n	.995	.99	.975	.95	.05	.025	.01	.005
18	11	11	12	13	24	25	26	26
19	11	12	13	14	24	25	26	27
20	11	12	13	14	25	26	27	28

m = 19

n	.995	.99	.975	.95	.05	.025	.01	.005
19	11	12	13	14	25	26	27	28
20	12	12	13	14	26	26	28	28

m = 20

n	.995	.99	.975	.95	.05	.025	.01	.005
20	12	13	14	15	26	27	28	29

For p > 0.50, the column probability is an upper limit for the table entry.
For p ≤ 0.50, the column probability is a lower limit for the table entry.

TABLE Q   Duncan's Multiple Comparison Factors

df	p	r=2	r=3	r=4	r=5	r=6	r=7	r=8	r=9
1	.05	18,0	18,0	18,0	18,0	18,0	18,0	18,0	18,0
1	.01	90,0	90,0	90,0	90,0	90,0	90,0	90,0	90,0
2	.05	6,09	6,09	6,09	6,09	6,09	6,09	6,09	6,09
2	.01	14,0	14,0	14,0	14,0	14,0	14,0	14,0	14,0
3	.05	4,50	4,50	4,50	4,50	4,50	4,50	4,50	4,50
3	.01	8,26	8,5	8,6	8,7	8,8	8,9	8,9	9,0
4	.05	3,93	4,01	4,02	4,02	4,02	4,02	4,02	4,02
4	.01	6,51	6,8	6,9	7,0	7,1	7,1	7,2	7,2
5	.05	3,64	3,74	3,79	3,83	3,83	3,83	3,83	3,83
5	.01	5,70	5,96	6,11	6,18	6,26	6,33	6,40	6,44
6	.05	3,46	3,58	3,64	3,68	3,68	3,68	3,68	3,68
6	.01	5,24	5,51	5,65	5,73	5,81	5,83	5,95	6,00
7	.05	3,35	3,47	3,54	3,58	3,60	3,61	3,61	3,61
7	.01	4,95	5,22	5,37	5,45	5,53	5,61	5,69	5,73
8	.05	3,26	3,39	3,47	3,52	3,55	3,56	3,56	3,56
8	.01	4,74	5,00	5,14	5,23	5,32	5,40	5,47	5,51
9	.05	3,20	3,34	3,41	3,47	3,50	3,52	3,52	3,52
9	.01	4,60	4,86	4,99	5,08	5,17	5,25	5,32	5,36
10	.05	3,15	3,30	3,37	3,43	3,46	3,47	3,47	3,47
10	.01	4,48	4,73	4,88	4,96	5,06	5,13	5,20	5,24
11	.05	3,11	3,27	3,35	3,39	3,43	3,44	3,45	3,46
11	.01	4,39	4,63	4,77	4,86	4,94	5,01	5,06	5,12
12	.05	3,08	3,23	3,33	3,36	3,40	3,42	3,44	3,44
12	.01	4,32	4,55	4,68	4,76	4,84	4,92	4,96	5,02
13	.05	3,06	3,21	3,30	3,35	3,38	3,41	3,42	3,44
13	.01	4,26	4,48	4,62	4,69	4,74	4,84	4,88	4,94
14	.05	3,03	3,18	3,27	3,33	3,37	3,39	3,41	3,42
14	.01	4,21	4,42	4,55	4,63	4,70	4,78	4,83	4,87
15	.05	3,01	3,16	3,25	3,31	3,36	3,38	3,40	3,42
15	.01	4,17	4,37	4,50	4,58	4,64	4,72	4,77	4,81
16	.05	3,00	3,15	3,23	3,30	3,34	3,37	3,39	3,41
16	.01	4,13	4,34	4,45	4,54	4,60	4,67	4,72	4,76
17	.05	2,98	3,13	3,22	3,28	3,33	3,36	3,38	3,40
17	.01	4,10	4,30	4,41	4,50	4,56	4,63	4,68	4,72
18	.05	2,97	3,12	3,21	3,27	3,32	3,35	3,37	3,39
18	.01	4,07	4,27	4,38	4,46	4,53	4,59	4,64	4,68
19	.05	2,96	3,11	3,19	3,26	3,31	3,35	3,37	3,39
19	.01	4,05	4,24	4,35	4,43	4,50	4,56	4,61	4,64
20	.05	2,95	3,10	3,18	3,25	3,30	3,34	3,36	3,38
20	.01	4,02	4,22	4,33	4,40	4,47	4,53	4,58	4,61
24	.05	2,92	3,07	3,15	3,22	3,28	3,31	3,34	3,37
24	.01	3,96	4,14	4,24	4,33	4,39	4,44	4,49	4,53
30	.05	2,89	3,04	3,12	3,20	3,25	3,29	3,32	3,35
30	.01	3,89	4,06	4,16	4,22	4,32	4,36	4,41	4,45
40	.05	2,86	3,01	3,10	3,17	3,22	3,27	3,30	3,33
40	.01	3,82	3,99	4,10	4,17	4,24	4,30	4,34	4,37
60	.05	2,83	2,98	3,08	3,14	3,20	3,24	3,28	3,31
60	.01	3,76	3,92	4,03	4,12	4,17	4,23	4,27	4,31
100	.05	2,80	2,95	3,05	3,12	3,18	3,22	3,26	3,29
100	.01	3,71	3,86	3,98	4,06	4,11	4,17	4,21	4,25
∞	.05	2,77	2,92	3,02	3,09	3,15	3,19	3,23	3,26
∞	.01	3,64	3,80	3,90	3,98	4,04	4,09	4,14	4,17

Reproduced from David B, Duncan, "Multiple range and multiple F tests",
BIOMETRICS 11: 1-42, 1955, With permission from the Biometrics Society,

TABLE Q Duncan's Multiple Comparison Factors (*Continued*)

df	p	r=10	r=12	r=14	r=16	r=18	r=20	r=50	r=100
1	,05	18,0	18,0	18,0	18,0	18,0	18,0	18,0	18,0
1	,01	90,0	90,0	90,0	90,0	90,0	90,0	90,0	90,0
2	,05	6,09	6,09	6,09	6,09	6,09	6,09	6,09	6,09
2	,01	14,0	14,0	14,0	14,0	14,0	14,0	14,0	14,0
3	,05	4,50	4,50	4,50	4,50	4,50	4,50	4,50	4,50
3	,01	9,0	9,0	9,1	9,2	9,3	9,3	9,3	9,3
4	,05	4,02	4,02	4,02	4,02	4,02	4,02	4,02	4,02
4	,01	7,3	7,3	7,4	7,4	7,5	7,5	7,5	7,5
5	,05	3,83	3,83	3,83	3,83	3,83	3,83	3,83	3,83
5	,01	6,5	6,6	6,6	6,7	6,7	6,8	6,8	6,8
6	,05	3,68	3,68	3,68	3,68	3,68	3,68	3,68	3,68
6	,01	6,0	6,1	6,2	6,2	6,3	6,3	6,3	6,3
7	,05	3,61	3,61	3,61	3,61	3,61	3,61	3,61	3,61
7	,01	5,8	5,8	5,9	5,9	6,0	6,0	6,0	6,0
8	,05	3,56	3,56	3,56	3,56	3,56	3,56	3,56	3,56
8	,01	5,5	5,6	5,7	5,7	5,8	5,8	5,8	5,8
9	,05	3,52	3,52	3,52	3,52	3,52	3,52	3,52	3,52
9	,01	5,4	5,5	5,5	5,6	5,7	5,7	5,7	5,7
10	,05	3,47	3,47	3,47	3,47	3,47	3,48	3,48	3,48
10	,01	5,28	5,36	5,42	5,48	5,54	5,55	5,55	5,55
11	,05	3,46	3,46	3,46	3,46	3,47	3,48	3,48	3,48
11	,01	5,15	5,24	5,28	5,34	5,38	5,39	5,39	5,39
12	,05	3,46	3,46	3,46	3,46	3,47	3,48	3,48	3,48
12	,01	5,07	5,13	5,17	5,22	5,24	5,26	5,26	5,26
13	,05	3,45	3,45	3,46	3,46	3,47	3,47	3,47	3,47
13	,01	4,98	5,04	5,08	5,13	5,14	5,15	5,15	5,15
14	,05	3,44	3,45	3,46	3,46	3,47	3,47	3,47	3,47
14	,01	4,91	4,96	5,00	5,04	5,06	5,07	5,07	5,07
15	,05	3,43	3,44	3,45	3,46	3,47	3,47	3,47	3,47
15	,01	4,84	4,90	4,94	4,97	4,99	5,00	5,00	5,00
16	,05	3,43	3,44	3,45	3,46	3,47	3,47	3,47	3,47
16	,01	4,79	4,84	4,88	4,91	4,93	4,94	4,94	4,94
17	,05	3,42	3,44	3,45	3,46	3,47	3,47	3,47	3,47
17	,01	4,75	4,80	4,83	4,86	4,88	4,89	4,89	4,89
18	,05	3,41	3,43	3,45	3,46	3,47	3,47	3,47	3,47
18	,01	4,71	4,76	4,79	4,82	4,84	4,85	4,85	4,85
19	,05	3,41	3,43	3,44	3,46	3,47	3,47	3,47	3,47
19	,01	4,67	4,72	4,76	4,79	4,81	4,82	4,82	4,82
20	,05	3,40	3,43	3,44	3,46	3,46	3,47	3,47	3,47
20	,01	4,65	4,69	4,73	4,76	4,78	4,79	4,79	4,79
24	,05	3,38	3,41	3,44	3,45	3,46	3,47	3,47	3,47
24	,01	4,57	4,62	4,64	4,67	4,70	4,72	4,74	4,74
30	,05	3,37	3,40	3,43	3,44	3,46	3,47	3,47	3,47
30	,01	4,48	4,54	4,58	4,61	4,63	4,65	4,71	4,71
40	,05	3,35	3,39	3,42	3,44	3,46	3,47	3,47	3,47
40	,01	4,41	4,46	4,51	4,54	4,57	4,59	4,69	4,69
60	,05	3,33	3,37	3,40	3,43	3,45	3,47	3,48	3,48
60	,01	4,34	4,39	4,44	4,47	4,50	4,53	4,66	4,66
100	,05	3,32	3,36	3,40	3,42	3,45	3,47	3,53	3,53
100	,01	4,29	4,35	4,38	4,42	4,45	4,48	4,64	4,65
∞	,05	3,29	3,34	3,38	3,41	3,44	3,47	3,61	3,67
∞	,01	4,20	4,26	4,31	4,34	4,38	4,41	4,60	4,68

TABLE R Newman-Keuls Multiple Comparison Factors

df	p	r=2	r=3	r=4	r=5	r=6	r=7	r=8	r=9	r=10	r=11
1	,05	18,0	27,0	32,8	37,1	40,4	43,1	45,4	47,4	49,1	50,6
	,01	90,0	135,	164,	186,	202,	216,	227,	237,	246,	253,
2	,05	6,09	8,3	9,8	10,9	11,7	12,4	13,0	13,5	14,0	14,4
	,01	14,0	19,0	22,3	24,7	26,6	28,2	29,5	30,7	31,7	32,6
3	,05	4,50	5,91	6,82	7,50	8,04	8,48	8,85	9,18	9,46	9,72
	,01	8,26	10,6	12,2	13,3	14,2	15,0	15,6	16,2	16,7	17,1
4	,05	3,93	5,04	5,76	6,29	6,71	7,05	7,35	7,60	7,83	8,03
	,01	6,51	8,12	9,17	9,96	10,6	11,1	11,5	11,9	12,3	12,6
5	,05	3,64	4,60	5,22	5,67	6,03	6,33	6,58	6,80	6,99	7,17
	,01	5,70	6,98	7,80	8,42	8,91	9,32	9,67	9,97	10,24	10,48
6	,05	3,46	4,34	4,90	5,30	5,63	5,90	6,12	6,32	6,49	6,65
	,01	5,24	6,33	7,03	7,56	7,97	8,32	8,61	8,87	9,10	9,30
7	,05	3,34	4,16	4,68	5,06	5,36	5,61	5,82	6,00	6,16	6,30
	,01	4,95	5,92	6,54	7,01	7,37	7,68	7,94	8,17	8,37	8,55
8	,05	3,26	4,04	4,53	4,89	5,17	5,40	5,60	5,77	5,92	6,05
	,01	4,75	5,64	6,20	6,62	6,96	7,24	7,47	7,68	7,86	8,03
9	,05	3,20	3,95	4,41	4,76	5,02	5,24	5,43	5,59	5,74	5,87
	,01	4,60	5,43	5,96	6,35	6,66	6,91	7,13	7,33	7,49	7,65
10	,05	3,15	3,88	4,33	4,65	4,91	5,12	5,30	5,46	5,60	5,72
	,01	4,48	5,27	5,77	6,14	6,43	6,67	6,87	7,05	7,21	7,36
11	,05	3,11	3,82	4,26	4,57	4,82	5,03	5,20	5,35	5,49	5,61
	,01	4,39	5,15	5,62	5,97	6,25	6,48	6,67	6,84	6,99	7,13
12	,05	3,08	3,77	4,20	4,51	4,75	4,95	5,12	5,27	5,39	5,51
	,01	4,32	5,05	5,50	5,84	6,10	6,32	6,51	6,67	6,81	6,94
13	,05	3,06	3,73	4,15	4,45	4,69	4,88	5,05	5,19	5,32	5,43
	,01	4,26	4,96	5,40	5,73	5,98	6,19	6,37	6,53	6,67	6,79
14	,05	3,03	3,70	4,11	4,41	4,64	4,83	4,99	5,13	5,25	5,36
	,01	4,21	4,89	5,32	5,63	5,88	6,08	6,26	6,41	6,54	6,66
15	,05	3,01	3,67	4,08	4,37	4,59	4,78	4,94	5,08	5,20	5,31
	,01	4,17	4,84	5,25	5,56	5,80	5,99	6,16	6,31	6,44	6,55
16	,05	3,00	3,65	4,05	4,33	4,56	4,74	4,90	5,03	5,15	5,26
	,01	4,13	4,79	5,19	5,49	5,72	5,92	6,08	6,22	6,35	6,46
17	,05	2,98	3,63	4,02	4,30	4,52	4,70	4,86	4,99	5,11	5,21
	,01	4,10	4,74	5,14	5,43	5,66	5,85	6,01	6,15	6,27	6,38
18	,05	2,97	3,61	4,00	4,28	4,49	4,67	4,82	4,96	5,07	5,17
	,01	4,07	4,70	5,09	5,38	5,60	5,79	5,94	6,08	6,20	6,31
19	,05	2,96	3,59	3,98	4,25	4,47	4,65	4,79	4,92	5,04	5,14
	,01	4,05	4,67	5,05	5,33	5,55	5,73	5,89	6,02	6,14	6,25
20	,05	2,95	3,58	3,96	4,23	4,45	4,62	4,77	4,90	5,01	5,11
	,01	4,02	4,64	5,02	5,29	5,51	5,69	5,84	5,97	6,09	6,19
24	,05	2,92	3,53	3,90	4,17	4,37	4,54	4,68	4,81	4,92	5,01
	,01	3,96	4,55	4,91	5,17	5,37	5,54	5,69	5,81	5,92	6,02
30	,05	2,89	3,49	3,85	4,10	4,30	4,46	4,60	4,72	4,82	4,92
	,01	3,89	4,45	4,80	5,05	5,24	5,40	5,54	5,65	5,76	5,85
40	,05	2,86	3,44	3,79	4,04	4,23	4,39	4,52	4,63	4,73	4,82
	,01	3,82	4,37	4,70	4,93	5,11	5,26	5,39	5,50	5,60	5,69
60	,05	2,83	3,40	3,74	3,98	4,16	4,31	4,44	4,55	4,65	4,73
	,01	3,76	4,28	4,59	4,82	4,99	5,13	5,25	5,36	5,45	5,53
120	,05	2,80	3,36	3,68	3,92	4,10	4,24	4,36	4,47	4,56	4,64
	,01	3,70	4,20	4,50	4,71	4,87	5,01	5,12	5,21	5,30	5,37
∞	,05	2,77	3,31	3,63	3,86	4,03	4,17	4,29	4,39	4,47	4,55
	,01	3,64	4,12	4,40	4,60	4,76	4,88	4,99	5,08	5,16	5,23

E. S. Pearson and H. O. Hartley (1966) Reproduced with permission from Biometrika, University College, Gower Street, London WC1E 6 BT.

TABLE R Newman-Keuls Multiple Comparison Factors (*Continued*)

df	p	r=12	r=13	r=14	r=15	r=16	r=17	r=18	r=19	r=20
1	.05	52.0	53.2	54.3	55.4	56.3	57.2	58.0	58.8	59.6
	.01	260.	266.	272.	277.	282.	286.	290.	294.	298.
2	.05	14.7	15.1	15.4	15.7	15.9	16.1	16.4	16.6	16.8
	.01	33.4	34.1	34.8	35.4	36.0	36.5	37.0	37.5	37.9
3	.05	9.95	10.15	10.35	10.52	10.69	10.84	10.98	11.11	11.24
	.01	17.5	17.9	18.2	18.5	18.8	19.1	19.3	19.5	19.8
4	.05	8.21	8.37	8.52	8.66	8.79	8.91	9.03	9.13	9.23
	.01	12.8	13.1	13.3	13.5	13.7	13.9	14.1	14.2	14.4
5	.05	7.32	7.47	7.60	7.72	7.83	7.93	8.03	8.12	8.21
	.01	10.70	10.89	11.08	11.24	11.40	11.55	11.68	11.81	11.93
6	.05	6.79	6.92	7.03	7.14	7.24	7.34	7.43	7.51	7.59
	.01	9.48	9.65	9.81	9.95	10.08	10.21	10.32	10.43	10.54
7	.05	6.43	6.55	6.66	6.76	6.85	6.94	7.02	7.10	7.17
	.01	8.71	8.86	9.00	9.12	9.24	9.35	9.46	9.55	9.65
8	.05	6.18	6.29	6.39	6.48	6.57	6.65	6.73	6.80	6.87
	.01	8.18	8.31	8.44	8.55	8.66	8.76	8.85	8.94	9.03
9	.05	5.98	6.09	6.19	6.28	6.36	6.44	6.51	6.58	6.64
	.01	7.78	7.91	8.03	8.13	8.23	8.33	8.41	8.49	8.57
10	.05	5.83	5.93	6.03	6.11	6.19	6.27	6.34	6.40	6.47
	.01	7.49	7.60	7.71	7.81	7.91	7.99	8.08	8.15	8.23
11	.05	5.71	5.81	5.90	5.98	6.06	6.13	6.20	6.27	6.33
	.01	7.25	7.36	7.46	7.56	7.65	7.73	7.81	7.88	7.95
12	.05	5.61	5.71	5.80	5.88	5.95	6.02	6.09	6.15	6.21
	.01	7.06	7.17	7.26	7.36	7.44	7.52	7.59	7.66	7.73
13	.05	5.53	5.63	5.71	5.79	5.86	5.93	5.99	6.05	6.11
	.01	6.90	7.01	7.10	7.19	7.27	7.35	7.42	7.48	7.55
14	.05	5.46	5.55	5.64	5.71	5.79	5.85	5.91	5.97	6.03
	.01	6.77	6.87	6.96	7.05	7.13	7.20	7.27	7.33	7.39
15	.05	5.40	5.49	5.57	5.65	5.72	5.78	5.85	5.90	5.96
	.01	6.66	6.76	6.84	6.93	7.00	7.07	7.14	7.20	7.26
16	.05	5.35	5.44	5.52	5.59	5.66	5.73	5.79	5.84	5.90
	.01	6.56	6.66	6.74	6.82	6.90	6.97	7.03	7.09	7.15
17	.05	5.31	5.39	5.47	5.54	5.61	5.67	5.73	5.79	5.84
	.01	6.48	6.57	6.66	6.73	6.81	6.87	6.94	7.00	7.05
18	.05	5.27	5.35	5.43	5.50	5.58	5.63	5.69	5.74	5.79
	.01	6.41	6.50	6.58	6.65	6.73	6.79	6.85	6.91	6.97
19	.05	5.23	5.31	5.39	5.46	5.53	5.59	5.65	5.70	5.75
	.01	6.34	6.43	6.51	6.58	6.65	6.72	6.78	6.84	6.89
20	.05	5.20	5.28	5.36	5.43	5.49	5.55	5.61	5.66	5.71
	.01	6.28	6.37	6.45	6.52	6.59	6.65	6.71	6.77	6.82
24	.05	5.10	5.18	5.25	5.32	5.38	5.44	5.49	5.55	5.59
	.01	6.11	6.19	6.26	6.33	6.39	6.45	6.51	6.56	6.61
30	.05	5.00	5.08	5.15	5.21	5.27	5.33	5.38	5.43	5.47
	.01	5.93	6.01	6.08	6.14	6.20	6.26	6.31	6.36	6.41
40	.05	4.90	4.98	5.04	5.11	5.16	5.22	5.27	5.31	5.36
	.01	5.76	5.83	5.90	5.96	6.02	6.07	6.12	6.16	6.21
60	.05	4.81	4.88	4.94	5.00	5.06	5.11	5.15	5.20	5.24
	.01	5.60	5.67	5.73	5.78	5.84	5.89	5.93	5.97	6.01
120	.05	4.71	4.78	4.84	4.90	4.95	5.00	5.04	5.09	5.13
	.01	5.44	5.50	5.56	5.61	5.66	5.71	5.75	5.79	5.83
∞	.05	4.62	4.68	4.74	4.80	4.85	4.89	4.93	4.97	5.01
	.01	5.29	5.35	5.40	5.45	5.49	5.54	5.57	5.61	5.65

TABLE S1   Dunnett's Multiple Comparison Factors

For use when the research hypothesis is that samples
are all greater or all less than the control.

df	p	r=2	r=3	r=4	r=5	r=6	r=7	r=8	r=9	r=10
5	.05	2,02	2,44	2,68	2,85	2,98	3,08	3,16	3,24	3,30
5	.01	3,37	3,90	4,21	4,43	4,60	4,73	4,85	4,94	5,03
6	.05	1,94	2,34	2,56	2,71	2,83	2,92	3,00	3,07	3,12
6	.01	3,14	3,61	3,88	4,07	4,21	4,33	4,43	4,51	4,59
7	.05	1,89	2,27	2,48	2,62	2,73	2,82	2,89	2,95	3,01
7	.01	3,00	3,42	3,66	3,83	3,96	4,07	4,15	4,23	4,30
8	.05	1,86	2,22	2,42	2,55	2,66	2,74	2,81	2,87	2,92
8	.01	2,90	3,29	3,51	3,67	3,79	3,88	3,96	4,03	4,09
9	.05	1,83	2,18	2,37	2,50	2,60	2,68	2,75	2,81	2,86
9	.01	2,82	3,19	3,40	3,55	3,66	3,75	3,82	3,89	3,94
10	.05	1,81	2,15	2,34	2,47	2,56	2,64	2,70	2,76	2,81
10	.01	2,76	3,11	3,31	3,45	3,56	3,64	3,71	3,78	3,83
11	.05	1,80	2,13	2,31	2,44	2,53	2,60	2,67	2,72	2,77
11	.01	2,72	3,06	3,25	3,38	3,48	3,56	3,63	3,69	3,74
12	.05	1,78	2,11	2,29	2,41	2,50	2,58	2,64	2,69	2,74
12	.01	2,68	3,01	3,19	3,32	3,42	3,50	3,56	3,62	3,67
13	.05	1,77	2,09	2,27	2,39	2,48	2,55	2,61	2,66	2,71
13	.01	2,65	2,97	3,15	3,27	3,37	3,44	3,51	3,56	3,61
14	.05	1,76	2,08	2,25	2,37	2,46	2,53	2,59	2,64	2,69
14	.01	2,62	2,94	3,11	3,23	3,32	3,40	3,46	3,51	3,56
15	.05	1,75	2,07	2,24	2,36	2,44	2,51	2,57	2,62	2,67
15	.01	2,60	2,91	3,08	3,20	3,29	3,36	3,42	3,47	3,52
16	.05	1,75	2,06	2,23	2,34	2,43	2,50	2,56	2,61	2,65
16	.01	2,58	2,88	3,05	3,17	3,26	3,33	3,39	3,44	3,48
17	.05	1,74	2,05	2,22	2,33	2,42	2,49	2,54	2,59	2,64
17	.01	2,57	2,86	3,03	3,14	3,23	3,30	3,36	3,41	3,45
18	.05	1,73	2,04	2,21	2,32	2,41	2,48	2,53	2,58	2,62
18	.01	2,55	2,84	3,01	3,12	3,21	3,27	3,33	3,38	3,42
19	.05	1,73	2,03	2,20	2,31	2,40	2,47	2,52	2,57	2,61
19	.01	2,54	2,83	2,99	3,10	3,18	3,25	3,31	3,36	3,40
20	.05	1,72	2,03	2,19	2,30	2,39	2,46	2,51	2,56	2,60
20	.01	2,53	2,81	2,97	3,08	3,17	3,23	3,29	3,34	3,38
24	.05	1,71	2,01	2,17	2,28	2,36	2,43	2,48	2,53	2,57
24	.01	2,49	2,77	2,92	3,03	3,11	3,17	3,22	3,27	3,31
30	.05	1,70	1,99	2,15	2,25	2,33	2,40	2,45	2,50	2,54
30	.01	2,46	2,72	2,87	2,97	3,05	3,11	3,16	3,21	3,24
40	.05	1,68	1,97	2,13	2,23	2,31	2,37	2,42	2,47	2,51
40	.01	2,42	2,68	2,82	2,92	2,99	3,05	3,10	3,14	3,18
60	.05	1,67	1,95	2,10	2,21	2,28	2,35	2,39	2,44	2,48
60	.01	2,39	2,64	2,78	2,87	2,94	3,00	3,04	3,08	3,12
120	.05	1,66	1,93	2,08	2,18	2,26	2,32	2,37	2,41	2,45
120	.01	2,36	2,60	2,73	2,82	2,89	2,94	2,99	3,03	3,06
∞	.05	1,64	1,92	2,06	2,16	2,23	2,29	2,34	2,38	2,42
∞	.01	2,33	2,56	2,68	2,77	2,84	2,89	2,93	2,97	3,00

Charles W. Dunnett (1955), Reprinted with permission from the Journal of the
American Statistical Association, 806 15th Street NW, Washington, DC 20005.

TABLE S2   Dunnett's Multiple Comparison Factors

For use when the research hypothesis is that some samples
are greater and some are less than the control.

df	p	r=2	r=3	r=4	r=5	r=6	r=7	r=8	r=9	r=10
5	.05	2.57	3.03	3.29	3.48	3.62	3.73	3.82	3.90	3.97
5	.01	4.03	4.63	4.98	5.22	5.41	5.56	5.69	5.80	5.89
6	.05	2.45	2.86	3.10	3.26	3.39	3.49	3.57	3.64	3.71
6	.01	3.71	4.21	4.51	4.71	4.87	5.00	5.10	5.20	5.28
7	.05	2.36	2.75	2.97	3.12	3.24	3.33	3.41	3.47	3.53
7	.01	3.50	3.95	4.21	4.39	4.53	4.64	4.74	4.82	4.89
8	.05	2.31	2.67	2.88	3.02	3.13	3.22	3.29	3.35	3.41
8	.01	3.26	3.77	4.00	4.17	4.29	4.40	4.48	4.56	4.62
9	.05	2.26	2.61	2.81	2.95	3.05	3.14	3.20	3.26	3.32
9	.01	3.25	3.63	3.85	4.01	4.12	4.22	4.30	4.37	4.43
10	.05	2.23	2.57	2.76	2.89	2.99	3.07	3.14	3.19	3.24
10	.01	3.17	3.53	3.74	3.88	3.99	4.08	4.16	4.22	4.28
11	.05	2.20	2.53	2.72	2.84	2.94	3.02	3.08	3.14	3.19
11	.01	3.11	3.45	3.65	3.79	3.89	3.98	4.05	4.11	4.16
12	.05	2.18	2.50	2.68	2.81	2.90	2.98	3.04	3.09	3.14
12	.01	3.05	3.39	3.58	3.71	3.81	3.89	3.96	4.02	4.07
13	.05	2.16	2.48	2.65	2.78	2.87	2.94	3.00	3.06	3.10
13	.01	3.01	3.33	3.52	3.65	3.74	3.82	3.89	3.94	3.99
14	.05	2.14	2.46	2.63	2.75	2.84	2.91	2.97	3.02	3.07
14	.01	2.98	3.29	3.47	3.59	3.69	3.76	3.83	3.88	3.93
15	.05	2.13	2.44	2.61	2.73	2.82	2.89	2.95	3.00	3.04
15	.01	2.95	3.25	3.43	3.55	3.64	3.71	3.78	3.83	3.88
16	.05	2.12	2.42	2.59	2.71	2.80	2.87	2.92	2.97	3.02
16	.01	2.92	3.22	3.39	3.51	3.60	3.67	3.73	3.78	3.83
17	.05	2.11	2.41	2.58	2.69	2.78	2.85	2.90	2.95	3.00
17	.01	2.90	3.19	3.36	3.47	3.56	3.63	3.69	3.74	3.79
18	.05	2.10	2.40	2.56	2.68	2.76	2.83	2.89	2.94	2.98
18	.01	2.88	3.17	3.33	3.44	3.53	3.60	3.66	3.71	3.75
19	.05	2.09	2.39	2.55	2.66	2.75	2.81	2.87	2.92	2.96
19	.01	2.86	3.15	3.31	3.42	3.50	3.57	3.63	3.68	3.72
20	.05	2.09	2.38	2.54	2.65	2.73	2.80	2.86	2.90	2.95
20	.01	2.85	3.13	3.29	3.40	3.48	3.55	3.60	3.65	3.69
24	.05	2.06	2.35	2.51	2.61	2.70	2.76	2.81	2.86	2.90
24	.01	2.80	3.07	3.22	3.32	3.40	3.47	3.52	3.57	3.61
30	.05	2.04	2.32	2.47	2.58	2.66	2.72	2.77	2.82	2.86
30	.01	2.75	3.01	3.15	3.25	3.33	3.39	3.44	3.49	3.52
40	.05	2.02	2.29	2.44	2.54	2.62	2.68	2.73	2.77	2.81
40	.01	2.70	2.95	3.09	3.19	3.26	3.32	3.37	3.41	3.44
60	.05	2.00	2.27	2.41	2.51	2.58	2.64	2.69	2.73	2.77
60	.01	2.66	2.90	3.03	3.12	3.19	3.25	3.29	3.33	3.37
120	.05	1.98	2.24	2.38	2.47	2.55	2.60	2.65	2.69	2.73
120	.01	2.62	2.85	2.97	3.06	3.12	3.18	3.22	3.26	3.29
∞	.05	1.96	2.21	2.35	2.44	2.51	2.57	2.61	2.65	2.69
∞	.01	2.58	2.79	2.92	3.00	3.06	3.11	3.15	3.19	3.22

Reproduced from Charles W. Dunnett, "New tables for multiple comparisons with a control
BIOMETRICS 20: 482-491, 1964. With permission from the Biometrics Society.

TABLE T Random Digits

1	2	3	4	5	6	7	8	9	10	11	12
8287	8317	4732	6499	1072	3982	4107	6530	9277	3296	8230	4170
3139	5047	3059	8467	4562	6428	8663	0354	9476	0753	5234	3178
8381	3789	0106	2703	7410	9892	7089	9994	6862	5465	0116	8456
4994	2285	0991	6143	0641	9602	4760	3886	7148	5275	7896	6949
9970	3259	6345	0202	9447	3480	6996	5248	1662	2809	0648	3678
3848	4217	2915	2725	0286	9919	1937	4858	1250	8871	3674	2053
1219	0028	7381	2415	7399	3624	1675	0343	2667	6475	7782	9851
8262	3022	5953	8534	3924	9656	7008	1941	6943	3414	7942	4374
7490	6897	6939	0896	0563	3221	3197	0342	2357	3492	3903	6985
2374	3862	2702	6679	7648	0168	0294	4734	7543	7775	5658	4650
8598	1612	2967	3400	9564	3110	7560	8148	5593	6340	7059	2623
6102	5931	5128	0575	0130	6730	7895	7062	4398	9911	8075	2787
9400	5215	1973	6321	6070	7940	3123	5730	6311	9570	6240	7802
0990	5940	0358	1779	1900	2548	5494	7504	4953	7256	8607	6742
2173	4933	5005	7927	4421	3090	7116	5265	2246	7116	5059	9860
9024	3488	1707	8863	8430	2578	3825	0496	3541	3431	7882	8992
2813	2533	7020	8634	3494	5791	2746	2955	6921	0222	1160	6128
2064	0552	6089	9672	6647	8591	7742	5231	2038	4737	9220	8740
3181	0608	8959	4960	1761	0689	7759	6030	3853	7150	5902	8060
6831	6932	7034	6799	8727	7715	9219	8115	5915	6442	7675	6087
8657	6897	6196	2582	6649	0584	5675	4806	0890	6369	4444	6911
1473	0894	8458	9225	2309	4799	7237	7196	3334	0626	0174	3015
4716	7499	2768	6156	8181	5280	1250	9303	7842	4797	5978	3928
1165	8942	5125	9541	8543	8840	3975	9918	1314	7194	2502	9015
1678	2121	3941	9815	0487	7885	0989	4886	8428	1214	7841	4071
7771	3461	6429	9612	0945	8391	9759	7729	8017	8021	0958	6988
9764	0874	7052	7901	9664	1304	9854	4156	6374	7586	3749	6015
0900	1169	1040	4508	4063	1519	7792	5387	4378	7548	0921	4477
5995	4183	3346	7955	1601	6051	6312	0624	8913	7641	6089	9564
5809	3106	5258	7740	4616	7613	8916	8991	4126	8468	4873	0149
3379	7428	0469	7207	9507	8334	5941	1419	8115	6337	5603	0713
4478	7015	5593	6440	6527	7716	9633	3292	6034	6257	8708	5977
4707	2264	8534	3711	3187	3528	5149	3148	0487	9051	9302	7003
6199	3837	7836	0829	9302	6889	2331	8100	7127	1861	0367	7546
0717	3451	9515	3249	0801	3598	7040	0670	6976	3631	5658	3695
5211	9362	3125	7095	3973	8968	1044	7445	0092	5597	8433	3937
5644	5639	2491	2851	5041	9389	9458	0499	5221	5756	1693	1018
8532	2879	1581	4541	4070	6129	2690	0413	4131	1398	7033	6810
9710	8840	4185	3864	4063	1838	6650	0581	3685	8224	0620	6407
3812	2853	6516	1011	7129	3020	5121	6605	3882	4741	2472	1541

TABLE T Random Digits (*Continued*)

1	2	3	4	5	6	7	8	9	10	11	12
7126	1868	4348	9643	9476	0759	8789	5199	3513	6463	9817	1644
2170	2412	5941	1200	9146	5832	7342	9909	6083	5474	5354	4475
3978	1479	4554	1819	4492	5713	6569	2844	1592	1449	6869	9965
8337	6772	2283	9840	5886	9488	8619	3750	6852	0133	8623	5944
2400	8397	3635	8813	8168	7639	5465	0749	3355	2767	5734	9246
5432	0430	4595	6428	9616	3352	0990	5412	8503	5173	7602	2213
5805	0577	1711	0848	1025	6450	2815	3259	6129	2693	2407	2481
1822	6806	2067	1802	5189	6911	4986	7995	5782	7930	6941	2472
2141	5352	3015	1133	9574	8862	7385	4508	3643	3207	5142	9485
1849	2300	9036	0714	2517	7711	7123	9456	8610	8519	4492	5603
6683	9844	7662	8341	9500	4674	3011	9042	3948	3902	5399	1711
8100	7542	7787	2457	2645	3179	9873	5781	7301	5297	1303	0070
4493	6026	1342	3205	4728	4939	8872	4094	9845	8394	1751	5148
0281	6983	7713	8489	7439	6433	2127	7710	7028	3135	2328	5470
1155	3086	3959	0173	2699	5539	3138	4526	4640	2482	7298	3421
1505	8882	0268	8924	4232	1601	6470	4326	6663	8218	7372	7493
5544	6813	6354	5650	9093	4529	6738	2510	3089	6480	0715	2830
4966	5315	1552	7896	7369	5294	9413	3063	0247	6033	5631	8213
3552	8966	9363	4181	1348	6873	2268	1048	9643	9685	4178	9881
8215	4745	4770	0062	7272	8561	9625	8366	5109	9059	4537	1553
6219	5134	3931	3002	3699	6810	4999	4898	5755	1484	7602	2431
1873	7600	1690	9556	8493	0275	2993	9637	5279	0512	3173	5675
1703	6456	5631	7991	3482	8990	3502	9230	5765	7567	2128	7490
8337	6982	6450	1858	8056	1482	6650	1004	3044	9883	2172	4206
3421	1814	2519	9172	1640	9616	3142	7046	3918	5570	2307	3019
7301	4453	2150	0685	5353	3848	4216	1759	9859	7294	1755	7661
5508	5346	9764	1298	7562	0035	1567	7012	4244	8086	9694	0158
3931	3841	0557	9872	4725	2946	1574	1094	7074	1520	8636	3694
6124	8919	1195	6750	9310	1609	0443	2136	2736	6983	7923	2328
4742	2888	7874	3860	1550	6219	5241	7474	7903	0817	3235	1902
1050	0895	9194	4098	2142	6398	2456	2225	4972	8567	2970	6134
1495	4291	6995	4830	4603	9857	6046	3490	3176	7880	7632	0641
7314	2528	4082	3355	3301	0515	4635	9557	8701	2225	4758	2418
3380	8166	6702	1039	4199	1822	6272	7817	9410	1383	7611	7974
7668	4136	4327	8764	9386	7362	1319	0025	4858	2094	8241	1088
9689	6482	1023	4563	7473	7063	4493	6334	4136	4225	7411	9995
9983	0369	9122	0848	1344	5102	5821	0855	5320	4481	8898	9374
4248	0383	7601	2325	4232	0859	7937	1025	5721	0541	0448	5068
1467	7435	4345	8178	4037	5809	3208	5874	1949	2077	7876	5537
2119	2366	8410	9801	2008	6521	3625	2102	2423	1583	6222	6915

Bibliography

Allen, David M., and Foster B. Cady: *Analyzing Experimental Data by Regression.* Belmont, Cal., Lifetime Learning Publications, 1982.

Badger, E. H. M.: The value and the limitations of high-speed turbo-exhausters for the removal of tar-fog from carburetted water-gas. *J Soc Chem Ind* 65:166–168, 1946.

Bartlett, M. S.: Properties of sufficiency and statistical tests. *Proc R Soc A* 160:268, 1937.

Bartlett, M. S.: The use of transformations. *Biometrics* 3:39–53, 1947.

Berkson, Joseph: Are there two regressions? *J Am Statist Assoc* 45:164–180, 1950.

Bishop, Yvonne M. M., Stephen E. Fienberg, and Paul W. Holland: *Discrete Multivariate Analysis: Theory and Practice.* Cambridge, Mass., The MIT Press, 1975.

Box, G. E. P., and S. L. Anderson: Permutation theory in the derivation of robust criteria and the study of the departures from assumption. *J R Statist Soc B* 17:1–26, 1955.

Brownlee, K. A.: *Statistical Theory and Methodology.* New York, John Wiley and Sons, 1965.

Burdette, Walter J., and Edmund A. Gehan: *Planning and Analysis of Clinical Studies.* Springfield, Ill., Charles C Thomas, 1970.

Cohen, Jacob: *Statistical Power Analysis for the Behavioral Sciences.* New York, Academic Press, 1969.

Conover, W. J.: *Practical Nonparametric Statistics.* New York, John Wiley and Sons, 1980.

Crow, Edwin L., Frances A. Davis, and Margaret W. Maxfield: *Statistics Manual.* New York, Dover Publications, 1960.

Doksum, Kjell: Robust procedures for some linear models with one observation per cell. *Ann Math Statist* 38:878–883, 1967.

Doolittle, M. H.: Method employed in the solution of normal equations and the adjustment of a triangulation. U.S. Coast and Geodetic Survey Report, pp. 115–120, 1878.

Duncan, David B.: Multiple range and multiple *F* tests. *Biometrics* 11:1–42, 1955.

Dunn, Olive Jean: Multiple comparisons among means. *J Am Statist Assoc* 56:52–64, 1961.

Dunn, Olive Jean: Multiple comparisons using rank sums. *Technometrics* 6:241–252, 1964.

Dunnett, Charles W.: A multiple comparison procedure for comparing several treatments with a control. *J Am Statist Assoc* 50:1096–1121, 1955.

Dunnett, Charles W.: New tables for multiple comparisons with a control. *Biometrics* 20:482–491, 1964.

Eisenhart, Churchill: The assumptions underlying the analysis of variance. *Biometrics* 3:1–21, 1947.

Eisenhart, Churchill, Millard W. Hastay, and W. Allen Wallis (eds.): *Selected Techniques of Statistical Analysis.* New York, McGraw-Hill, 1947.

Fisher, Ronald A.: Applications of "Student's" distribution. *Metron* 5:90–104, 1926.

Fisher, Ronald A.: Comment on the notes by Neyman, Bartlett, and Welch in this journal. *J R Statist Soc B* 19:179, 1957.

Fisher, Ronald A.: *Statistical Methods and Scientific Inference,* 3d ed. New York, Hafner, 1973.

Fleiss, Joseph L.: *Statistical Methods for Rates and Proportions.* New York, John Wiley and Sons, 1973.

Friedman, Milton: The use of ranks to avoid the assumption of normality implicit in the analysis of variance. *J Am Statist Assoc* 32:675–701, 1937.

Galton, Francis: *Natural Inheritance.* London, Macmillan, 1889.

Gibbons, Jean Dickinson: *Nonparametric Statistical Inference.* New York, McGraw-Hill, 1971.

Hart, B. I.: Significance levels for the ratio of the mean square successive difference to the variance. *Ann Math Statist* 13:445–447, 1942.

Hodges, J. L., Jr., and E. L. Lehmann: Estimates of location based on rank tests. *Ann Math Statist* 41:783–801, 1963.

Hollander, Myles, and Douglas A. Wolfe: *Nonparametric Statistical Methods.* New York, John Wiley and Sons, 1973.

Hunt, George A.: A training program becomes a clinic. *Ind Qual Cont* 4-04:25–27, 1948.

Iman, Ronald L., and James M. Davenport: Approximations of the critical region of the Friedman statistic. *Commun Statist Theor Meth* A9:571–595, 1980.

Johnson, Norman L., and Samuel Kotz: *Discrete Distributions.* Boston, Houghton Mifflin, 1969.

Kendall, Maurice G.: A new measure of rank correlation. *Biometrika* 30:81–93, 1938.

Kendall, Maurice G.: *Rank Correlation Methods.* New York, Hafner, 1955.

Kendall, Maurice G., and B. Babington Smith: The problem of *m* rankings. *Ann Math Statist* 10:275–287, 1939.

Kendall, Maurice G., and William R. Buckland: *A Dictionary of Statistical Terms.* Edinburgh, Oliver and Boyd, 1957.

Keppel, Geoffrey: *Design and Analysis: A Researcher's Handbook.* Englewood Cliffs, N.J., Prentice-Hall, 1973.

Keuls, M.: The use of the Studentized range in connection with an analysis of variance. *Euphytica* 1:112–122, 1952.

Kimball, A. W.: Errors of the third kind in statistical consulting. *J Am Statist Assoc* 52:133–142, 1957.

Koopmans, Lambert H.: *Introduction to Contemporary Statistics.* Boston, Duxbury Press, 1981.

Kruskal, William H.: A nonparametric test for the several sample problem. *Ann Math Statist* 23:525–540, 1952.

Kruskal, William H., and W. Allen Wallis: Use of ranks in one-criterion variance analysis. *J Am Statist Assoc* 47:583–621, 1952.

Lehmann, E. L.: Asymptotically nonparametric inference: An alternative approach to linear models. *Ann Math Statist* 34:1494–1506, 1963.

Lilliefors, Hubert W.: On the Kolmogorov-Smirnov test for normality with mean and variance unknown. *J Am Statist Assoc* 62:399–402, 1967.

Mann, H. B., and D. R. Whitney: On a test of whether one of two random variables is stochastically larger than the other. *Ann Math Statist* 18:50–60, 1947.

Mannweiler, G. B.: Report of subcommittee AO4.33. *Proc Am Soc Test Mater* 72:44–63, 1972.

Massey, Frank J., Jr.: The Kolmogorov-Smirnov test for goodness of fit. *J Am Statist Assoc* 46:68–78, 1951.

Miller, Rupert G., Jr.: *Simultaneous Statistical Inference.* New York, McGraw-Hill, 1966.

Moses, Lincoln E.: Query: Confidence limits from rank tests. *Technometrics* 7:257–260, 1965.

Nagel, Ernest: *The Structure of Science.* New York, Harcourt, Brace, Jovanovich, 1961.

Nemenyi, Peter B.: Distribution-free multiple comparisons. Ph.D. thesis, Princeton University, 1963.

Newman, D.: The distribution of the range in samples from a normal population, expressed in terms of an independent estimate of standard deviation. *Biometrika* 31:20–30, 1939.

Ostle, Bernard: *Statistics in Research.* Ames, Iowa, Iowa State College Press, 1954.

Owen, D. B.: *Handbook of Statistical Tables.* Reading, Mass., Addison-Wesley, 1962.

Pearson, E. S., and John Wishart: *Student's Collected Papers.* Cambridge, England, Cambridge University Press, 1958.

Pearson, E. S., and H. O. Hartley: *Biometrika Tables for Statisticians,* Vol. I. Cambridge, England, Cambridge University Press, 1966.

Pearson, E. S., and H. O. Hartley: *Biometrika Tables for Statisticians,* Vol. II. Cambridge, England, Cambridge University Press, 1972.

Pratter, Nilon H.: Estimate gasoline yield from crudes. *Petroleum Refiner* 35-05:236–238, 1956.

Pratt, John W.: Robustness of some procedures for the two-sample location problem. *J Am Statist Assoc* 59:665–680, 1964.

Rijkoort, P. J.: A generalization of Wilcoxon's test. *Indag Math* 14:394–404, 1952. Correction. 15:407, 1953.

Satterthwaite, F. E.: An approximate distribution of estimates of variance components. *Biometrics Bull* 2:110–114, 1946.

Scheffè, Henry: A method for judging all contrasts in the analysis of variance. *Biometrika* 40:87–104, 1953.

Sen, Pranab Kumar: On nonparametric simultaneous confidence regions and tests for the one criterion analysis of variance problem. *Ann Inst Statist Math* 18:319–336, 1966.

Shapiro, S. S., and M. B. Wilk: An analysis of variance test for normality (complete samples). *Biometrika* 52:591–611, 1965.

Shapiro, S. S., and M. B. Wilk: Approximations for the null distribution of the W statistic. *Technometrics* 10:861–866, 1968.

Shapiro, S. S., M. B. Wilk, and Mrs. H. J. Chen: A comparative study of various tests for normality. *J Am Statist Assoc* 63:1343–1372, 1968.

Siegel, Sidney: *Nonparametric Statistics for the Behavioral Sciences.* New York, McGraw-Hill, 1956.

Snedecor, George W.: *Analysis of Variance and Covariance.* Ames, Iowa, Collegiate Press, 1934.

Spearman, C.: The proof and measurement of association between two things. *Am J Psychol* 15:72–101, 1904.

Stevens, S. S.: On the theory of scales of measurement. *Science* 103:677–680, 1946.

Stevens, W. L.: Distribution of groups in a sequence of alternatives. *Ann Eugen* 9:10–17, 1939.

Swed, Frieda S., and Churchill Eisenhart: Tables for testing randomness of grouping in a sequence of alternatives. *Ann Math Statist* 14:66–87, 1943.

Thiel, H.: A rank-invariant method of linear and polynomial regression analysis, I. *Proc Kon Med Akad v Wetensch A* 53:386–392, 1950.

Thiel, H.: A rank-invariant method of linear and polynomial regression analysis, II. *Proc Kon Med Akad v Wetensch A* 53:521–525, 1950.

Thiel, H.: A rank-invariant method of linear and polynomial regression analysis, III. *Proc Kon Med Akad v Wetensch A* 53:1397–1412, 1950.

Tukey, John: The problem of multiple comparisons. Unpublished dittoed manuscript, 1953.

Upton, Graham J. G.: *The Analysis of Cross-Tabulated Data.* New York, John Wiley and Sons, 1978.

von Neumann, John: Distribution of the ratio of the mean square successive differences to the variance. *Ann Math Statist* 12:367–395, 1941.

von Neumann, J., R. H. Kent, H. R. Bellinson, and B. I. Hart: The mean square successive difference. *Ann Math Statist* 12:153–162, 1941.

Wald, A., and J. Wolfowitz: On a test whether two samples are from the same population. *Ann Math Statist* 11:147–162, 1940.

Walker, Helen M., and Joseph Lev: *Statistical Inference.* New York, Holt, Rinehart and Winston, 1953.

Wallis, W. Allen: The correlation ratio for ranked data. *J Am Statist Assoc* 34:533–538, 1939.

Walsh, John E.: Some significance tests for the median which are valid under very general conditions. *Ann. Math. Statist.* 20:64–81, 1949.

Welch, B. L.: The significance of the difference between two means when the population variances are unequal. *Biometrika* 29:350–362, 1937.

Welch, B. L.: The generalization of Student's problem when several different population variances are involved. *Biometrika* 34:28–35, 1947.

Werner, George: The effect of type of capping material on the compressive strength of concrete cylinders. *Proc Am Soc Test Mater* 58:1166–1186, 1958.

Wilcoxon, Frank: Individual comparisons by ranking methods. *Biometrics* 1:80–83, 1945.

Wilks, S. S.: Certain generalizations in the analysis of variance. *Biometrika* 24:471–494, 1951.

Yates, F.: Contingency tables involving small numbers and the chi-squared test. *J R Statist Soc* 1 [Suppl]:217–235, 1934.

Index